NEW TECHNOLOGIES AND THE LAW IN WAR AND PEACE

Policymakers, legislators, scientists, thinkers, military strategists, academics and all those interested in understanding the future want to know how twenty-first-century scientific advances should be regulated in war and peace. This book tries to provide some of the answers. Part I summarises some important elements of the relevant law. In Part II, individual chapters are devoted to cyber capabilities, highly automated and autonomous systems, human enhancement technologies, human degradation techniques, the regulation of nanomaterials, novel naval technologies, outer space, synthetic brain technologies beyond artificial intelligence and biometrics. Part III notes important synergies that emerge between the different technologies and legal provisions (existing and proposed), assesses notions of convergence and of composition in international law and provides some concluding remarks. The new technologies, their uses and their regulation in war and peace are presented to the reader, who is invited to draw conclusions.

WILLIAM H. BOOTHBY is an Adjunct Professor at the La Trobe University, Melbourne and a retired Air Commodore. He retired in 2011 as Deputy Director of Legal Services for the Royal Air Force. He is the author of several leading texts on the law of armed conflict, including *Weapons and the Law of Armed Conflict* (2nd edn, 2016) and *The Law of War: A Detailed Assessment of the US Department of Defense Law of War Manual* (Cambridge, 2018) with Professor Wolff Heintschel von Heinegg. He has been a member of the Groups of Experts that addressed *Direct Participation in Hostilities* and that produced the *HPCR Manual on International Law Applicable to Air and Missile Warfare* (Cambridge, 2013), the *Tallinn Manual on the International Law Applicable to Cyber Warfare* (Cambridge, 2013) and the *Leuven Manual on the International Law Applicable to Peace Operations* (Cambridge, 2017). He teaches in Australia, Denmark and Geneva.

New Technologies and the Law in War and Peace

Edited by
WILLIAM H. BOOTHBY
La Trobe University, Melbourne

CAMBRIDGE
UNIVERSITY PRESS

CAMBRIDGE
UNIVERSITY PRESS

University Printing House, Cambridge CB2 8BS, United Kingdom

One Liberty Plaza, 20th Floor, New York, NY 10006, USA

477 Williamstown Road, Port Melbourne, VIC 3207, Australia

314–321, 3rd Floor, Plot 3, Splendor Forum, Jasola District Centre, New Delhi – 110025, India

79 Anson Road, #06–04/06, Singapore 079906

Cambridge University Press is part of the University of Cambridge.

It furthers the University's mission by disseminating knowledge in the pursuit of education, learning, and research at the highest international levels of excellence.

www.cambridge.org
Information on this title: www.cambridge.org/9781108497534
DOI: 10.1017/9781108609388

First published 2019

Printed and bound in Great Britain by Clays Ltd, Elcograf S.p.A.

A catalogue record for this publication is available from the British Library.

Library of Congress Cataloging-in-Publication Data
NAMES: Boothby, William H., editor.
TITLE: New technologies and the law in war and peace / edited by William H. Boothby, Geneva Centre for Security Policy.
DESCRIPTION: Cambridge, United Kingdom ; New York, NY, USA : Cambridge University Press, [2019]
IDENTIFIERS: LCCN 2018035150 | ISBN 9781108497534 (hardback : alk. paper) | ISBN 9781108740128 (pbk. : alk. paper)
SUBJECTS: LCSH: Military weapons–Law and legislation. | War (International law) | Technological innovations–Law and legislation. | Military art and science–Technological innovations.
CLASSIFICATION: LCC KZ5624 .N49 2019 | DDC 341.6–dc23
LC record available at https://lccn.loc.gov/2018035150

ISBN 978-1-108-49753-4 Hardback
ISBN 978-1-108-74012-8 Paperback

Contents

Contributors

It will be appreciated that this is a book that tackles a wide range of complex technologies, each of which is the subject of diverse international law and/or policy initiatives, domestic legislation, human rights implications and expert writings. I am therefore privileged and grateful that the contributing authors, all experts in their various fields, have been prepared to contribute chapters on particular technologies. Without their valuable contributions the preparation of the present volume would have simply been impossible. I therefore offer them my heartfelt thanks for all their hard work in preparing the chapters and for all their help in taking the project forward. With their contributions, I am able to offer the reader a comprehensive and in-depth treatment of this most challenging subject.

In the rest of this section, therefore, brief biographical notes will be offered in relation to myself and each of the contributing authors.

Harry Aitken is an Australian lawyer who holds a Bachelor of Laws (Hons I) and Bachelor of Arts from the Australian National University. He was law clerk to Judge James Crawford at the International Court of Justice as part of the Traineeship Program. He has also worked as a litigation lawyer in the disputes group of a major Australian law firm. He is currently an LLM Candidate at the University of Cambridge, where he is on a Cambridge Australia Allen Scholarship.

Air Commodore William H. Boothby (Retd) retired from the Royal Air Force Legal Branch as Deputy Director of Legal Services in July 2011. He holds a doctorate from the European University Viadrina, Frankfurt (Oder) in Germany. His publications include *The Law of Targeting* (2012), *Conflict Law* (2014), *Weapons and the Law of Armed Conflict* (2nd edn, 2016) and *The Law of War* (with Professor Wolff Heintschel von Heinegg, 2018). He has been a member of the Groups of Experts that addressed Direct Participation in

Hostilities, that produced the HPCR *Manual of the Law of Air and Missile Warfare* and that produced the *Tallinn Manual on the Law of Cyber Warfare*. He teaches at the Australian National University, at La Trobe University, at the University of Southern Denmark and at the Geneva Centre for Security Policy.

Diana M. Bowman is the Associate Dean for International Engagement in the Sandra Day O'Connor College of Law and the Associate Director for Students in the School for the Future of Innovation in Society, Arizona State University. Diana also holds visiting appointments in the Faculty of Law, KU Leuven (Belgium) and the Faculty of Law, University of New England (Australia).

Dr Melissa de Zwart is Professor and Dean of Law of the Adelaide Law School, University of Adelaide, and Deputy Director of the Research Unit on Military Law and Ethics. Prior to joining academia, Professor de Zwart was the Legal Manager, Corporate Legal Services, of CSIRO, Australia's largest science and engineering research organisation, where she was responsible for advising on the protection and commercialisation of leading-edge technology. She has published widely on issues relating to technology and regulation, including digital copyright, internet regulation, cybersecurity, surveillance, privacy, whistleblowing, social media, online communities and gaming platforms, commercialisation and space law. She is a member of the Advisory Council of the Space Industry Association of Australia and Director of the Woomera Manual on the International Law of Military Space Operations.

David P. Fidler is the James Louis Calamaras Professor of Law at the Indiana University Maurer School of Law and is Adjunct Senior Fellow for Cybersecurity and for Global Health at the Council on Foreign Relations. His books include *The Snowden Reader* (editor and contributor, 2015), *India and Counterinsurgency: Lessons Learned* (co-editor and contributor, 2009), *Biosecurity in the Global Age* (co-author with Lawrence O. Gostin, 2008) and *SARS, Governance and the Globalization of Disease* (2004). He holds degrees in international relations and law from the University of Oxford and in law from Harvard Law School.

Dr Heather A. Harrison Dinniss is a senior lecturer at the Centre for International and Operational Law at the Swedish Defence University. Her research and publications focus on the impact of modern warfare on international humanitarian law; in particular, on emerging military technologies such as cyber warfare, advanced and autonomous weapons systems and the legal aspects of human enhancement techniques on members of the armed forces. She is the author of *Cyber War and Laws of War* (Cambridge

University Press, 2012). She has served as a member of advisory groups to the Swedish government on both autonomous weapon systems and cyber operations. A member of the International Law Association's Study Group on Cyber Terrorism and International Law (2014–2016) since 2016, she has been involved as co-editor and a core expert for the project to establish a Manual on International Law Applicable to Military Uses of Outer Space (MILAMOS) and remains a core expert on the Woomera Manual on the International Law of Military Space Operations.

Professor Dr Wolff Heintschel von Heinegg holds the Chair of Public Law, especially Public International law, European Law and Foreign Constitutional Law at the European University Viadrina in Germany, where he has been Dean of the Law Faculty and University Vice President. He has held teaching posts at a number of universities and institutions, has been the Rapporteur of the International Law Association Committee on Maritime Neutrality and has taken leading roles in the German Society of Military Law and the Law of War, the International Institute of Humanitarian Law, the International Society for Military Law and the Law of War and the Cooperative Cyber Defence Centre of Excellence. He has been a member of the International Groups of Experts that produced the *San Remo Manual on International Law Applicable to Armed Conflicts at Sea*, the *Manual on the Law Applicable to Air and Missile Warfare* (2010) and the Tallinn Manuals on the *International Law Applicable to Cyber Warfare* (2013) and to *Cyber Operations* (2016). He publishes widely on legal matters.

Kobi Leins is a PhD candidate at the Law School at the University of Melbourne. Leins earned her Bachelor of Arts (Hons) and Bachelor of Laws at Monash University (Australia). In 2003, she was admitted to practice as a Barrister and Solicitor of the Supreme Court of Victoria, Australia. Her professional experience includes working for the United Nations, the International Committee of the Red Cross, and as a commercial litigator. Leins is currently undertaking research on whether the use of nanotechnology-enhanced or -based weapons is prohibited or limited in an armed conflict, with a focus on the adequacy of existing weapons review requirements for nanomaterials. Leins is more broadly interested in the governance and implications of new and emerging technologies.

Rob Mclaughlin is Professor of Military and Security Law, and Director of the Australian Centre for the Study of Armed Conflict, at UNSW Canberra (Australian Defence Force Academy). Prior to this he taught at the Australian National University (2011–2017), and was head of the United Nations Office on

Drugs and Crime's Global Maritime Crime Programme (2012–2014). Before becoming an academic, he served in the Royal Australian Navy (RAN) for more than two decades, as both a seaman officer and a legal officer. His final postings in the RAN were as Director of Operations and International Law, and Director of the Naval Legal Service.

Hitoshi Nasu is Professor of International Law at the University of Exeter, with expertise in public international law, particularly in the fields of international security law and the law of armed conflict. He holds bachelor's and master's degrees in political science from Aoyama Gakuin University and a master's degree and a PhD in law from the University of Sydney. He is the author of *International Law on Peacekeeping: A Study of Article 40 of the UN Charter* (2009) and co-editor of *Human Rights in the Asia Pacific Region: Towards Institution Building* (with Ben Saul, 2011), *Asia-Pacific Disaster Management: Comparative and Socio-legal Perspectives* (with Simon Butt and Luke Nottage, 2013), *New Technologies and the Law of Armed Conflict* (with Rob McLaughlin, 2014) and *Legal Perspectives on Security Institutions* (with Kim Rubenstein, 2015).

Ioana Maria Puscas is Research Officer at the Geneva Centre for Security Policy (GCSP), Geneva. She has master's degrees from the Graduate Institute of International and Development Studies (Geneva) and the Central European University (Budapest). Prior to her work at the GCSP, she was a trainee in the Operations Division at the North Atlantic Treaty Organisation (NATO) HQ in Brussels.

Preface

This book tackles the challenging topic of advances in technology by examining how the law seeks to cope with novel developments. If the practical issues raised by scientific and technological progress in numerous fields of research are complex, and that is undoubtedly the case, matters become even more difficult when the intricacies of different elements of international law are thrown into the mix. By seeking to consider how the various legal principles and rules address the applications, or potential applications, of new and emerging technologies in both the military and civil contexts and by asking whether either can learn from the other, this book adds numerous additional levels of complexity and challenge to an already tricky topic. And yet, that was the challenge that the editor set himself, and the following pages are the result.

The intent in preparing this volume has been to lay before the reader an explanation of the kinds of legal challenge that new and emerging technologies present, and to show how, in differing contexts, legal rules and initiatives for the generation of new law, guidance, codes of practice, etc., strive to address the challenges that have been identified. What emerges is not a blueprint that is capable of general application in all contexts. Rather, the discussion shows that it is the challenges posed by a particular technology or capability when applied in specific circumstances that will have to be the focus of appropriate provision.

However, it is not for the editor nor for the contributors to tell the reader which lessons can be learned, which approaches or solutions are to be preferred and which should be avoided. The role of the editor and of the contributors is to lay the relevant facts, rules and information before the reader while it is for the readers to draw their own conclusions. In that spirit, the editor very much hopes that readers will find worthwhile insights in this book, insights that will be of use to policymakers, lawyers, technologists, scientists, academics and indeed to all with an interest in where scientific endeavour is taking us, what the consequences seem likely to be and how the law might profitably address such changes.

It would have been folly in the extreme to seek to meet the challenges associated with preparing a book of this nature alone. Indeed, some will consider that addressing such a topic at all is something of a tall order. Be that as it may, the editor has carefully selected a group of distinguished contributors, each of them authoritative in a variety of fields and is indebted to each and every one of them for all their hard work in putting together and refining the chapters of this book.

Grateful thanks are also offered to Simon Phippard for his helpful advice in relation to Chapter 6, to Jay Jenkins for helping with the reviewing of Chapter 10, to Kobi Leins for her helpful and insightful suggestions throughout the project and to Melissa de Zwart for taking on the Outer Space chapter with less than an ideal period of notice. Thanks also go to Cambridge University Press for agreeing to publish the book, and to copy-editor Sean Connolly and all in the production team for turning our humble writings into the impressive volume you have before you.

Abbreviations

AMW Manual	*Manual on International Law Applicable to Air and Missile Warfare, Program on Humanitarian Policy and Conflict Research,* Harvard University, March 2010
API	Protocol Additional to the Geneva Conventions of 12 August 1949, and Relating to the Protection of Victims of International Armed Conflicts, Geneva, 8 June 1977, 1125 UNTS 3
CCW	Convention on Prohibitions or Restrictions on the Use of Certain Conventional Weapons Which May be Deemed to be Excessively Injurious or to Have Indiscriminate Effects, Geneva, 10 April 1981, 1342 UNTS 137
DoD Law of War Manual	Department of Defense Law of War Manual, Office of General Counsel, US Department of Defense, June 2015
ECHR	European Convention for the Protection of Human Rights and Fundamental Freedoms, 4 November 1950, 213 UNTS 222

ECtHR	European Court of Human Rights
ENMOD	Convention on the Prohibition of Military or Any Other Hostile Use of Environmental Modification Techniques, 10 December 1976, 1108 UNTS 151
GCI	Convention (I) for the Amelioration of the Condition of the Wounded and Sick in Armed Forces in the Field, Geneva, 12 August 1949, 75 UNTS 31
GDPR	European Union, General Data Protection Regulation, Regulation 2016/679, adopted on 27 April 2016, became effective on 25 May 2018
ICRC	International Committee of the Red Cross
ICRC Customary Law Study, or CIHL	J.-M. Henckaerts and L. Doswald-Beck, *Customary International Humanitarian Law*, ICRC, 2005 – Volume 1: Rules
ICRC Commentary	Y. Sandoz, C. Swinarski and B. Zimmermann, eds., Commentary on the Additional Protocols of 8 June 1977 to the Geneva Conventions of 12 August 1949, ICRC, 1987
ICT	information and communication technologies
Nuclear Weapons Case	International Court of Justice, Legality of the Threat or Use of Nuclear Weapons (Advisory Opinion) [1996] ICJ Rep 226
Oslo Manual	*Oslo Manual on Select Problems of the Law of Armed Conflict: Rules and Commentaries* (to be published)
Rome Statute	Statute of the International Criminal Court, Rome, 17 July 1998, 2187 UNTS 90

Tallinn Manual	M. N. Schmitt, ed., *Tallinn Manual on the International Law Applicable to Cyber Warfare*, Cambridge University Press, 2013
Tallinn Manual 2.0	M. N. Schmitt and L. Vihul, eds., *Tallinn Manual 2.0 on the International Law Applicable to Cyber Operations*, Cambridge University Press, 2017
UK Manual	*The Manual of the Law of Armed Conflict, UK Ministry of Defence*, Oxford University Press, 2004

PART I

The Legal Background

1

Introduction

WILLIAM H. BOOTHBY

1.1 INTRODUCTORY REMARKS

We live in times of very rapid technological advance in many fields of science. We are witnessing scientific revolutions that will radically change many aspects of the way in which ordinary people live their lives. What has happened in the field of digital technology is an example from the past that serves to show how novel developments can rapidly change traditional ways of doing things across numerous fields of activity. The digital revolution is an expression that trips off the tongue with almost casual ease, and yet for those 'of a certain age', like the editor of this volume, the now increasingly pervasive employment of, and dependence upon, digital electronics in virtually every sphere of life amounts in truth to a revolution compared to the methods that preceded it. To an author who, as a qualified lawyer, was required to prepare first drafts of his legal advices in long hand in his early days in the Royal Air Force Legal Branch and whose advices in later years were routinely dictated to and typed by professional secretaries, the irresistible march of the personal computer brought with it an absolute and permanent transformation of ways of working. Bulky case files gave way to data records, paper communications gave way to emails, the pace of work quickened markedly and the secretaries were no more.

If the workplace was dramatically altered forever, almost all other fields of activity followed suit, and one rather doubts that those who initiated the first faltering steps towards electronic communication could, in their wildest imaginings, have conceived how commercial, industrial, professional, academic and daily life would all be changed to such a considerable degree.

The reference in the opening paragraph to 'those of a certain age' is deliberate, because for many, perhaps a majority of the population, aspects of modern digital technology have been evident in everyday life to a greater or lesser extent throughout their lives. It is those for whom that is not the case that will appreciate most keenly the breadth of effect that these technologies have brought.

These thoughts, or something rather like them, were forming in the present editor's mind while he was preparing the second edition of *Weapons and the Law of Armed Conflict*.[1] He appreciated that while the subject matter of that book, as reflected in its title, was indeed in its own right a worthy topic for a book, it nevertheless only dealt with a singular aspect of the technologies that it addressed, namely their employment in connection with warfare. The thought started to develop that emerging technologies that are used in warfare often also have applications, perhaps their origins, in the non-military, peacetime environment. In the past, the focus and motivating factor driving scientific research forward into new technologies has often been the development of new and improved weaponry or military ways of doing things, with the clear purpose of securing a technological advantage over prospective or actual enemies.

In more recent times, it has frequently been the large-scale peacetime, consumer markets that have inspired technological development and progress, and the product of those improvements has tended to be adapted to military use where this was feasible and seen to be useful. Either way, whether it was a recognised military need or anticipated commercial gains that motivated the initial research, a number of emerging technologies have ended up being, in this sense, 'dual use'.

Sometimes new technologies are applied to discharge similar functions in the military/armed conflict setting as in the peacetime/commercial setting. On other occasions, the technology is used in significantly different ways and to achieve fundamentally different purposes. Sometimes the same technology is used for both similar and different purposes.

Realising that this is the case led to the thought that the bodies of law that address these emerging technologies will often differ as between the military/armed conflict sphere and the peacetime/commercial sphere. Where the former is concerned, it was clear that it would be the law of armed conflict that would take precedence when the technology is employed in connection with hostilities associated with an armed conflict. Nevertheless, in certain applications even military use of novel technologies would be regulated by other bodies of law. In the peacetime/commercial context, however, it was obvious that domestic law, human rights law and some other elements of

[1] W. H. Boothby, *Weapons and the Law of Armed Conflict*, 2nd edn (Oxford: Oxford University Press, 2016).

international law would regulate the circumstances in which, and the ways in which, the new or emerging technology could lawfully be used. These recognitions, however, did not really address the more intriguing challenges: it is, let us face it, relatively easy to assert in fairly general terms that the bodies of law mentioned in the previous sentence will apply. It is altogether more challenging to consider, in the context of an emerging, incompletely understood technology what domestic law rules appear to be relevant, what if any legislation, domestic or international, is being produced to address the challenges posed by the new way of doing things and to what extent the existing legal rules apply. It is, moreover, difficult to determine what the future seems likely to look like in terms of further technological evolution and in terms of changes in regulations and other legal and policy provision to address evolving challenges posed by these scientific changes.

So, it was with these inchoate thoughts in mind that attention shifted back in favour of getting on with the completion the second edition of *Weapons and the Law of Armed Conflict*. Revisiting what at that stage was still a rather nebulous notion after the 'Weapons Law' manuscript was safely in the hands of the publisher, it became clear that turning such vague concepts into anything resembling a coherent book was going to be a considerable challenge – yet the issues involved seemed to be fascinating and there was the inescapable thought that there were potentially valuable lessons to be learned here. It was at that point, however, that following an invitation to write a chapter critiquing part of the US Department of Defense *Law of War Manual*, the editor took on the daunting task of co-authoring with Professor Wolff Heintschel von Heinegg a commentary addressing the whole of that most lengthy and important legal document. Ideas associated with new and emerging technologies therefore 'took a back seat' for the best part of a further year until that manuscript had also been safely despatched to its publisher.

With those two major diversions out of the way, the opportunity now presented itself to try to take forward the ideas outlined in the opening paragraphs of the present chapter. It was immediately evident that it would not be possible to consider all emerging technologies. There are too many and to do so would likely produce an unwieldy text that would inevitably lack the required sharpness of focus. So an attempt has been made to select specific technologies that seem to be relevant in both the military/armed conflict and the peacetime/commercial spheres and that involve interesting issues where plans or measures for their regulation are concerned.

Many of the new technologies that have been chosen have applications across numerous fields of human activity. Here again a choice has had to be made. To seek to discuss all such applications, or potential applications, would likely result in chapters that would be excessively lengthy and potentially confusing. It is, of course, not possible to predict in which respects legal and practical challenges associated with these new technologies and methods will arise. Sometimes they will pose challenges that differ as between those different fields; sometimes the challenges are similar. Sometimes the aspects that have been discussed in this book will indeed be the focus of future concern and regulation; sometimes that focus will lie elsewhere for reasons that may not, at the time of writing, have been evident, foreseeable or foreseen. Be that as it may, rightly or wrongly the technologies reflected in Chapters 5–14 of this book are the ones that have been chosen and it is hoped that the resulting analysis is thought to be illuminating and worthwhile.

A considerable effort has been made to try to ensure that the technological and legal explanations that are given are both as accurate and up to date as possible as at the date of completion of the present manuscript, namely 15 April 2018. This is, however, a rapidly changing field. In a sense, if some aspect of regulation has changed, or if the technology has further matured since a particular chapter was written, that is somewhat tangential to the main purpose of the book. Indeed, the idea that spurred the writing of this book is that there are, perhaps, lessons to be learned by considering how both the law and policy attempt to deal with the diverse applications of these evolving technologies and the legal, security and other challenges that these developments pose. By giving an account of how the relevant legislatures, governments and other institutions, international and national, were then approaching the task of addressing those challenges, the intended points and the basis for comparison will emerge.

The purpose of the book is to reassure, where reassurance seems to be indicated, to draw attention to matters of concern where that seems appropriate, and above all to try to build into the discussion a recognition that what is being discussed in the context of conflict situations may have relevance in the peacetime environment, and vice versa. It is the collective view of the contributors participating in this endeavour that to examine the issues posed by rapid technological advance from a singular standpoint, such as the law of armed conflict, would be to omit vitally important perspectives reflected in other bodies of law and would thus risk producing a potentially distorted view. However, by looking at how the law grapples with new technologies in somewhat different contexts, it is possible that a deeper understanding, for example of what the law can and cannot do for us, will emerge. Certainly, the emerging technologies and their significance will be placed in a more comprehensive context which, alone, may help readers to devise their own appreciation of the relative merits and drawbacks of what is going on in the scientific arena and of the regulatory or analogous measures that are being taken or that are required.

1.2 THE SOURCES OF TECHNOLOGICAL AND SCIENTIFIC INNOVATION

As was noted earlier, in the past it was often said that new technologies for peacetime applications tended to emerge from military applications. According to this appreciation, it was the pressing need to find technological solutions to problems on the battlefield that yielded new scientific insights which were then adapted to peacetime use. The development of radar during World War II and its subsequent evolution as one of the chief cornerstones of modern, safe civil aviation is but one among many examples of such a process. Perhaps this 'military invention – subsequent civilian adaptation' phenomenon had at least as much to do with the traditionally large expenditures that were devoted to armament and the conduct of military affairs, particularly in the lead up to and during the conduct of two world wars.

In much more recent times, we have witnessed somewhat diminishing military budgets coinciding with rapid growth in consumer spending. Arguably, it is nowadays the peacetime applications that tend to drive invention and refinement of new technologies that are then applied, where possible and appropriate, to military requirements.

By asking whether there are similarities in the challenges that new technologies pose in the military and peacetime environments and by trying to identify in which disciplines such similarities reside, this book tries to determine how those challenges are being identified, clarified and addressed; if there are inadequacies in the way the challenges are being addressed; what the implications of those inadequacies might be; what problems the identified technologies of the future seem likely to bring as they mature; and whether existing law helps us as we consider how to deal with these new products of science. In short, is this a process of technical and scientific advance that we face unassisted by existing legal provision, or do existing law, domestic and international, and established policy and doctrine include rules, norms and principles that can point the way towards methods of coping with the challenges that novel technologies seem destined to pose?

1.3 THE PURPOSE OF THE BOOK

In a sense, for the purposes of the discussion in this book, it does not really matter whether a technology first emerged to satisfy consumer demand or to meet military need. The feature that is of most interest where the analysis in this book is concerned is the tendency for there to be a crossover between military and consumer uses of a number of pivotal technologies. It is that very duality of potential application that leads one to consider how law will develop to regulate the application of the respective technologies in these two distinct spheres of activity: the military on the one hand and the peacetime/civilian/consumer on the other.

What will be the challenges in formulating such regulation, what will be the purposes that it is designed to achieve, how do those challenges and purposes compare as between the different technologies and as between the military/armed conflict sphere and the peacetime/civilian/consumer sphere and what lessons can be learned from the diverse legal/policy approaches to addressing these various issues? It is the central purpose of this book to try to shed some light on these matters. No claim is made that the book answers all relevant questions, nor that it represents any kind of blueprint on how to approach the control of new technologies within society. Indeed, if anything, the book will tend to show that 'one size does not fit all' in the sense that regulatory measures must be designed with the specific technology, application and relevant surrounding circumstances in mind.

So, in working out how best to adjust legal and policy arrangements to take account of new ways of doing things, the peculiarities of the new science or method, the particularities of the context in which the new science or method is to be employed, the framework of relevant existing law and policy and the goals that any new law or policy is being designed to achieve are among the more important factors that must be taken into account. At best, this book aims to represent some 'food for thought'. By reflecting on how we are seeking to address these matters at the time of writing, trends can be identified, ideas as to improved approaches may emerge and both the individual technological/scientific advances themselves and the legal and policy responses to each of them can, usefully it is hoped, be placed into a broader context.

The authors who have contributed to this book are therefore laying before the reader factual material relating to the technologies in question and their appreciation of the regulations, policies and initiatives that are being applied to those technologies in selected applications or, in the case of 'over the horizon' capabilities, that seem to be indicated by their characteristics and by their likely uses. It is then for the readers to develop their own appreciation assisted, it is hoped, by the discussions and analyses in the following pages. So this book seeks to draw attention to some at least of the factors that seem to be relevant in reaching any such conclusions on these important issues.

1.4 STRUCTURE OF THE BOOK

The discussion will be framed by reference to particular emerging technologies each of which will be the subject of a dedicated chapter. Each chapter will seek to explain the technology concerned in terms that are as accessible as possible. As will be appreciated when the particular technologies are listed in the next section of the present chapter, formulating such explanations is by no means an easy task. There are, for example, controversies as to the meanings to be given to significant terms and there will be differences of view as to how the relevant existing law should be interpreted and applied. Nevertheless, each such technology will be described and the way in which law of armed conflict principles and rules apply to its potential use in an armed conflict setting will be explained.

As a distinct matter, the legal implications of particular applications of the technology in certain fields within the peacetime/civilian/consumer environment will be explored with a view to identifying where there are differences in the uses to which the technology is put in armed conflict and in peace – where there are similarities – and whether the differences translate into distinctions in the legal and policy provision that is made in relation to such uses of the technology. The purpose of the discussion is not to delve into the theoretical and philosophical underpinnings of comparative law. Rather, the objective is to look in a practical way at the legal challenges that these new technologies pose in fundamentally different situational settings and to note where there are similarities and where there are differences.

The book is divided into three parts. Part I comprises Chapters 1–4 and introduces the subject matter of the book, explains its purposes and gives a general explanation of the core legal principles and rules that address emerging military technologies in the military/armed conflict and in certain peacetime/commercial spheres. The purpose of this part is to give the reader a sufficient understanding of aspects of the applicable law to support a proper consideration of some of the legal challenges that will be addressed in the ensuing chapters. Part II, comprising Chapters 5–14, discusses each of the chosen technologies, applying a chapter to each such technology.[2] Part III draws the discussion in the preceding chapters together, considers notions that have been described as 'convergence' by referring to the technologies themselves and to the applicable law, and then seeks to draw some conclusions.

[2] It will, however, be noted that two chapters are devoted to human enhancement.

1.5 AN INTRODUCTION TO THE CHAPTERS

1.5.1 *Part I*

Chapter 2 will look at how the acceptability of military uses of new technologies is determined. It will do this by examining the international law principles and rules that differentiate between lawful and unlawful weapons and methods of warfare and will draw attention to the obligation of all States to determine whether new weapons they acquire and new ways of fighting that they adopt match the international law rules by which they are bound. This will provide the first part of the framework against which the analysis of particular technologies will be set.

Chapter 3 recognises that a comprehensive discussion of the law as it will apply to novel military activities in armed conflict must not be limited to the law as to the weapons that states may lawfully employ but must also consider the legal rules that regulate the actual conduct of hostilities. In heavily summarised form, therefore, Chapter 3 explains the core principles and rules of international law that regulate hostile operations in warfare.

Chapter 4 completes Part I of the book by dealing with applications of military technologies outside the scope of armed conflict and, thus, outside the regulatory regime of the law of armed conflict. The chapter addresses the issue through the useful lens of the domestic law of Australia and cites examples to illustrate the applicability of relevant domestic and human rights law. Reference is made to other, particular international law rule sets that apply to the employment of military technologies outside armed conflict, pointing out that these may also apply alongside the law of armed conflict. Particular mention is made of status of forces arrangements, rules as to liability and agency, export control obligations, some privacy-related regulations and review requirements related to acquisition of military technologies.

The particular regulations that apply when a specific technology is devoted to military or non-military peacetime activities will tend to depend on the peculiarities of that technology, on the kinds of activity for which it is to be employed and on relevant surrounding circumstances. Accordingly, in each of Chapters 5–14, existing, proposed or suggested peacetime regulatory arrangements are discussed by reference to the technology being addressed in that chapter and by reference to the peacetime applications of that technology that are being considered.

1.5.2 *Part II*

Chapter 5 starts the process of looking at specific technologies by addressing the use of cyberspace, a new environment in which hostilities will potentially be conducted. Initiatives, respectively, by International Groups of Experts and under the auspices of the UN General Assembly aimed at identifying the applicable rules are discussed and the conclusions so far reached by those processes are summarised and compared. Aspects of the positions taken by Russia and China are noted and proposals for a Code of Conduct are explained. In the peacetime/commercial context, the policy and legal challenges posed by hacking and by the Internet of Things are considered, e.g., by reference to the law applicable to human rights and to data protection.

Chapter 6 takes the analysis further by looking at emerging automation and autonomy technologies. Much has been discussed and written on these topics, particularly as they apply to the conduct of daily life and of warfare. The chapter proposes an interpretation of the respective terms, assesses the legal challenges that their use in warfare implies, considers some potential peacetime applications and determines how law and policy will regulate such uses. The use of autonomous surgery techniques, the rules and guidance on the use of autonomy in civil aviation and the emerging plans for testing and ultimate use of driverless cars on the public roads are the chosen peacetime applications that are considered.

Chapter 7 examines human enhancement technologies. After differentiating between enhancement and therapy and summarising the history of military human enhancement, the chapter explains many of the techniques that are currently available. Enhancing physical capabilities, cognitive performance, pharmacological approaches, neuro-enhancements including implanted neural interfaces, memory restoration and neurostimulation are among the techniques that are explained, citing in each case relevant examples. Some applicable rules of the law of armed conflict are considered, ethical challenges and opportunities are addressed and the discussion is set against the context of the use of enhancements in society, particularly in the sporting and employment settings.

Chapter 8 continues the exploration of issues raised by human enhancement technologies. Building on the discussion in the previous chapter, this contribution begins by examining the question of whether and under what circumstances we might consider that individuals who enhance their natural abilities might be considered something other than human – and what that might mean for their treatment under the law. Biochemical enhancement, cybernetic technologies such as brain–machine interfaces and advances in prosthetic technologies all have the capacity to alter and augment the human experience and raise interesting challenges for the law. This chapter looks specifically at the application of the law of armed conflict (international humanitarian law) in relation to these techniques and the effects of human rights law in an age of enhanced humans – whether they be civilian or military personnel. Clear synergies also exist with the discussions in Chapter 13 on brain–machine interfaces. Attention is given in the final section to questions as to the adequacy of the current rights frameworks and as to the distinction between national and international systems.

Chapter 9 deals with the techniques that can be employed to degrade the performance of human beings. Recognising that the idea of human degradation is nothing new, the chapter defines the meaning of the term and points out the various contexts in which such technologies might be employed. Attention readily focuses on the use of incapacitants and the use by the Russians of a substance based on Fentanyl to deal with the 2002 Moscow Theatre siege is analysed. Other degradation technologies are explained and prohibited means of human degradation are noted. The lawfulness of human degradation is considered by reference to the law as to weapons, targeting and human rights and conclusions are reached.

Issues raised by the creation and potential uses of purposefully engineered nanoparticles (ENPs) are the focus of Chapter 10. The international governance frameworks respectively applying to commercial and military employment of ENPs are considered, the idea of the nanoscale is explained and the lack of agreed definitions is noted. The application of weapons law to nanotechnology substances is considered and attention is paid to governance, e.g., of the hostile use of nanotechnologies. The implicit questions as to whether such governance is adequate in general and whether in particular the law of armed conflict weapon review obligation is a sufficient mechanism to address the issues raised are also addressed. Some conclusions and recommendations are then offered.

Chapter 11 deals with novel developments in the technologies employed in naval warfare. While many such developments are adequately regulated under the established rules, e.g., as to weapons and targeting, attention naturally focuses on the legal issues that arise in relation to unmanned maritime systems and underwater infrastructure, to which the rest of the chapter is devoted. Taking each of these in turn, the relevant law is teased out, proper attention being given to the legal status and lawful operation of such systems and infrastructures. Gaps in the law are identified and peacetime and armed conflict law are both addressed. Suitable conclusions are drawn as to the need for state action to develop the law in relation to each of the technologies.

Chapter 12 looks at the evolution of the international law framework in relation to outer space, pointing out that early outer space law focused on the prevention of the deployment of weapons of mass destruction in outer space and on the regulation of its military and scientific use by States. After outlining and assessing the significance of the outer space treaties and related provisions, the chapter considers new activities in space, including those involving commercial actors providing communications, Internet, tourist and mining capabilities. The dual use nature of relevant technologies is noted and measures to promote future security in outer space, including the development of an international manual, are assessed.

Chapter 13 assesses developing notions of an artificial brain and of brain computer interfaces, noting the very considerable technological challenges involved. Recognising that this is an 'over the horizon' technology, the chapter explains the possibilities and clarifies the range of its contemplated applications in peace and war. Inevitably there is a degree of well-informed speculation associated with some aspects of the discussion, but important linkages emerge between these brain-related technologies and human enhancement, cyber operations, autonomous capabilities and the Internet of Things. The chapter spells out the numerous important and intractable issues that are raised by these brain-associated technologies and draws important conclusions.

Chapter 14 evaluates the use in warfare and in peacetime of biometrics technologies and the legal issues involved. Three peacetime applications of the technology are selected and are considered in some detail. The chapter recognises the potential for some such applications to be employed in both the military/armed conflict and the peacetime, internal security contexts and notes the legal implications. Some legal considerations specifically associated with the peacetime employment of biometrics are identified, including those relating to the lawful processing of the personal data that some forms of biometric testing are likely to generate. Some tentative conclusions are reached and this chapter concludes Part II of the book.

1.5.3 *Part III*

Chapter 15 brings the book to a close. The chapter starts by recalling some of the significant conclusions taken from the chapters of Part II. In a second, important section of the chapter, notions of convergence and composition are considered. Comment in the past has centred on the idea of a convergence between the jurisdictions of different kinds of international court. The chapter discusses the suggestion that international law should be understood as a composite whole, not as a selection of different bodies of law that are independent of one another. Similar notions of convergence or composition involving at least some of the technologies themselves are also reflected upon, and the folly of seeking to consider the implications of a single technology in isolation is made plain. Thereafter, some reflections are offered on the book as a whole in the form of concluding remarks.

1.6 READERSHIP

This book is intended to have a relatively broad appeal and to be useful. It is intended to indicate to scientists and those involved in technological development and in the implementation of new systems some of the ways in which the law may be expected to impact on their projects. By showing where some of the legal challenges can be expected to lie, it is hoped that project directors and those taking research forward will be able better to anticipate legal problem areas and to set in place timely coping arrangements. The book will be of interest to those addressing the challenges associated with the peacetime application of novel ways of doing things and will, of course, have great relevance to those developing and implementing novel military systems. Legislators, persons involved in treaty negotiations, those developing international policy and guidance, members of think tanks, NGOs and all who have an interest in regulating the products of science will find the following pages useful, as will academics in the legal, war studies, applied technology and other fields. Finally, anyone interested in how the law of the past and of the present might sensibly be adjusted to cope with the technological challenges of the future is invited to read on.

1.7 HOW TO USE THIS BOOK

There is no single proper way of using a book of this nature. While some hardy souls may choose to read it from cover to cover, others will prefer to dip into particular topics of interest and then, when suitably intrigued, to read further, and perhaps more expansively than originally intended. An index has been included to assist those who prefer to use it as a book of reference, but throughout an attempt has been made to render the text as readable as possible for the benefit of the 'cover to cover' reader. As time passes following the date of publication, some of the legislation, codes of practice, policy initiatives, negotiation positions, etc. will inevitably be overtaken by events but, as hinted earlier in this chapter, that is not really the point. The book is a snapshot taken at a particular moment in time that shows how emerging technologies were then being addressed and which considers the associated implications. Those implications, and the lessons that the reader draws from them, are what the book is really all about, and they have an enduring quality and value that is not diminished by subsequent events and developments.

2

Regulating New Weapon Technologies

WILLIAM H. BOOTHBY

2.1 INTRODUCTION

There will always be a tension between emerging technology and the established legal order. Law consists of rules that are designed to regulate behaviour and frequently do this by differentiating between the acceptable and the unacceptable, prohibiting the latter. Scientific research is concerned with exploring the hitherto unknown, for example in order to develop novel solutions to established challenges. There is always, therefore, going to be a certain tension between the rules that are based on what is already understood and the process of expanding the boundaries of knowledge and, thereby, of activity. Is new activity already prohibited by existing rules, or were those rules formed in ignorance of the new technology and are they rendered irrelevant by it? Does law need to change to address dangers that the new technology is considered to pose, and are we sure that we understand enough about the capabilities of the new technology to be able to formulate rules that maximise the advantages to be derived from it while minimising the dangers it is assessed as posing? These are just some of the questions that underpin the debates associated with the technologies addressed in this book.

In the present chapter, we will examine how the legal acceptability of new weapon technologies is determined. We will look at the compliance and review processes that international law requires, shall assess the extent to which they are actually implemented, shall summarise the relevant rules and shall assess in process, not substance, terms how the review requirements should be applied to the new technologies covered by the present book. We will not be applying the relevant rules of weapons law to the technologies. That will be undertaken in the later chapters by reference to the characteristics of each technology.

2.2 REVIEWING NEW WEAPON TECHNOLOGIES UNDER INTERNATIONAL LAW

There is a duty placed by international law upon all States to consider whether the use of a new weapon or weapon technology would, in some or all circumstances, breach the international law rules that apply to that State. The requirement is implied, meaning that it can be inferred from numerous other legal rules that bind States. These other legal rules include, but are not necessarily limited to, the 1868 St Petersburg Declaration,[1] Article 1 of Hague Convention II of 1899 and Article 1 of Hague Convention IV of 1907.[2] Article 1 common to the Geneva Conventions 1949 specifies the obligation of States to respect, and ensure respect, for the Conventions, and the language of a number of weapon treaties makes it clear that States are subject to a continuing obligation to satisfy themselves that the weapons they acquire and use comply with the legal rules by which they are bound. The ICRC view is that '[t]he requirement that the legality of all new weapons, means and methods of warfare be systematically assessed is arguably one that applies to *all* States'[3] and the present author agrees that this implied obligation requires all States legally to review all new weapons.

[1] Note in particular the final paragraph of the Declaration Renouncing the Use, in Time of War, of Explosive Projectiles Under 400 Grammes Weight, St Petersburg 11 December 1868 (St Petersburg Declaration) in which States party 'reserve to themselves to come hereafter to an understanding whenever a precise proposition shall be drawn up in view of future improvements which science may effect in the armament of troops, in order to maintain the principles which they have established, and to conciliate the necessities of war with the laws of humanity.'

[2] 'The contracting Powers shall issue instructions to their armed land forces which shall be in conformity with the Regulations respecting the laws and customs of war on land, annexed to the present Convention.' Consider, e.g., W. Hays Parks, 'Conventional Weapons and Weapons Reviews', 8 *Yearbook of International Humanitarian Law* (2005) 55 at pages 55–57 and note that an important State not party to API, the United States, legally reviews the new weapons that it acquires and has done so for many years.

[3] K. Lawand, 'A Guide to the Legal Review of New Weapons, Means and Methods of Warfare: Measures to Implement Article 36 of Additional Protocol I of 1977', *ICRC* (2006) at page 4; *AMW Manual*, Rule 9; *Tallinn Manual*, Rule 48(a), the Commentary to which cites Article 1 common to the Geneva Conventions of 1949. See also the *UK Manual*, paras 6.20–6.20.1, the *United States Naval Commanders' Handbook on the Conduct of Naval Operations*, NWP 1–14, para 5.3.4; *Canadian Department of National Defence, Joint Doctrine Manual Law of Armed Conflict at the Operational and Tactical Levels*, 13 August 2001, para 530 and *Bundesministerium der Verteidigung, Law of Armed Conflict Manual*, ZDv 15/2 dated May 2013, para 405.

States that are party to API[4] are subject to a rather more extensive obligation set forth in Article 36 of that treaty. They are obliged '[i]n the study, development, acquisition or adoption of a new weapon, means or method of warfare [. . .][5] to determine whether its employment would, in some or all circumstances, be prohibited by th[e] Protocol or by any other rule of international law applicable to the High Contracting Party.' So, under Article 36 weapon reviews must take place when a weapon or method of warfare[6] is being studied and when it is being developed. 'Study' for these purposes seems to refer to the study of the weaponisation of a technology, that is the assessment of how a technology can be adapted or applied so as to cause death, injury, damage or destruction in the context of an armed conflict. 'Development', as that term is used in Article 36 of API, includes the actual creation of a weapon and its improvement, refinement and testing[7] while 'acquisition' of weapons involves obtaining them, whether from commercial undertakings, from other States or otherwise; the 'adoption' of a method of warfare seems to contemplate a State or its armed forces deciding to use a weapon or method of warfare in military operations.

Although this is a treaty law obligation by which the overwhelming majority of States are bound, relatively few States are known to have formalised systems for the legal review of new weapons[8] and even fewer States are known actually to conduct legal reviews of new methods of warfare.

At the time of writing, there are 174 States party to API,[9] all of which are treaty bound to comply with the Article 36 requirement to review the lawfulness of new weapons and methods of warfare.

2.3 THE CUSTOMARY PRINCIPLES OF WEAPONS LAW

In the next few sections of this chapter we will summarise the law of weaponry with a view to identifying the criteria that should be applied when reviewing new weapon technology and which accordingly will be applied in the

4 Protocol Additional to the Geneva Conventions of 12 August 1949, and Relating to the Protection of Victims of International Armed Conflicts, Geneva, 8 June 1977 (API).

5 Lawand, 'A Guide to the Legal Review', at page 5.

6 A 'method of warfare' essentially consists of any way of conducting hostilities, such as aerial bombardment, blockade, siege or cyber operations.

7 Consider I. Daoust, R. Coupland and R. Ishoey, 'New Wars, New Weapons? The Obligation of States to Assess the Legality of Means and Methods of Warfare' (2002) 846 *IRRC* 345, 348.

8 Lawand, 'A Guide to the Legal Review', at page 5.

9 Source www.icrc.org viewed on 5 April 2018.

later chapters to assess the acceptability of the scientific developments considered there.

We should start by looking at three fundamental customary principles in the law of weaponry. In 1874 the authors of the Brussels Declaration[10] suggested that '[t]he laws of war do not recognize in belligerents an unlimited power in the adoption of means of injuring the enemy'. This notion was repeated in the Hague Regulations of 1899 and 1907[11] and in Article 35(1) of API. The further principles and rules set forth below represent the limitations alluded to in the Declaration, Regulations and Protocol. So the first principle is that international law prescribes that there must be limits; the law then specifies what those limits are.

The second principle prohibits the employment of weapons, projectiles and material and methods of warfare of a nature to cause superfluous injury or unnecessary suffering. This is a cardinal principle of the law of armed conflict[12] described in the Hague Regulations, 1907 as a prohibition of the employment of 'arms, projectiles or material calculated to cause unnecessary suffering'.[13] More recently, and in the opinion of the present author more authoritatively, the rule is described as prohibiting the employment of 'weapons, projectiles and material and methods of warfare of a nature to cause superfluous injury or unnecessary suffering'.[14] It is a customary rule that binds all States in relation to international and non-international armed conflicts.[15] The rule first appeared in the preamble to the St Petersburg Declaration of 1868.[16]

Applying the principle involves comparing the wounding effect, injury and other suffering consequent on the use of the weapon with an un-stated comparator. The general aim in using a weapon is to achieve a military advantage or purpose during armed conflict, so the superfluous injury/unnecessary

[10] Project of an International Declaration concerning the Laws and Customs of War, Brussels, 27 August 1874 (Brussels Declaration), Article 12.

[11] Regulations Respecting the Laws and Customs of War on Land annexed, respectively, to Hague Convention II of 29 July 1899 and Hague Convention IV of 18 October 1907, Article 22.

[12] International Court of Justice, Legality of the Threat or Use of Nuclear Weapons (Advisory Opinion) (1996) ICJ Reports 226, 35 ILM 809 (the *Nuclear Weapons* case), paras 74–87.

[13] Regulations annexed to Hague Convention IV, 18 October 1907, Respecting the Laws and Customs of War on Land, Article 23(e).

[14] API, Article 35(2) and for corresponding language in the Rome Statute of the International Criminal Court, Rome, 17 July 1998, see article 8(2)(b)(xx).

[15] J.-M. Henckaerts and L. Doswald-Beck, 'Customary International Humanitarian Law, Vol 1: Rules' *ICRC* (2005) (Customary Law study), 237, Rule 70.

[16] St Petersburg Declaration, preambular paras 3–6; G D Solis, The Law of Armed Conflict (CUP, 2011) at pages 269–272.

suffering test must of necessity involve comparing the nature and scale of the generic military advantage to be anticipated from the weapon in the application(s) for which it is designed to be used, with the pattern of injury and suffering associated with the normal intended use of the weapon.[17]

It follows from this that the principle requires an evaluation of the suffering, injury and military utility associated with the intended use of the weapon. These phenomena are hard to measure and difficult to compare. It should be remembered that a weapon is not rendered unlawful merely because it causes severe injury, suffering or loss of life. It is the injury or suffering inevitably caused by the weapon in its normal or designed circumstances of use that is the relevant factor and this must be disproportionate to its military purpose or utility for the rule to be broken. When making that assessment, consideration must be given to comparable lawful weapons in current use.[18]

This unnecessary suffering principle is self-evidently addressing the effects of a weapon on personnel. Weapons the designed effects of which are limited to objects, for example, are unlikely to breach this principle, unless of course the impact of the weapon on the relevant object is intended to have a consequent effect of causing superfluous injury or unnecessary suffering to persons.

The third principle of weapons law prohibits the employment of weapons that are of a nature to be indiscriminate. This indiscriminate weapons principle derives from the principle of distinction. The latter principle requires parties to the conflict at all times to distinguish between combatants and civilians and between military objectives and civilian objects and only to direct their military operations against combatants and military objectives.[19] The drafters of the Hague Draft Rules of Aerial Warfare, a text that was never adopted by States in treaty form and that is not a direct source of law, included at Article 24(3) a draft rule that would have prohibited indiscriminate attacks.[20]

[17] Consider W. J. Fenrick, 'The conventional weapons convention: A modest but useful treaty', 279 *IRRC* 498 (1990) 500 and W. Hays Parks, 'Means and Methods of Warfare, Symposium Issued in Honour of Edward R. Cummings', 38 *George Washington International Law Review* (2006) 511, 536 at note 25.

[18] See Hays Parks, 'Means and Methods of Warfare'.

[19] See F. Lieber, 'Instructions for the Government of Armies of the United States in the Field, 24 April 1863' (The Lieber Code), Articles 14, 15 and 22; the Preamble to the St Petersburg Declaration; Brussels Declaration, Articles 12, 13, 15 and 17; and Hague Regulations, 1907 Articles 23(g), 25 and 27. The modern formulation of the principle of distinction is to be found in API, Articles 48, 51 and 52.

[20] Rules of Aerial Warfare, The Hague, 1923, Article 24(3) asserted that in cases where objectives cannot be bombarded without the indiscriminate bombardment of the civilian population, 'the aircraft must abstain from bombardment'.

The first appearance of a treaty rule prohibiting both indiscriminate attacks and indiscriminate weapons was in 1977[21] with the adoption of API. Article 51(4) of that treaty prohibits indiscriminate attacks which it defines as:

> '(a) those which are not directed at a specific military objective;
>
> (b) those which employ a method or means of combat which cannot be directed at a specific military objective; or
>
> (c) those which employ a method or means of combat the effects of which cannot be limited as required by th[e] Protocol;
>
> and [which], consequently, in each such case, are of a nature to strike military objectives and civilians or civilian objects without distinction.'[22]

Sub-paragraphs (b) and (c) set forth the indiscriminate weapons principle. It is those sub-paragraphs that prohibit weapons that, either because they cannot be directed at a specific military objective or because their effects cannot be limited, essentially do not distinguish as required by the principle of distinction.[23] Accordingly, any new technology weapon is going to be unlawful if, when used in its normal or designed circumstances, it cannot be directed at a specific military objective, or if its effects cannot be reasonably restricted to the target, and if as a result its nature is to strike lawful targets, whether in the form of objects or persons, and protected persons and objects without distinction. Of course, any weapon is capable of indiscriminate use. The indiscriminate weapons principle is concerned with the inherent nature or characteristics of the weapon and not with its use on a particular occasion.

So, if when a new weapon technology is being studied or developed, it is clear that likely uses of it will have incidental injurious effects on civilians or incidental damaging effects on civilian objects, a legal review of the weapon technology should draw attention to this and to the consequent need that commanders receive targeting law advice before employing such weapons in the relevant circumstances. The legal advice at the study and developmental stages is likely also to suggest that the technology should be developed if possible with a view to minimising or obviating entirely such adverse incidental effects on civilians and civilian objects. The weapon review may also need

[21] Consider Hague Regulations 1907, Article 27 and Hague Convention IX, 1907, Article 5, neither of which addresses indiscriminate weapons explicitly.

[22] The discrimination rule in Article 51(4) of API is an important element in compliance with the principle of distinction.

[23] Consider the *Nuclear Weapons* case, 257: 'States must never make civilians the object of attack and must consequently never use weapons that are incapable of distinguishing between civilian and military targets.'

to consider the alternative methods of achieving the same generic military purpose to try to determine whether the new technology is more or less discriminating than current methods for achieving that relevant military purpose.

Like the superfluous injury principle, the customary indiscriminate weapons principle binds all States irrespective of whether they are party to API and applies in both international and non-international armed conflicts.[24]

2.4 WEAPONS AND THE ENVIRONMENT

The two kinds of weapons law rule that relate to the environment deal, respectively, with the causing of damage to the environment and with the use of the environment as a weapon. We will deal first with the rules that relate to damage caused to the natural environment by military activities in armed conflict. API has two rules on this subject that do not have customary status and which do not therefore bind States that are not party to API. The customary law rule which binds all States characterises the natural environment as a civilian object and thus accords it the same protection from attacks as is enjoyed by all civilian objects. The natural environment must not, therefore, be made the object of attack, unless the relevant part of it becomes a military objective, and must not be the subject of wanton destruction.[25] The reference in this rule is to direct damage to the natural environment, whether as the intended or as the incidental outcome of a military operation.[26] It follows from this that when developing new weapons all States must have due regard to the impact they may be expected to have on the natural environment.[27]

States that are party to API are bound by Article 35(3) which prohibits the use of methods or means of warfare which are intended or may be expected to cause widespread, long-term and severe damage to the natural environment. Article 55 prohibits such methods or means that thereby prejudice the health

[24] ICRC Customary Law Study report, vol. 1, Rule 71 at p. 244. See also, in relation to international armed conflicts, Rome Statute, Article 8(2)(b)(xx).

[25] AMW *Manual*, Rules 88 and 89; *Tallinn Manual*, Rule 83(a). The *US Department of Defense Law of War Manual* (2015), para 6.10.3.1 notes the US rejection of Articles 35(3) and 55 of API and expresses the view that 'use of such weapons is prohibited only if their use is clearly excessive in relation to the concrete and direct overall military advantage anticipated'. While the way in which the DoD Manual expresses the point can be criticised, it is clear that the USA regards environmental damage as only prohibited when it breaches proportionality.

[26] Consider Y. Dinstein, *The Conduct of Hostilities under the Law of International Armed Conflict*, 3rd edn, (2016), 237–238.

[27] AMW *Manual*, Rule 89.

or survival of the population.[28] The terms 'widespread, long-term and severe' are cumulative such that if any one such factor is not present, the rule will not have been breached. API contains no definition of 'widespread, long-term and severe', but only grave damage will breach the rule.[29] A number of States when ratifying API made statements excluding the application of, inter alia, this rule to nuclear weapons.[30]

The UN Environmental Modification Convention (ENMOD) prohibits States party from engaging in military or any other hostile use of environmental modification techniques having widespread, long-lasting or severe effects as the means of destruction, damage or injury to any other State party to that Convention.[31] The term 'environmental modification techniques' means 'any technique for changing – through the deliberate manipulation of natural processes – the dynamics, composition or structure of the Earth, including its biota, lithosphere, hydrosphere and atmosphere, or of outer space '.[32] There are therefore a number of factors that must be present before the ENMOD prohibition is breached, namely:

1. the dynamics, composition or structure of one of the listed items must have been changed;
2. natural processes must have been manipulated to achieve such an effect;
3. the manipulation must have been deliberate;
4. the use of the technique must have been military or hostile in nature;

[28] API, Article 55(1).

[29] 'The time or duration required (i.e. long-term) was considered by some to be measured in decades. References to twenty or thirty years were made by some representatives as being a minimum. Others referred to battlefield destruction in France in the First World War as being outside the scope of the prohibition... It appeared to be a widely shared assumption that battlefield damage incidental to conventional warfare would not normally be proscribed by this provision. What the article is primarily directed to is thus such damage as would be likely to prejudice, over a long term, the continued survival of the civilian population or would risk causing it major health problems.'; Rapporteur's Report CDDH/215/Rev.1 para 27, reported in ICRC Commentary, para 1454.

[30] Consider, for example, statement (a) made by the UK on ratification of API on 28 January 1998. Other NATO States made similar statements.

[31] Convention on the Prohibition of Military or any Other Hostile Use of Environmental Modification Techniques, Geneva, 2 September 1976 (ENMOD), Article I.

[32] ENMOD, Article II. The term 'widespread' has been interpreted as 'encompassing an area on the scale of several hundred square kilometres', while 'long-lasting' would involve 'lasting for a period of months, or approximately a season' and effects are 'severe' if they involve 'serious or significant disruption or harm to human life, natural and economic resources or other assets.' Conference Understanding relating to Article I, available at the ICRC treaty database at www.icrc.org.

5. widespread, long-lasting or severe effects must have resulted;
6. those effects must have been the means of causing destruction, damage or injury to another State; and
7. both States, that is the State using the technique and the victim State, must be party to the Convention.

ENMOD only applies when the destructive or injurious effect is suffered by another State, so it only effectively applies to international armed conflicts between States party. Environmental modification would include an attempt to modify the weather by either increasing or reducing rainfall in an area in order to bring about floods or drought[33] and the results of environmental modification may include earthquakes, tsunamis, a disturbance in the ecological balance of a region, changes in weather and climate patterns.[34] So any new weapon technology that is used to initiate environmental modification that produces the prohibited damage and effects is likely to breach the treaty rule.

2.5 WEAPONS LAW RULES RELATING TO SPECIFIC WEAPON TYPES

International law includes rules that prohibit the use of certain weapons or weapon technologies. Some weapon prohibitions apply only to States that are party to a relevant treaty and attention is drawn to this aspect where it arises.

2.5.1 *Poisons and Gases*

The long-established prohibition of the use of poison or poisoned weapons was noted in the Lieber Code,[35] the Brussels Declaration and the Oxford Manual.[36] It is set forth as a treaty rule in the Hague Regulations of 1899 and 1907.[37] It is a customary law prohibition that binds all States in both international and non-international armed conflicts.

Poison or poisoned weapons comprise weapons whose primary, or even exclusive, effect is to poison or asphyxiate.[38] The Rule prohibits any substance

33 P. J. Rowe, *Defence, the Legal Implications* (London: Brassey's 1987) at page 117.
34 Conference Understanding relating to Article II, available at the ICRC treaty database at www.icrc.org.
35 Lieber Code, Article 16.
36 Brussels Declaration, Article 12 and Oxford Manual, Article 8(a).
37 Hague Regulations 1907, Article 23(a).
38 ICJ, *Nuclear Weapons* Case, para 111.

that is used to aggravate a wound. It also prohibits the poisoning of wells, pumps and rivers from which the enemy draws water supplies.[39]

2.5.2 *Expanding Bullets*

The Hague Declaration 3 of 1899[40] prohibits 'the use of bullets which expand or flatten easily in the human body, such as bullets with a hard envelope which does not entirely cover the core or is pierced with incisions'. This treaty prohibition reflects customary law and thus applies to all States in relation to the conduct of hostilities in international armed conflicts. In non-international armed conflicts, a similar customary rule is subject to certain limited exceptions. These may arise, e.g., in situations, such as hostage-taking, when it is necessary instantly to disable the targeted individual to protect civilians or in situations where use of a conventional bullet involves the risk of ricochet putting civilians in danger. If, in such relatively limited non-international armed conflict situations, the use of expanding ammunition yields an additional military utility, e.g., by facilitating compliance with distinction and discrimination, such use would not therefore occasion superfluous injury/unnecessary suffering and would not, on this basis, be regarded as breaching the customary law rule.[41]

2.5.3 *Exploding Bullets Designed for Anti-Personnel Use*

Under the St Petersburg Declaration, the employment during international armed conflict of projectiles below 400 grammes weight that are either explosive or charged with fulminating or inflammable substances was prohibited. Later technical developments rendered the 400-gramme limit obsolete and this was noted in the 1923 draft Hague Rules of Aerial Warfare.[42] However, State practice shows that explosive or incendiary bullets designed solely for use against personnel are prohibited under customary law. The likely

39 H. Lauterpacht (ed), *Oppenheim's International Law* Vol II (Reissue of 1952 edn, Longmans 1955) available at www.gutenburg.org/files/41047/41047-h/41047-h.htm) 340; ICRC Customary Law Study Report, vol. 1, 253 and API Commentary, para 1419.

40 Declaration Concerning Expanding Bullets, The Hague, 29 July 1899.

41 *UK Manual*, para 6.9 and note 32. Note the resolution adopted on 10 June 2010 by the Kampala Review Conference for the Rome Statute of the International Criminal Court, RC/Res 5 adopted at the 12th Plenary Meeting, which makes it clear that the crime of using expanding bullets in a non-international armed conflict is only committed 'if the perpetrator employs the bullets to uselessly aggravate suffering or the wounding effect upon the target of such bullets, as reflected in customary international law'; preambular para 9.

42 Draft Hague Rules of Aerial Warfare, 1923, Article 18(2).

basis for this is that a solid round would achieve the relevant military purpose, so using an explosive round in such circumstances will cause injury for which there is no corresponding military purpose.[43]

2.5.4 *Asphyxiating Gas*

The use of asphyxiating, poisonous or other gases, all analogous liquids, materials or devices and bacteriological methods of warfare is prohibited. Hague Declaration 2 of 1899 addressed the use of asphyxiating gas in warfare but did not prevent its widespread use during World War I. In large measure because of the World War I experience, the Geneva Gas Protocol[44] was adopted in 1925. The Protocol prohibited the use in war of asphyxiating, poisonous or other gases, of all analogous liquids, materials or devices and of bacteriological methods of warfare. This prohibition applies to all States in connection with both international and non-international armed conflicts. Accordingly, the development of any new weapons technology in the form of a gas, an analogous liquid, a material or a device that is designed to asphyxiate[45] or to poison is explicitly prohibited by this rule which is a customary law rule binding on all States in both international and non-international armed conflicts. The prohibition of bacteriological methods of warfare will be subsumed within the discussion of the Biological Weapons Convention below.

2.5.5 *Fragmentation Weapons*

The first Protocol adopted under the Conventional Weapons Convention[46] provides that 'it is prohibited to use any weapon the primary effect of which is to injure by fragments which in the human body escape detection by x-rays'.[47] To come within this prohibition, a new weapon technology would need to disperse fragments, those fragments would need to be non-detectable by the technology referred to in the treaty, namely x-rays, and injury by such fragments would need to be the primary effect of the weapon.

[43] See *UK Manual*, para 6.10.1. This implies that the prohibition is based on the customary superfluous injury and unnecessary suffering principle as opposed to the treaty rule.

[44] Protocol for the Prohibition of the Use in War of Asphyxiating, Poisonous or Other Gases, and of Bacteriological Methods of Warfare, Geneva, 17 June 1925.

[45] 'Asphyxia' is defined as 'a condition arising when the body is deprived of oxygen, causing unconsciousness or death'; *Concise Oxford English Dictionary*, 11th edn (2006), 77.

[46] Convention on Prohibitions or Restrictions on the Use of Certain Conventional Weapons Which may be Deemed to be Excessively Injurious or to Have Indiscriminate Effects, Geneva, 10 October 1980 (CCW).

[47] Protocol on Non-Detectable Fragments, Geneva, 10 October 1980.

2.5.6 *Laser Weapons*

The fourth Protocol to the Conventional Weapons Convention[48] prohibits the employment of 'laser weapons specifically designed, as their sole combat function or as one of their combat functions, to cause permanent blindness to unenhanced vision, that is, to the naked eye or to the eye with corrective eyesight devices'.[49] The Protocol distinguishes between laser weapons that are specifically designed as a combat function to cause permanent blindness and '[b]linding as an incidental or collateral effect of the legitimate military employment of laser systems, including laser systems used against optical equipment'. The distinction is important because the latter, incidental or collateral kind of blinding 'is not covered by the prohibition of this Protocol'.[50] It is important to note, however, that the treaty requires States party, when employing laser systems, to 'take all feasible precautions to avoid the incidence of permanent blindness to unenhanced vision', noting that such precautions should include training and other practical measures.[51]

So, the mere fact that a weapon based on novel technology could cause incidental or collateral permanent blindness does not per se render that novel technology weapon unlawful. Furthermore, a novel technology weapon that is designed as a combat function to cause temporary blindness or to dazzle does not breach the Protocol. However, when using any such weapons, everything practicable, including training and precautions at the time of use, must be done to avoid causing permanent blindness. Moreover, laser or similar equipment designed to cause damage to other objects or vehicles or to perform measurement, range-finding or other military tasks but that have no combat function to cause permanent blindness are not prohibited by this treaty rule.

2.5.7 *Chemical Weapons*

The Geneva Gas Protocol 1925 did not prohibit the possession of chemical weapons. The Chemical Weapons Convention 1993[52] is an arms control treaty in the sense that it does not limit itself to prohibiting the use of chemical

[48] Protocol on Blinding Laser Weapons, Geneva, 13 October 1995 (Protocol IV).

[49] CCW, Protocol IV, Article 1. Permanent blindness means irreversible and uncorrectable loss of vision which is seriously disabling with no prospect of recovery. Serious disability is equivalent to visual acuity of less than 20/200 Snellen measured using both eyes; Protocol IV, Article 4.

[50] CCW, Protocol IV, Article 3.

[51] CCW, Protocol IV, Article 2.

[52] Convention on the Prohibition of the Development, Production, Stockpiling and Use of Chemical Weapons and on Their Destruction, Paris, 13 January 1993 (Chemical Weapons Convention).

weapons but also bans the development, production, acquisition, stockpiling or retention of chemical weapons or their direct or indirect transfer to anyone. States party are prohibited from preparing to use them and must not assist, encourage or induce anyone to do any of these things.[53]

For the purposes of this rule, chemical weapons are toxic chemicals, their precursors and equipment and munitions connected with their use. The rule does not apply to such substances if they are intended for purposes not prohibited under the Convention, provided that the types and quantities must be consistent with such purposes.[54] Toxic chemicals are any chemicals whatever their origin and however or wherever produced which, through their chemical action on life processes can cause death, temporary incapacitation or permanent harm to humans or animals.[55] A 'precursor' is any chemical reactant which takes part, at any stage, in the production, by whatever method, of a toxic chemical.[56]

The purposes that the Convention does not prohibit are industrial, agricultural, research, medical, pharmaceutical or other peaceful purposes; protective purposes, namely those purposes directly related to protection against toxic chemicals and to protection against chemical weapons; military purposes not connected with the use of chemical weapons and not dependent on the use of the toxic properties of chemicals as a method of warfare; and law enforcement, including domestic riot control purposes.[57] 'Law enforcement' refers to the enforcement of domestic law.[58]

The Chemical Weapons Convention defines riot control agents as 'chemicals not listed in a Schedule to the Treaty which can produce rapidly in humans sensory irritation or disabling physical effects which disappear within a short time following termination of exposure'.[59] It is lawful to use such weapons during an armed conflict for law enforcement, including domestic riot-control purposes. It is, however, prohibited to use them as a method of warfare.

The Convention applies to international and non-international armed conflicts and includes an important verification annex.[60] The prohibitions

53 Chemical Weapons Convention, Article I(1).
54 For the precise definition see Chemical Weapons Convention, Article II(1).
55 Chemical Weapons Convention, Article II(2).
56 Chemical Weapons Convention, Article II(3).
57 Chemical Weapons Convention, Article II(9).
58 D. P. Fidler, 'The Meaning of Moscow: "Non-Lethal"' Weapons and International Law in the 21st Century', (2005) 859 *IRRC* 525 at pages 540–4.
59 Chemical Weapons Convention, 1993, Article II(7).
60 The Chemical Weapons Convention Annex on Implementation and Verification includes complex provisions as to national declarations and verification arrangements.

as to use and possession of chemical weapons, and probably the remaining activities prohibited by the Convention, now reflect customary law[61] and so bind all States, whether they are party to the Convention or not.

2.5.8 *Biological Weapons*

Although the 1925 Geneva Gas Protocol banned bacteriological warfare, the treaty did not prohibit the stockpiling of bacteriological weapons. Article 1 of the Biological Weapons Convention[62] prohibits the development, production, stockpiling, acquisition and retention of microbial or other biological agents or toxins whatever their origin or method of production, of types and in quantities that have no justification for prophylactic, protective or other peaceful purposes, and weapons or equipment for their use for hostile purposes or in armed conflict.[63] The Fourth Review Conference of the Biological Weapons Convention took place in 1996 and agreed that the Convention also effectively prohibits the use of such weapons,[64] although States party to the Geneva Gas Protocol are prohibited from using biological weapons by that provision anyway.

The numerous activities prohibited by the Convention show that it is also an arms control treaty and its prohibitions apply to both international and non-international armed conflicts. Almost all militarily significant States are party to the Convention and the general practice of States establishes that the prohibitions on possession and use of such weapons, and almost certainly the other activities prohibited by the Convention, reflect customary law and thus bind all States in relation to both international and non-international armed conflicts.

The position in relation to both chemical and biological weapons is therefore clear. It is unlawful for any State, whether it is party to the Conventions or not, to equip itself for or undertake a chemical or biological attack. No State is permitted to develop, produce, stockpile, acquire, retain or use any kind of new technology that uses, or that in any way contributes to the use, of toxic chemicals or their precursors or microbial or other biological agents or toxins for hostile purposes or in an armed conflict. This prohibition will also apply, for example, to such use of synthetic biological substances, bacteria, viruses and incapacitating chemical substances.

[61] Note that there are now 192 States party to the Chemical Weapons Convention, source www.icrc.org viewed on 5 April 2018.

[62] The Convention on the Prohibition of the Development, Production and Stockpiling of Bacteriological (Biological) and Toxin Weapons and on their Destruction, 1972, was opened for signature on 10 April 1972 (Biological Weapons Convention).

[63] Biological Weapons Convention, Article I.

[64] *UK Manual*, 104, note 8.

2.5.9 *Incendiary Weapons*

Some weapons law rules restrict the circumstances in which the relevant class of weapon may lawfully be used. Protocol III[65] to the Conventional Weapons Convention makes that kind of provision in relation to incendiary weapons. These are defined as 'any weapon or munition which is primarily designed to set fire to objects or to cause burn injury to persons through the action of flame, heat, or a combination thereof, produced by a chemical reaction of a substance delivered on the target'.[66]

Under the Protocol '[i]t is prohibited in all circumstances to make any military objective located within a concentration of civilians[67] the object of attack by air-delivered incendiary weapons'.[68] The Protocol also prohibits making military objectives located within a concentration of civilians the object of attack using incendiary weapons other than air-delivered incendiary weapons 'except when such military objective is clearly separated from the concentration of civilians and all feasible precautions are taken with a view to limiting the incendiary effects to the military objective and to avoiding, and in any event to minimizing, incidental loss of civilian life, injury to civilians and damage to civilian objects.'[69]

In addition, the Protocol prohibits making forests or other kinds of plant cover the object of attack using incendiary weapons except when such natural elements are used to cover, conceal or camouflage combatants or other military objectives, or have themselves become military objectives.[70]

Accordingly, the first question to consider when evaluating relevant novel weapons technologies is whether the capability in question comes within the definition of an incendiary weapon. If it does, the next important issue to consider is the delivery system. If it amounts to air delivery, then the prohibition in Article 2(2) of the Protocol will apply. If it is not an air delivery system, the restrictions in Article 2(3) of the Protocol will be relevant.

65 Protocol on Prohibitions or Restrictions on the Use of Incendiary Weapons (Protocol III), Geneva, 10 October 1980.

66 Protocol III article 1(1). The definition notes that such weapons can take a variety of forms, but excludes munitions with incidental incendiary effects, such as tracers or illuminants, smoke or signaling systems and combined effects munitions in which the incendiary effect is designed to be used against objects not persons.

67 A 'concentration of civilians' may be permanent or temporary, and can include inhabited parts of cities, towns, villages, camps, columns of refugees or groups of nomads; Protocol III, Article 1(2).

68 Protocol III, Article 2(2).

69 Protocol III, Article 2(3).

70 Protocol III, Article 2(4).

So, if a weapon system employing novel technology were to involve the deployment of an incendiary weapon against a target, the way in which the weapon is fired to the target would determine whether the weapon system is subject to the prohibition of attacking military objectives located within concentrations of civilians using air-delivered incendiaries. If some method were to be used other than air delivery, and this might consist of an entirely novel method, the prohibition in Article 2(2) of the Protocol would not apply. However, Article 2(3) would only permit the use of such a weapon against a military objective located within a concentration of civilians if the conditions specified in that paragraph of the treaty are satisfied.[71]

2.5.10 *Nuclear Weapons*

International law does not specifically prohibit, nor does it explicitly permit, the use of, or threats to use, nuclear weapons. It is the general principles of international law and the rules of the law of armed conflict that determine whether any such use or threat is lawful. When it addressed these matters in July 1996, the International Court of Justice came to the conclusion that there is in neither customary nor in conventional international law any specific authorisation of the threat or use of nuclear weapons and that there is in neither customary nor conventional international law any comprehensive and universal prohibition of the threat or use of nuclear weapons as such.

In its judgment, the Court noted that a threat to use, or use of, nuclear force that is contrary to Article 2(4) of the UN Charter and that fails to meet the requirements of Article 51 is unlawful and that such a threat or use should be compatible with the law of armed conflict and with treaties and other undertakings dealing expressly with nuclear weapons. Having acknowledged that a threat or use of nuclear weapons would generally be contrary to the law of armed conflict, the Court stated that it could not 'conclude definitively whether the threat or use of nuclear weapons would be lawful or unlawful in an extreme circumstance of self defence, in which the very survival of a state would be at stake.'[72]

[71] I.e. that the military objective is clearly separated from the concentration of civilians, that all feasible precautions are taken with a view to limiting the incendiary effects to the military objective and to avoiding and in any event minimising incidental loss of civilian life, injury to civilians and damage to civilian objects.

[72] *Nuclear Weapons* case, para 97.

This non liquet judgment by the Court has been the subject of academic criticism[73], but the legal view expressed in the judgment represents the position of the Court unless and until it reconsiders the matter. It can therefore be concluded that the possession of nuclear weapons as such does not breach any specific rule under the law of armed conflict. States are however bound by any other relevant treaties to which they are party such as the Nuclear Non-Proliferation Treaty.

API was ratified by a number of NATO States on the basis of a stated understanding that the new rules introduced by the treaty would not apply to nuclear weapons.[74] The statements did not specify which rules were considered to be covered by these declarations on ratification, but they are generally recognised as including the environmental protection rules in Articles 35(3) and 55 and the prohibition on taking reprisals against civilians and civilian objects. It has been suggested that they also include the indiscriminate weapons principle.[75] The proportionality and discrimination rules[76] do, however, apply to any use of nuclear weapons. The lawfulness of taking the reprisal action prohibited by API will depend on the position taken by the relevant State when ratifying that treaty.[77]

States that do not have nuclear weapons are bound under Article II of the Non-Proliferation Treaty not to receive the transfer from any transferor whatsoever of nuclear weapons or nuclear explosive devices or of control over them, and not to seek or receive any assistance in the manufacture of such weapons or explosive devices.[78]

[73] See for example T. L. H. McCormack, 'A Non Liquet on Nuclear Weapons – The ICJ Avoids the Application of General Principles of International Humanitarian Law' (1997), 316 *IRRC* 76 and C. Greenwood, 'The Advisory Opinion on Nuclear Weapons and the Contribution of the International Court to International Humanitarian Law' (1997), 316 *IRRC* 65, 73.

[74] See, for example, statement (a) made by the UK on ratification of API on 28 January 1998, first statement made by Belgium on ratification of API on 20 May 1986, first statement by Italy on ratification of API on 27 February 1986 and statement 2 made by France on ratification of API on 11 April 2001.

[75] F. Kalshoven, 'Arms, Armaments and International Law (1985-II)' 191 *RdC* 183–341, at 236 and 287.

[76] API, Article 51(5)(b) and 51(4) respectively.

[77] Consider in that regard the statement made by the UK on ratification of API on 28 January 1998, statement (m).

[78] Treaty on the Non-Proliferation of Nuclear Weapons, New York, 1 July 1968. Consider also the Treaty on the Prohibition of Nuclear Weapons, New York, 7 July 2017, which will enter into force 90 days after the 50th ratification. Article 4 provides for international verification of the elimination of nuclear weapons-related facilities; Article 2 requires declarations by States party as to their ownership, possession or control of nuclear weapons, whether they host such weapons and as to the elimination of them. Article 1 contains the core obligations not to develop, test, produce, manufacture, acquire, possess or stockpile nuclear weapons, not to

2.5.11 *Mines, Booby-Traps and Other Devices*

Protocol II[79] and Amended Protocol II[80] are concerned with the use in armed conflict of mines,[81] booby traps and other devices.[82] This part of the discussion will start by looking at the weapons prohibited by these two treaties and will then note the restrictions that the treaties impose. It should be noted, however, that the law relating to such weapons is extensive and complex and the present discussion is only intended to amount to an overview.

Article 1 of Protocol II defines booby-traps as 'any device or material which is designed, constructed or adapted to kill or injure and which functions unexpectedly when a person disturbs or approaches an apparently harmless object or performs an apparently safe act'.[83] Article 6 of the Protocol prohibits the use of booby-traps in the form of apparently harmless portable objects if they are specifically designed and constructed to contain explosive material and to detonate when they are disturbed or approached.[84] Protocol II also prohibits the use of booby-traps in any way attached to or associated with:

> 'internationally recognised protective emblems, signs or signals;
> sick, wounded or dead persons;
> burial or cremation sites or graves;
> medical facilities, medical equipment, medical supplies or medical transportation;
> children's toys or other portable objects or products specially designed for the feeding, health, hygiene, clothing or education of children;

directly/indirectly transfer or receive them, not to use or threaten to use them, not to assist, encourage or induce such actions and not to permit stationing of them on a State party's territory. The treaty is not yet in force and the States that are permanent members of the UN Security Council seem unlikely to ratify. Currently 5 States have ratified the treaty; source www.icanw.org viewed on 4 March 2018.

79 Adopted in Geneva on 10 October 1980.

80 Adopted in Geneva on 3 May 1996.

81 'Mine means any munition placed under, on or near the ground or other surface area and designed to be detonated or exploded by the presence, proximity or contact of a person or vehicle.'; Protocol II, Article 2(1). There are minor and relatively inconsequential amendments to the equivalent definition in Amended Protocol II, Article 2(1).

82 'Other devices means manually-emplaced munitions and devices designed to kill, injure or damage and which are actuated by remote control or automatically after a lapse of time.'; Protocol II, Article 1(3). The Amended Protocol II definition of 'other devices' is: 'Other devices means manually-emplaced munitions and devices including improvised explosive devices designed to kill, injure or damage and which are actuated manually, by remote control or automatically after a lapse of time.'; Amended Protocol II, Article 2(5).

83 Protocol II, Article 1(2).

84 Protocol II, Article 6(1)(a).

food or drink;
kitchen utensils or appliances except in military establishments, military locations or military supply depots;
objects clearly of a religious nature;
historic monuments, works of art or places of worship which constitute the cultural or spiritual heritage of peoples;
animals or their carcasses.'[85]

Amended Protocol II specifically prohibits the following munitions:

'mines, booby-traps or other devices which employ a mechanism or device specifically designed to detonate the munition by the presence of commonly available mine detectors as a result of their magnetic or other non-contact influence during normal use in detection operations';[86]

'a self-deactivating mine equipped with an anti-handling device that is designed in such a manner that the anti-handling device is capable of functioning after the mine has ceased to be capable of functioning';[87]

anti-personnel mines that do not incorporate in their construction a material or device that enables the mine to be detected by commonly-available technical mine detection equipment and that provides a response signal equivalent to a signal from 8 grammes or more of iron in a single coherent mass;[88]

remotely delivered[89] anti-personnel mines which do not comply with certain self-destruction and self-deactivation requirements;[90]
certain remotely-delivered mines other than anti-personnel mines;[91]
booby-traps or other devices in the form of apparently harmless portable objects which are specifically designed and constructed to contain explosive material.[92]

For States that are party to the Ottawa Convention, anti-personnel landmines are entirely prohibited weapons. The Convention defines 'anti-personnel

85 Protocol II, Article 6(1); Amended Protocol II, Article 7(1).
86 Amended Protocol II, Article 3(5).
87 Amended Protocol II, Article 3(6).
88 Amended Protocol II, Article 4 and Technical Annex, para 2(a). This restriction applies to anti-personnel mines produced after 1 January 1997. Mines produced before that date must either incorporate into their construction or have attached prior to their emplacement in a manner that is not easily removable a material or device with identical characteristics; Amended Protocol II, Technical Annex para 2(b).
89 For the definition of 'remotely delivered', see Amended Protocol II, Article 2(1).
90 Amended Protocol II, Article 6(2). For the requirements, see Amended Protocol II, Technical Annex, para3(a) taken with (b).
91 Amended Protocol II, Article 6(3).
92 Amended Protocol II, Article 7(2).

mine' as 'a mine designed to be exploded by the presence, proximity or contact of a person and that will incapacitate, injure or kill one or more persons. Mines designed to be detonated by the presence, proximity or contact of a vehicle as opposed to a person, that are equipped with anti-handling devices, are not considered anti-personnel mines as a result of being so equipped'.[93]

The Ottawa Convention is an arms control treaty in the sense that it provides that States party must never use, develop, produce, otherwise acquire, stockpile, retain or transfer anti-personnel mines or assist, encourage or induce, in any way, anyone to do so. The Convention makes detailed provision as to the destruction of anti-personnel mines in stockpiles and in mined areas and limits the retention of anti-personnel mines for training purposes. The Convention also includes an extensive compliance regime.[94] The Convention prohibitions and requirements apply to international and non-international armed conflicts but do not yet reflect customary law.

2.5.12 *Cluster Munitions*

Another arms control treaty, the Cluster Munitions Convention[95] obliges States party never to use, develop, produce, otherwise acquire, stockpile, retain or transfer cluster munitions or to assist, encourage or induce anyone to do any of these things.[96]

The Convention defines 'cluster munition' as 'a conventional munition that is designed to disperse or release explosive sub-munitions each weighing less than 20 kilograms and includes those explosive sub-munitions'. The definition excludes: munitions or sub-munitions designed to dispense flares, smoke, pyrotechnics or chaff; munitions designed exclusively for an air defence role; munitions or sub-munitions designed to produce electrical or electronic effects; munitions that, in order to avoid indiscriminate area effects and the risks posed by unexploded sub-munitions, contain fewer than 10 explosive sub-munitions each of which weighs more than 4 kilograms, is designed to detect and engage a single target object, is equipped with an electronic self-destruction mechanism and is equipped with an electronic self-deactivating feature.[97] Explosive sub-munitions consist of conventional munitions which, in order to perform their

[93] Ottawa Convention, Article 2(1).
[94] See Ottawa Convention 1997, Article 8.
[95] Convention on Cluster Munitions, Dublin, 30 May 2008 (Cluster Munitions Convention).
[96] Cluster Munitions Convention, Article 1(1).
[97] Cluster Munitions Convention, Article 2(2).

task, are dispersed or released by a cluster munition and are designed to function by detonating an explosive charge prior to, on or after impact.[98]

Article 21 of the Convention requires States party to encourage States that are not party to participate in the Convention, obliges a State that is party to notify the governments of States that are not party of its Convention obligations but explicitly permits States party or their military personnel or nationals to 'engage in military cooperation and operations with States not party to th[e] Convention that might engage in activities prohibited to a State party'.[99] This provision does not permit a State party itself to acquire, transfer or use cluster munitions, neither may the State party specifically request the use of cluster munitions where the choice of munitions used is within its exclusive control.[100]

The use of certain weapons is subject to treaty law restrictions. Both Protocol II and Amended Protocol II to CCW restrict the use of mines, booby-traps and other devices. Some restrictions simply re-state targeting law rules that apply to the use of all weapons and it is not therefore necessary to repeat them here. The following restrictions under Protocol II reflect established doctrine and are customary in nature. The Amended Protocol II restrictions only bind States that are party to that treaty.

Article 4 of Protocol II prohibits the use of mines other than remotely-delivered mines, booby-traps and other devices in any city, town, village or other area containing a similar concentration of civilians in which combat between ground forces is not taking place or does not appear to be imminent unless either they are placed on or in the close vicinity of the military objective belonging to or under the control of an adverse party, or measures are taken to protect civilians from their effects such as, for example, the posting of warning-signs, the posting of sentries, the issue of warnings or the provision of fences.[101]

Article 5 of Protocol II limits the use of remotely-delivered mines to areas that are or contain military objectives and requires that their location can be accurately recorded and that an effective neutralising or remotely-controlled mechanism is used on each such mine.[102] Article 7 of Protocol II requires that parties to the conflict record the location of all pre-planned minefields that

98 Cluster Munitions Convention, Article 2(3). The treaty defines the notions of self-destruction and self-deactivation.

99 Cluster Munitions Convention, Article 21(3).

100 Cluster Munitions Convention, Article 21(4).

101 'Remotely delivered mines' means, for the purposes of Protocol II, any mine delivered by artillery, rocket, mortar or similar means or dropped from an aircraft.

102 See Protocol II to CCW, Article 5(1).

they lay and of all areas in which they have made large-scale and pre-planned use of booby-traps.

Amended Protocol II, Article 3, makes a party to the conflict that employs mines, booby-traps or other devices responsible to clear, remove, destroy or maintain them as specified later in the Treaty. Practical precautions must be taken.[103] Article 5 of the same treaty prohibits use of anti-personnel mines other than remotely-delivered mines[104] if they are not designed and constructed according to criteria set forth in the treaty.[105] Weapons that do not comply with those requirements may be used by a State party to Amended Protocol II if the treaty's conditions as to use and subsequent action are met.[106]

Some weapons of this sort are excluded from the restrictions referred to in the previous paragraph. This exclusion applies to weapons which propel fragments in a horizontal arc of less than 90° and which are placed on or above the ground. The maximum permitted use of such weapons is seventy-two hours, they must be located in immediate proximity to the military unit that emplaced them and the area must be monitored by military personnel to ensure the effective exclusion of civilians.[107]

Remotely delivered mines may only be used if they are recorded in accordance with the Technical Annex and remotely delivered mines other than anti-personnel mines must, to the extent feasible, be equipped with an effective self-destruction or self-neutralisation mechanism and have a back-up self-deactivation feature which is designed so that the mine will no longer function as a mine when the mine no longer serves the military purposes for which it was placed in position. Amended Protocol II prohibits the use of booby-traps or other devices in any town, village or other area containing a similar concentration of civilians and in which combat between ground forces is not taking place or does not appear to be imminent, unless either they are placed on or in the close vicinity of a military objective; or, measures are taken to protect civilians from their effects, for example, by the posting of warning-sentries, the issuing of warnings or the provision of fences.[108]

[103] As to the factors to consider, see Amended Protocol II, Article 3(10).

[104] 'Remotely delivered mine' is defined in Article 2(2), Amended Protocol II.

[105] See Amended Protocol II, Technical Annex, para 3(a) and (b).

[106] For these requirements, see Amended Protocol II, Article 5(2).

[107] Amended Protocol II, Article 5(6).

[108] Amended Protocol II, Article 7(3).

2.5.13 *Non-Lethal Weapons*

Weapons law applies to non-lethal weapons in just the same way as it applies to lethal weapons.[109] 'Non-lethal' is not a designation provided for in the law of armed conflict and non-lethal weapons are not limited to those that have zero lethality; indeed, lethality will inevitably in part depend on how the weapon is used. Nevertheless, fatal or permanent injury should generally be less frequent than is to be expected from the use of conventional lethal weapons. Importantly, when conducting a weapon review of a non-lethal weapon, the same legal criteria must be applied as in the case of lethal weaponry.

2.6 THE CRITERIA THAT DETERMINE THE LAWFULNESS OF A WEAPON TECHNOLOGY

The criteria to be applied in order to determine the lawfulness of a novel weapon technology comprise the customary principles, the environment-related rules and the rules relating to specific weapon technologies, all of which have been discussed in this chapter. The customary principles and the rules that have been identified as having customary status will bind all States. The remaining treaty-based rules will bind the States that are party to the relevant treaties, subject to any reservations or other declarations that those States may have made on ratification. So a weapon review under either Article 36 of API or the implied rule, both of which were discussed earlier, will address:

whether the weapon system is of a nature to cause superfluous injury or unnecessary suffering;

whether the weapon system is indiscriminate by nature;

for States that are party to API, whether the weapon system is intended, or may be expected, to cause widespread, long-term and severe damage to the natural environment;

for States that are not party to API, whether the use of the weapon is going to be consistent with the State's obligation to have due regard to the natural environment;

109 'Non-lethal weapons are weapons which are explicitly designed and developed to incapacitate or repel personnel, with a low probability of fatality or permanent injury, or to disable equipment with minimal undesired damage or impact on the environment'; NATO Policy on Non-Lethal Weapons dated 27 September 1999.

for States that are party to ENMOD, whether the use of the weapon would involve environmental modification techniques of the sort and involving the consequences prohibited by that Convention; and
whether there are ad hoc weapons law rules that apply to the weapon.

When a State undertakes a weapon review involving a novel technology, it should apply these criteria to the technology in question, taking into account the general circumstances in which it is intended to use the weapon, means or method of warfare.

There is no prescribed form for a weapon review and there is no particular procedure that the weapon review process must adopt. Sometimes, advice to an appropriate commander, whether oral or written, may, depending on the circumstances, be sufficient.[110]

Accordingly any State that is acquiring a weapon that involves novel technologies including those discussed in this volume, and any API State party that is studying or developing such a weapon or that is adopting a novel method of warfare must conduct a legal review of that weapon or method to determine whether its employment is prohibited or restricted by the principles and rules discussed in this chapter.

Before concluding this section, one additional point should be made. If the novel weapon technology or system that is being reviewed is autonomous in the sense that it is the weapon system that undertakes target identification, recognition, designation and attack decision-making without reference to a human operator, it will be necessary when such a weapon system is being reviewed for those conducting the review to be satisfied that the targeting law rules, including the distinction principle and the discrimination, proportionality and precaution rules, can be applied. This issue is likely to apply to autonomous weapon systems, to certain cyber weapon systems and perhaps to certain other novel weapon technologies. The issue is discussed more fully in Chapter 6.

2.7 LESSONS AS TO THE REGULATION OF EMERGING TECHNOLOGIES

Two additional points should be made before this chapter ends. The first relates to a clause included in the preambles to the 1899 Hague Convention II and the 1907 Hague Convention IV and known as the Martens Clause. In its 1907 iteration, the clause reads as follows: 'Until a more complete code of the

[110] *Tallinn Manual*, commentary accompanying Rule 48, para 3.

laws of war has been issued, the high contracting Parties deem it expedient to declare that, in cases not included in the Regulations adopted by them, the inhabitants and the belligerents remain under the protection and the rule of the principles of the law of nations, as they result from the usages established among civilized peoples, from the laws of humanity, and the dictates of the public conscience.'[111] Contrary to what some may seek to argue, the present author takes the view that the Martens Clause, the laws of humanity and the dictates of the public conscience do not individually or collectively constitute additional criteria or standards against which the lawfulness of means or methods of warfare must be judged.[112] The clause does however remind the reader that if specific rules on a particular matter are lacking, customary principles continue to apply, and this will be just as much the case when considering the lawfulness of novel weapon technologies such as are discussed in this book as in any other context.

The second point, alluded to in the course of the previous chapter, is that not all rules of the law of weaponry have stood the test of time, and this is a point that should be borne in mind when legal provisions to deal with emerging technologies are being considered. At the time of the first Hague Peace Conference, 1899, the possibility that technology would permit the employment of explosive weapons from the air was already causing concern. The delegates to that conference adopted Declaration 1, which prohibited for a period of five years 'the launching of projectiles and explosives from balloons, or by other methods of a similar nature'.[113] The Declaration came into force in 1900 and the inclusion of the term 'other methods of a similar nature' illustrates the drafting problems posed when seeking to achieve appropriately comprehensive provision at a time of rapid technological advance. As the present author has noted elsewhere, '[t]he drafters and negotiators of the treaty sought to balance the legal requirement for precision of expression, particularly where the scope of the treaty was concerned, with the recognition that while technical advance may render tightly drafted language obsolete, overly flexible text may catch technologies and weapons not seen, at the time of the negotiation, as meriting prohibition or restriction'.[114]

111 Paragraph 8 of the Preamble to Hague Convention IV, 1907.

112 Dinstein, *The Conduct of Hostilities*, 13–15.

113 Declaration IV, 1 to Prohibit, for the Term of Five Years, the Launching of Projectiles and Explosives from Balloons, and Other Methods of Similar Nature, The Hague, 29 July 1899.

114 W. H. Boothby, *Weapons and the Law of Armed Conflict*, 2nd edn (Oxford: Oxford University Press, 2016), 15.

The Declaration was replaced by Hague Declaration XIV of 1907[115] the operative part of which prohibited 'the discharge of projectiles and explosives from balloons or by other new methods of a similar nature'. Informed commentators were already expressing the view, as soon as seven years after the adoption of the later Declaration, that developments in aerial warfare had rendered it redundant.[116] So what we had here was grave concern about the dangers that the new technology posed, concern incidentally that was more than amply borne out by ensuing events. The concern inspired the adoption of legal measures to address that concern at a time when the relevant technology had yet to mature. It was, arguably, only when the military opportunities presented by the emerging technology had later become evident that the balance of advantage between humanitarian concern and military utility was seen to tilt in favour of military utility. This kind of experience will cause some experts to be nervous about proposing international law rules, particularly in the form of prohibitions, before the relevant technology has achieved sufficient maturity for its advantages and drawbacks to be reasonably understood.

It could be argued that this is not the only example of misconceived, or perhaps inadequately conceived, legal provision of this sort. As noted earlier in the chapter, the first weapons law treaty of the modern era, namely the St Petersburg Declaration 1868, prohibited the employment of 'any projectile of a weight below 400 grammes, which is either explosive or charged with fulminating or inflammable substances'. Already by 1923 the jurists who prepared the 1923 Draft Hague Air Rules agreed, in Article 18(2), that the use of tracer, incendiary or explosive projectiles by or against aircraft is not prohibited and that this applies equally to States that are party to the 1868 Declaration. Technological advance had rendered the weight limit no longer sustainable.

A different issue arose with asphyxiating gas. Here again we must go back to the 1899 First Hague Peace Conference and to a time when gas weapons were seen as being the novel products of advances in science. Poison was already prohibited. That prohibition had been confirmed in treaty form in 1899 with the adoption of the prohibition of poisons and poisoned weapons in Article 23 (a) of the Regulations annexed to Convention II of that year. It was considered however that there was an element of perfidy associated with the use of gas weapons and that customary principles of humanity should be applied to their use. It was also considered, however, that decisive weapons might shorten wars

[115] Declaration (XIV) Prohibiting the Discharge of Projectiles and Explosives from Balloons, The Hague, 18 October 1907. The Declaration was stated to apply until the close of the Third Peace Conference, but no such Conference took place.

[116] J. M Spaight, *Aircraft in War* (London, Macmillan 1914), 30.

and thus reduce casualties so there was no outright ban. In Declaration IV, 2 of 1899 'on the one hand, the traditional norms are made applicable to poison gas, but the act of making poisonous and deleterious gases the subject of a separate document rather than incorporating them in the regulations (for the conduct of war on land) constituted their recognition as a novel type of weapon'.[117] Nevertheless, the Declaration did prohibit 'the use of projectiles the sole object of which is the diffusion of asphyxiating or deleterious gases'. At the time of the adoption of the Declaration, such technology had not been used in combat.[118] The Declaration clearly did not prevent the use of such gases in the trenches of the Western Front during World War I, with all of its attendant misery and suffering. Arguably, it was the reference in the 1899 text to 'sole object' that opened the loophole to permit the lawful use of weapons that combined gas diffusion with some other object, say by means of an explosion.

Be all of that as it may, the point of interest in relation to the present discussion is that here we had a failure to ban a novel technology which, when used in battle, was seen to have unacceptable effects. Only a few years after the conclusion of the war, a comprehensive ban on the use of such technologies was achieved by the adoption of the 1925 Geneva Gas Protocol. Whether the deletion of the 'sole object' language from the 1899 text would have prevented the World War I attacks is an open question. It is however clearly important that emerging weapon technologies that have unacceptable humanitarian consequences should be the subject of timely and adequate prohibitions or, as appropriate, restrictions.

The difficulty that this discussion identifies is in getting the timing and content of the legal arrangements right. Law is notoriously slow to respond to technological advance in weaponry. The fear is that if this sluggishness persists, some of the technologies of the future will pose humanitarian challenges that dwarf those seen from 1915 to 1918.

[117] J. P. Zanders, 'International Norms Against Chemical and Biological Warfare: An Ambiguous Legacy', (2003) 8(2) *JCSL* 391, 407–408.

[118] R. L. O'Connell, Of Arms and Men (Oxford: Oxford University Press, 1989), 238. The first recorded use took place on the afternoon of 22 April 1915 at the Ypres salient; L. F. Haber, *The Poison Cloud: Chemical Warfare in the First World War* (Oxford: Oxford University Press, 1986), 34.

3

The Law on the Conduct of Hostilities

WILLIAM H. BOOTHBY

3.1 INTRODUCTION

Any discussion of how international law addresses new weapon technologies must of necessity consider two distinct aspects, namely the law that regulates the weapon technologies as such and, as a discrete matter, the law that applies to targeting, i.e. to the actual use of weapons in general in armed conflict. Having addressed the former topic in Chapter 2, it is the purpose of this chapter to summarise the latter body of law.

The scientists researching new weapon technologies, the engineers developing them into viable weapon systems, those whose job it is to test and procure such weapons and ultimately the armed forces and units that will employ them are all seeking to ensure that the particular required operational tasks that have been identified are in fact going to be successfully fulfilled by the weapon system that this process of research, development and production generates. It is that aspect that renders an understanding of the law on the conduct of hostilities of such central importance, and in the present chapter attention will accordingly focus on the aspects of that part of the law that appear to be of greatest relevance to the topic at hand.

Military units and individuals whose task it is to use weapons will reasonably assume that the weapons issued to the unit and the planned ways in which they will be used comply with the international law rules by which the State is bound. That should be a reasonable assumption if the relevant State has developed and implemented a system that legally reviews new weapons in accordance with the arrangements explained in Chapter 2. The weapon review process will inform those studying the weaponisation of a new or emerging technology, and those who develop the weapon, which legal constraints apply to the designed or intended uses of such a weapon.

However, there is another significant relevance of targeting law to the present discussion that must not be ignored. A number of the technologies that will be discussed in the following chapters exclude or erode the control of a human being over the weapon system or at least significantly alter the relationship between human being and weapon. When considering the lawfulness of a traditional weapon such as a mine, a rifle, bullets, bombs or missiles, attention is not focused on the ability to implement targeting law when using the weapon system. This is because there is a human operator centrally involved in attack decision-making, and that human operator is the agency through whom decisions required by targeting law can be taken. If, however, the human operator is removed, e.g., because the weapon system is autonomous, that human applier of targeting law is by definition also removed. In those circumstances, it becomes vitally important to determine whether, absent a human operator of the weapon system, the decisions that the law of targeting requires will nevertheless be made. That may involve determining whether persons fulfilling other roles relating to the mission can make sufficiently effective targeting-law decisions, for example before the mission commences. If, however, that is not the case and it is the weapon system that, acting autonomously, decides, e.g., who or what is to be attacked, how, when and using what weapon, if in other words the human decision-maker is removed from the attack decision, it becomes important to determine whether the weapon system itself is capable of making the assessments and of taking the decisions that the law of targeting requires.[1] If we are to be able to assess this aspect by reference to the technologies for which it is relevant, we need to have identified the targeting law rules that seem to involve decisions that currently require human consideration.

The law of targeting[2] as it applies to the United Kingdom will be the basis for the present discussion, but most of the rules discussed below are customary in nature and therefore bind all States irrespective of their participation or otherwise in particular treaties. Targeting law essentially consists of fundamental customary law principles, such as humanity, distinction and military necessity, some additional customary rules[3] and some rules that only bind

[1] For the basis of the US rejection of this approach, and for the present author's criticism of the stated US position, see Chapter 6, note 27.

[2] See M. Sassoli, 'Legitimate Targets of Attacks Under International Humanitarian Law', *HCPR Policy Brief* (January 2003); W. Fenrick, 'The Law Applicable to Targeting and Proportionality after Operation Allied Force: A View from the Outside' (2000) *YIHL* 51; and W. H. Boothby, *The Law of Targeting* (Oxford: Oxford University Press, 2012).

[3] See ICRC Customary International Law Study, Vol. 1, particularly Parts I–III. See also Memorandum for Assistant General Counsel (International), Office of the Secretary of

States that are party to specific treaties. Much of the modern law of targeting is to be found in Articles 48–67 of API and will apply subject to any statements or reservations that a State party may have entered when it ratified the treaty. The United Kingdom ratified API on 28 January 1998 subject to numerous statements so in the following discussion the rules in the treaty will be interpreted subject to those statements, to the extent that they are relevant.

The majority of the targeting law rules set forth in API have the status of customary law and therefore bind all States in the language used in that treaty. Rules provided for in API that are not customary bind the overwhelming majority of States given that, at the time of writing, 174 States are party to the treaty, but do not bind States that are not party, such as the United States and Israel. Appropriate reference will be made in the following discussion to the ICRC Customary Law Study[4]. While the ICRC Study does not in the present author's view constitute a formal source of law, it represents what the contributing Experts collectively considered customary humanitarian law to comprise and is therefore deserving of suitable attention.

3.2 APPLICATION OF THE API TARGETING LAW RULES

The targeting rules in this chapter apply to attacks[5] from land, sea, or air platforms that may affect the civilian population, individual civilians, or civilian objects on land. The rules also apply, as a matter of treaty law, to attacks from the sea or from the air against objectives on land but do not otherwise affect armed conflict at sea or in the air. However, to the extent that the rules explained below are identified as being customary, they do in the opinion of the present author apply to the conduct of hostilities in general and irrespective of the environment in which those hostilities take place.[6]

Defense, 1977 Protocols Additional to the Geneva Conventions: Customary International Law Implications, 8 May 1986 and M. J. Matheson, 'The United States Position on the Relation of Customary International Law to the 1977 Protocols Additional to the 1949 Geneva Conventions' (1987) 2 *AU JILP* 419. While this chapter states the law applicable to the UK, the cited documents and other assessments of the ICRC Customary Law Study Report serve to clarify the customary rules that bind the UK and all other States.

4 J.-M. Henckaerts and L. Doswald-Beck, 'Customary International Humanitarian Law (2005), Volume 1 Rules', *ICRC*, 2005.

5 For a discussion of the meaning of 'attack' as used in API, Article 49 (1), see W. Hays Parks, 'Air War and the Law of War' (1990) 32 (1) *The Air Force Law Review* 1, 113–116.

6 Note API, Article 49(3).

3.3 THE BASIC PRINCIPLE

The parties to an armed conflict must at all times distinguish between the civilian population and combatants and between civilian objects and military objectives and must accordingly direct their operations only against military objectives.[7] This, taken together with the prohibitions, mentioned below, of making civilians or civilian objects the object of attack, constitutes the principle of distinction, a principle that lies at the core of the law of targeting. The principle is customary in nature and binds all States.

3.4 SOME IMPORTANT DEFINITIONS

To understand the legal framework that regulates targeting, it is important to appreciate the meanings given in international law to some pivotal terms. In some respects these meanings differ as between States that are party to API and those that are not. For States that are party to API, Article 43 makes it clear that the armed forces are regarded as including 'all organized armed forces, groups and units which are under a command responsible to [a Party to the conflict] for the conduct of its subordinates' whether or not that Party's government or authority is recognised. The armed forces must be 'subject to an internal disciplinary system which, *inter alia*, shall enforce compliance with the rules of international law applicable in armed conflict'. Article 43(2) then makes it clear that members of the armed forces of a Party to the conflict, other than medical and religious personnel, are combatants and thus have the right to participate directly in hostilities. By contrast, a civilian is any person who does not belong to the categories of person referred to in Article 4A(1), (2), (3) and (6) of the Third Geneva Convention and Article 43 of API.

Some States not party to API, in particular the United States, do not consider this a customary rule and essentially use the term combatant in the sense of describing the persons who actually take part in the fighting, as opposed to those who have the legal right to do so. Accordingly, for the USA for example, the right to take part in hostilities applies to armies, and to militias and volunteer corps fulfilling the four criteria in Article 1[8] of the

7 API, Article 48. See A. P. V. Rogers, *Law on the Battlefield*, 3rd edn (Manchester: Manchester University Press, 2012), 7–17 and Boothby (2012), 60–2 and 77–117.

8 The four criteria are that the militia or corps is commanded by a person responsible for his subordinates, that its members have a fixed, distinctive sign recognisable at a distance, that they carry their arms openly and that they conduct their operations in accordance with the laws and customs of war.

Hague Regulations 1907. Both traditions recognise members of a *levee en masse* as having combatant status.[9]

It follows from the API provisions that, at least for States that are party to that treaty, there are three classes of person, namely (1) combatants who have the right to fight in an international armed conflict, who have prisoner of war status if captured during such a conflict and who are exempt from prosecution for their lawful acts of war; (2) medical and religious personnel who collectively are classed as 'non-combatants' and (3) civilians who are generally protected by the law of armed conflict unless they take a direct part in the hostilities.

Where there is doubt whether a person is a civilian, he must be considered a civilian.[10] However, this rule only arises where the person making the assessment considers that there is substantial doubt after he has considered the information from all sources that is reasonably available to him at the relevant time. This doubt rule does not override the commander's duty to protect the safety of troops under his command or to preserve his military situation in conformity with other rules in API.[11]

3.5 THE PROTECTION OF CIVILIANS

The civilian population consists of all persons who are civilians.[12] The presence within the civilian population of persons who are not civilians does not deprive the population of its civilian character.

The civilian population and individual civilians must enjoy general protection against dangers arising from military operations.[13] It follows from this that the civilian population and individual civilians shall not be made the object of attack. Moreover, acts or threats of violence the primary purpose of which is to spread terror among the civilian population are prohibited. Individual civilians enjoy this protection unless and for such time as they take a direct part in

9 Article 2 of the Hague Regulations provides that '[t]he inhabitants of a territory which has not been occupied, who, on the approach of the enemy, spontaneously take up arms to resist the invading troops without having had time to organize themselves in accordance with Article 1, shall be regarded as belligerents if they carry arms openly and if they respect the laws and customs of war.' This is the recognised definition of a *levee en masse* and being regarded as belligerents is to be equated with having combatant status in the API sense.

10 API, Article 50(1).

11 Statement (h) made by the UK on ratification of API on 28 January 1998.

12 API, Article 50(2).

13 Consider Rule 22(2) and (3) of the Hague Draft Rules of Aerial Warfare 1923 and see Hays Parks, 'Air War and the Law of War' (1990), 34–35.

hostilities.[14] The use therefore of suicide bombers or other methods of attack that have the purpose of killing civilians or that have the primary purpose of causing terror among the civilian population is a clear breach of this cardinal international law principle.

3.6 THE CLASSIFICATION OF OBJECTS

All objects are either lawful targets, i.e. military objectives, or protected objects. Civilian objects, meaning all objects that are not military objectives, must not therefore be made the object of attack.[15] This prohibition only applies to attacks *directed at* items which are not military objectives. It does not address the issue of collateral damage arising from an attack on a military objective.[16] That issue is dealt with in the next section when we discuss indiscriminate attacks. Equally, it does not address attacks that were directed at a military objective but that, whether because of human error, technical malfunction or some other intervening reason, failed to hit the intended target.

Military objectives,[17] where objects are concerned, are ' those objects which by their nature, location, purpose, or use make an effective contribution to military action and whose total or partial destruction, capture, or neutralization, in the circumstances ruling at the time, offers a definite military advantage '.[18] A specific area of land is capable of being a military objective if its location, etc. is such that total or partial destruction, capture, or neutralisation of it in the circumstances ruling at the time offers a definite military advantage.[19] In case of doubt as to whether an object normally dedicated to civilian purposes is being used to make an effective contribution to military action, it shall be presumed not to be so used.[20]

[14] API, Article 50(1)–(3).

[15] API, Article 52(1).

[16] Statement (j)(2) made by the UK on ratification of API on 28 January 1998.

[17] For the application of this widely accepted, arguably customary, definition in relation to effects based operations see M. N. Schmitt, 'War, Technology and International Humanitarian Law', *Harvard University HPCR Occasional Paper Series* (Summer 2005), 48 and J. M. Meyer, 'Tearing Down the Façade: A Critical Look at the Current Law on Targeting the Will of the Enemy and Air Force Doctrine' (2001) 51 *Air Force Law Review* 143. Professor Dinstein argues that the definition 'leaves a lot to be desired' and would favour a composite definition combining an abstract statement with a non-exhaustive catalogue of specific illustrations; Y. Dinstein, *The Conduct of Hostilities under the Law of International Armed Conflict*, 3rd edn (Cambridge: Cambridge University Press, 2016), 103–104.

[18] API, Article 52(2). See further Rogers, *Law on the Battlefield*, 97–124.

[19] Statement (j)(1) made by the UK on ratification of API on 28 January 1998.

[20] API, Article 52(3).

3.7 INDISCRIMINATE ATTACKS

Indiscriminate attacks[21] are prohibited[22] and are defined as those attacks which are not directed at a specific military objective, or which employ a method or means of combat which cannot be so directed, or which employ a method or means of combat whose effects cannot be limited as required by API and which, in each case, are of a nature to strike military objectives and civilians or civilian objects without distinction.[23]

It is clear from this widely accepted formulation of the rule that the first element represents a targeting law rule and focuses on the nature of the attack as such. The second and third elements, however, that prohibit methods and means that cannot be directed at a specific military objective and the use of methods or means the effects of which cannot be limited as required by API clearly constitute rules of weapons law as the focus in each is on the characteristics of the relevant weapon. Unusually API includes in the operative part of the treaty examples of attacks that would breach the first, or targeting law element in the rule.

The examples of indiscriminate attacks given in API are as follows:

- bombardments by any method or means which treat as a single military objective a number of clearly separated and distinct military objectives located in a city, town, village, or other area containing a similar concentration of civilians or civilian objects; and
- attacks which may be expected to cause incidental loss of civilian life, injury to civilians, damage to civilian objects, or a combination thereof, which would be excessive in relation to the concrete and direct military advantage anticipated.[24]

The former example is frequently cited as prohibiting carpet bombing and similar methods, and the latter example, which as the wording demonstrates focuses on notions of excessiveness, is widely known as the proportionality rule. Neither rule will usually determine the legal acceptability of a new weapon that employs traditional technologies in which a human operator

[21] See Rogers, *Law on the Battlefield*, 27–31 and Y. Dinstein, 'The Laws of Air, Missile and Nuclear Warfare' (1997) 27 *Israel Yearbook on Human Rights* 1, 6–9.

[22] Tony Rogers explains that the rule of distinction, and possibly the rule of proportionality, is violated if an attack is indiscriminate: Rogers, *Law on the Battlefield*, 27.

[23] API, Article 51(4).

[24] API, Article 51(5). For a more detailed discussion of the proportionality rule, see Rogers (2012), 21–27, and see Hays Parks, 'Air War and the Law of War' 197–198 for a sobering account of the limitations that apply in practice to the realisation of proportionality in attacks.

controls the weapon. However, and as indicated in the first section of this chapter, if for example the operation of a new technology weapon does not require a human operator or diminishes that person's role, these rules that at present presuppose human involvement and decision-making must be complied with by some other means, and the person considering the legal acceptability of the weapon or weapon technology must therefore consider how in its intended circumstances of use these targeting law rules are to be implemented.[25]

3.8 PARTICULARLY PROTECTED OBJECTS AND PERSONS

Military and specified civilian medical units and establishments must be respected and protected at all times and must not be made the object of attack. Medical units must not be used to shield military objectives from attack; indeed they shall be so sited that attacks against military objectives do not imperil their safety.[26] Fixed and mobile military and civilian medical units do not lose their protection unless they are used to commit, outside their humanitarian function, acts harmful to the enemy. Even then, protection may only cease after a warning has been given setting, wherever appropriate, a reasonable time limit and after such warning has gone unheeded.[27]

Geneva Convention I and Articles 21–31 of API make specific provision for the protection of medical transports, including but not limited to hospital ships, medical aircraft and medical transportation by land: 'Medical personnel exclusively engaged in the search for, or in the collection, transport or treatment of the wounded or sick, or in the prevention of disease, staff exclusively engaged in the administration of medical units and establishments, as well as chaplains attached to the armed forces, shall be respected and protected in all circumstances.'[28] API extends this requirement of respect and protection to civilian medical personnel.[29]

25 For assessments of the practical application of the proportionality rule, see A. E. Wall, 'Legal and Ethical Lessons of NATO's Kosovo Campaign', *US Naval War College International Law Studies*, Vol. 78, Parts III and IV.

26 API, Article 12(1) and (4).

27 GCI, Article 21 and API, Article 13(1). Certain acts are not considered harmful to the enemy, namely, in the case of civilian medical units for example, that personnel of the unit are equipped with light personal weapons for their own defence or that of the wounded and sick in their charge; that the unit is guarded by a picket, by sentries or by an escort; that small arms and ammunition taken from the wounded and sick have not yet been handed to the proper service and are found in the unit; and that members of the armed forces or other combatants are in the unit for medical reasons; API, Article 13(2).

28 Geneva Convention I, Article 24.

29 API, Article 15(1).

Acts of hostility against historic monuments, works of art, or places of worship constituting the cultural or spiritual heritage of peoples are prohibited[30] unless those places are used unlawfully for military purposes, in which case they will lose protection from attacks directed against such unlawful military uses.[31] The Hague Cultural Property Convention 1954 requires that States party respect cultural property located within their own territory and in the territory of other States party by refraining 'from any use of the property and its immediate surroundings or of the appliances in use for its protection for purposes which are likely to expose it to destruction or damage in the event of armed conflict; and by refraining from any act of hostility directed against such property'.[32]

Starvation of civilians as a method of warfare is prohibited.[33] Attacking, destroying, removing or rendering useless objects that are indispensable to the survival of the civilian population such as foodstuffs, agricultural land, crops, livestock, drinking water installations and supplies and irrigation works is prohibited[34] unless the purpose of such an attack, etc. is other than the denial of sustenance to the civilian population or to the adverse party.[35] This proviso goes rather further than that in API, which states that the rule does not apply to objects used by an adverse party solely as sustenance for members of its armed forces or in direct support of military action, although no attack on such objects is permissible if it may be expected to leave the civilian population with such inadequate food or water as to cause starvation or force its movement.[36] There is, however, an exception provided for in API that has the effect of permitting a scorched-earth policy based on imperative military necessity if it is limited to national territory that it under the control of the party to the conflict employing scorched earth.[37]

API prohibits the employment of weapons, methods or means of warfare which are intended, or may be expected, to cause widespread, long-term and

[30] API, Article 53. For a comprehensive definition of cultural property and for the extensive law of armed conflict rules as to its protection, see the Convention for the Protection of Cultural Property in the Event of Armed Conflict, The Hague, 14 May 1954 and its two Protocols; see also Rogers (2012), 175–208.

[31] Statement (k) made by the UK on ratification of API on 28 January 1998.

[32] Hague Cultural Property Convention, Article 4. Numerous more detailed provisions are made in the Convention, and in its two Protocols and in the associated Regulations, including for example arrangements for enhanced protection.

[33] API, Article 54(1).

[34] API, Article 54(2).

[35] Statement (l) made by the UK on ratification of API on 28 January 1998.

[36] API, Article 54(3).

[37] API, Article 54(5).

severe damage to the natural environment.[38] In an additional provision, Article 55 of API requires that care be taken in warfare to avoid the degree of environmental damage prohibited by Article 35(3) and further prohibits methods or means of warfare which by causing such damage thereby prejudice the health or survival of the population. States which are not party to API are not bound by these provisions but there is a customary rule that binds all States and which prohibits the wanton destruction of the natural environment.[39]

States party to API are subject to particular rules that protect installations containing dangerous forces, namely dams, dykes and nuclear electrical generating stations, which shall not be made the object of attack, even if they are military objectives, if such attack may cause the release of dangerous forces and consequent severe losses among the civilian population.[40] Other military objectives located in the vicinity of such works or installations are also protected from attack if the attack may have the same results.[41] This protection only ceases if, in the case of a dam or dyke, it is used for other than its normal function in regular, significant, and direct support of military operations and if such attack is the only feasible way to terminate that support. In the case of a nuclear electrical generating station, the protection only ceases if the power station provides electric power in regular, significant, and direct support of military operations and if such attack is the only feasible way to terminate that support. For military objectives located in the vicinity of such installations, protection only ends if they are used in regular, significant and direct support of military operations.[42] If an attack on such works or installations is decided upon, precautions must be taken including all feasible precautions to prevent the release of the dangerous forces.[43]

Parties to an armed conflict must endeavour to avoid locating military objectives in the vicinity of such works and installations. Arrangements to defend the installations from attack are permitted provided their use remains a defensive response to attacks against the installation and provided weapons are

38 API, Article 35(3). For the rules protecting the environment in respect of attacks, see Rogers (2012), 209–237.

39 *AMW Manual*, rule 88. '"Wanton" means that the destruction is the consequence of a deliberate action taken maliciously. In other words, it is an action that cannot be justified by considerations of imperative military necessity'; *AMW Manual*, para 2 of the Commentary to rule 88.

40 See W. Hays Parks, 'Air War and the Law of War', 202–218 for a critical discussion of this rule in API.

41 API, Article 56(1).

42 API, Article 56(2).

43 API, Article 56(3).

limited to 'weapons capable only of repelling hostile action' against the installation.[44] Special signs are provided for in API to assist an opposing party to identify such installations.[45]

In a statement made on ratification of API,[46] the United Kingdom declared it:

> '|.|.|.| cannot undertake to grant absolute protection to installations which may contribute to the opposing Party's war effort, or to the defenders of such installations, but will take all due precautions in military operations at or near the installations ... in the light of the known facts, including any special marking which the installation may carry, to avoid severe collateral losses among the civilian populations; direct attacks on such installations will be launched only on authorisation at a high level of command.'

3.9 PRECAUTIONS IN ATTACK

The law of armed conflict requires that certain precautions be taken by attackers in order to try to ensure that those persons and objects entitled to protection under the law receive it. Article 57 of API lists the precautions that are required in attack.[47] The first requirement, which reflects and is arguably also part of the principle of distinction, is that constant care shall be taken in the conduct of military operations to spare the civilian population, civilians and civilian objects.[48]

Article 57 then prescribes additional precautions that must be taken by those who plan or decide upon an attack, namely:

- to do everything feasible[49] to verify that the objectives to be attacked are not civilians, civilian objects, or subject to special protection but are military objectives (meaning lawful targets) whose attack is not prohibited by API;
- to take all feasible precautions in the choice of means and methods of attack to avoid, and in any case to minimise, incidental loss of civilian life, injury to civilians, or damage to civilian objects and

44 API, Article 56(5).

45 API, Article 56(7).

46 Statement (n) made by the UK on ratification of API on 28 January 1998.

47 For a more comprehensive discussion see the UK Manual, paras 5.32–5.33.5, Rogers, *Law on the Battlefield*), 125–159 and Boothby, *The Law of Targeting* (2012), 118–140.

48 API, Article 57(1).

49 Statement (b) made by the UK on ratification of API on 28 January 1998 notes the UK understanding that the term 'feasible' here means 'that which is practicable or practically possible, taking into account all circumstances ruling at the time, including humanitarian and military considerations.'

- to refrain from deciding to launch an attack which may be expected to cause excessive incidental loss of civilian life, injury to civilians, or damage to civilian objects in relation to the military advantage the attack is anticipated to yield.

The article goes on to require that an attack shall be cancelled or suspended if it becomes apparent that the objective is not a military one, or is subject to special protection, or that the attack may be expected to result in excessive injury or death of civilians or damage or destruction of civilian objects in relation to the anticipated military advantage. However, when ratifying API, the United Kingdom stated its understanding that the obligation to comply with this rule only extends to those who have the authority and practical possibility to cancel or suspend the attack, an understanding that may have implications for some of the new weapon technologies considered later.[50]

Effective advance warning must be given of attacks which may affect the civilian population,[51] unless circumstances do not permit.[52] The final precautionary rule in Article 57 stipulates that where, in order to obtain a similar military advantage, it is possible to choose between military objectives, the objective to be chosen shall be that the attack on which may be expected to cause the least danger to civilian lives and to civilian objects.[53]

Novel weapon technologies or methods of warfare may render it appropriate to consider taking particular precautions that are additional to those prescribed by existing targeting law. Some technologies may be such that certain precautions take on a special importance. Other novel technologies may require entirely new kinds of precaution to be considered. This is only likely to become apparent when the testing of the new technology or method is under way.

While many of the precautionary rules set forth in Article 57 of API are generally accepted by States in the terms in which they are drafted, thus have customary status and are binding on all States, that is not necessarily the case with all such rules and it is specifically known, for example, that the US does not accept the final rule mentioned in this section.[54]

[50] Statement (o) made by the UK on ratification of API on 28 January 1998. Those persons, moreover, necessarily have to reach decisions on the basis of their assessment of the information from all sources which is reasonably available to them at the relevant time; UK statements on ratification, statement (c).

[51] It is generally accepted that 'affect' here means that members of the civilian population would be injured or killed.

[52] API, Article 57(2). Those circumstances may include military and humanitarian factors.

[53] API, Article 57(3).

[54] DoD Manual, footnote 303 on page 241.

3.10 NON-DEFENDED LOCALITIES, DEMILITARISED ZONES AND CIVIL DEFENCE

Article 59 of API prohibits the attack of non-defended localities. These are inhabited places open for occupation by an adverse party, declared by a party to an international armed conflict and which are in or near a zone where armed forces are in contact. The zone must comply with the stipulations in Article 59(2) but the parties to the conflict may agree to establish non-defended localities that do not comply with the Article 59(2) requirements. Article 59 contains a number of detailed provisions that are not set forth here. It should, however, be noted that Article 60 prohibits the extending of military operations to zones on which the parties to the conflict have conferred by agreement the status of demilitarised zone where such extension breaches the agreement.[55] Again, Article 60 prescribes conditions to be fulfilled by such a zone and other obligations that must be complied with.

Articles 61–67 of API deal with the protected status of civil defence establishments, facilities and personnel.

3.11 HUMAN SHIELDS AND PRECAUTIONS AGAINST THE EFFECTS OF ATTACKS

Human shielding is unlawful. Accordingly, '[t]he presence or movements of the civilian population or individual civilians shall not be used to render certain points or areas immune from military operations, in particular in attempts to shield military objectives from attacks or to shield, favour or impede military operations.' In particular, the 'Parties to the conflict shall not direct the movement of the civilian population or individual civilians in order to attempt to shield military objectives from attacks or to shield military operations'.[56] While human shielding is therefore absolutely prohibited, the violation by the enemy of this or other 'prohibitions shall not release the Parties to the conflict from their legal obligations with respect to the civilian population and civilians, including the obligation to take the precautionary measures provided for in Article 57'.[57] All of this clearly demonstrates that while human shielding of military objectives is unlawful, the fact that it has taken place does not allow the attacking party simply to ignore the

55 API, Articles 59 and 60 contain more detailed provisions, which it is not necessary to repeat here.

56 API, Article 51(7).

57 API, Article 51(8).

precautionary rules that bind it. However, human shielding activities may render it more difficult for the attacking party to offer effective protection to the civilian population notwithstanding that required precautions are taken. There is also some controversy over the value to be allocated to those who are shielding a military objective on a voluntary basis when the proportionality rule is being considered.[58]

Any act of human shielding constitutes a clear breach, but by no means the only potential example of a breach, of the obligation on both parties to the conflict to take precautions against the effects of attacks, sometimes inaccurately referred to as 'passive precautions'. Article 58 of API requires the parties to the conflict 'to the maximum extent feasible' to do the following:

> '(a) without prejudice to Article 49 of the Fourth Convention, [to] endeavor to remove the civilian population, individual civilians and civilian objects under their control from the vicinity of military objectives;
>
> (b) [to] avoid locating military objectives within or near densely populated areas;
>
> (c) [to] take the other necessary precautions to protect the civilian population, individual civilians and civilian objects under their control against the dangers resulting from military operations.'[59]

What is feasible by way of such precautions will inevitably depend on a variety of factors. These will include what is known as to how attacks are liable to be undertaken by the adverse party; what is known about the objects and/or persons that it will choose to attack; and whether the anticipated targets can realistically be isolated from civilian objects, networks or centres of population. In determining what is feasible, the precautions that can realistically be taken and that can sensibly be expected to protect civilians and civilian objects should be the focus of consideration.

3.12 CUSTOMARY LAW RULES AS TO TARGETING

The ICRC Customary Law Study Report reflects the obligation to distinguish between civilians and combatants in Rules 1–6; the requirement to

[58] As to human shielding generally see AMW Manual, rule 45 and associated Commentary. As to the controversies associated with the evaluation of voluntary human shields, see para 6 of the Commentary.

[59] API, Article 58. For a comprehensive discussion of the precautions that are required against the effects of attacks, see Rogers (2012), 160–174 and Hays Parks, 'Air War and the Law of War', 57–58 and 149–168.

distinguish between civilian objects and military objectives in Rules 7–10; the prohibition of indiscriminate attacks at Rules 11–13; the rule of proportionality at Rule 14; the rules as to precautions in attack at Rules 15–21 and at Rules 22–24, the requirement to take precautions against the effects of attacks. It should be recalled that the text of the ICRC Customary Humanitarian Law Study report does not constitute a formal source of customary law in its own right. Rather, the text represents the views of its contributing experts as to what those rules comprise, views which States may or may not accept but which they should take carefully into account.

So the rules that are in fact customary and precisely how they should be expressed will be a matter of opinion and experts' views may differ. Nevertheless, the principle of distinction is core to the law and binds all States as noted earlier.[60]

It is for States to determine the customary rules that they recognise both as to subject and content. Customary law exists, however, independently of the recognition of particular States. The findings of the ICRC Study, and the writings of experts, will assist States in determining the customary rules that apply to them. However, States and their practice form customary law in the sense that it is the general practice of States, whether in acting or in refraining from acting, that is combined with *opinio juris*, namely the national acceptance that a particular act or omission is required by law, to form customary law rules. Accordingly, any formulation of generally accepted customary law must take as its starting point the interpretations of States and must only articulate rules that have the demonstrable and near universal support of States in the suggested terms.

A number of the 174 States that are party to API[61] have made national statements of interpretation when ratifying the treaty. As the number of States party to API is so large, it is always going to be sensible, in determining what the customary law is in relation to targeting, to use the exact language of Articles 48–67 of API where this is generally accepted. Where the declared positions of States that are not party to API deviate from the API language, it is suggested that the customary law rules must only deviate from that language to the degree necessary to accommodate the positions necessary to achieve the general acceptance that is required for a customary rule to exist. The result of such a process will usually produce a customary law rule that is less stringent or restrictive than the corresponding rule in API.

60 ICJ, Advisory Opinion on the Legality of the Threat or Use of Nuclear Weapons, Judgment on 8 July 1996, para 78.

61 Source www.icrc.org viewed on 5 April 2018.

It is sufficient for the present purposes to note that the United States has expressed official concern about and disagreement with aspects of the elements of the ICRC Customary Law Study Report.[62] The UK interpretation of targeting law is largely given in Chapter 5 of the UK Manual and the interpretations of other States are set forth in military manuals issued by them.[63]

3.13 REPRISALS

The leading thinker on the issue of reprisals, Frits Kalshoven, suggested the following 'working definition', namely 'an act in breach of a specific rule of the law of armed conflict, directed by one belligerent party against another party with a view to inducing the latter to stop violating this particular or some other rule of this body of law (reprisal in kind, or not in kind, respectively)'.[64] In the UK Manual a reprisal is described as 'an extreme measure of coercion', noting that the threat or use of reprisal action may be the only practical way of inducing an adverse party to desist from unlawful conduct. Reprisals are never lawful if directed against the wounded, sick, and shipwrecked; medical personnel and chaplains; medical units, establishments, and transports; against prisoners of war; or against protected persons and their property.[65]

Additional Protocol 1 also prohibits reprisals against civilians and the civilian population; civilian objects; historic monuments, works of art or places of worship which constitute the cultural or spiritual heritage of peoples; objects indispensable to the survival of the civilian population such as foodstuffs, crops, livestock, drinking water installations and supplies, and irrigation works; the natural environment; and works or installations containing dangerous

[62] Letter by John B. Bellinger III, legal adviser to the State Department and William J. Haynes II, Department of Defense General Counsel, at www.defenselink.mil/news/newsarticle.aspx?id=3308.

[63] Consider, for example, the Canadian Manual, Law of Armed Conflict at the Operational and Tactical Levels, B-GJ-005–104/FP-021, issued 13 Aug 2001; Australian Defence Doctrine Publication 06.4, Law of Armed Conflict, issued 11 May 2006; US DoD Law of War Manual dated June 2015 as amended; German Law of Armed Conflict Manual 2013 (Joint Service Regulation (ZDv) 15/2) issued May 2013; French Manuel de Droit des Conflits Armés (2012); and Norway Manual I krigens folkerett (2013).

[64] F. Kalshoven, 'Belligerent Reprisals Revisited' (1990) 21 *Netherlands Yearbook of International Law* 43; see also US DoD Law of War Manual, June 2015, para 18.18.

[65] UK Manual, para 16.18, citing as authorities GC1, Article 14; GC2, Article 16; GC3, Article 13; GC4, Article 33; and API, Article 20; Tallinn Manual, Rule 46 and associated Commentary and see US DoD Law of War Manual, June 2015, para 18.18.3.2.

forces, namely dams, dykes and nuclear electrical generating stations.[66] States that are not party to API are not, generally speaking, bound by this prohibition, and if a State that is party to API made a reservation in this regard, its obligations must be construed by reference to that reservation.

On ratifying API, the United Kingdom made a statement in the following terms:

> 'The obligations of Articles 51–55 are accepted on the basis that any adverse party against which the United Kingdom might be engaged will itself scrupulously observe these obligations. If an adverse party makes serious and deliberate attacks, in violation of Article 51 or Article 52 against the civilian population or civilians or against civilian objects, or, in violation of Articles 53, 54 and 55, on objects or items protected by those Articles, the United Kingdom will regard itself as entitled to take measures otherwise prohibited by the Articles in question to the extent that it considers such measures necessary for the sole purpose of compelling the adverse party to cease committing violations under those Articles, but only after formal warning to the adverse party requiring cessation of the violations has been disregarded and then only after a decision taken at the highest level of government. Any measures thus taken by the United Kingdom will not be disproportionate to the violations giving rise thereto and will not involve any action prohibited by the Geneva Conventions of 1949 nor will such measures be continued after the violations have ceased. The United Kingdom will notify the Protecting Powers of any such formal warning given to an adverse party, and if that warning has been disregarded, of any measures taken as a result.'[67]

This means that reprisals taken in accordance with the statement are permissible by and against the United Kingdom; requests to take such action must be submitted by commanders and commanders-in-chief to the Ministry of Defence and require approval at Cabinet level.[68]

[66] UK Manual para 16.19, citing API, Articles 51(6), 52(1), 53(c), 54(4), 55(2), and 56; and see Rule 47 of the Tallinn Manual which reflects the same prohibitions.

[67] Statement (m) made by the UK on ratification of API on 28 January 1998; see also para 11 of the statements on ratification of API made on 11 April 2001 by France by reference to Article 51 (8) which has the apparent effect of maintaining the option of employing reprisals to protect its civilian population, available at www.icrc.org/applic/ihl/ihl.nsf/vwTreatiesByCountrySelected.xsp?xp_countrySelected=FR&nv=4. The United States rejects API's provisions on reprisals; US DoD Law of War Manual, June 2015, para 18.18.3.4.

[68] Tallinn Manual, Commentary accompanying Rule 47, para 2, which explains that the prohibition on reprisals against civilians is limited to States party to API that have not reserved on the point; UK Manual, para 16.19.2.

3.14 USE OF WEAPONS AS A REPRISAL

There is no general prohibition at international law of the use of unlawful weapons, or of unlawful uses of weapons not previously discussed in this section, as a reprisal. In general terms, 'if any enemy employs illegal weapons against a State, the victim may resort to the use of weapons which would otherwise be unlawful in order to compel the enemy to cease its prior violation'.[69]

If a treaty prohibits reprisals, States party will only be permitted to use such reprisals if they entered a reservation in suitable terms at the time of ratification. By way of illustration, no such statement was made when the United Kingdom ratified Protocols I, II, and III to CCW. Accordingly, the prohibition on reprisals in Article 3(2) of Protocol II binds the United Kingdom not to use mines, booby-traps, or other devices against the civilian population or individual civilians by way of reprisal.

Of course, this issue only arises in relation to weapons that have either been prohibited as such or the lawful circumstances of use of which have been restricted by treaty or customary law. Where a novel weapon technology has not yet been the subject of treaty provision and where its use does not involve the breach of established law, the question of reprisals as a justification for its employment does not arise, simply because no illegality is involved.

3.15 CONCLUSION

The rules of targeting law, briefly discussed in the present chapter, provide an important part of the context against which the utility of emerging weapon technologies will be considered. Moreover, when human control is either diminished or removed, an important determinant of the legal acceptability of the weapon system will be the ability, or otherwise, to apply the targeting law rules that we have been discussing notwithstanding that human control has been diminished or extinguished. The following chapters will seek to show how the relevant rules apply to the technology in question.

69 USAF Judge Advocate General's Department, *Air Force Operations and the Law, A Guide for Air and Space Forces* (2002), 47.

4

Non-LOAC-Governed Deployment of Military Technologies: Some Regulatory Touchstones

ROB MCLAUGHLIN

4.1 INTRODUCTION

Militaries routinely deploy technology in non-LOAC-governed contexts. After an unconscionably slow start in relation to 'monitoring and surveillance [where] there has not been parallel progress in UN operations, despite a commercial revolution in sensor technology',[1] UN Peace Operations Forces in the eastern Democratic Republic of the Congo have recently employed unmanned aerial vehicles for surveillance in situations not governed by the law of armed conflict (LOAC).[2] Other UN agencies are similarly seeking to employ unmanned aerial vehicle (UAV) technology in conflict contexts. As UNICEF recently observed, UAV technology could be a game-changer in humanitarian assistance operations in three ways:

> First, drones can leapfrog over broken infrastructure in places where developed transportation networks or roads do not exist, carrying low-weight supplies. Second, UAVs can be used for remote sensing, such as gathering imagery and data, in the wake of natural disasters like mudslides, to locate where the damage is and where the affected peoples are ... Third, drones can extend WiFi connectivity, from the sky to the ground, providing refugee camps or schools with access to the Internet.[3]

Similarly, intelligence, surveillance and reconnaissance (ISR) technologies are today employed to provide 'battlespace awareness to successfully plan

[1] A. W. Dorn, *Keeping Watch: Monitoring, Technology and Innovation in UN Peace Operations* (Tokyo: United Nations University Press, 2011), 2.

[2] M. Tafirenyika, 'Drones are Effective in Protecting Civilians – Herve Ladsous', *Africa Renewal*, April 2016, at www.un.org/africarenewal/magazine/april-2016/drones-are-effective-protecting-civilians.

[3] 'Does Drone Technology Hold Promise for the UN?', *UN News*, 6 September 2017, at https://news.un.org/en/story/2017/09/564452-feature-does-drone-technology-hold-promise-un.

and conduct future operations', very much as they provide current intelligence for LOAC-governed operations.[4] Further, 'dual use' technology and capabilities are often managed and deployed by specialist military forces for domestic security and law enforcement purposes. For example, the Australian Signals Directorate (ASD), a Defence agency,[5] is responsible for signals intelligence (SIGINT) collection for both deployed military and domestic security operations, and is legislatively mandated to that effect.[6] The legal and ethical consequences of employment of telecommunications and cyber interception capabilities in domestic law enforcement and security contexts[7] can thus become correspondingly more complicated where those technologies and capabilities are military – not least because of the different legislative, authorisation, and attitudinal foci required of military electronic warfare (EW) units, personnel, and capabilities when employed in such roles. When, for example, Australian Defence Force (ADF) EW and ISR personnel are employed on warlike operations overseas (such as communications interception and surveillance for pattern of life assessments to support targeting), their conduct is primarily governed by LOAC (international law) and the executive power (domestic law);[8] when engaged in the same activities within Australia for national security purposes, the governing legal regime(s) engaged are a significantly more complicated mix of interrelated legislative authorisations, obligations, and

4 Australia, 'Intelligence, Surveillance and Reconnaissance, Space, Electronic Warfare and Cyber Security', *Defence White Paper*, 2016, at www.defence.gov.au/Whitepaper/docs/ISR-Cyber.pdf.

5 https://asd.gov.au/about/index.htm.

6 Australia, Intelligence Services Act (Commonwealth) 2001, Section 7; the 'functions' of ASD include, inter alia: '(c) to provide material, advice and other assistance to Commonwealth and State authorities on matters relating to the security and integrity of information that is processed, stored or communicated by electronic or similar means': '(e) to provide assistance to Commonwealth and State authorities in relation to: (i) cryptography, and communication and computer technologies; (ii) other specialised technologies acquired in connection with the performance of its other functions and (iii) the performance by those authorities of search and rescue functions'; and '(f) to co-operate with and assist bodies referred to in section 13A [including the Australian Security Intelligence Organisation – ASIO] in accordance with that section…', at www.legislation.gov.au/Details/C2017C00360.

7 S. Bronitt and J. Stellios, 'Regulating Telecommunications Interception and Access in the Twenty-first Century: Technological Evolution or Legal Revolution?' (2006) 24 *Prometheus: Critical Studies in Innovation* 413.

8 See, inter alia, Division 268 Criminal Code Act 1995 (Commonwealth), at www.legislation.gov.au/Details/C2017C00408; s 61, Commonwealth of Australia Constitution Act 1900, at www.legislation.gov.au/Details/C2013Q00005.

immunities.[9] It is therefore useful for any assessment of the relationship between law and military technology to also consider, as a baselining exercise, a general appreciation of the types of regulatory issues that arise in such situations, if only as a counter-point to understanding how regulation of such technologies occurs within the context of LOAC.

4.1.1 *Outline*

In this chapter, I shall briefly survey a range of non-LOAC based regulatory issues that routinely affect the deployment of military technologies in both non-LOAC and LOAC-governed situations. To this end, I shall commence with a general recapitulation of the non-LOAC bodies of law, both domestic and international, which might apply to such employment of military technology. Following this, I shall briefly describe some examples of specific regulatory regimes that apply to the employment of military technology, noting that many of these regimes also affect the employment of these capabilities in LOAC-governed contexts, albeit in often moderated or mediated ways. These regimes include regulations relating to such issues as licences, acquisition, export control, privacy, and liability. The unsurprising conclusion of this assessment is that although the use of military technologies in armed conflict is primarily governed by LOAC, there are a range of 'peacetime' and 'baseline' legal regimes that are relevant to military capabilities which retain some persistent regulatory power even where LOAC is the applicable lex specialis. This phenomenon is in many ways merely an exemplar of a broader theme in LOAC application and interpretation, in that as militarily and operationally relevant non-LOAC regimes achieve greater specificity external to the lex specialis of LOAC, this regulatory precision is increasingly fed back into LOAC as second or third order specificity employed to fill perceived lacunae, or flesh out generalities, within the lex specialis of LOAC.

9 For example, depending upon the specific context and the organisational/agency affiliations engaged, the relevant legislation could extend to a mixture of (non-exhaustively): The 'call out' provisions in Part IIIAAA of the Defence Act 1903, at www.legislation.gov.au/Details/C2017C00342; the function, authorisation, and intelligence cooperation and sharing limitations within the Intelligence Services Act 2001, at www.legislation.gov.au/Details/C2017C00360; the cooperation and intelligence sharing and authorisation scheme within the Australian Security Intelligence Organization Act 1979, at www.legislation.gov.au/Details/C2016C01133; and the access and exemption provisions within the Telecommunications (Interception and Access) Act 1979, at www.legislation.gov.au/Details/C2017C00308.

4.2 GENERAL NON-LOAC REGULATORY REGIMES THAT APPLY TO THE EMPLOYMENT OF MILITARY TECHNOLOGY

4.2.1 *Domestic Regimes*

There are two fundamental, indeed paradigmatic regimes of domestic law that routinely apply to the employment of military technologies outside of LOAC-governed contexts. The first regime is 'criminal law', which is 'conventionally defined as a set of legal norms for determining the conditions under which individuals may be held liable to punishment'.[10] In order to explore this regime, I shall use the case study of Australia. The second regime is domestically implemented human rights law, which can differ from international human rights law in nuanced but significant ways; specific examples from the UK and Canadian contexts offer some indicators of this broader point.

4.2.1.1 Criminal Law

Criminal law is centrally concerned with the domestic regulation of conduct, states of affairs, or consequences that attract criminal responsibility.[11] One of the hallmark aspirations of criminal law is, consequently, to 'provide a clear and unambiguous statement of fundamental principles of criminal responsibility', ensuring – for consistency is essential - that any caveats to this general and universal applicability 'will be rare and the intention of the legislature to override its provisions, in those rare instances, will be explicit and unmistakable'.[12] One component of consistency and 'democracy' in criminal law is that the sanctionable 'wrong' is generally defined as universally applicable, and any exceptions, exclusions, or justifications are then defined expressly and explicitly.[13] Thus, in most democracies, the executive, its agents and all other subjects are equally bound by the criminal law. In Australia, this universality of application is expressed in a variety of ways:

[10] S. Bronitt and B. McSherry, *Principles of Criminal Law*, 3rd edn (Sydney: Lawbook Co, 2010), 6.

[11] In the Australian context, this is made explicit in Chapter 2 of the Criminal Code Act 1995 – see generally, S. Odgers, *Principles of Federal Criminal Law*, 2nd edn (Sydney: Lawbook Co, 2010), 2–3.

[12] *The Commonwealth Criminal Code: A Guide for Practitioners*, Commonwealth of Australia, Attorney-General's Department, Canberra, March 2002, 3.

[13] See generally, G. Williams, *Textbook of Criminal Law*, 2nd edn (London: Stevens and Son, 1983), at, inter alia, 27 et seq.

– Jurisdictional provisions with no executive caveats or carve outs;[14]
– Express regulation of executive activity (such as with the conduct and operations of intelligence agencies to which military capabilities often contribute);[15]
– Express provisions for the civil control of military capabilities, and the subjection of military operations to judicial and other forms of independent oversight;[16]
– Legal defences and excuses which are underpinned by the assumption that the offence-creating law applies to the executive, its agents, and its capabilities (including the military and its technology), such as Criminal Code Act, Section 10.5 'Lawful authority' ('A person is not criminally responsible for an offence if the conduct constituting the offence is justified or excused by or under a law');[17]
– Explicit 'binding of the Crown';[18] and
– The provision of express and legislatively limited immunities, such as the Intelligence Services Act 2001, Section 14 immunity for ASD officials – including, where relevant, ADF members in ASD;[19] and

[14] Criminal Code Act 1995. Part 2.7.
[15] For example, Intelligence Services Act 2001, Sections 13 and 13A on cooperation, and s 7 on the functions of the civil service and Australian Defence Force staffed Australian Signals Directorate.
[16] For example, Defence Act 1903, Part II in relation to 'Control and Administration', and Part VIIIB, in relation to the Inspector-General of the Australian Defence Force; Inspector-General of Intelligence and Security Act 1986, Section 8, in relation to the intelligence agency inquiry functions of the Inspector-General of Intelligence and Security (IGIS), whose bailiwick includes the Defence Intelligence Organisation (DIO).
[17] Criminal Code Act 1995, Section 10.4.
[18] For example, in relation to the Telecommunications (Interception and Access) Act 1979, Section 4: 'Act binds the Crown. This Act binds the Crown in right of the Commonwealth, of each of the States, of the Australian Capital Territory and of the Northern Territory'.
[19] Intelligence Services Act 2001, Section 14: 'Liability for certain acts. (1) A staff member or agent of an agency is not subject to any civil or criminal liability for any act done outside Australia if the act is done in the proper performance of a function of the agency.

(2) A person is not subject to any civil or criminal liability for any act (whether done inside or outside Australia) if:

(a) the act is preparatory to, in support of, or otherwise directly connected with, overseas activities of the agency concerned; and
(b) the act: (i) taken together with an act, event, circumstance or result that took place, or was intended to take place, outside Australia, could amount to an offence; but (ii) in the absence of that other act, event, circumstance or result, would not amount to an offence and
(c) the act is done in the proper performance of a function of the agency'.

judicial pronouncements to the effect that the executive is not immune from the law of the land.[20]

Domestic criminal law thus clearly holds consequences for the employment of military technologies at many levels of consideration, including the use of force, the tactical employment of technologies, and the regulatory constraints upon intelligence operations. An example of each will assist in demonstrating this point. So, in relation to the use of force, the employment of lethal technologies in national security operations is in general always subject to the same or similar criminal law limitations as those that are applicable to police operations in general. It is clear from points made during the inquest proceedings arising from the Lindt cafe siege in Martin Place, Sydney from 15 to 16 December 2014[21] that the use of lethal force is generally limited in criminal law to situations of self-defence, i.e. the defence of self or of others from the threat of imminent death or serious harm. This limitation is equally applicable to the use of military technologies and capabilities with lethal effects in domestic counter-terrorism operations by ADF units.[22] In that case, the Coroner specifically observed: 'I can readily appreciate why individual officers might be inclined to take a cautious approach to interpreting their powers. Their careers and even their own liberty could hinge on the later concurrence by others in the criminal justice system that their resort to deadly

[20] See, for example, *A v Hayden* [1984] HCA 67, dealing with the conduct, during a hostage rescue exercise, of the Australian Secret Intelligence Service (ASIS), which provoked Justice Mason to assert that, 'For the future, the point needs to be made loudly and clearly, that if counter-espionage activities involve breaches of the law they are liable to attract the consequences that ordinarily flow from breaches of the law' (para 2 of his judgment). Justice Murphy, in the same case, made the eviscerating observation that, 'The Executive power of the Commonwealth must be exercised in accordance with the Constitution and the laws of the Commonwealth. The Governor-General, the Federal Executive Council and every officer of the Commonwealth are bound to observe the laws of the land. If necessary, constitutional and other writs are available to restrain apprehended violations and to remedy past violations.
I restate these elementary principles because astonishingly one of the plaintiffs asserted through counsel that it followed from the nature of the executive government that it is not beyond the executive power, even in a situation other than war, to order one of its citizens to kill another person. Such a proposition is inconsistent with the rule of law. It is subversive of the Constitution and the laws. It is, in other countries, the justification for death squads. . .' (at para 3 of his judgment).

[21] State Coroner of NSW, *Inquest into the deaths arising from the Lindt café siege: Findings and Recommendations*, May 2017, ch14, especially paras 5–47, at www.lindtinquest.justice.nsw.gov .au/Documents/findings-and-recommendations.pdf.

[22] See, inter alia, R.McLaughlin, 'The Use of Lethal Force by Military Forces on Law Enforcement Operations – Is There a "Lawful Authority"?' (2009) 37:3 *Federal Law Review* 441.

force was justified'.[23] Any similar use of force by ADF members, employing lethal military technologies – even when under special 'call out' orders in accordance with Part IIIAAA of the Defence Act 1903 – is still subject to use of force limitations broadly based on the justification of self-defence.[24] An example of the second limitation is the tactical use of military technologies in domestic security operations. In the Northern Ireland context, for example, the necessity for employing force protection technologies designed for armed conflict in domestic security operations was a topic of conscious operational, reputational, medical, and legal consideration.[25] The use of such military technologies in this manner poses increasing legal challenges because, though necessarily both routine and essential,[26] operations using such technologies are also perennially controversial.[27] The third example, regulatory constraints upon intelligence operations, is also well illustrated by a case in point drawn from the context of British operations during 'the Troubles' in Northern Ireland. This example relates to the liability of undercover intelligence and military agents, and their handlers, for murder and torture under the ordinary criminal law.[28] Whilst, in the 1970s and 1980s, such operations were predominantly human intelligence (HUMINT)-based, such intelligence-led security

[23] *Inquest into the deaths arising from the Lindt café siege: Findings and Recommendations*, ch 14, para 45.

[24] Defence Act 1903, Section 51T Use of reasonable and necessary force; see D.Letts and R. McLaughlin, 'Call-out Powers for the Australian Defence Force in an Age of Terrorism: Some Legal Implications' (2017) No 85 *Australian Institute of Administrative Law Forum* 63, 75–79.

[25] See, for example, W. Miller, 'The British Experience in Northern Ireland: A Model for Modern Peacemaking Operations', US Army Command and General Staff College, Fort Leavenworth, 1993, 40, at www.dtic.mil/dtic/tr/fulltext/u2/a264419.pdf. One example is the use of plastic bullets/baton rounds: see, for example, E. Phoenix, 'NI State Papers: Concern over Use of Plastic Bullets', *BBC*, 28 December 2013, at www.bbc.com/news/uk-northern-ireland-25425779; E. Sweeney, '"Troubles" legacy: Derry used as testing ground for plastic bullets', *Derry Journal*, 26 April 2016, at www.derryjournal.com/news/troubles-legacy-derry-used-as-testing-ground-for-plastic-bullets-1-7348893; A. Ritchie and J. Gibbons, 'Letter to Editor: Plastic bullets in Northern Ireland' (1990) 301(6764) *British Medical Journal* 1332.

[26] See, inter alia, US National Research Council, *An Assessment of Non-Lethal Weapons Science and Technology* (Washington, DC: The National Academies Press, 2003), at for example 50–63.

[27] For example, the issue of 'drones': Birmingham Policy Commission, *The Security Impact of Drones: Challenges and Opportunities for the UK*, October 2014, 48–52 in relation to, for example, complicity in intelligence sharing and employment outside of armed conflict contexts, 59 as to 'lowering the threshold on the use of force', and 76–77 as to privacy, at www.birmingham.ac.uk/Documents/research/policycommission/remote-warfare/final-report-october-2014.pdf.

[28] See, for example, S. O'Neill, 'British spy in IRA arrested on murder and torture charges', *The Times*, 31 January 2018, 5.

operations today are increasingly driven by signals intelligence (SIGINT) and technological requirements, employing bivalent technologies and capabilities (that is, they are useful and employable in both armed conflict and domestic security contexts) which are often maintained and deployed primarily or in part by a state's military forces.

4.2.1.2 Domestic Human Rights Law

That domestic human rights law carries with it regulatory implications for the employment of military technology is today a statement of the obvious. The experience of the UK in relation to military materiel acquisition and deployment, and the availability and use of military technology, is an ongoing matter before the courts, where issues as to liability arising from the deployment of thin-skinned snatch Land Rovers, or the non-deployment of certain targeting and identification technology in Challenger II tanks, remains the subject of ongoing tensions between human rights obligations and the doctrine of combat immunity.[29] However, domestic human rights law can also affect the employment of military technology in other ways. One example can be extrapolated from the *Khadr* series of cases[30] in relation to the application of the *Canadian Charter of Human Rights*.[31] In this case, the involvement of Canadian intelligence and foreign service members in questioning Omar Khadr, a Canadian minor held in the US detention facility at Guantanamo Bay, Cuba, in circumstances where their questioning contributed to his ongoing detention, was held to have enlivened certain *Canadian Charter* rights:

> It is reasonable to infer from the uncontradicted evidence before us that the statements taken by Canadian officials are contributing to the continued detention of Mr Khadr, thereby impacting his liberty and security interests. In the absence of any evidence to the contrary... we conclude on the record before us that Canada's active participation in what was at the time an illegal

[29] *Smith and others* v. *Ministry of Defence* [2013] UKSC 41.

[30] *Canada (Justice)* v. *Khadr* [2008] 2 SCR 125; *Canada (Prime Minister)* v *Omar Ahmed Khadr* [2010] 1 SCR 44; Nathan Denette, 'Supreme court rejects federal government's argument that Omar Khadr is an adult offender', *National Post*, 14 May 2015, at http://nationalpost.com/news/canada/supreme-court-rejects-tory-argument-that-omar-khadr-was-adult-offender; L. McGregor, 'Are Declaratory Orders Appropriate for Continuing Human Rights Violations? The Case of Khadr v Canada' (2010) 10:3 *Human Rights Law Review* 487; K. Roach, 'The Supreme Court at the Bar of Politics: The Afghan Detainee and Omar Khadr Cases' (2010) 28 *National Journal of Constitutional Law* 115.

[31] Constitution Act (1982) Canada, Part I: Canadian Charter of Rights and Freedoms, article 7 'Everyone has the right to life, liberty and security of the person and the right not to be deprived thereof except in accordance with the principles of fundamental justice.'

regime has contributed and continues to contribute to Mr Khadr's current detention, which is the subject of his current claim.[32]

And further:

> This conduct establishes Canadian participation in state conduct that violates the principles of fundamental justice. Interrogation of a youth, to elicit statements about the most serious criminal charges while detained in these conditions and without access to counsel, and while knowing that the fruits of the interrogations would be shared with the US prosecutors, offends the most basic Canadian standards about the treatment of detained youth suspects.[33]

The question to ask, in relation to the employment of military technology, is whether the same conclusions as to the applicability of the *Canadian Charter* would follow if the contribution to continued detention had not been physical presence and questioning by Canadian officials, but rather the provision of key SIGINT or other intelligence material collected via Canadian military EW capabilities. It is difficult to see why not.

4.2.2 *International Legal Regimes*

Depending upon the context, certain international law rule sets routinely apply to non-LOAC-governed employment of military technology. Indeed, many of these same regimes will also continue to apply in parallel with LOAC when the context is one of armed conflict, noting of course that the principle of lex specialis must be borne in mind when considering precisely (and on a case by case basis) how such regimes apply. Five examples will illustrate this point: The COLREGS, or the 'Rules of the Road' for maritime navigation; the legal regime relating to cyber operations below the threshold of a UN Charter Article 51 'armed attack'; the law relating to attribution of naval mines to particular States; the customary international law encapsulated in the Articles on State Responsibility; and the Law of the Sea Convention 1982, and its parallel customary rules and predecessor treaties.

The COLREGS are the universally accepted rules for maritime navigation,[34] and they apply to military technology, including warships.[35] In the

[32] *Canada (Prime Minister)* v *Omar Ahmed Khadr* [2010] 1 SCR 44 at para 21.

[33] Ibid., para 25.

[34] Convention on the International Regulations for Preventing Collisions at Sea 1972 (COLREGs), 'Rule 1 (a): 'These Rules shall apply to all vessels upon the high seas and in all waters connected therewith navigable by seagoing vessels.'

[35] For example: US Navy, *Report on the Collision between USS FITZGERALD (DDG 62) and Motor Vessel ACX CRYSTAL, and Report on the Collision between USS JOHN S MCCAIN*

recent Philippines–China arbitral award (the 'South China Sea arbitration') the application of both the general LOSC obligations in relation to safe navigation and safety at sea, and the COLREGS was once again assessed in terms of its regulatory effects upon and application to military and state law enforcement technology, including warships and state vessels.[36]

The second example of a general regulatory scheme which routinely governs military technology, whilst simultaneously taking little cognisance of the military status of the technology and its custodians, is cyber capabilities when employed outside of armed conflict in situations that fall below the UN Charter Article 51 'armed attack' threshold. As the Tallinn Manual 2.0 explains in Rule 15, 'Cyber operations conducted by organs of a State, or by persons or entities empowered by domestic law to exercise elements of governmental authority, are attributable to the State.'[37] The Tallinn Manual 2.0 Commentary then elaborates as follows:

> Attribution to a State occurs in a number of circumstances. The clearest case is when State organs, such as the military or intelligence agencies, commit the wrongful acts. For instance, all cyber activities of US Cyber Command, the Netherlands Defence Cyber Command, the French Network and Information Security Agency (ANSSI), the Estonian Defence League's Cyber Unit, the People's Liberation Army cyber unit, and Israel's Unit 2800 are fully attributable to the respective States.[38]

In situations governed by such fundamental, responsibility apportioning, legal regimes – particularly those which are concerned with the regulation of counter-measures[39] – the fact that the particular technology employed in the operation is maintained and deployed by armed forces is largely irrelevant: The applicable non-LOAC rules generally assume that the relevant obligation in practice falls upon the State in general. This is consistent with the

(DDG 56) and Motor Vessel ALNIC MC, 23 October 2017, §8.1, 8.2 (USS FITZGERALD), §1.2, 7.2 (USS JOHN S MCCAIN), at http://s3.amazonaws.com/CHINFO/USS+Fitzgerald+and+USS+John+S+McCain+Collision+Reports.pdf; M. Troy, 'USS *John McCain* Destroyer: Who's At Fault When Ships Collide?', *ABC News*, 21 August 2017, at www.abc.net.au/news/2017-08-21/uss-john-mccain-destroyer:-whos-at-fault-when-ships-collide/8827516.

[36] PCA Case N° 2013–19 In the Matter of the South China Sea Arbitration before an Arbitral Tribunal Constituted Under Annex VII to the 1982 United Nations Convention on the Law of the Sea between the Republic of the Philippines and the People's Republic of China, Award, 12 July 2016, paras 1060–1109, at www.pcacases.com/web/sendAttach/2086.

[37] M. Schmitt, et al. (eds.), *Tallinn Manual 2.0 on the International Law Applicable to Cyber Operations* (Cambridge: Cambridge University Press, 2017), 87.

[38] Ibid.

[39] UN General Assembly, A/RES/56/83 Responsibility of States for internationally wrongful acts, 12 December 2001, Articles 49–53.

obligations encapsulated in Articles 4–9 of the Articles on State Responsibility. To this end, the particularised status of the discrete State agent involved – military, police, or public servant – is on the whole irrelevant as to the nature and existence of the obligation (although it is not irrelevant for the purposes of international criminal law, or for domestic legal accountability mechanisms). This is quite different in both practice and consequence to many LOAC obligations relating to the employment of military technology, which in reality fall more directly and particularly upon the commander and/or the military force as the employer of the technology – as, for example, with the 1977 Additional Protocol I Article 57 'precautions in attack', which impose a 'very heavy burden of responsibility…on military commanders'.[40]

A third example of the regulatory reach of non-military, non-LOAC, regulatory schemes in relation to military technologies is the law governing the use of naval mines outside armed conflict[41] – in particular the legal assessment for attribution of national responsibility in relation to the deployment of this particular military technology in non-LOAC-governed situations. The ICJ has dealt with this question directly on several occasions.[42] The *Corfu Channel* case involved Albanian responsibility for notifications as to the presence of mines in its waters, regardless of the fact that it did not lay those mines,[43] in a context outside of 'war' (given that the 1949 Geneva Conventions and the new

[40] Protocol Additional to the Geneva Conventions of 12 August 1949, and relating to the Protection of Victims of International Armed Conflicts (Protocol I), 1977, article 57; Commentary: Additional Protocol I, ICRC, Geneva, 1987, para 2187.

[41] D. Letts, 'Naval Mines: Legal Considerations in Armed Conflict and Peacetime' (2016) 98/102 *International Review of the Red Cross* 543; D. Stephens and M.Fitzpatrick, 'Legal Aspects of Contemporary Naval Mine Warfare' (1999) 21:4 *Loyola of Los Angeles International and Comparative Law Journal* 553.

[42] The *Nicaragua* case is less on point here, as it was clearly within the context of an armed conflict in which the USA was implicated – ICJ, *Case Concerning Military and Paramilitary Activities in and Against Nicaragua (Nicaragua* v. *United States of America) (Merits)* [1986] ICJ Reps 14, para 292(6): 'Decides that the United States of America, by failing to make known the existence and location of the mines laid by it, referred to in subpara (6) hereof, has acted in breach of its obligations under customary international law in this respect…'.

[43] ICJ, *Corfu Channel* case, Judgment, [1949] ICJ Reps 4, pp15, 19–22: 'From all the facts and observations mentioned above, the Court draws the conclusion that the laying of the minefield which caused the explosions on October 22nd, 1946, could not have been accomplished without the knowledge of the Albanian Government…'; see generally, R. McLaughlin, 'Dangerous Waters and International Law: The Corfu Channel Case, Warships, and Sovereignty Irritants', in K. Bannelier, T. Christakis and S. Heathcote (eds.) *The ICJ and the Evolution of International Law: The enduring impact of the Corfu Channel case* (Abingdon: Routledge, 2012), ch 10; A.Lalaj *Burning Secrets of the Corfu Channel Incident* (Woodrow Wilson International Center Cold War History Project Working paper #70, September 2014), at for example, 20–24.

concept of 'armed conflict' were at this stage still a few months from finalisation). The factually similar, but differently decided, *Oil Platforms* case involved a mine strike against the USS *Samuel B. Roberts*. Although it occurred within the context of the Iran–Iraq War, the mining of the US warship actually preceded the very short armed conflict between the USA and Iran (Op Praying Mantis)[44]. The ICJ ultimately dealt with the attribution of responsibility for the deployment of the naval mines in terms of governance regimes external to LOAC, finding in this case that the mines could not be attributed to Iran.[45]

A fourth example of a persistent regulatory scheme which always governs the deployment of military technology, yet owes little to LOAC, is the 'nationalising' effect of (for example) the Articles on State Responsibility upon the deployment of military technology for LOAC-governed ISR. The regulatory effect of the Articles on State Responsibility (or more correctly, of the customary international law that stands behind them) does not cease or abdicate the legal field simply because LOAC is also engaged – this legal regime persists in parallel even in armed conflict situations. One example is the employment of State A military technology, for example an unmanned aerial vehicle, by State B military personnel deployed to State A to assist in such an operation. In such cases the contextual and jurisdictional fact that LOAC may be applicable to the operation does not displace the separate legal assessment required of the attribution to both States A and B of any wrongful consequences of the employment of State A military technology by State B military personnel.[46]

The fifth example of a persisting regulatory scheme is the Law of the Sea Convention 1982, and its counterpart customary and predecessor rules. The law of the sea does not cease to regulate the deployment and

[44] See, inter alia, J. B. Perkins, 'Operation Praying Mantis: The Surface View' (1989) 115:5 *US Naval Institute Proceedings* 1035, at www.usni.org/magazines/proceedings/1989-05/operation-praying-mantis-surface-view.

[45] ICJ, *Oil Platforms (Islamic Republic of Iran* v *United States of America)*, Judgment, [2003] ICJ Reps 161, paras 23–26, 44–71 et seq; see also R. McLaughlin and D. Stephens, 'International Humanitarian Law in the Maritime Context: Conflict Characterization in Judicial and Quasi-Judicial Contexts', in D. Jinks, J.Maogoto and S. Solomon (eds.) *Applying International Humanitarian Law in Judicial and Quasi-Judicial Bodies: International and Domestic Aspects* (The Hague: TMC Asser Press, 2014), 103.

[46] R. McLaughlin, 'State Responsibility for Third Country Deployed / Embedded Military Personnel Engaged in Armed Conflict', *EJIL Talk!*, 10 September 2015, at www.ejiltalk.org/state-responsibility-for-third-country-deployed-embedded-military-personnel-engaged-in-armed-conflict/.

employment of military technology simply because it is military technology, and regardless – albeit with some lex specialis nuance – of the fact that the law of naval warfare and the law of neutrality may also simultaneously operate as between the concerned parties. As the San Remo Manual on the International Law Applicable to Armed Conflict at Sea notes, the scheme of maritime zones and associated passage regimes established in the LOSC 1982 does not cease to apply in times of armed conflict; indeed, the questions are more about how the law of naval warfare might interact with, limit, and be limited by, for example, the transit passage, Archipelagic Sea Lanes passage, and innocent passage regimes.[47] Additionally, the status at sea of military technology such as unmanned surface and sub-surface vessels remains relevant regardless of the parallel applicability of LOAC.[48] Thus the sovereign immune status[49] of military technologies remains vital to assessing the deployment and employment of military technologies in all legal contexts. This is the case in relation to entire vessels such as the USS *Pueblo*, seized in January 1968 by North Korean forces in waters which on any Law of the Sea compliant reading were outside the DPRK territorial sea, although a parallel question regarding the 1953 Armistice persists[50]. It is also the case in relation to specific items of equipment, such as submersible hydrographic 'gliders'.[51]

[47] L. Doswald-Beck (ed.) *San Remo Manual on International Law Applicable to Armed Conflicts at Sea* (Cambridge/San Remo: Cambridge University Press/International Institute of Humanitarian Law, 1995), Rules 27–33.

[48] R.McLaughlin, 'Unmanned Naval Vehicles at Sea: USVs, UUVs, and the Adequacy of the Law' (2011) 21:2 *Journal of Law Information and Science* 100; W. Heintschel von Heinegg, 'Unmanned Maritime Systems: Does the Increasing Use of Naval Weapon Systems Present a Challenge for IHL?', in W. Heintschel von Heinegg, R. Frau, and T. Singer (eds.), *Dehumanization of Warfare: Legal Implications of New Weapon Technologies* (New York: Springer, 2018), 119.

[49] Law of the Sea Convention 1982, Article 32.

[50] G. Aldrich, 'Questions of International Law raised by the Seizure of the USS *Pueblo*' (1969) 63 *Proceedings of the American Society of International Law at its Annual Meeting (1921–1969)* 2; S. Radchenko, *The Soviet Union and the North Korean Seizure of the USS Pueblo: Evidence from Russian Archives*, Woodrow Wilson International Center Cold War History Project Working paper #47, 7 July 2011, at (for example) 11, and Document 20.

[51] See, for example, M.Ryan and D. Lamothe, 'Pentagon: Chinese Naval Ship Seized an Unmanned US Underwater Vehicle in South China Sea', *Washington Post*, 17 December 2016, at www.washingtonpost.com/news/checkpoint/wp/2016/12/16/defense-official-chinese-naval-ship-seized-an-unmanned-u-s-ocean-glider/?utm_term=.2dfia5ad523d.

4.3 SOME SPECIFIC REGULATORY ISSUES WITH IMPLICATIONS FOR EMPLOYMENT OF MILITARY TECHNOLOGY

4.3.1 *Licensing and Employment of Military Technologies*

In non-LOAC-governed situations the military forces and physical military technology of one State may be deployed in the territory of another State, usually with the consent of that territorial State, on the basis of an appropriate UN Charter, Chapter VII authorisation from the UN Security Council. A 'Status of Forces Arrangement' (SOFA) or instrument of similar effect will often be entered into between the consenting/authorising entity (the territorial State or the UN) and the deploying State. Such arrangements routinely canvas matters that touch upon the deployment and employment of military technologies, including use of the electromagnetic spectrum, immunity from local laws as to the use of technologies such as driverless vehicles or UAV, and the licensing and use of military medical technologies.[52] Such regulatory arrangements are not generally required in situations where LOAC applies, such as occupation,[53] or the conduct of hostilities in the territory of an adversary State. However, these regulatory provisions will often be necessary in a situation governed by the law of non-international armed conflict where the deploying State is employing its forces and military technology to assist the territorial State against an adversary non-State armed group.[54]

4.3.2 *Liability and Agency*

Rules on liability, agency, and the remedies applicable to the consequences of the use of military technology often leverage – or hinge upon – the

[52] See, for example, *Draft Model United Nations SOFA*, 1990, Articles 11 (communications systems), 13 (vehicle, aircraft, and vessel registration and licensing), 38–39 (permits and licenses, including for weapons); see the commentary on the Draft Model UN SOFA in B. Oswald, H. Durham, and A. Bates (eds.), *Documents on the Law of UN Peace Operations*, (Oxford: Oxford University Press, 2010), 34–38.

[53] For example, 1949 Geneva Convention IV, Article 56, which authorises the occupying power to establish, in certain circumstances, Article 18 protected civilian hospitals, and to employ/assign Article 20 protected civilian hospital staff, independent of the normal processes and control of the 'competent organs of the occupied State…'.

[54] See, for example, Agreement between the North Atlantic Treaty Organization and the Islamic Republic of Afghanistan on the Status of NATO Forces and NATO personnel conducting mutually agreed NATO-led activities in Afghanistan, 30 September 2014, e.g. Article 5(6) on respect for (but not being bound by) local work, health and safety laws; Article 8 on vehicles, vessels and aircraft; Article 10(2)-(4) on communications and use of the radio spectrum; Article 16 on driving and professional licenses, at www.nato.int/cps/ic/natohq/official_texts_116072.htm?selectedLocale=en.

jurisdictionally relevant rules for the liability of the Crown/State/Executive within domestic law. However, these rules can often continue to apply to military technologies even during armed conflict, where the particular liability instance at issue has a limited nexus to the armed conflict. Thus, when HMAS *Adelaide*, a Royal Australian Navy cruiser, collided during the dark hours with a merchant vessel (*Coptic*) off the coast of eastern Australia during World War II, the Commonwealth was held liable for the damage to the merchant vessel caused by the negligent seamanship of the crew of the cruiser. In this case, both vessels were, pursuant to orders, steaming without lights to reduce the potential for detection by enemy vessels. *Coptic* was sailing a Navy-assigned passage plan to simplify the surface picture for Australian and Allied forces through knowledge of which vessels were expected and which were suspicious and thus warranted further investigation. All of these actions were authorised by regulations and were clearly designed to support the prosecution of the war effort. Yet the High Court found that the circumstances of the collision were sufficiently removed from a direct combat or near to combat situation that the normal rule on combat immunity from liability did not apply, and that the negligent seamanship of the naval unit rendered the Commonwealth liable to the owners of *Coptic* for the damages suffered.[55]

4.3.3 *Export Control Regimes*

Control regimes regulating the export of sensitive military or dual use technologies come in a variety of guises. The Wassenaar Arrangement, for example, is a non-treaty level arrangement that was

> ... established in order to contribute to regional and international security and stability, by promoting transparency and greater responsibility in transfers of conventional arms and dual-use goods and technologies, thus preventing destabilising accumulations. The aim is also to prevent the acquisition of these items by terrorists.[56]

This regime is one of several that deal with the vexed issue of export control and dual use technology. The latter is, for example, defined by the EU as 'items, including software and technology, which can be used for both civil and military purposes ... [including] all goods which can be used for both

55 *Shaw Savill and Albion Co Ltd v. Commonwealth* (1940) 66 CLR 344.

56 *Wassenaar Arrangement on Export Controls for Conventional Arms and Dual-Use Goods and Technologies*, 1995, Wassenaar Arrangement Secretariat, Vienna, at www.wassenaar.org/wp-content/uploads/2015/06/WA-DOC-17-PUB-001-Public-Docs-Vol-I-Founding-Documents.pdf.

non-explosive uses and assisting in any way in the manufacture of nuclear weapons or other nuclear explosive devices'.[57] The Arms Trade Treaty 2013, by contrast, is an example of an international convention embodying legally binding commitments and obligations – in this case, in relation to trade in certain forms of military technology:[58]

The object of this Treaty is to:

- Establish the highest possible common international standards for regulating or improving the regulation of the international trade in conventional arms;
- Prevent and eradicate the illicit trade in conventional arms and prevent their diversion;

for the purpose of:

- Contributing to international and regional peace, security and stability;
- Reducing human suffering; and
- Promoting cooperation, transparency and responsible action by States Parties in the international trade in conventional arms, thereby building confidence among States Parties.[59]

A third form of regulatory regime affecting the export of military technology is illustrated by UN Security CouncilR 1540 (2004), authorised in accordance with the UNSC's Chapter VII powers. This binding resolution was targeted at the supply of certain military technologies of concern to non-State armed groups, and to this end mandated that

> all States shall take and enforce effective measures to establish domestic controls to prevent the proliferation of nuclear, chemical, or biological weapons and their means of delivery, including by establishing appropriate controls over related materials and to this end shall:

57 Council Regulation (EC) No 428/2009 of 5 May 2009 setting up a Community regime for the control of exports, transfer, brokering and transit of dual-use items, Article 2(1), at http://eur-lex.europa.eu/legal-content/EN/TXT/PDF/?uri=CELEX:02009R0428–20161116&from=EN; see also the 'Guidance' provided by the UK government in relation to the 'strategic export control list' and the special rules applicable to 'military or dual use goods, services or technology', at www.gov.uk/guidance/export-military-or-dual-use-goods-services-or-technology-special-rules; see similarly, the Australian 'Defence and Strategic Goods List', at www.defence.gov.au/ExportControls/DSGL.asp.

58 Arms Trade Treaty 2013, Article 2(1): 'This Treaty shall apply to all conventional arms within the following categories: (a) Battle tanks; (b) Armoured combat vehicles; (c) Large-calibre artillery systems; (d) Combat aircraft; (e) Attack helicopters; (f) Warships; (g) Missiles and missile launchers; and (h) Small arms and light weapons.'

59 Ibid., Article 1.

(a) Develop and maintain appropriate effective measures to account for and secure such items in production, use, storage or transport;
(b) Develop and maintain appropriate effective physical protection measures;
(c) Develop and maintain appropriate effective border controls and law enforcement efforts to detect, deter, prevent and combat, including through international cooperation when necessary, the illicit trafficking and brokering in such items in accordance with their national legal authorities and legislation and consistent with international law;
(d) Establish, develop, review and maintain appropriate effective national export and trans-shipment controls over such items, including appropriate laws and regulations to control export, transit, trans-shipment and re-export and controls on providing funds and services related to such export and trans-shipment such as financing, and transporting that would contribute to proliferation, as well as establishing end-user controls; and establishing and enforcing appropriate criminal or civil penalties for violations of such export control laws and regulations.[60]

Such international regulatory regimes concerned with controlling the export of sensitive military and dual use technologies are often directly reflected in the domestic law of exporter States, where regulatory measures are designed to address both international obligations and domestic national security concerns. However, the 'domestication' pathway adopted is often slave to the iterative process by which such controls have evolved in tandem with both the technologies of concern, and the interdiction and control mechanisms available. Taking Australia as an example, the cohort of legislation that affects the export control regime includes: Customs Act 1901 (in particular the military end-user provisions in Part VI, Division 1AA of that Act);[61] Defence Act 1903; Charter of the United Nations Act 1945; *Weapons of Mass Destruction (Prevention of* Proliferation) Act 1995; Criminal Code Act 1995;

[60] UNSCR 1540 on Non-Proliferation of Weapons of Mass Destruction, 28 April 2004, operative para 3.

[61] Customs Act 1901 (Commonwealth), at, for example, Section 112BA: 'Notice prohibiting export. (1) If:

(a) the Defence Minister suspects that, if a person (the first person) were to export particular goods to a particular place or to a particular person, the goods would or may be for a military end use that would prejudice the security, defence or international relations of Australia; and
(b) the goods are not prohibited exports under section 112; the Defence Minister may give the first person a notice prohibiting the first person from exporting the goods to the particular place or particular person...', at www.legislation.gov.au/Details/C2017C00411.

Autonomous Sanctions Act 2011; Defence Trade Controls Act 2012 and Maritime Powers Act 2014.

4.3.4 *Privacy*

Privacy is a matter of fundamental regulatory concern in national law and international human rights law.[62] However, in many legal systems, privacy regulations do not prima facie apply to certain military and security uses of data. For example, clause 16 of EU Regulation (EU) 2016/679 provides that

> This Regulation does not apply to issues of protection of fundamental rights and freedoms or the free flow of personal data related to activities which fall outside the scope of Union law, such as activities concerning national security. This Regulation does not apply to the processing of personal data by the Member States when carrying out activities in relation to the common foreign and security policy of the Union.[63]

In other jurisdictions, privacy regulations continue to apply in toto to certain types of military data collection technologies and activities, or in a mediated form which often involves oversight by a separate body cognisant of, and authorised in accordance with, the appropriate security considerations.[64] The Australian Defence Intelligence Organisation, for example, is exempt from the routine operation of the Australian Privacy Principles,[65] but is nevertheless required to abide by Ministerial 'Guidelines to Protect the Privacy of Australian Persons'.[66] The Australian Signals Directorate (ASD), in

62 For example, International Covenant on Civil and Political Rights 1966, Article 17.

63 Regulation (EU) 2016/679 of the European Parliament and of the Council of 27 April 2016 on the protection of natural persons with regard to the processing of personal data and on the free movement of such data, and repealing Directive 95/46/EC (General Data Protection Regulation), at http://eur-lex.europa.eu/legal-content/EN/TXT/HTML/?uri=CELEX:32016R0679&from=EN.

64 The Australian Department of Defence, for example, has a specialised policy on this topic: DI (G) OPS 13–16 Defence policy relating to the collection of biometric data for intelligence purposes (03 October 2012), which is referenced in Re-thinking Systems of Inquiry, Investigation, Review and Audit in Defence, Annex I: Current mandated legal support available to Defence under Defence policy and instructions, at www.defence.gov.au/Publications/Reviews/SystemsInquiry/docs/Annex_I_Current_mandated_legal_support_available_to_Defence_under_Defence_policy_and_instruction.pdf.

65 Contained in Schedule 1 to the Privacy Act 1988, at www.legislation.gov.au/Series/C2004A03712.

66 Defence Intelligence Organisation, 'Guidelines to Protect the Privacy of Australian Persons', 16 December 2012: 'The purpose of these guidelines is to ensure that DIO preserves the privacy of Australian persons while ensuring the proper performance of the organisation's intelligence functions. . .', at www.defence.gov.au/dio/privacy-rules.shtml.

accordance with section 15 of the Intelligence Services Act,[67] is likewise required to operate in accordance with Ministerial guidelines that provide an attenuated but operationally coherent system for privacy regulation of – in this case – Australian Signals Directorate employed military data collection, storage, and transmission technologies.[68]

A key recent example of the implications of the interaction of privacy regulatory regimes with military technology is biometric data capture, storage, and transmission.[69] 'Biometric technologies', can be described as those which 'take automated measurements of certain physiological or behavioural traits for purposes of human recognition'.[70] These technologies have considerable utility when used, e.g., by and for military purposes to achieve 'identity dominance', that is, 'where military forces have the distinct ability to separate friend from foe by linking people to their previous identities and past terrorist or criminal activities'.[71] They are obviously significant contributors to ISR, detention and targeting operations. One recent study of the use of US biometric data collection capabilities as a component of identity operations in Iraq asserts that improvements to the forensic capacities of the Iraqi legal system, and to the accuracy of identification of detainees, were positive consequences of the deployment of the technology.[72] The military collection, storage, and transmission of biometric data (and also of cyber identity data) of both one's own force and of any adversary/potential adversary force, is still subject to overarching privacy and information assurance requirements.[73] Nevertheless, the implications of military biometric data collection

[67] Intelligence Services Act 2001, Section 15: 'Rules to protect privacy of Australians. (1) The responsible Minister in relation to ASIS, the responsible Minister in relation to AGO and the responsible Minister in relation to ASD, must make written rules regulating the communication and retention by the relevant agency of intelligence information concerning Australian persons.' (2) In making the rules, the Minister must have regard to the need to ensure that the privacy of Australian persons is preserved as far as is consistent with the proper performance by the agencies of their functions. . .', at www.legislation.gov.au/Details/C2017C00360.

[68] Australian Signals Directorate, 'Rules to Protect the Privacy of Australians', 02 October 2012, at www.asd.gov.au/publications/broadcast/20121002-privacy-rules.htm.

[69] See Chapter 14.

[70] J.Woodward, Jr, 'Using Biometrics to Achieve Identity Dominance in the Global War on Terrorism' (September–October 2005) LXXXV No.5 *Military Review* 30 at 31.

[71] Ibid., 34.

[72] W. Buhrow, *Biometrics in Support of Military Operations: Lessons from the Battlefield* (Boca Raton: Taylor and Francis, 2017), 64–65.

[73] See, for example, *US Department of Defense Directive 8500.0.1E – Information Assurance (IA)*, 24 October 2002, certified current as of 23 April 2007, para 4.8.2, at http://dodcio.defense.gov/Portals/0/Documents/DIEA/850001p.pdf; on Cyber, Identity and Information Assurance (CIIA), see US Department of Defense, *Deputy Assistant Secretary of Defense for Cyber,*

technologies for privacy and the associated challenges for regulation are fundamental. As *Scientific American* reported in 2014, 'this shift to biometric-enabled security creates profound threats to commonly accepted notions of privacy and security. It makes possible privacy violations that would make the National Security Agency's data sweeps seem superficial by comparison.'[74] A specific illustration of the challenges, perceived or actual, posed by such military technologies and/or their military use is offered by a report by *the Guardian* on US biometric data collection in Afghanistan:

> With each iris and fingertip scanned, the device gave the operator a steadily rising percentage chance that the goat herder was on an electronic 'watch list' of suspects. Although it never reached 100%, it was enough for the man to be taken to the nearest US outpost for interrogation.
>
> Since *the Guardian* witnessed that incident, which occurred near the southern city of Kandahar earlier this year, US soldiers have been dramatically increasing the vast database of biometric information collected from Afghans living in the most war-torn parts of southern and eastern Afghanistan.
>
> The US army now has information on 800,000 people, while another database developed by the country's interior ministry has records on 250,000 people.
>
> It is the sort of operation that would horrify civil liberties campaigners in the west, but there has been little public debate in Afghanistan. Kitted out with handheld devices that contain a camera to scan eyes and an electronic pad to take fingerprints, US soldiers have been collecting huge amounts of biometric data, with little oversight from the Afghan government.[75]

Recent moves to place the US Army's 'classified Defense Cross-Domain Analytical Capability, a database storing various kinds of security-relevant information' in the 'cloud',[76] have also generated significant expressions of concern regarding privacy vulnerabilities and the potential for misuse by authorised partner forces or governments.[77]

Identity, and Information Assurance Strategy, August 2009, at http://dodcio.defense.gov/Portals/0/Documents/DoD_IA_Strategic_Plan.pdf.

74 'Biometric Security Poses Huge Privacy Risks: Without explicit safeguards, your personal biometric data are destined for a government database', *Scientific American*, 01 January 2014, at www.scientificamerican.com/article/biometric-security-poses-huge-privacy-risks/.

75 See, for example, J. Boone, 'US Army Amasses Biometric Data in Afghanistan', *The Guardian*, 27 October 2010, at www.theguardian.com/world/2010/oct/27/us-army-biometric-data-afghanistan.

76 S. Ackerman, 'Army's Fingerprint and Iris Scan Databases Head for the Cloud', *Wired*, 30 May 2013, at www.wired.com/2013/05/army-biometrics-cloud/.

77 For example: A. Estes, 'The Pentagon is Putting its Big Brother Data Collection in the Cloud', *Gizmodo*, 01 June 2013, at www.gizmodo.com.au/2013/06/the-pentagon-is-putting-its-big-brother-data-collection-in-the-cloud/.

4.3.5 *Review and Acquisition*

A final regulatory regime applicable to military technology – one that is in fact bounded by LOAC, but which applies outside of immediate armed conflict situations – is that relating to weapons review and acquisition.[78] The indicative obligation is that in Article 36 of API, relating to the conduct of reviews for new weapons: 'In the study, development, acquisition or adoption of a new weapon, means or method of warfare, a High Contracting Party is under an obligation to determine whether its employment would, in some or all circumstances, be prohibited by this Protocol or by any other rule of international law applicable to the High Contracting Party'.[79] This obligation, although expressed as a component of LOAC, necessarily pre-dates any armed conflict arising,[80] in the same way that the obligation in Common Article 1 of the Geneva Conventions 1949 to respect and ensure respect of the Conventions applies equally to States both party to, and outside of, any particular armed conflict.[81] Other 'weapons law' LOAC instruments which entail direct regulatory effects for military technology outside of situations of armed conflict include the Convention on Certain Conventional Weapons[82] and its Protocols, the Ottawa Convention (concerning anti-personnel landmines),[83] and the Oslo Convention (Cluster Munitions).[84]

4.4 CONCLUSION

Although LOAC is a lex specialis regime with respect to military use of technologies, there are a range of enduring, non-LOAC, governance

78 See Chapter 2.

79 API, Article 36.

80 ICRC, *A Guide to the Legal Review of New Weapons, Means and Methods of Warfare*, Rev. Nov 2006, 23: 'The temporal application of Article 36 is very broad. It requires an assessment of the legality of new weapons at the stages of their "study, development, acquisition or adoption". This covers all stages of the weapons procurement process, in particular the initial stages of the research phase (i.e. conception, study), the development phase (i.e. development and testing of prototypes) and the acquisition phase (including "off-the-shelf" procurement).'

81 See, inter alia, *Commentary of 2016 to Convention (I) for the Amelioration of the Condition of the Wounded and Sick in Armed Forces in the Field. Geneva, 12 August 1949*, ICRC, 2016, paras 119–120, at https://ihl-databases.icrc.org/applic/ihl/ihl.nsf/Comment.xsp?action=openDocument&documentId=72239588AFA66200C1257F7D00367DBD.

82 Convention on Prohibitions or Restrictions on the Use of Certain Conventional Weapons which may be deemed to be Excessively Injurious or to have Indiscriminate Effects 1980.

83 Convention on the Prohibition of the Use, Stockpiling, Production and Transfer of Anti-Personnel Mines and on Their Destruction 1997.

84 Convention on Cluster Munitions 2008.

frameworks and issues which continue to operate in relation to military technologies both during and outside situations of armed conflict. These non-LOAC regimes – although they are often targeted at technology generally, and thus capture military technology either incidentally, or consciously but in a capability and contextually bespoke manner – must thus be considered when thinking about military acquisition and use of technology. As the interface and crossover between civil and military development, deployment, and employment of such technologies increases, 'dual use' will likely presage yet deeper concerns for coherence between the regulation of analogous civil and military technologies. It is probable, consequently, that a tendency to regulatory convergence will take increasing prominence – particularly in the civil-military 'security' arena. However, this trend will also serve to further underline, and emphasise, the parallel need to maintain a robust degree of military particularity when regulating the deployment and employment of military technology within the lex specialis framework provided by LOAC.

PART II

Applying the Law to Some New Technologies

5

Cyber Capabilities

WILLIAM H. BOOTHBY

5.1 INTRODUCTION

It has become increasingly apparent that cyberspace is likely to join the land, sea, air and outer space domains as yet another environment in which hostilities can be undertaken. We have already seen its employment, inter alia, to commit crime, to perpetrate identity thefts, to steal personal and other data[1] including data of a commercial nature,[2] to interfere with the proper operation of governmental websites, allegedly to interfere with the political electoral process of a super-power, and to obtain intellectual property and intelligence information on a substantial scale. Against such a background, it would be a brave commentator who would argue that the use of cyber capabilities in warfare is either impossible or unlikely. Indeed, the establishment of the US Cyber Command,[3] of the Cyber Security Centre in the United Kingdom,[4] of

[1] The notion of 'data' is central to the discussion in this chapter. It is therefore, perhaps, useful to give some idea of the meaning of the notion. A useful dictionary definition is 'facts and statistics used for reference or analysis' and in the philosophical context, 'things known or assumed as facts, making the basis of reasoning.' Alternatively and more immediately relevantly for the purposes of the following discussion, data is 'the quantities, characters, or symbols on which operations are performed by a computer'; *Concise Oxford English Dictionary*, 11th edn (2006), 364. So data consists of the factual materials on the basis of which computer functions operate.

[2] Note, e.g., Report by Price Waterhouse Coopers and BAe Systems, Operation Cloud Hopper, April 2017 available at www.pwc.co.uk/cyber-security/pdf/cloud-hopper-report-final-updated .pdf reporting a systematic hacking operation by China-based APT10 with an unprecedented web of global victims.

[3] Full operational command was achieved on 31 October 2010; see www.stratcom.mil/Media/ Factsheets/Factsheet-View/Article/960492/us-cyber-command-uscybercom/. Note the reported allocation by the US of 90% of its cyber expenditure to offensive efforts; J. Menn, 'A Scramble at Cisco Exposes Uncomfortable Truths about US Cyber Defense', Reuters, 30 March 2017, available at http://uk.reuters.com/article/us-usa-cyber-defense-idUKKBN17013U.

[4] The UK National Cyber Security Centre was formally opened on 14 February 2017. For its role, see www.ncsc.gov.uk. As to the establishment of a Ukrainian Cyber Security Centre, see

a cyber command in Germany,[5] plans for the development of cyber forces in China[6] and India[7] and the acknowledgement of an offensive cyber capability in the Australian Cyber Security Strategy document[8] are among the developments that show that States perceive such a development as sufficiently probable to justify the making of policy and tangible arrangements.[9]

5.2 LAW, CYBER OPERATIONS AND THE LEGAL STATUS OF THE TALLINN MANUALS

In the present chapter the temptation to talk in general terms about cyber warfare will be resisted. Instead, the opportunity will be taken to review the current state of the law in relation to the procurement and employment of cyber warfare capabilities. In that endeavour, we will refer to a number of authoritative statements of the law that do not necessarily coincide with one another.

It was in 2009 that the NATO Cooperative Cyber Defence Centre of Excellence (CCD COE) in Tallinn, Estonia, invited an International Group of Experts, acting in their personal capacities, to produce a Manual on the law relating to cyber warfare. Such International Manuals had previously been produced by similarly acting International Groups of Experts and dealt, e.g.,

Interfax-Ukraine, Ukraine with NATO's support prepares creation of cyber security centre, KyivPost, 5 April 2017, available at www.kyivpost.com/ukraine-politics/ukraine-natos-support-prepares-creation-cyber-security-centre.html.

5 Germany's cyber command was launched on 5 April 2017, S. Saeed, 'German Cyber Security Chief: Army attacked over 284,000 times this year', *Politico*, 3 April 2017, available at www.politico.eu/article/german-cybersecurity-chief-army-attacked-over-284000-times-this-year/.

6 G. Webster, 'Observations on China's New International Cyberspace Cooperation Strategy', Lawfare blog, 7 March 2017, available at www.lawfareblog.com/observations-chinas-new-international-cyberspace-cooperation-strategy. For the text of the strategy (in translation), see T. Shaohui (ed.), 'International Strategy of Cooperation on Cyberspace', 1 March 2017, available at http://news.xinhuanet.com/english/china/2017–03/01/c_136094371.htm.

7 R. Panditl, 'New Joint Doctrine Stresses Space, Cyber, Special Ops in War-Fighting', *Times of India*, 25 April 2017, available at http://timesofindia.indiatimes.com/india/new-joint-doctrine-stresses-space-cyber-special-ops-in-war-fighting/articleshow/58365512.cms.

8 Australia's Cyber Security Strategy (2016), page 28, available at https://cybersecuritystrategy.dpmc.gov.au/assets/img/PMC-Cyber-Strategy.pdf.

9 Consider also R. Emmott, 'NATO Mulls "Offensive Defence" with Cyber Warfare Rules', Reuters, 30 November 2017, available at www.reuters.com/article/us-nato-cyber/nato-mulls-offensive-defense-with-cyber-warfare-rules-idUSKBN1DU1G4; for the assessment that Russia 'has a highly advanced offensive cyber program', see D. R. Coats, 'Worldwide Threat Assessment of the US Intelligence Community', 11 May 2017, available at www.dni.gov/files/documents/Newsroom/Testimonies/SSCI%20Unclassified%20SFR%20-%20Final.pdf.

with the international law applicable to armed conflicts at sea[10] and the international law applicable to air and missile warfare.[11] In the present chapter we will refer to the Tallinn Manual 2.0.[12] The original Tallinn Manual was published in 2013[13] and limited itself to a consideration of the international law relating to cyber warfare. The Tallinn Manual 2.0, also published under the aegis of the CCD COE, has a considerably broader scope, covering as it does the law that applies to cyber operations in general, not just cyber warfare. Accordingly, references in the present chapter will be made to this text that was published in 2017.

The legal status of texts such as the Tallinn Manual and Tallinn Manual 2.0 needs to be established clearly at the commencement of this discussion. Neither constitutes a recognised source of international law, yet both base themselves on what the participating experts consider the current state of international law on their respective topics to be. When defining the law to be applied by the International Court of Justice, the Statute of that Court lists as the recognised sources of international law: 'a. international conventions, whether general or particular, establishing rules expressly recognized by the contesting states; b. international custom, as evidence of a general practice accepted as law' and 'c. the general principles of law recognized by civilized nations'. The Statute goes on to refer to 'judicial decisions and the teachings of the most highly qualified publicists of the various nations, as subsidiary means for the determination of rules of law'.[14] This is increasingly being regarded as a somewhat conservative approach. Indeed, the decisions of judicial bodies and the writings of authoritative publicists not only provide useful indications as to the existence and form of such rules but also frequently point the way to the formation of such rules, whether by general and consistent subsequent State practice that forms customary rules, by providing the inspiration for subsequent treaty provision or by expressing positions that are generally adopted in the discourse and that are accepted by States in general as reflecting the law.

Customary law, as referred to in Article 38(1)(b) of the ICJ Statute, comprises the actions that States take or the things that States refrain from doing in the belief, in either such case, that they are so behaving pursuant to a legal

[10] *San Remo Manual on International Law applicable to Armed Conflicts at Sea*, 1994.

[11] *HPCR Manual on the International Law applicable to Air and Missile Warfare*, 2010.

[12] M. N. Schmitt and L. Vihul (eds.), *Tallinn Manual 2.0 on the International Law applicable to Cyber Operations* (2017) (Tallinn Manual 2.0).

[13] M. N. Schmitt (ed.) Tallinn Manual on the International Law applicable to Cyber Warfare, (2013) (*Tallinn Manual*).

[14] Statute of the International Court of Justice, San Francisco, 26 June 1945, Article 38(1).

obligation.[15] It is the generality of this State practice and of the associated belief in a legal obligation, or *opinio juris*, that forms a rule that is binding under international law irrespective of whether a particular State may or may not be a party to a specific treaty reflecting such a rule. While there is a great deal more that could be said about the formation, significance and termination of rules of customary law, the above suffices for the actual purposes of the present discussion.

A treaty or convention, being expressions which for the present purposes can be regarded as synonymous, consists of 'an international agreement concluded between States in written form and governed by international law, whether embodied in a single instrument or in two or more related instruments and whatever its particular designation'.[16] Accordingly, it is only two or more States that can make treaties, and there must be one or more documents recording the agreement between them.

The sources of law that we have identified make it clear that States are in the forefront of international law formation, whether by negotiating and adopting treaty texts or by employing State practice that forms the basis of customary law. There is at the time of writing no treaty provision dealing explicitly with the conduct of cyber hostilities in warfare. The question that naturally arises is whether customary law rules of more general application, such as the customary law rules addressing the conduct of attacks, can be applied in the cyber context. This chapter argues that they can indeed. This conclusion is based in part on the findings of the International Court of Justice in the Nuclear Weapons Advisory Opinion,[17] on the Martens Clause as expressed in the Preamble to Hague Convention IV of 1907,[18] on Article 1(2) of API,[19] and on a common sense interpretation of the concepts of attack, of weapon systems and other applicable notions in international law. There is,

[15] *UK Manual*, paras 1.12 and 1.12.1; DoD Manual, Section 1.8.

[16] Vienna Convention on the Law of Treaties, 1969, Article 2(1)(a).

[17] Legality of the Threat or Use of Nuclear Weapons (Advisory Opinion) (1996) ICJ Reports 226; 35 ILM 809, (Nuclear Weapons case), para 86.

[18] Para 8 of the Preamble to Hague Convention IV of 1907 provides as follows: 'Until a more complete code of the laws of war has been issued, the high contracting Parties deem it expedient to declare that, in cases not included in the Regulations adopted by them, the inhabitants and the belligerents remain under the protection and rule of the principles of the law of nations, as they result from the usages established among civilized peoples, from the laws of humanity, and the dictates of the public conscience'.

[19] Article 1(2) of API provides as follows: 'In cases not covered by this Protocol or by other international agreements, civilians and combatants remain under the protection and authority of the principles of international law derived from established custom, from the principles of humanity and from the dictates of public conscience.'

however, an absence of international consensus on these matters, which renders it necessary to consider what the writings of experts tell us and what status at law those writings have.

With that in mind, what legal status do the original Tallinn Manual and Tallinn Manual 2.0 assert for themselves? The original Tallinn Manual makes it clear that it 'is not an official document, but is only the product of a group of independent experts acting solely in their personal capacity. The Manual does not represent the views of the NATO CCD COE, its supporting nations, or NATO ... participation in the International Group of Experts by individuals with official positions in their own countries must not be interpreted as indicating that the Manual represents the views of those countries.'[20] Similar sentiments are expressed in Tallinn Manual 2.0.[21] It is therefore abundantly clear that the two Tallinn Manuals do not come within the three sources of international law listed in Article 38. States were not involved with them, so they constitute neither treaties nor State practice. In ICJ Statute terms, they are writings of highly respected jurists and are therefore likely to come within Article 38(1)(d). Consideration of the ICJ Statute would therefore suggest that they are subsidiary means to enable the applicable rules of law to be determined but they also have the capacity to inspire future general understanding as to the rules of law that should be applied in the relevant circumstances. Whether they do in fact lead to such a future understanding will depend on whether the propositions set forth in the Manuals achieve the requisite, i.e. general, level of acceptance by States. Self-evidently, this capacity to form the basis of emerging law may be diminished, even extinguished, if for example a publication like the Tallinn Manual or Tallinn Manual 2.0 attracts such opposition among certain States, particularly subject matter-relevant States, that its statements of law can no longer be regarded as reflecting a general international view.

In the next three sections, the extent and detail of the rules found by the Tallinn experts will be summarised in respect of two specific topic areas, namely the *jus ad bellum* and the *jus in bello*. Only some of the rules found within each such topic will be addressed and the commentaries associated with the rules will not be generally considered. The purpose is not to assess whether the Tallinn Manual correctly states the law but to demonstrate the level of detail reflected in the Manual's rules and to show the broad legal positions that the rules demonstrate. Reference throughout the next two sections will be to the Rules as stated in Tallinn Manual 2.0.

[20] *Tallinn Manual*, 11.

[21] *Tallinn Manual* 2.0, 2.

5.3 THE TALLINN MANUALS – *JUS AD BELLUM*

Taking the *jus ad bellum* first, the Manual reflects Article 2.4 of the UN Charter by finding that 'a cyber operation that constitutes a threat or use of force against the territorial integrity or political independence of any State, or that is in any other manner inconsistent with the purposes of the United Nations, is unlawful.'[22] Adopting the 'scale and effects' approach employed by the International Court of Justice,[23] the Manual then finds that 'a cyber operation constitutes a use of force when its scale and effects are comparable to non-cyber operations rising to the level of a use of force'.[24] The employment of the word 'comparable' avoids any suggestion that identical scale and effects to those resulting from non-cyber operations would be required and the Commentary to Rule 69 helpfully puts forward factors that States are likely to consider in determining whether a particular cyber operation amounts to a use of force. Rule 70 makes the point that a cyber operation, or threatened cyber operation, constitutes an unlawful threat of force 'when the threatened action, if carried out, would be an unlawful use of force'. Rule 71 then provides that a 'State that is the target of a cyber operation that rises to the level of an armed attack may exercise its inherent right of self-defence' and indicates that whether a cyber operation constitutes an armed attack depends on its scale and effects. The requirement that the use of force by a State acting in self-defence be necessary and proportionate is stated in Rule 72,[25] and the requirements as to imminence and immediacy are set forth in Rule 73. Collective self-defence is addressed in Rule 74, noting the requirement that the victim State must have requested such collective action which must be within the scope of its request. The reporting measures provided for in Article 51 of the UN Charter are then applied by Rule 75.

[22] *Tallinn Manual 2.0*, Rule 68.

[23] Military and Paramilitary Activities in and against Nicaragua, Nicaragua v United States, 1986, ICJ, 14 (27 June), para 195.

[24] *Tallinn Manual 2.0*, Rule 69.

[25] Consider in this regard reports that in August 2017 a petrochemical plant in Saudi Arabia was the target of a cyber operation designed to sabotage the company's operations and to trigger an explosion; N. Perlroth and C. Krauss, 'A Cyber Attack in Saudi Arabia Had a Deadly Goal: Experts Fear Another Try', *New York Times*, 15 March 2018, available at www.nytimes.com/2018/03/15/technology/saudi-arabia-hacks-cyberattacks.html. Of course, to be classed as a use of force, such an event would need to be attributable to a State; to be classed as an armed attack, it would need to meet the required criteria, including as to seriousness of the consequences. Note also K. Donaldson, 'The UK Would Respond to a Russian Cyber Attack with The Weapon of Its Own Choice', *Bloomberg*, 18 December 2017, available at www.bloomberg.com/news/articles/2017-12-18/u-k-would-respond-to-russian-cyberattack-with-weapon-of-choice, reporting remarks attributed to the UK's national security adviser.

Section 15 of Tallinn Manual 2.0 deals with collective security. Rule 76 is the Rule that is of particular relevance to the present discussion and notes the power of the UN Security Council, if it determines that a cyber operation constitutes a threat to the peace, breach of the peace or act of aggression, to authorise non-forceful measures in response. Alternatively, if it considers such measures to be inadequate, the Rule notes that the Security Council may decide upon forceful measures, including cyber measures. This Rule is essentially reflecting Articles 39, 41 and 42 of Chapter VII of the UN Charter. The section concludes with rules addressing regional organisations and peace operations.

5.4 TALLINN MANUALS – *JUS IN BELLO*

The Manual's treatment of *jus in bello* starts with a useful assertion in Rule 80 that: 'Cyber operations executed in the context of an armed conflict are subject to the law of armed conflict'. Paragraph 28d of the 2015 UN GGE Report (discussed below), paragraph 72 of the NATO Wales Summit Declaration 2014, certain reports to the UN Secretary General associated with the UN GGE work on Information and Telecommunications in the context of International Security and an EU General Affairs Council Meeting document are cited in support of this proposition. The applicability to cyber operations of the geographical limitations imposed by international law is specified in Rule 81 and the recognised classifications of international and non-international armed conflicts are reflected in Rules 82 and 83 respectively.

It is not the purpose of this section to reproduce all of the *in bello* rules found in the Tallinn Manuals. Rather, in the immediately following paragraphs some of the Manuals' rules that apply to the conduct of hostilities will be summarised, after which its rules applying to cyber weapons will be briefly discussed. As before, the detailed discussion of each rule in the associated Commentaries will not be addressed as the purpose of this part of the discussion is only to illustrate to the reader the granularity of the rules that the participating Experts agreed regulate the conduct of cyber hostilities under existing international law.[26]

The Tallinn Manual 2.0 tackles the important issue of the meaning of the term 'cyber attack' in Rule 92 where that expression is defined as 'a cyber

[26] '*Tallinn Manual* 2.0 is intended as an objective restatement of the *lex lata*'; *Tallinn Manual* 2.0, 3.

operation, whether offensive or defensive, that is reasonably expected to cause injury or death to persons or damage or destruction to objects'. By interpreting the notion of 'act of violence' as used in the API Article 49(1) definition of attack as involving events that have violent consequences, meaning death, injury, damage or destruction, the Manual establishes the logic according to which the experts found that the international law rules relating to attack can sensibly be applied to cyber operations the results of which are the same as those associated with more conventional attacks in armed conflicts. If the reasonable expectation is that the cyber operation will generate the same outcome as, say, the firing of a rifle or the detonation of a bomb, there would indeed seem to be little justification for treating the two kinds of event in a significantly different manner. Accordingly, the Manual, sensibly finds in the next rule that the 'principle of distinction applies to cyber attacks'.[27] It would, frankly, have been hard to reconcile any contrary finding with the Martens Clause, discussed earlier, with the ICJ Nuclear Weapons Advisory Opinion, also discussed earlier and, for that matter, with common sense. The application of the principle of distinction is then developed as follows. Rule 94 provides that the 'civilian population as such, as well as individual civilians, shall not be the object of cyber attack." Rules 96 and 97 address the persons who may be made the lawful object of cyber attacks and Rules 99 and 100 make the important distinction for cyber warfare purposes between civilian objects and military objectives.

The prohibition of indiscriminate attacks of the kind referred to in Article 51 (4)(a) of API forms the substance of Rule 111. The example of indiscriminate attacks given in Article 51(5)(a) of API is adapted to the cyber domain as a discrete Rule in the following terms:

> 'A cyber attack that treats as a single target a number of clearly discrete cyber military objectives in cyber infrastructure primarily used for civilian purposes is prohibited if to do so would harm protected persons or objects.'[28]

The rule of proportionality then becomes Rule 113 in terms that are identical to those given in Article 51(5)(b) of API subject to the addition of the term 'cyber'.

The Manual also specifically addresses the important issue of precautions. The constant care rule in Article 57(1) of API is accurately re-stated as a requirement that '[d]uring hostilities involving cyber operations, constant care shall be taken to spare the civilian population, individual civilians, and civilian

[27] *Tallinn Manual* 2.0, Rule 93.
[28] *Tallinn Manual* 2.0, Rule 112, applying API Article 51(5)(a).

objects.'[29] The individual precautions required of attackers by API are then applied in the cyber context as follows:

> 'Those who plan or decide upon a cyber attack shall do everything feasible to verify that the objectives to be attacked are neither civilians nor civilian objects and are not subject to special protection.'[30]
>
> 'Those who plan or decide upon a cyber attack shall take all feasible precautions in the choice of means or methods of warfare employed in such an attack, with a view to avoiding, and in any event to minimising, incidental injury to civilians, loss of civilian life, and damage to or destruction of civilian objects.'[31]
>
> 'Those who plan or decide upon attacks shall refrain from deciding to launch any cyber attack that may be expected to cause incidental loss of civilian life, injury to civilians, damage to civilian objects, or a combination thereof, which would be excessive in relation to the concrete and direct military advantage anticipated.'[32]
>
> 'For States Parties to Additional Protocol I, when a choice is possible between several military objectives for obtaining a similar military advantage, the object to be selected for cyber attack shall be that the attack on which may be expected to cause the least danger to civilian lives and to civilian objects.'[33]
>
> 'Those who plan, approve, or execute a cyber attack shall cancel or suspend the attack if it becomes apparent that: (a) the objective is not a military one or is subject to special protection; or (b) the attack may be expected to cause, directly or indirectly, incidental loss of civilian life, injury to civilians, damage to civilian objects, or a combination thereof that would be excessive in relation to the concrete and direct military advantage anticipated.'[34]
>
> 'Effective advance warning shall be given of cyber attacks that may affect the civilian population unless circumstances do not permit'.[35]

The Manual includes specific rules dealing with numerous other related aspects of the conduct of hostilities such as the precautions that are required against the effects of attacks, the general and special protection of particular classes of person and object, perfidy and the prohibition of using certain signs

[29] *Tallinn Manual* 2.0, Rule 114, applying API Article 57(1).

[30] *Tallinn Manual* 2.0, Rule 115, applying API Article 57(2)(a)(i).

[31] Tallinn Manual 2.0, Rule 116, applying API Article 57(2)(a)(ii).

[32] Tallinn Manual 2.0, Rule 117, applying API Article 57(2)(a)(iii).

[33] Tallinn Manual 2.0, Rule 118, applying API Article 57(3) and expressed so as to recognise that Article 57(3) is not considered to have customary law status.

[34] *Tallinn Manual* 2.0, Rule 119, applying API Article 57(2)(b).

[35] *Tallinn Manual* 2.0, Rule 120, applying API Article 57(2)(c).

and indicia. It immediately becomes clear that the Tallinn Manual process has achieved a very considerable degree of granularity in terms of the number and specificity of the rules that it finds to apply in relation to the conduct of cyber hostilities.

Turning to the Tallinn Manual rules that were found by the International Group of Experts to apply to cyber weapons, the Manual starts by discussing the means and methods of cyber warfare and cites the ICJ Nuclear Weapons case observation that 'the established principles and rules of humanitarian law... appl[y] to all forms of warfare, and to all kinds of weapons, those of the past, those of the present and those of the future'.[36] The International Group of Experts is reported to have concluded, sensibly it can be argued, that 'the general rules that determine the legality of weapons will also determine the lawfulness of cyber methods and means of warfare'.[37] The Manual then makes the point that the relevant rules apply to methods or means that a State develops or procures and to those over which it acquires control including by cyber means, i.e. by employing them as if they were its own.

The Manual defines 'means of cyber warfare' as 'cyber weapons and their associated cyber systems', while 'methods of cyber warfare' consist of 'the cyber tactics, techniques and procedures by which hostilities are conducted'.[38] Cyber weapons are then defined as 'cyber means of warfare that are used, designed, or intended to be used to cause injury to, or death of, persons or damage to, or destruction of, objects, that is, that result in the consequences required for qualification of a cyber operation as an attack.'[39] The notion will therefore, as the Commentary to Rule 103 goes on to explain, extend to any cyber device, material, instrument, mechanism, equipment, or software used, designed or intended to be used to conduct a cyber attack. This must be a sensible interpretation. After all, it would defy logic to accept that a conventional munition used, designed or intended to cause injury or damage during an armed conflict is a weapon, but that a cyber tool used, designed or intended to generate similar injurious or damaging effects is not.[40]

36 *Nuclear Weapons* case, para 86.
37 *Tallinn Manual* 2.0, 451.
38 *Tallinn Manual* 2.0, Rule 103.
39 *Tallinn Manual* 2.0, para 2 of the Commentary accompanying Rule 103.
40 Complications can, however, arise. Consider for example the development of a cyber capability involving the spoofing of GPS signals so as to cause the geographical location of a ship to be falsely stated on its GPS navigational equipment. If such technology is used with the explicit purpose of causing the targeted vessel to suffer a collision and thus to be damaged or sunk, it is arguable that the use of the capability in this way constitutes an attack and that the equipment so used taken together constitutes a weapon system. However, there are those who would likely consider that such a capability can better be described as a method of

If the use of a cyber tool for cyber espionage has the effect of causing damage to an adversary during an armed conflict, that would still fall to be considered an attack[41] and the tool that is so employed would therefore, logically, be treated as a weapon. It follows from this approach that a cyber intelligence gathering operation, for example, that causes damage to the targeted network, node or computer or that causes injury or damage to the customers of that network, node or computer, would be regarded as an attack and that the cyber device, material, instrument, mechanism, equipment and software so employed would collectively be treated on this basis as a weapon system and thus as a means of warfare.

The Manual draws a distinction between means of cyber warfare and the cyber infrastructure connecting the target with the computer system being used to undertake the attack. The cyber infrastructure is not, the Manual asserts, a means of warfare 'because an object must be in the control of an attacking party to comprise a means of warfare'.[42] If, as is undoubtedly the case, a weapon system comprising a conventional rifle and the associated bullets is not interpreted as also including the air through which the bullet will pass on its way to the assigned target, so also the cyber infrastructure through which the cyber tool passes on its way to the targeted computer or network does not logically thereby become part of the means of warfare. Where cyber methods of warfare are concerned, these are stated to go wider than operations amounting to attacks, and thus would include cyber operations to interfere with enemy communications capabilities.[43]

It is by classifying the relevant devices, materials, instruments, mechanisms, equipment or software as weapons, weapon systems or means of warfare that the Manual is logically able to apply the principles and rules of weapons law to them. Rule 104 therefore prohibits means or methods of cyber warfare that are of a nature to cause superfluous injury or unnecessary suffering.[44] The Manual gives the example of Internet-addressable pacemaker devices, noting that while it would be lawful by cyber means to take control of such a device

warfare; see D. Hambling, 'Ships Fooled into GPS Spoofing Attack Suggest Russian Cyberweapon', *New Scientist Daily News*, 10 August 2017, available at www.newscientist.com/article/2143499-ships-fooled-in-gps-spoofing-attack-suggest-russian-cyberweapon/.

[41] *Tallinn Manual* 2.0, Commentary to Rule 92, para 14.

[42] *Tallinn Manual* 2.0, para 3 of the Commentary to Rule 103. This is an interpretation that may not survive the introduction of autonomous cyber weapon capabilities. Autonomous technologies are discussed in Chapter 6. Perhaps control is not the relevant factor. Arguably the relevant point is simply that the cyber infrastructure in general is not being incorporated by the attacker as part of the weapon system.

[43] *Tallinn Manual* 2.0, para 5 of the Commentary to Rule 103.

[44] *Tallinn Manual* 2.0, Rule 104 on page 453.

fitted to a combatant either to kill the combatant or render him *hors de combat*, stopping his heart and then reviving him multiple times before finally killing him would occasion suffering serving no military purpose and thus would breach the principle.[45] The Manual also finds a rule prohibiting the employment of 'means or methods of cyber warfare that are indiscriminate by nature.'[46] Adapting closely the language of API, Article 51(4), the Rule defined indiscriminate means and methods of cyber warfare as those which 'cannot be: (a) directed at a specific military objective, or (b) limited in their effects as required by the law of armed conflict and consequently are of a nature to strike military objectives and civilians and civilian objects without distinction.' The Commentary clarifies that the first element of the Rule is directed at 'shots in the dark', meaning a means or method where 'it is impossible to predict with any reasonable certainty whether it will strike a specific military objective rather than cyber infrastructure protected by the law of armed conflict'.

The ICJ has described both of these principles, i.e. the superfluous injury principle and the indiscriminate weapons principle, as 'cardinal' and 'intransgressible' and it is clear that the ICJ interprets them as applying to means and methods of injuring the enemy irrespective of the technology that is used. Rules 103 and 104 show how legal effect can be given to that stated ICJ position in the context of cyber capabilities. Any State that rejects that approach would seem implicitly to be rejecting the frequently cited and authoritative ICJ position and, arguably, the clear meaning of Article 1(2) of API.

The Commentary to Rule 105 explains that the second part of the Rule would prohibit 'cyber weapons that create an uncontrollable chain of events.' Malware that inevitably and harmfully spreads into civilian networks in a manner beyond the control of the attacker is given as an example.[47]

45 *Tallinn Manual* 2.0, para 6 of the Commentary to Rule 104.

46 *Tallinn Manual* 2.0, Rule 105 on page 455.

47 Tallinn Manual 2.0, para 4 of the Commentary to Rule 105. Effects are not harmful for these purposes if they constitute mere inconvenience or annoyance. Likewise, malware that infects civilian networks but that limits its injurious or damaging impact to military objectives is not rendered indiscriminate by that non-damaging collateral effect; ibid., para 5. For an example of an apparently indiscriminate cyber operation, consider the NotPetya computer worm that targeted systems in Ukraine but also reportedly affected banks, shipping and a chocolate factory in Australia; P. Polityuk, 'Ukraine Points Finger at Russian Security Services in Recent Cyber Attack', Reuters, available at www.reuters.com/article/us-cyber-attack-ukraine-idUSKBN19M39P; see also M. N. Schmitt and J. Biller, 'The NotPetya Cyber Operation as a Case Study in International Law', *EjilTalk*, 11 July 2017. The NotPetya operation has now formally been attributed, e.g., by the US, to Russia; White House Press Secretary Statement

The inclusion in the Manual of Rule 106 addressing cyber booby-traps illustrates the need to consider ad hoc weapons law rules by reference to particular cyber weapons. The ad hoc rules may be relevant because the particular cyber weapon may take the form of a weapon prohibited or restricted by ad hoc weapons law rules, e.g., a cyber booby-trap, or because a planned cyber operation has the effect of taking control of an enemy network and thereby of exercising control over an enemy weapon system to which weapons law rules that bind the hacking State apply. Thus, if a hacking operation were to have the effect that control is exercised over, say, a remotely piloted aircraft carrying cluster munitions, a State party to the Cluster Munitions Convention[48] would need to ensure that it fulfils its treaty obligations in relation to any cluster munitions that such a hacking operation places under its control.

The Manual takes the position that weapon review obligations apply to cyber means of warfare, and in the case of States party to API, to cyber methods of warfare.[49] The associated Commentary suggests that the requirement extends to any cyber weapon acquired or used by a State, to cyber weapons designed as such and procured by States, to cyber weapons designed by the armed forces to exploit vulnerabilities and to malicious software not originally developed for military purposes but that is subsequently acquired by States for use in armed conflict.[50]

It is therefore evident from the preceding paragraphs that the Tallinn Manual finds detailed rules to be applicable to the resort to the use of cyber force and to the manner in which such force may lawfully be employed in an armed conflict, and that it achieves these rules by applying to the cyber environment widely accepted legal principles and rules relating to non-cyber activity. Some commentators might argue that what is going on here is the application in the cyber context of rules by analogy. This is not, however, the case. An analogy is defined as 'an agreement or correspondence in certain respects between things otherwise different'. 'Analogous' is defined as 'having analogy; bearing some correspondence or resemblance; similar in certain circumstances or relations (with, to)'.[51] The discussion of whether the notion of cyber attack is included within the idea of attack, whether the notion of cyber weapon is included within the notion of weapon, whether the notion

dated 15 February 2018, available at www.whitehouse.gov/briefings-statements/statement-press-secretary-25/.

48 Cluster Munitions Convention, Dublin, 3 December 2008.

49 *Tallinn Manual* 2.0, Rule 110 at 464.

50 *Tallinn Manual* 2.0, para 3 of the Commentary to Rule 110.

51 The *Chambers Dictionary*, 2003, 48.

of cyber use of force is included within the notion of use of force, and whether the notion of cyber armed attack is included within the notion of armed attack is not based on the application of analogy because the respective notions are not considered to be different from one another. Rather, the discussion consists of determining whether the latter notions actually include the former. This involves determining whether the two concepts are the same as one another, not whether an analogy based on difference can properly be drawn. Any critique of the Tallinn Manual approach will therefore have to address whether the cyber notion is indeed included within the more general notion, not on whether a suggested analogy is sensible or appropriate. So, it can be argued that a cyber operation that has damaging or injurious consequences is an attack and the malware, systems and hardware used to undertake such an operation form part of a weapon system. On this basis, it can properly be concluded that the existing law as to the conduct of hostilities actually applies to cyber warfare.

5.5 UN GROUP OF GOVERNMENTAL EXPERTS DISCUSSIONS

Now that we have considered the approach taken by a Group of Experts drawn from what are essentially like-minded States,[52] we should consider the work of differently constituted but more geographically balanced Groups of Governmental Experts who have been discussing these matters under UN General Assembly authority. The United Nations General Assembly has, since 2004 and by resolution, established five Groups of Governmental Experts (UN GGE) the purpose of which has been, as the titles of their reports suggest, to examine Developments in the Field of Information and Telecommunications in the Context of International Security.[53] The Groups produced Reports in 2010, 2013 and 2015 and a fifth Group undertook such work in 2016 and 2017. The fifth Group failed to achieve consensus on a draft report.[54]

[52] See the 'Tallinn Manual 1.0 International Group of Experts and Other Participants', as listed at pages xix–xxii in *Tallinn Manual 2.0*.

[53] See para 4 of UN General Assembly Resolution 66/24 of 2012 and para 4 of UN General Assembly Resolution 68/243 of 2014.

[54] See M. G. Markoff, 'Explanation of Position at the Conclusion of the 2016–2017 Group of Governmental Experts on Developments in the Field of Information and Telecommunications in the Context of International Security', US Department of State, 23 June 2017, available at www.state.gov/s/cyberissues/releasesandremarks/272175.htm and for a careful analysis of the implications of the outcome of the 2016–2017 UN GGE process, see E. Tikk and M. Kerttunen, 'The Alleged Demise of the UN GGE: An Autopsy and Eulogy', Cyber Policy Institute, 2017, available at http://cpi.ee/wp-content/uploads/2017/12/2017-Tikk-Kerttunen-Demise-of-the-UN-GGE-2017–12–17-ET.pdf.

The Experts who produced the 2013 and 2015 Reports were taken from, respectively, fifteen and twenty geographically diverse States. The 2016–2017 Group comprised Experts from twenty-four States. This aspect is reflected in the products of the two processes. Discussions among the like-minded inevitably produce a greater level of consensus and, thus, of granularity in the rules that are found to exist. Discussions among a more diverse group will, equally inevitably, produce text that must take account of differences of perspective and/or of opinion, and will therefore generate texts that are less granular and less detailed. This is certainly what we find when comparing the Tallinn Manual texts with the UN GGE reports.

The Reports are expressed as a series of recommendations addressed to States and, in broad terms, set forth: recommended norms, rules and principles for responsible behaviour by States; recommendations on confidence building measures and the exchange of information; recommendations on capacity building measures, international cooperation and assistance in Information and Communication Technology (ICT) security; and recommendations on how international law applies to the use of ICTs.

In the following paragraphs some of the very many pertinent points set forth in the 2013 and 2015 Reports are summarised. The purpose of this discussion is to show how far this inclusive, geographically balanced process has got in agreeing on the norms, rules and principles, on the confidence building measures and on the other arrangements that are required to ensure international security in this field. Some general conclusions in that regard will be offered at the end of the discussion.

It will be immediately apparent that the focus of the UN GGE work is not quite the same as that of the Tallinn experts. Whereas Tallinn was specifically concerned with identifying rules of law and explaining how they may be considered to apply in the cyber context, the UN GGE focus includes recommended national and international action and approaches, including cooperative measures and best practice, with a view to the fostering of ICT security and good behaviours in the ICT environment. Nevertheless, where there are observable and relevant parallels between certain UN GGE recommendations and rules in the Tallinn Manual, attention is drawn to these in the associated footnotes below.

Section I of the 2013 Report notes the economic and social benefits brought by ICTs and their possible use for legitimate or malicious purposes, draws attention to the potential for concealment and that 'attribution to a specific perpetrator can be difficult thus creating impunity'.[55] After referring to the increasingly acute threats to individuals, businesses, national infrastructure and governments from State and non-State actors and the roles of proxies in malicious ICT activity, the 2013 Report made specific mention, e.g., of the link with terrorism, the potential for an erosion of trust in commerce and damage to national security. Pointing to the need for cooperative action against such threats, Section II of the 2013 Report suggests that consideration should be given to cooperative measures that could enhance international peace, stability and security, such as 'common understandings on the application of relevant international law and derived norms, rules and principles of responsible behaviour of States.'[56] There is then a proposal in the 2013 Report that the UN should play a leading role in promoting dialogue and the valuable work of other international organisations is noted. The application of norms derived from existing international law relevant to the use of ICTs by States is seen as being essential to reduce risks to international peace, security and stability.[57]

Section III of the 2013 Report puts forward a number of propositions, some of which are given below:

International law, in particular the UN Charter, is applicable[58];

State sovereignty and associated norms and principles apply to State ICT activities[59] and to jurisdiction over ICT infrastructure in their territory[60];

State ICT security action must go 'hand in hand with respect for human rights in the Universal Declaration of Human Rights';

States should intensify cooperation against criminal or terrorist use of ICT; and

55 Report of the Group of Governmental Experts on Developments in the Field of Information and Telecommunications in the Context of International Security, A/68/98 dated 24 June 2013 (2013 Report), at 6.

56 2013 Report, at 7. While the GGE recognized the merit of cooperative measures with a view to achieving this common understanding, it is, in view of the failure of the 2016–2017 GGE process to achieve consensus, an open question whether, when and how such an understanding will emerge.

57 At face value, some observers may interpret this as an endorsement of the *Tallinn Manual* approach, but which norms are seen as being applicable in this sense is left open.

58 This would, in general terms, lead one to conclude that the *jus ad bellum* rules in the UN Charter are seen by the UN GGE as applying to ICT activities. To the extent that the *jus ad bellum* rules in the *Tallinn Manual* are reflective of UN Charter provision, and they are, one would imagine that the UN GGE participants would support those Tallinn rules but perhaps the 'devil' here, to coin a phrase, lies in the detail.

59 This would seem to support in part the finding of *Tallinn Manual* 2.0, Rule 2.

60 Consider *Tallinn Manual* 2.0, Rules 8–13, which, however, lie outside the scope of the present chapter and therefore are not discussed in the preceding section.

States must meet obligations regarding internationally wrongful acts attributable to them[61] and must not use proxies to commit such acts.

Section IV of the 2013 Report recommends confidence-building measures and these can be summarised as follows:

The voluntary exchange of views and information on national strategies, policies, best practices, decision-making processes and measures to improve international cooperation;
The creation of bilateral, regional and multilateral consultative frameworks;
Information sharing on ICT security incidents;
Information exchanges between national Computer Emergency Response Teams; and
Enhanced mechanisms for law enforcement cooperation.

Section V of the 2013 Report emphasises the importance of providing 'technical and other assistance to build capacities in ICT security'[62], particularly in developing countries, and lists suggested measures to that end.

The 2015 Report seems to have taken the conclusions reached by the 2013 Group somewhat further. In the following paragraphs some of the apparently more significant points in that report are summarised. Section II identifies a number of widely appreciated cyber threats, noting that the use of ICTs in future conflicts between States is becoming more likely, that the most harmful attacks using ICTs include those targeted against infrastructure and associated information systems of a State and that the risk of them is real and serious; that terrorist use of ICT including for attacks is an increasing possibility that threatens international peace and security[63]; and that numerous factors such as the diversity of malicious non-State actors and the speed at which malicious ICT actions can occur increase risk, such that States are rightfully concerned.

Section III of the 2015 Report deals with norms, rules and principles. Observing that voluntary, non-binding norms of responsible State

[61] This somewhat opaque formulation is to be contrasted with the clear *Tallinn Manual* 2.0 assertion, based on highly persuasive authority, that a State bears international responsibility for cyber-related acts attributable to that State that constitute a breach of an international legal obligation.

[62] 2013 Report, p.10.

[63] Note for example B. Riley-Smith, et al., 'Airports and Nuclear Power Stations on Terror Alert As Government Officials Warn of Credible Cyber Threat', *Daily Telegraph*, 2 April 2017, available at www.telegraph.co.uk/news/2017/04/01/airports-nuclear-power-stations-terror-alert-government-officials/.

behaviour[64] can reduce risks to international peace, security and stability, norms are stated to reflect the expectations of the international community and to set standards for responsible State behaviour. The Report puts forward voluntary, non-binding norms for consideration by States, among which are the following:

States should cooperate in developing and applying measures to increase stability and security in the use of ICTs and to prevent harmful or threatening practices;

States should not knowingly allow their territory to be used for internationally wrongful acts using ICTs;[65]

States should consider how best to cooperate to exchange information, assist each other and prosecute terrorist and criminal use of ICTs;

States, in ensuring secure use of ICTs, should respect specified Resolutions on human rights on the Internet and on the right to privacy;

States should not conduct or knowingly support ICT activity contrary to international law that intentionally impairs use and operation of critical infrastructure serving the public;[66]

States should take appropriate measures to protect their critical infrastructure from ICT threats;[67]

[64] It is, of course, the voluntary, non-binding nature of the norms identified here that most immediately distinguishes them from the legally binding rules of *lex lata* that the *Tallinn Manual* experts found to exist.

[65] Consider the *Tallinn Manual 2.0*, Rule 6 which states the general principle on due diligence in the following terms: 'A State must exercise due diligence in not allowing its territory, or territory or cyber infrastructure under its governmental control, to be used for cyber operations that affect the rights of, and produce serious adverse consequences for, other States.'

[66] But note, for example, the Kaspersky report on Lazarus Group allegedly linked to the North Korean government and reporting cyber espionage and cyber sabotage operations, including attacks on banks and financial manipulation; A. O. Kaspersky 'Lab, Lazarus under the Hood', 3 April 2017, available at https://securelist.com/blog/sas/77908/lazarus-under-the-hood/; see also R. Sherstobitoff, 'Hidden Cobra Targets Turkish Financial Sector With New Bankshot Implant', 8 March 2018, available at https://securingtomorrow.mcafee.com/mcafee-labs/hidden-cobra-targets-turkish-financial-sector-new-bankshot-implant/. Hidden Cobra is a hacking group allegedly linked to the North Korean government. Note also the reports that link North Korea with the WannaCry ransomware attacks that crippled parts of the UK National Health Service and other organisations around the world; G. Corera, 'NHS Cyber attack was "launched by North Korea"', BBC News, 16 June 2017, available at www.bbc.co.uk/news/technology-40297493 and consider the assessment that Russia is becoming more aggressive in its cyber espionage effort.

[67] Note, however, discussion of Russian state-sponsored cyber operations, or the risk thereof, against critical national infrastructure; US-CERT Alert, TA-18–074A, Russian Government Cyber Activity Targeting Energy and Other Critical Infrastructure Sectors, 16 March 2018; M. Townsend and T. Helm, 'Threat of Russian Cyber Reprisal Puts UK Finance, Power and

States should respond to appropriate requests for assistance by another State or to mitigate malicious ICT activity emanating from their territory, taking due account of sovereignty;
States should seek to ensure the integrity of the supply chain; and
States should encourage responsible reporting of ICT vulnerabilities and should share information on available remedies.

Section IV of the 2015 Report emphasises the value of confidence building measures in strengthening international peace and security and includes among recommended voluntary measures the following:

identifying points of contact at policy and technical levels to address serious ICT incidents;
employing bilateral, regional, sub-regional and multilateral consultation mechanisms to enhance confidence building and reduce the risk from ICT incidents of misperception, escalation and conflict;
transparency at similar levels to increase confidence or inform future work including sharing national views and information on listed matters including vulnerabilities and
voluntary sharing of national views on which infrastructure is considered critical and on national efforts to protect it. Mention is made under this heading of national laws and policies to protect data and cross-border cooperation to address critical infrastructure vulnerabilities that transcend national borders.

The Section suggests that additional specified confidence-building measures to strengthen cooperation should be considered and Section V proposes voluntary measures to help build capacity to secure ICTs in countries where such help is needed.

Section VI of the 2015 Report addresses how international law applies to the use of ICTs. The Report identifies five principles as of central importance, namely:
sovereign equality;[68]

Water on High Alert', *The Guardian*, 18 March 2018, available at www.theguardian.com/world/2018/mar/17/uk-finance-power-water-on-high-alert-threat-russian-cyber-reprisal-grows. Consider also in this respect D. Wroe, 'New sabotage laws for cyber attacks on Australia's critical infrastructure', *Sydney Morning Herald*, 8 December 2017, available at www.smh.com.au/federal-politics/political-news/new-sabotage-laws-for-cyber-attacks-on-australias-critical-infrastructure-20171207-h000c2.html.

68 *Tallinn Manual* 2.0, para 1 of the Commentary associated with rule 3 expressly bases the idea of external sovereignty as reflected in Rule 3 on the sovereign equality of States and cites the 2015 UN GGE report in support.

peaceful settlement of international disputes so that international peace, security and justice are not endangered;[69]

refraining from the threat or use of force against the territorial integrity or political independence of any State or otherwise inconsistently with the purposes of the UN;[70]

respect for human rights and fundamental freedoms; and

non-intervention in the internal affairs of other States.[71]

The Report observes that State sovereignty and international norms and principles that flow from it 'apply to the conduct by States of ICT-related activities and to their jurisdiction over ICT infrastructure within their territory.'[72]

Six 'non-exhaustive views on how international law applies to the use of ICTs by States' are then set forth, and these are:

'States have jurisdiction over the ICT infrastructure located within their territory';[73]

In using ICTs, States must observe State sovereignty, sovereign equality, peaceful settlement of disputes and non-intervention in the internal affairs of other States. 'Existing obligations under international law are applicable to State use of ICTs'[74] and States must comply with their obligations to respect and protect human rights;

[69] *Tallinn Manual* 2.0, Rule 65a, points out that 'States must attempt to settle their international disputes involving cyber activities that endanger international peace and security by peaceful means.' Rule 65b provides that '[i]f States attempt to settle international disputes involving cyber activities that do not endanger international peace and security, they must do so by peaceful means.' It can therefore be seen that the Tallinn Manual and the UN GGE adopted similar approaches which, in both cases, were coherent with the UN Charter.

[70] *Tallinn Manual* 2.0, Rule 68, expresses the point as a clear prohibition and considerably more detail is achieved in that rule, in rules 69 and 70 and in the associated Commentaries, where the ideas of a use of force and a threat of force and related matters are clarified.

[71] *Tallinn Manual* 2.0, Rule 66 stipulates that a 'State may not intervene, including by cyber means, in the internal or external affairs of another State.' Here again therefore there is a common approach as between the Tallinn experts and the UN GGE. It is just that the Tallinn experts achieve greater detail both in their rules and by virtue of the associated Commentaries.

[72] 2015 Report, 12.

[73] *Tallinn Manual* 2.0, Rule 9 recognises, inter alia, that a State may exercise territorial jurisdiction over cyber infrastructure and persons engaged in cyber activities on its territory. Again, the Tallinn Manual rule goes very much wider than the somewhat hesitant UN GGE recommendation, but there is a clear parallel between the two.

[74] This is of course a very general statement. It neither specifies which rules of international law are being referred to, nor does it indicate that all relevant rules of international law are seen as being covered by the statement. As it stands, therefore, the statement might be regarded as broadly to be welcomed, but of doubtful practical value.

The Group noted the inherent right of States to take measures consistent with international law as recognised in the UN Charter;[75]

'The Group notes the established international legal principles, including, where applicable, the principles of humanity, necessity, proportionality and distinction';[76]

States must not use proxies to commit internationally wrongful acts using ICTs[77] and should seek to ensure that their territory is not used by non-State actors to do so.

The indication that an ICT activity was launched or otherwise originates from the territory or ICT infrastructure of a State may be insufficient in itself to attribute the activity to that State.

5.6 WHICH IS TO BE PREFERRED, TALLINN OR GGE?

It is against this background of slow but perceptible progress that the reported failure of the 2016–2017 GGE to achieve consensus must be judged. It is evident that the Group as a whole did not feel able to achieve greater specificity in the norms, rules and other measures that it had previously recommended to States in somewhat general terms. Clearly, the significant differences in the ways in which some of the participating States approach cyber matters will explain the absence of consensus. The US Deputy Coordinator for Cyber Issues, in an Explanation document issued on 23 June 2017, notes that some participants in the GGE continue to contend that it is premature to specify which bodies of international law apply to States' use of

[75] The observations in the immediately preceding footnote would apply similarly here. The Report does not indicate which measures under the UN Charter are being referred to. One might reasonably infer that measures by way of self-defence and measures undertaken by States under the authority of a UN Security Council resolution would likely be included. Whether that is right and which other measures are or are not included is not explicitly stated.

[76] 2015 Report, at 13. That the UN GGE merely 'noted' the established international legal principles neither suggests that the UN GGE specifically found those principles to apply in relation to ICT activities, nor that they found the principles to be inapplicable. The fact that the UN GGE merely noted the principles does not therefore affect the applicability of the principles either way. As noted earlier, the Martens Clause, the previously cited extract from the ICJ Nuclear Weapons Advisory Opinion, clear legal logic and the findings of the Tallinn experts all lead to the firm conclusion that the customary principles of international law do indeed apply to ICT/cyber operations undertaken by States and probably also to such operations undertaken by non-State actors. To hold otherwise would be to establish a legal vacuum for which legal justification would be hard to find.

[77] Consider for example *Tallinn Manual* 2.0, Rules 16 and 17 which respectively deal with the attribution of cyber operations by organs of a State that are placed at the disposal of another State and with the attribution of cyber operations by non-State actors.

ICTs, much less to state how such bodies of law apply. Suggesting that some such participants 'seem to want to walk back progress made in previous GGE reports', Michele Markoff expresses the 'unfortunate conclusion that those who are unwilling to affirm the applicability of these international legal rules and principles believe their states are free to act in or through cyberspace to achieve their political ends with no limits or constraints on their actions', a conclusion which is indeed dangerous, insupportable and that must be rejected.[78]

So, the high points in the achievements of the UN GGE process currently consist of the 2013 and 2015 Reports, being documents which, as we have seen do not go into anything like enough detail to give States a clear set of widely accepted rules as to what action is, or respectively is not, acceptable in cyberspace. It is significant that the GGE Reports express many, but by no means all, of their propositions in rather general terms, particularly when it is recalled that those propositions only have the status of recommendations. It is both significant and unsatisfactory that the Group only felt able to 'note' the principle of distinction, the application of which to military ICT operations during armed conflicts ought not to be controversial, particularly in light of the ICJ comments in the Nuclear Weapons case noted earlier. When it made the relevant remark, it should be noted that the ICJ emphasised the broader context by noting that the 'intrinsically humanitarian character of the legal principles. . .which permeates the entire law of armed conflict and applies to all forms of warfare and to all kinds of weapons, those of the past, those of the present and those of the future'.[79] This would reinforce the appreciation that the cardinal, intransgressible principle of distinction clearly applies to the use, for example, of ICT as a weapon in the course of an armed conflict.[80] So, what are we to make of the evident reluctance of the GGE to do anything other than 'note' the existence of the principle? Noting its existence does not of course constitute in any form a denial of its relevance nor of its applicability. Equally, it does not constitute a clear and unambiguous endorsement of that applicability. It seems, at best, to reflect an unhelpfully agnostic view of the matter.

If it is accepted that the principle of distinction does indeed apply to cyber operations conducted as part of an armed conflict, the customary rules of the law of targeting that give practical expression to that principle, e.g., such of Articles 48–60 of API as reflect customary law, will all therefore logically apply to military ICT operations in armed conflict. However, that the 2015 GGE

78 M.G. Markoff, 'Explanation of Position'.
79 *Nuclear Weapons* case, para 86.
80 *Nuclear Weapons* case, paras 78 and 79.

only 'noted' the principle of distinction and that the 2016–2017 Group could not achieve consensus on anything more specific, seem to illustrate the position taken by certain States that starting with the existing law and assessing how it can be applied to ICT operations fails adequately to reflect the fundamental differences between ICT activities and more conventional military operations.[81] Moreover, the very notion of 'cyber weapons' has hitherto proved controversial,[82] so unsurprisingly the existing GGE reports do not refer to any rules that would distinguish between lawful and unlawful ICT weapons.

It is known that Russia and China consider that sovereignty extends to ICT infrastructure located in a State's territory, ICT activities undertaken there and to that portion of cyberspace that is located within such territory.[83] While all States will have an interest in restricting certain kinds of activity on the Internet[84] or access to certain materials there,[85] it is evident that certain States

81 Clear impression gained during discussions involving certain Russian officials at which the author was present during 2016. Note comments of the Russian Foreign Minister, Sergey Lavrov, that an international convention is what is required to combat cybercrime; T. Kanunnikova, 'Russia Offers to Draft International Convention on Combating Cybercrime', 23 March 2017, available at http://russianconstruction.com/news-1/26888-russia-offers-to-draft-international-convention-on-combating-cybercrime.html but note, e.g., Lithuanian complaints of cyber and information attacks from Russia; L. Weymouth, 'Russia is a threat…to all of Europe', *Foreign Policy*, 24 March 2017, available at http://foreignpolicy.com/2017/03/24/russia-lithuania-nato-grybauskaite-putin-trump-interview/; reports from Germany of cyber attacks by the Russian group APT28 in 2016, e.g., A. Shalal, 'Germany Blocked Russian Hacking Attacks in 2016', Reuters, 24 March 2017, available at http://uk.reuters.com/article/uk-germany-elections-russia-idUKKBN16V2FJ.

82 Chinese representatives made statements objecting to the notion during a meeting at which the author was present in Geneva in advance of the commencement of the 2016/2017 GGE process.

83 The Chinese International Strategy of Cooperation on Cyberspace, 1 March 2017, available at http://news.xinhuanet.com/english/china/2017–03/01/c_136094371_2.htm, Chapter II section 2 states: 'National governments are entitled to administer cyberspace in accordance with law. They exercise jurisdiction over ICT infrastructure, resources and activities within their territories, and are entitled to protect their ICT systems and resources from threat, disruption, attack and destruction so as to safeguard citizens' legitimate rights and interests in cyberspace.' It is the first sentence that reflects the Chinese assertion of sovereignty over cyberspace as such.

84 Note, for example, that on 20 March 2017 the Supreme Court of India directed the Indian government, Internet majors and intermediaries to stop the sharing of videos of sexual offences on social media; 'SC Asks Govt, Others To Mull Ways to Check Sex Offence Videos', *India Today*, 20 March 2017, available at http://indiatoday.intoday.in/story/sc-asks-govt-others-to-mull-ways-to-check-sex-offence-videos/1/908579.html. For a discussion of the Chinese view in relation to sovereignty, see Z. Huang and K. Mačák, 'Towards the International Rule of Law in Cyberspace: Contrasting Chinese and Western Approaches', 16 *Chinese Journal of International Law* (2017) 271, paras 44–51, and on the present point see in particular para 48.

85 Blocking access to websites may, depending on the circumstances and the extent of the blocking activity, lead to freedom of expression concerns; see, for example, Egypt extends its assault on freedom of expression by blocking dozens of websites UN experts warn, Office of the

wish to impose more extensive restrictions on the information to which their populations and persons within their territories may have access and may wish to stop such access when they deem it expedient to do so.[86] As to the practicalities of such cyber operations, Sigmar Stadlmeier points out that 'neutralising the domain name servers of a network does indeed render all members of the affected network *incommunicado* until the damage is repaired ... Without their service neither incoming nor outgoing messages can be routed. The effect of taking out the DNSs exactly equals jamming a whole radio net or blowing up a main network node.'[87] Accordingly, the difference of view between the Western States and Russia and China in this regard seems to have more to do with the extent of the restrictions on access that should be imposed and on the reasons for such restrictions, and rather less to do with the principle that certain restrictions are likely to be required.

An arguably more significant difference of perspective between what might be characterised as the 'like-minded States' such as the USA, UK, Canada and Australia on the one hand and, e.g., Russia and China on the other relates to the more fundamental question of the adequacy of current international law. So, for example, while Russia and China see the development of *lex specialis*, in the form of treaty provision, to address Information and Communications

United Nations High Commissioner for Human Rights, 30 August 2017, available at www.ohchr.org/EN/NewsEvents/Pages/DisplayNews.aspx?NewsID=22008&LangID=E. Where the UK is concerned, for example, see, e.g., M. Jackson, 'UK Government Confirm Move to Force ISPs into Blocking "Adult" Sites', *ISP Review*, 21 November 2016, available at www.ispreview.co.uk/index.php/2016/11/uk-government-confirm-move-force-isps-blocking-adult-sites.html. Access to certain websites may be blocked pursuant to a UK High Court order; see, e.g., Sky, Websites we've blocked under order of the High Court, available at www.sky.com/help/articles/websites-blocked-under-order-of-the-high-court. Consider also the power of a constable to require the blocking of terrorism-related material on the Internet under Section 3 of the Terrorism Act 2006 (2006 c.11) and note M. Pham, 'Vietnam Urges Firms to Stop Youtube and Facebook Ads in Protest over Fake Content', Reuters, 16 March 2017, available at www.reuters.com/article/us-vietnam-google-idUSKBN16N110.

86 Consider in this respect the apparent blocking by China of access by its citizens to Facebook, Twitter and, apparently, now Pinterest; D. Ngo, 'Pinterest is now blocked by China's Great Firewall', C/Net, 16 March 2017, available at www.cnet.com/uk/news/pinterest-blocked-in-china-great-firewall/ and for an assessment of China's approach to online censorship, see L. Ruan, 'Internet Censorship: how China does it', *The Strategist*, 9 October 2017, available at www.aspistrategist.org.au/internet-censorship-how-china-does-it/. There is evidence that Chinese censorship is also affecting, e.g., Australian citizens in Australia; see H. Hanson, 'China's Censorship Fingers Reach into Our Cyberspace', *The Weekend Australian*, 24 June 2017, available at www.theaustralian.com.au/news/inquirer/chinas-censorship-fingers-reach-into-our-cyberspace/news-story/846944042e1e4227782611979a211cae.

87 S. Stadlmeier, 'CPU and Keyboard: Weapons of Mass Disruption?' In W. Heintschel von Heinegg, R. Frau and T. Singer (eds.), *Dehumanization of Warfare* (Cham, Switzerland: Springer 2018), 147, 155.

Technologies as essential, the like-minded generally consider that existing international law, including the law of armed conflict, is the appropriate way of addressing cyber challenges.[88] After pointing out the fragility and apparent erosion of the Western, like-minded approach and noting the growing support for Russia-inspired calls for international agreement to be the basis for the way forward, Eneken Tikk and Mika Kerttunen argue for a progressive Western initiative, a Cyber Marshall Plan building robust national capacities and unprecedented international transfers of ICTs to promote a climate of wealth, health and security and to counter Russian demonstrations of vulnerability through its own national operations. They argue persuasively for consistency of approach among the like-minded and for shared national interests and objectives to be identified across continents. They perceptively point out that the Western argument that existing international law is sufficient to address cyber challenges 'can be seen as an attempt to avoid restraint' and suggest that the West will have difficulty in convincing the international community of its stand against the negotiation of new treaty law. The paper concludes that the West needs to be prepared for the possibility of treaty negotiations but could sensibly move forward with Russia on developing principles for international cyber security.[89]

One is left with the impression that it is these fundamental national differences of approach to cyber/ICT policy questions that have led to the unfortunate inability of the UN GGE to make progress in 2017. Whether certain States do indeed see a narrow, national benefit in an absence of agreement as to the application of existing law is, and will remain, a matter of opinion. It is, however, clear that the fundamentally opposed views of important States on these matters do not necessarily preclude the application of customary law status to the Tallinn Manual approach to applying existing rules of customary law to cyber warfare activities. While some may argue that there is an absence of general practice accepted as law such as is required to regard such an approach as customary, there is an arguably more powerful

88 E. Tikk and M. Kerttunen, 'The Alleged Demise of the UN GGE: An Autopsy and Eulogy', Cyber Policy Institute, 2017, 21 and see Z. Huang and K. Mačák, 'Towards the International', para 21.

89 E. Tikk and M. Kerttunen, 'Cyber Treaty is Coming: Что делать?', Cyber Policy Institute (2018), 5, available at http://cpi.ee/wp-content/uploads/2018/02/Cyber-Treaty-is-Coming-Tikk-Kerttunen.pdf and see E. Tikk and M. Kerttunen, 'Russians in the Red Zone: Send in the White Knights', Cyber Policy Institute (2018), 5 where a third Russian objection to the Western view is noted, namely that ICTs do not provide economic prosperity and in fact erode societal stability.

argument that cyber uses of force, cyber armed attacks, cyber attacks and cyber weapons are simply novel forms of the established phenomena of use of force, armed attack, attack and weapon and that the law that applies to the established phenomena applies equally to them in their cyber forms. So, for example, the argument would be that the Tallinn Manuals do not just show how established international law principles and rules could be applied in cyberspace; where cyber hostilities are concerned, they also for the most part show the principles and rules that actually apply.

Developing this point somewhat, some additional remarks should be made. While Article 38 of the ICJ Statute recognises other materials, such as Experts' writings and judicial decisions, as being indicative of the existence of rules, but not as forming a source as such of the rules so indicated, numerous commentators argue for a more inclusive approach to the definition of sources, pointing to the highly authoritative nature of much judicial and scholarly work.[90] So, while traditionalists would class the Tallinn Manual 2.0 and the GGE Reports as, at best, being suggestive of where the law is on these matters, they both reflect the views of their respective sets of authors as to the state of the law, and both fall to be considered in determining the legal position on any particular relevant issue.

The alternative interpretation, to which the present author subscribes, argues that the position expressed by the International Court of Justice in the Nuclear Weapons Advisory Opinion, the application by the Martens Clause as found in Hague Convention IV and in Article 1(2) of API and common-sense interpretations of relevant notions of use of force, armed attack, attack and weapon all lead to the conclusion that the established principles of international law apply in the cyber context. It would follow from this that the customary rules of international law that are based on those principles also apply with equal force. To apply the principles but not to apply the rules that necessarily flow from those principles simply would not make sense.

The mere fact that certain States do not presently adopt the Tallinn approach does not of course render that approach wrong as such. Likewise, the failure of the UN GGE to do more than 'note' the principle of distinction does not as such call into question the applicability of the principle or of the associated rules. It just means that those Groups of Experts chose not to go

[90] Cassandra Steer points to the roles of participants, e.g., in international criminal justice proceedings, as makers of the law and argues for a policy-oriented, dynamic approach to recognising legal rules; C. Steer, *Translating Guilt: Identifying Leadership Liability for Mass Atrocity Crimes* (The Hague: Asser Press, 2017) 66–68 and 76.

into more detail. There is the possibility that some participants in the GGE process are declining at the moment to accept the application of certain rules for fear of lending apparent legitimacy to actions in cyberspace of which they disapprove or which for political reasons they prefer not to dignify with the finding of rules. That would not per se exclude the application of such rules.

It is at this point worth noting that at its meeting in Lucca, Italy on 11 April 2017 the G7 Group of States made a Declaration on Responsible States [*sic*] Behaviour in Cyberspace.[91] Having called upon States to be guided by the UN GGE Reports, the Declaration restates many of the principles agreed by the UN GGEs, notes the circumstances in which resort may be made to countermeasures, that attribution must accord with international law and calls on States publicly to set forth their views on how existing international law applies to State activities in cyberspace. Emphasising the importance of confidence building measures and of communication channels among States for crisis management, the Declaration supports the promotion of voluntary non-binding norms of responsible State behaviour, repeating items (a)–(k) of paragraph 13, Section III, of the 2015 GGE Report. The Declaration then adds a final norm suggesting inter alia that 'no country should conduct or support ICT-enabled theft of intellectual property with the intention of providing competitive advantages to companies or commercial sectors.'

Given that there is no treaty provision dealing with cyber hostilities and that States seem to be taking somewhat contrary positions on some core matters, there are those who would argue that there are in reality no detailed international rules of cyber warfare beyond what the GGE has agreed.[92] In the preceding paragraphs of this Chapter, however, a strong argument has been put forward that currently understood international law applicable to uses of force, to armed attacks, to attacks and to weapons applies equally in the cyber context.

5.7 PROPOSED INTERNATIONAL CODE OF CONDUCT

It is at this point that attention should shift to alternative norm-creating processes some of which have already been discussed in the context of the UN GGE Reports. Another example of such a proposal is reflected in a letter

[91] The full text of the Declaration is available at www.esteri.it/mae/resource/doc/2017/04/declaration_on_cyberspace.pdf.

[92] There are, however, other initiatives looking at these matters, as noted in M. Schmitt and L. Vihul, 'International Cyber Law Politicized: The UN GGE's Failure to Advance Cyber Norms', *Just Security*, 30 June 2017.

sent on 9 January 2015 by the UN Permanent Representatives of China, Kazakhstan, Kyrgyzstan, the Russian Federation, Tajikistan and Uzbekistan to the UN Secretary General.[93] Referring to an earlier draft Code of Conduct on matters of information security submitted in 2011 to the UN General Assembly by the Russian Federation, China, Tajikistan and Uzbekistan, the letter submits a version of the code of conduct that was evidently revised taking into account the comments and suggestions received in relation to the earlier text. Annexed to the letter is a draft of a voluntary International Code of Conduct for Information Security that is stated to be open to all States. The purpose of the Code of Conduct is stated to be to identify the rights and responsibilities of States, to promote responsible behaviour, to enhance cooperation, to establish a peaceful, secure and open information environment, and to ensure that the use of these technologies does not run counter to international peace and security.[94]

Paragraph 2 of the Code then sets forth thirteen pledges which, stated briefly, are as follows:

To comply with the UN Charter and universally recognised norms governing international relations, e.g., respect for sovereignty, territorial integrity, respect for human rights and fundamental freedoms and for cultural diversity;

Not to use ICTs and networks for activities contrary to the maintenance of international peace and security or 'to interfere in the internal affairs of other States or with the aim of undermining their political, economic and social stability';

To cooperate in combatting criminal and terrorist activities that use ICTs and networks and in curbing dissemination of information that incites terrorism, separatism or extremism or that inflames hatred on ethnic, racial or religious grounds;

To endeavour to ensure the supply chain security of ICT goods and services for purposes listed in the Code;

'To reaffirm the rights and responsibilities of all States in accordance with the relevant norms and rules regarding legal protection of their information space and critical information infrastructure against damage resulting from threats, interference, attack and sabotage';

93 United Nations General Assembly, Document A/69/723, dated 13 January 2015.

94 International Code of Conduct for Information Security ('the Code'), para 1.

That individuals' rights in the offline environment must also be protected online; to fully respect rights and freedoms in the information space, e.g., to seek, receive and impart information taking into account that ICCPR Article 19 attaches to that right duties and responsibilities. It may be subject to certain restrictions provided for in law and necessary: for the rights and reputations of others; and for the protection of national security, public order, public health or morals;

That all States must play the same role in and carry equal responsibility for international governance of the Internet, its security, continuity and stability of operation and its development;

That States must cooperate with interested parties to encourage deeper understanding by all elements of society of their responsibility to ensure information security including by creating an information security culture and by supporting efforts to protect critical information infrastructure;

To develop confidence building measures to increase predictability and reduce likelihood of misunderstanding and the risk of conflict. Examples are given;

To assist developing countries in their efforts to enhance capacity building on information security and to close the digital divide;

To bolster bilateral, regional and international cooperation and promote a prominent role for the United Nations and

To settle any dispute arising from the application of the Code by peaceful means and to refrain from the threat or use of force.

These are the topics addressed in the Code, not always in the precise language selected here. The list does, however, convey the general scope of the Code and its broad focus. States in general have not however signified their support for the Code, so it too represents a text that does not have the status of law but that suggests ways of addressing perceived challenges in cyberspace.

It is also worth mentioning that industry has been calling for action. Thus, and to cite but one example, Microsoft published a statement in February 2017 suggesting the need for a digital Convention the core proposals of which were: that there should be no targeting of tech companies, the private sector or critical infrastructure; that private sector efforts to detect, contain, respond to and recover from events be assisted; that vulnerabilities be reported to vendors rather than exploiting them; that there be restraint in developing cyber weapons; that offensive operations should be limited and a commitment to non-proliferation.[95] In a recently published undated policy paper, Microsoft proposes the formation of an international cyber attack attribution organisation.[96] It remains to be seen how much traction this initiative gains among States.

Finally, it should be noted that international consultations among States, for example under the auspices of the Hague Process,[97] are likely to be helpful in fostering the achievement of greater international consensus as to applicable principles, rules and norms. Other useful examples of such consultations among States include the Global Commission on the Stability of Cyberspace[98] and annual Cycon meetings[99] which usefully explore contemporary issues and advance understanding. Such meetings are frequently attended inter alia by government officials but do not usually involve direct inter-State engagement.

These are not the only contemporary processes under which these issues are being considered. Engagement at the regional level,[100] and diplomatic, official and academic engagements on a 'State to State' or 'group of States to State' level[101] can clarify mutual understanding of the issues involved and thus achieve progress in the finding of norms. The outcomes of such discussions may not necessarily be made public, so such processes may not directly lead to the formation of customary rules of law. Nevertheless, processes that enhance mutual international understanding in this field are to be welcomed.

95 B. Smith, 'The need for a Digital Geneva Convention', Microsoft Blog, 14 February 2017, available at https://blogs.microsoft.com/on-the-issues/2017/02/14/need-digital-geneva-convention/#sm.00001e7re9mi1sd80wxcuwmn80906.

96 The document is available at https://mscorpmedia.azureedge.net/mscorpmedia/2017/04/Policy-Paper-Attribution-Organization.pdf.

97 More than fifty States and international organisations participated in a series of meetings sponsored by the Dutch Ministry of Foreign Affairs which gave them the opportunity to provide input to the Tallinn Manual 2.0 Experts with regard to cyber operations by States and by non-State actors; M. N. Schmitt, Tallinn Manual 2.0 on the International Law of Cyber Operations: What it is and isn't, Just Security, 9 February 2017, available at www.justsecurity.org/37559/tallinn-manual-2-0-international-law-cyber-operations/.

98 Note that the Global Commission met at Las Vegas on 27 July 2017 as described at https://cyberstability.org/news/645–2/.

99 The Ninth International Conference on Cyber Conflict, or Cycon, took place in Tallinn, Estonia on 31 May–2 June 2017.

100 Consider the consultations leading to the adoption of the European Union's Cyber Security Strategy, as to which see www.consilium.europa.eu/en/policies/cyber-security/. As to the importance of such consultations in Asia, consider T. M. Wettach, 'Rebalancing Cyber Security in Asia', *The Diplomat*, 16 December 2016, available at http://thediplomat.com/2016/11/rebalancing-cybersecurity-in-asia/.

101 Consider, for example, track 1, track 1.5 and track 2 discussions, e.g., between the US, Australia, the UK and the EU and China.

5.8 RANGE OF USES OF CYBER TECHNOLOGIES OUTSIDE HOSTILITIES

Much of the discussion to date has centred on cyber activities associated with conflict, armed or otherwise. When, however, we start to consider cyber activities outside conflict, it becomes immediately apparent that a vast array of kinds of operation may take place, involving a similarly broad selection of legal issues addressed by disparate areas of the law. Generally speaking it will be a mixture of domestic and human rights law that will often be relevant.

To take one example, there is a very lively debate that has been going on for some time as to where the balance lies between notions of free speech, privacy and security on the Internet.[102] State agencies have the sovereign responsibility to take appropriate action to maintain the security of their citizens. In the modern era of social media communications this will necessarily involve achieving as clear and timely an understanding as possible of what terrorists, criminals and others seeking to perpetrate violence are planning.[103] Those same citizens have the legal right under human rights law to respect for their private lives.[104] Numerous social media communications systems are equipped with 'end to end encryption' that, it has been claimed, renders security services monitoring either impossible or at least difficult.[105] As with

[102] Consider in this regard the proposals being considered in Germany that would require technology and car companies to provide German security services with digital access to cars, phones, computers and other devices; German government wants 'backdoor' access to every digital device: Report, *The Local*, 1 December 2017, available at www.thelocal.de/20171201/german-government-wants-backdoor-access-to-every-digital-device-report.

[103] Note the report that forty-one of Pakistan's sixty-four banned terrorist groups are active on Facebook; J. Haque and O. Bashir, 'Banned Outfits in Pakistan Operate Openly on Facebook', 13 June 2017, *Dawn*.

[104] Consider, for example, the ruling of the Supreme Court of India that individual privacy is a fundamental right; S. Mohanty and R. Bhatia, 'India Court's Privacy Ruling Is Blow to Government', Reuters, 24 August 2017, available at www.reuters.com/article/us-india-court-privacy/indian-courts-privacy-ruling-is-blow-to-government-idUSKCN1B40CE?utm_campaign=trueAnthem:+Trending+Content&utm_content=599e6b6204d3015a2054937c&utm_medium=trueAnthem&utm_source=twitter. Citizens are of course concerned that their personal data should be securely held by those to whom they have entrusted it. Note that in 2016 in China 6.05 billion pieces of personal information were leaked according to a report issued by the Chinese Internet security company, Qihoo 360; J. Jie, 'Personal Information Leakage on the Rise in China', *People's Daily Online*, 31 March 2017, available at http://en.people.cn/n3/2017/0331/c90000-9197748.html. Note, however, L. Caiyu, 'China's First Quantum Information Network Will Prevent Hacking: Expert', *Global Times*, 9 July 2017.

[105] See reported comments by the UK Home Secretary in G. Rayner, 'WhatsApp Accused of Giving Terrorists A Secret Place to Hide as it Refuses To Hand over Private Messages', *Daily Telegraph*, 27 March 2017, available at www.telegraph.co.uk/news/2017/03/26/home-secretary-amber-rudd-whatsapp-gives-terrorists-place-hide/ and note K. McCann, 'Government

so many issues of this nature, the challenge lies in achieving a satisfactory balance between these apparently but not necessarily competing interests. By breaking encryption, security services are given potential access to information which ought, perhaps, to be protected. On the other hand, if terrorist incidents can be prevented, human life is thereby saved. These issues have been played out in the legal dispute between Apple Corp. and the United States Federal Bureau of Investigation,[106] but there are suggestions that the law is something of a blunt instrument in dealing with this sort of issue. The tech companies are, it seems, doing more than may immediately be apparent to work with governmental agencies towards achieving that 'hard to find' balance between personal privacy and public security.[107]

In the UK, it is the Investigatory Powers Act 2016[108] that addresses these issues by determining the extent to which certain investigatory powers may be used to interfere with privacy. The Act provides for duties in relation to, and protections of, privacy, for example by providing for offences and penalties relating to the unlawful interception of communications, or the unlawful obtaining of communications data. The Act sets out the circumstances, including under a warrant, in which it is lawful to intercept communications[109] and in which the obtaining of communications data is lawful under a warrant, and deals with the obtaining and treatment of such data.[110] The retention of communications data, warrants covering the interference with

"blocked" from Accessing Twitter Data to Help Spot Terrorist Plots', *The Telegraph*, 25 April 2017, available at www.telegraph.co.uk/news/2017/04/25/government-blocked-accessing-twitter-data-help-spot-terrorist/ but consider N. Lomas, 'Facebook Expands Its Hate-fighting Counterspeech Initiative in Europe', 23 June 2017, available at https://techcrunch.com/2017/06/23/facebook-expands-its-hate-fighting-counterspeech-initiative-in-europe/.

[106] The issue arose because Syed Farook, who undertook attacks in San Bernadino in December 2015, owned an iPhone which the FBI wished to unlock. The counter argument was that if the FBI could unlock the iPhone in question, nothing could stop them from doing likewise to many others; for further details of the case, see A. Kharpal, 'Apple vs FBI: All You Need to Know', 29 March 2016, available at www.cnbc.com/2016/03/29/apple-vs-fbi-all-you-need-to-know.html.

[107] Note that the G20 States committed themselves to work with the private sector, communication service providers and administrators of particular applications, to fight exploitation of the Internet and social media for terrorist purposes while respecting human rights; filtering, detecting and removing relevant content are seen as critical. G20 States encouraged collaboration with industry to provide lawful and non-arbitrary access to information where necessary for the protection of national security; G20 Statement 17/1955, G20 Leader's Statement on Countering Terrorism, Hamburg, 7 July 2017, para 20.

[108] 2016 c.25. The text of the legislation is available at www.legislation.gov.uk/ukpga/2016/25/contents/enacted.

[109] See Investigatory Powers Act 2016, part 2 and part 6, chapter 1.

[110] See Investigatory Powers Act 2016, part 3 and chapter 2 of part 6.

equipment, and warrants dealing with bulk personal datasets are dealt with in further parts of the Act.[111] So the relationship between personal privacy and public security, to which reference was made at the end of the previous paragraph, has been set forth in the UK in a substantial piece of legislation that comprises 272 sections and ten schedules.

It should not be forgotten, however, that domestic law, such as the 2016 Act, must also be interpreted consistently with applicable human rights, such as the right to respect for private and family life. So far as the UK is concerned, this is most relevantly expressed in Article 8 of the European Convention: '(1) Everyone has the right to respect for his private and family life, his home and his correspondence. (2) There shall be no interference by a public authority with the exercise of this right except such as is in accordance with the law and is necessary in a democratic society in the interests of national security, public safety or the economic well-being of the country, for the prevention of disorder or crime, for the protection of health or morals, or for the protection of the rights or freedoms of others.' Clearly, the 2016 Act addresses the 'prescribed by law' aspect, but the relevant issues under Article 8 will tend to be whether the interference is for one of the listed purposes and whether it is necessary for that purpose.

The posting of what has come to be known as 'fake news'[112] and of radical and other subversive material on websites operated by organisations such as Google is also posing challenges for security agencies. The challenge lies for example in the capacity of such materials to radicalise individuals with all the potential consequences that we have seen, such as terrorist activity and the like. There is evidence that Google and Twitter, for example, have been taking some steps to remove certain material from their websites.[113] UK Government ministers have, in the aftermath of the March 2017 terrorist attack

[111] Investigatory Powers Act 2016, Section 1.

[112] See, for example, J. Kosslyn and C. Yu, 'Fact Check Now Available in Google Search and News round the World', Google Blog, 7 April 2017, available at https://blog.google/products/search/fact-check-now-available-google-search-and-news-around-world/ and D. Thomas, 'Facebook to Tackle Fake News with Education Campaign', BBC News, 6 April 2017, available at www.bbc.com/news/technology-39517033.

[113] See The *Daily Telegraph*, 'Google Removes 640 Videos From Youtube Promoting Terrorism', 18 June 2012, available at www.telegraph.co.uk/technology/google/9337993/Google-removes-640-videos-from-YouTube-promoting-terrorism.html. However, see also A. Topping and others, 'Google is "profiting from hatred" say MPs in row over adverts', *The Guardian* 17 March 2017 and for Google's response, see P. Schindler, 'Expanded safeguards for advertisers', 21 March 2017, available at https://blog.google/topics/ads/expanded-safeguards-for-advertisers/. Twitter has issued a transparency report for the second half of 2016 that includes information as to the removal of terrorism-promoting material, available at https://transparency.twitter.com and see B. Popken, 'Twitter Deleted 200,000 Russian Troll Tweets', NBC News, 14 February 2018,

in Westminster, argued that more could and should be done in this regard, and some of the largest Internet companies, including Google, have agreed to create tools to identify and remove terrorist propaganda.[114] This raises the question whether a 'naming and shaming' process will induce the Internet companies to do that which blunt domestic law levers[115] may or may not achieve. It also leads one to wonder whether some kind of technical solution will be the more appropriate way ahead.[116]

As alluded to earlier in the chapter, there are suggestions that Russian hackers intruded into US Democratic Party communication systems in the run-up to the 2016 US Presidential Election.[117] Information gathering by

available at www.nbcnews.com/tech/social-media/now-available-more-200-000-deleted-russian-troll-tweets-n844731.

[114] C. Hope, 'Google: We'll make new tools to tackle terrorists online', The *Daily Telegraph*, 30 March 2017, available at www.telegraph.co.uk/news/2017/03/30/google-facebook-agree-set-new-industry-board-accelerate-removal/.

[115] Note, for example, that social media firms face fines in Germany of up to €50m if they persistently fail to remove illegal content from their sites; Germany approves plans to fine social media firms up to €50m, *The Guardian*, 30 June 2017.

[116] Consider reports as to the use of artificial intelligence to identify terrorist websites with a view to them being taken down; L. Dearden, 'ISIS Videos Targeted by Artificial Intelligence That Can Detect Propaganda Before It's Uploaded', The *Independent*, 13 February 2018, available at www.independent.co.uk/news/uk/home-news/isis-videos-artificial-intelligence-propaganda-ai-home-office-islamic-state-radicalisation-asi-data-a8207246.html. The draft European Commission Recommendation for tackling illegal material online, dated February 2018, is available at https://edri.org/files/illegal_content_ec_draft_recommendation_201802.pdf.

[117] US Office of the Director of National Intelligence, Background to 'Assessing Russian Activities and Intentions in Recent US Elections: The Analytic Process and Cyber Incident Attribution, 6 January 2017: 'We assess Russian President Vladimir Putin ordered an influence campaign in 2016 aimed at the US Presidential Election. Russia's goals were to undermine public faith in the US democratic process, denigrate Secretary Clinton, and harm her electability and potential presidency… When it appeared to Moscow that Secretary Clinton was likely to win the election, the Russian influence campaign began to focus more on undermining her future presidency…. We assess with high confidence that Russian Military intelligence … used the Guccifer 2.0 persona and DCLeaks.com to release US victim data obtained in cyber operations publicly and in exclusives to media outlets and relayed material to WikiLeaks'; pages ii and iii. For the evidence of the then FBI Director on the matter, see 'Full Transcript: FBI Director James Comey Testifies on Russian Interference in 2016', *Washington Post*, 20 March 2017, available at www.washingtonpost.com/news/post-politics/wp/2017/03/20/full-transcript-fbi-director-james-comey-testifies-on-russian-interference-in-2016-election/?utm_term=.6b6ef8d4c68d. As to suggestions that the 2017 Macron French Presidential Election campaign was targeted by a cyber espionage group linked with Russian military intelligence see E. Auchard, 'Macron campaign was target of cyber attacks by spy-linked group', Reuters, 24 April 2017, available at www.reuters.com/article/us-france-election-macron-cyber-idUSKBN17Q200. As to interference in Canadian electoral processes, see Canada Communications Security Establishment, Cyber Threats to Canada's Democratic Process, 7 June 2017, available at www.cse-cst.gc.ca/sites/default/files/cse-cyber-threat-assessment-e.pdf.

foreign security agencies is of course nothing new.[118] The mere process of gaining access to a political party's communications systems is the kind of activity that one would confidently imagine that certain Western security agencies have undertaken in the past. It must, in this regard, be borne in mind that States bear international responsibility for cyber-related acts that are attributable to the State and that constitute a breach of an international legal obligation. Cyber espionage is not something that is regulated by international law, so the mere act of spying on the Democratic Party's communications, even if undertaken on the instructions of the Russian Government, would not per se be internationally wrongful.[119]

It is clear that the 'Democratic Party hack' is, at the time of writing, the subject of on-going investigation by a number of US agencies. The discussion in the present paragraph will therefore proceed on a hypothetical basis. Let us imagine that hackers are acting under the direction and control of the government of State R or that their activities are in whatever manner attributable to State R. The question to consider is whether actions actually taken by the hackers and/or by the State R government amount to an unlawful intervention into the internal affairs of State U. This is because no State may intervene by cyber means in the internal or external affairs of another State.[120] To constitute intervention, the activity must relate to the internal or external affairs of the target State and must be coercive in nature.[121] It seems likely that coercive action 'designed to influence outcomes in, or conduct with respect to, a matter reserved to a target State' would likely come within the rule. It would seem highly likely that seeking to undermine public faith in the political processes of the target State, seeking to harm the electoral prospects of one of the two candidates in the target State's governmental election and seeking to harm the potential presidency of that candidate, taken together, does indeed constitute unlawful intervention, and would therefore amount to an internationally wrongful act, or series of acts. Arguably, it is the release of the relevant information with the suggested intent that actually constitutes the intervention.

Finally, it must be emphasised that whether such activity is in fact attributable to a State government, whether that activity goes beyond mere espionage, and if so, whether it can be shown to have been intended to influence an

[118] Consider, for example, C. Andrew and V. Mitrokhin, The Mitrokhin Archive: The KGB in Europe and the West (Alan Lane: The Penguin Press, 1999).

[119] *Tallinn Manual* 2.0, Rule 14 and para 6 of the Commentary to the same Rule.

[120] *Tallinn Manual* 2.0, Rule 66.

[121] *Tallinn Manual* 2.0, para 6 of Commentary to Rule 66.

outcome in a matter internal to the target State would need to be definitively determined.

Much hacking activity is directed at commercial or criminal purposes, not political ones.[122] Hacking for political purposes also, however, is likely to constitute a breach of the domestic law of the affected State.[123] In the US there has been a relatively recently disclosed case of the hacking of Yahoo and the associated theft of large amounts of personal data.[124] Where the UK is concerned, the Computer Misuse Act 1990[125] renders it an offence if a person 'causes a computer to perform any function with intent to secure access to any program or data held in any computer, or to enable any such access to be secured; [if] the access he intends to secure, or to enable to be secured, is unauthorised; and [if] he knows at the time when he causes the computer to perform the function that this is the case.'[126] Unauthorised access with intent to commit or to facilitate the commission of a further offence is criminalised by Section 2 of the Act. Theft or fraud are of course examples of the further offences that might be relevant. Committing an unauthorised act with the intention of impairing the operation of a computer or program or to hinder access to data are offences under Section 3 of the Act and unauthorised access that causes serious damage, or whereby there is a serious risk of such damage, is covered by Section 3ZA of the Act. While the relevant legislation goes into much more detail than this, the basic point emerges that the Act does address the kinds of activity reported in respect of the Yahoo and other high profile hacks. There are of course evidential and other issues surrounding the practical possibility of identifying, investigating and taking proceedings against

[122] Note, for example, the reported hacking by the Russian hacking group APT28 (Fancy Bear) of IAAF medical records; Press Release, IAAF victim of cyber attack, 3 April 2017, available at www.kyivpost.com/ukraine-politics/ukraine-natos-support-prepares-creation-cyber-security-centre.html.

[123] Note that on 16 February 2018 an indictment was handed down by the District Court for the District of Columbia in the United States charging numerous Russian entities and individuals with conspiracy offences arising from alleged interference in the US Presidential election 2016; Case 1:18-cr-00032-DLF, available at www.justice.gov/file/1035477/download.

[124] S. Thielman, 'Yahoo Hack: 1bn Accounts Compromised by Biggest Data Breach in History', *The Guardian* 15 December 2016, available at www.theguardian.com/technology/2016/dec/14/yahoo-hack-security-of-one-billion-accounts-breached.

[125] 1990 c. 18.

[126] Computer Misuse Act 1990, section 1(1). As to Singapore's recently amended Computer Misuse and Cyber Security Act, see K Kwang, Changes to Singapore's Cybercrime Law Passed, Channel News Asia, 3 April 2017, available at www.kyivpost.com/ukraine-politics/ukraine-natos-support-prepares-creation-cyber-security-centre.html. See also K. Abuya, 'Kenya's Computer and Cybercrimes Bill 2016 Gets Approval', *techweez*, 7 April 2017, available at www.techweez.com/2017/04/07/computer-cybercrimes-bill-2017-kenya-approved/.

those responsible, but those issues are distinct from the legislation's coverage of relevant activity.

The crimes committed by using the hacked personal data would of course be charged and punished in accordance with the legislation that criminalises such activities. The mere fact that data stolen as a result of a cyber hack is used, e.g., to defraud a bank into transferring money from a customer's account to the criminal does not preclude the charge of theft or fraud, depending on the precise circumstances, being preferred against the perpetrator.[127] Reportedly in China a plan has been adopted to establish a cyberspace court which would accept filings electronically, would try cases via livestream and would hear only E-commerce and Internet related cases.[128]

Intellectual property theft is an activity that has attracted significant comment in both the UK and the US, featuring in the 2016 US Presidential Campaign. It is an activity that is not peculiar to the Internet nor is it a post-Internet invention. Nevertheless it is clear that the existence of the Internet, and the lodging of intellectual property materials in such a way that they can be accessed, often improperly, via the Internet, has had the effect that intellectual property theft now has a somewhat higher profile than may hitherto have been the case. In the US there have been accusations of systemic and extensive such theft by China,[129] proceedings before the courts have been commenced[130] and the phenomenon gave rise to discussions and an agreement between the then US President and the President of China.[131] Yet again, even supposing identifiable criminal activity can be proved against named persons, there are profound difficulties that at best impede and often preclude

[127] Consider the G20 decision to 'jointly fight cyber attacks on the global banking system' following the 2016 '$81 million heist of the Bangladesh central bank's account'; B. Koranyi, 'G20 to Jointly Fight Bank Sector Hacking', Reuters, 17 March 2017, available at http://uk.reuters.com/article/us-g20-germany-cyber-idUKKBN16O2F0.

[128] S. Xia, 'China Establishes Its First Cyber Court in Hangzhou: Thank You Alibaba', *China Law Blog*, 16 August 2017, available at www.chinalawblog.com/2017/08/china-establishes-its-first-cyber-court-in-hangzhou-thank-you-alibaba.html.

[129] J. W. Miller, 'US Steel Accuses China of Hacking', *Wall Street Journal*, 28 April 2016, available at www.wsj.com/articles/u-s-steel-accuses-china-of-hacking-1461859201.

[130] D. E. Sanger and N. Perlroth, '6 Chinese Men Indicted in Theft of Code from US Tech Companies', *New York Times*, 19 May 2015, available at www.nytimes.com/2015/05/20/technology/6-chinese-men-indicted-in-theft-of-code-from-us-tech-companies.html?_r=0.

[131] E. Rosenfeld, 'US-China Agree Not to Conduct Cybertheft of Intellectual Property', CNBC, 25 September 2015, available at www.cnbc.com/2015/09/25/us-china-agree-to-not-conduct-cybertheft-of-intellectual-property-white-house.html. See also US Department of Justice, 'Nine Iranians Charged With Conducting Massive Cyber Theft Campaign on Behalf of the Islamic Revolutionary Guard Corps', Department of Justice Office of Public Affairs, 23 March 2018, available at www.justice.gov/opa/pr/nine-iranians-charged-conducting-massive-cyber-theft-campaign-behalf-islamic-revolutionary.

the taking of legal action. Whether achieving mutual understandings between the relevant political leaders better achieves the desired goal of proper behaviour on both, indeed on all, sides is an open question the answer to which is also hard to verify.

So, with the scale and extent of 'questionable' cyber activity in mind, we will select for consideration in the next section one specific aspect of the peacetime cyber environment that is likely in the coming years to be of increasing importance, namely the 'Internet of Things'.

5.9 THE INTERNET OF THINGS

The Internet of Things is a rapidly developing feature associated with a wide range of novel consumer products. It raises a number of potential legal issues and of particular legal interest are those issues concerned with its likely implications for established notions of data protection and the right to privacy. This discussion starts with a brief description of the Internet of Things (IOT) sufficient only to ground the analysis of the relevant issues. The rules applicable in the European Union that regulate the processing of personal data are noted and European understandings as to the nature and scope of the human right to privacy are given. We will then address how the characteristics of the IOT may be expected to challenge or, as the case may be, to be challenged by the identified data protection and human rights rules and thereafter some conclusions will be suggested.

Before launching into a discussion of some of the legal issues raised by the IOT, we need to consider what exactly the notion means. Anyone researching the topic will readily discover that there are any number of definitions but for present purposes, rather than fix the analysis by reference to a hard and fast definitional text, it suffices to state descriptively that the IOT involves everyday physical objects such as fridges, mobile phones, medical equipment, washing machines, television sets, central heating controls, air conditioning controls or vehicle components. 'Smart' objects of this nature are integrated into the information sphere and it is possible to interact with them via the Internet and to exert control over them by that means. This process of interaction generates data and it is the security of, and potential for exploitation of, such data that raises the issues discussed in this and the next section.

This is what seems to be meant by this phenomenon. Perhaps the potential to use the IOT to generate adverse effects was most recently and clearly shown in the Dyn Corp attack. Briefly, in October 2016 much of the Internet in the US was reportedly brought down by a new weapon called the Mirai botnet. 'The victim was the servers of Dyn, a company that controls much of the internet's domain name system (DNS) infrastructure. It was hit on 21 October and remained under sustained assault for most of the day, bringing down sites including Twitter, the Guardian, Netflix, Reddit, CNN and many others in Europe and the US.' Interestingly, unlike previous distributed denial of service operations that have employed botnets based on computers and similar devices, 'the Mirai botnet is largely made up of so-called 'internet of things' devices such as digital cameras and DVR players.'[132] It is important to bear these circumstances of the Dyn attack in mind when we consider the importance of IOT security in the following section.

5.10 INTERNET OF THINGS DATA ISSUES

In October 2014 the UK Information Commissioner's office responded to a consultation paper put out by the UK communications regulator, Ofcom, addressing the promotion of investment and innovation in the IOT. The Information Commissioner (IC) discussed whether Internet-enabled devices produce personal data. The importance of that question stems from the EU-derived legislation in the UK that regulates the processing, a concept interpreted broadly, of personal data.[133] Basing himself on the idea that 'data protection law is concerned with information that identifies or is reasonably likely to identify an individual (directly or indirectly)', the IC concluded that while an organisation collecting and using information from, e.g., a smart phone is fully subject to data protection law[134] the position is less clear for less

[132] N. Woolf, 'DDoS Attack That Disrupted Internet Was Largest of Its Kind in History, Experts Say', *The Guardian*, 26 October 2016, available at www.theguardian.com/technology/2016/oct/26/ddos-attack-dyn-mirai-botnet.

[133] 'Processing' is defined for the purposes of the Regulation as 'any operation or set of operations which is performed on personal data or on sets of personal data, whether or not by automated means, such as collection, recording, organisation, structuring, storage, adaptation or alteration, retrieval, consultation, use, disclosure by transmission, dissemination or otherwise making available, alignment or combination, restriction, erasure or destruction;' European Union, General Data Protection Regulation, Regulation 2016/679, adopted 27 April 2016 and became effective on 25 May 2018 (GDPR), Article 4(2). As to the requirement in practice for service providers such as web operators to treat all IP addresses, including dynamic ones, as personal data see CCD-COE, T Minárik and A. Garcia, 'CJEU Determines Dynamic IP Addresses Can Be Personal Data But Can Also Be Processed for Operability purposes', reporting on *Patrick Breyer* v. *Bundesrepublik Deutschland*, CJEU C-582–14 dated 19 October 2016.

[134] Consider, for example, controversies associated with the apparent sharing of personal data by WhatsApp with its parent company Facebook; J. Fioretti, 'French Privacy Watchdog Raps WhatsApp over Facebook Data Sharing', Reuters, 18 December 2017, available at www.reuters.com/article/us-whatsapp-privacy-france/french-privacy-watchdog-raps-whatsapp-over-facebook-data-sharing-idUSKBN1EC285.

'personal' devices such as washing machines or TV sets that all members of a household use autonomously. He considered however, that only if an individual is identified, or is reasonably likely to be identified, will the information become personal data and thus subject to the processing restrictions. Importantly, the IC recognised that privacy risks from IOT devices would be likely to come from linking or matching data with other datasets.

The IC suggested that the limited physical interfaces that IOT devices are liable to have as compared, e.g., with conventional PCs may reduce the opportunity to notify consumers of data processing and to obtain consent. He therefore recommended the adoption of industry standards as to the fair, transparent and necessary collection and use of data. Matters are rendered rather more complex by the possible multiplicity of individuals and companies involved in handling the data.[135] The IC therefore recommends that privacy be considered at the design stage, by addressing data minimisation, data retention periods and data accuracy. 'Privacy by default' arrangements would, e.g., require certain data processes to be activated by the user. If adequate software updates are not put in place, there is the danger that the IOT device will become increasingly insecure, that the user's personal data could be put at risk and that an attacker could compromise the device.[136] Publicly available databases of such vulnerabilities already existed in 2014, so producing security updates is important. Also noted in the response document are the risks that limited interfaces pose for certain encrypted communications and that software lifecycles are potentially shorter than the lifespan of an IOT device.

In a recent press release, the Information Commissioner's Office reported on the findings of a study that 6 in 10 IOT devices do not sufficiently inform their customers how their personal information is being collected, used and disclosed. The ICO Head of Enforcement reportedly said: 'Companies making these devices need to be clear how they're protecting customers. We would encourage companies to properly consider the privacy impact on individuals before they go to market with their

[135] IC Response document, page 4 of 8.

[136] See, e.g., M. Burgess, 'Wikileaks Drops "Dark Matter": part two of its Vault 7 CIA leaks', *Wired*, 24 March 2017, available at www.wired.co.uk/article/cia-files-wikileaks-vault-7. Consider the comments of Neal Ziring on the importance of the cyber-security industry sharing information on vulnerabilities, emphasising the need for a shared public–private framework, e.g., to roll out patches at machine speed to individual researchers, industry and government, i.e. there is a need to establish the infrastructure, practices and standards for shared responses; C. Bing, 'NSA Technical Director: Sharing Hacker Information Isn't Enough, We Need a Shared Response', 30 March 2017, *Cyberscoop*, available at www.cyberscoop.com/nsa-technical-director-sharing-hacker-information-isnt-enough-need-shared-response/.

product and services.'[137] The UK Department for Digital, Culture, Media and Sport has recently published a report setting forth proposals to improve security associated with the IOT. The report identifies two main risks arising from the IOT, namely: '(1) consumer security, privacy and safety is [sic] being undermined by the vulnerability of individual devices; and (2) the wider economy faces an increasing threat of large-scale cyber attacks launched from large volumes of insecure IOT devices.' The report suggests that these risks need to be addressed through joint government and industry action, that this must happen urgently and that this is important 'because the Government has a duty of care to UK citizens to help ensure that they can access and use the Internet safely.' Interestingly, the report proposes a move away from focus on consumers protecting themselves by securely configuring their own devices and towards ensuring that strong security is built in by design. Accordingly the report proposes a draft Code of Practice directed at manufacturers of consumer IOT products and providers of associated services.[138]

The current regulation of these matters at European level is to be found in the General Data Protection Regulation.[139] This is a complex and lengthy piece of European legislation. Core to the regulation are the data protection principles which, in summary form, involve requirements that personal data processing be lawful, transparent and fair; that it be collected for an explicit and lawful purpose and that it should not be further processed for an inconsistent purpose; that the data be adequate, relevant and limited to what is necessary; that personal data be accurate, kept up to date and that reasonable steps be taken to ensure that inaccurate personal data is rectified without delay; that the data is kept in a form that enables the identification of data subjects for no longer than the purpose for which they were processed requires; and that the personal data are processed in a manner that ensures appropriate security of the personal data.[140]

[137] ICO Press Release 'Privacy regulators study finds Internet of Things shortfalls', 22 September 2016.

[138] UK Department for Digital, Culture, Media and Sport, Secure by Design: Improving the cyber security of consumer Internet of Things Report, available at www.gov.uk/government/uploads/system/uploads/attachment_data/file/686089/Secure_by_Design_Report_.pdf.

[139] Regulation (EU) 2016/679 of 27 April 2016, available at http://eur-lex.europa.eu/legal-content/EN/TXT/PDF/?uri=CELEX:32016R0679&from=EN.

[140] 2016/679, Article 5. It should be recalled that the European Court of Justice has found that, confronted with data on the Internet that appears unfair, one-sided or wrong, an individual has the so-called 'right to be forgotten', i.e. the right, e.g., for Google to delete information that is inadequate, irrelevant or no longer relevant; A. Travis and C. Arthur, 'EU Court Backs Right To Be Forgotten: Google Must Amend Results on Request', *The Guardian*, 13 May

Personal data is any data relating to an identified or identifiable natural person and the processing of such data includes any operation or set of operations performed on personal data, by automated means or otherwise, including collecting it, organising it, recording it, structuring it, storing it, adapting it, altering it, ordering it, retrieving it, consulting it, using it, disclosing it, disseminating it, making it available, combining it, erasing it or destroying it.[141]

While there are other grounds on which the processing of personal data may be lawful, arguably the most important such ground is that the data subject, i.e. the person referred to in the data, consents to the particular processing.

Special rules apply to the processing of personal data that reveals racial or ethnic origins, political opinions, religious or philosophical beliefs, trade union membership, genetic data, biometric data, and data relating to health or sex life or sexual orientation. Processing of such data is prohibited by Article 9, but there are several exceptions to this ban. The exceptions include explicit consent of the data subject (if the State's law allows such consent), necessary processing for the data controller's employment law obligations and rights, protection of the vital interests of the data subject or another person where the data subject is incapable of giving his consent, very limited processing by certain non-profit organisations for limited purposes and with specified safeguards, or that the processing relates to data manifestly made public by the data subject or is necessary to establish, exercise or defend legal claims. The Article provides for limited medical and public interest-related exemptions.[142]

The Regulation clarifies the obligations arising from the requirement for transparency, and makes detailed provision as to, e.g., the information that must be provided when data is being collected from a data subject, the

2014, available at www.theguardian.com/technology/2014/may/13/right-to-be-forgotten-eu-court-google-search-results.

[141] 2016/679, Article 4(1) and (2).

[142] At the time of writing, controversy surrounds suggestions that there may have been transfers of personal data relating to very large numbers of people between Facebook and a company called Cambridge Analytica; H. Osborne and J. Elgot, 'MPs Summon Mark Zuckerberg and Accuse Facebook of Misleading Them', *The Guardian*, 20 March 2018, available at www.theguardian.com/uk-news/2018/mar/20/officials-seek-warrant-to-enter-cambridge-analytica-hq. Consider also concerns expressed in response to Facebook's plans to introduce opt-in arrangements for the use of facial recognition technology in the European Union and Canada to enable Facebook users to be identified from photographs and videos. The system would reportedly seem to be intended to facilitate the application of name tags to images, to detect when a scammer is attempting to use a stolen photo for their profile and to enable new 'friends' suggestions to be offered; L. Kelion, 'Facebook Seeks Facial Recognition Consent in EU and Canada', BBC, 18 April 2018, available at www.bbc.com/news/technology-43797128.

information to be given when personal data is not obtained from a data subject, the right of the data subject to confirmation of whether personal data relating to him is being processed and, if so, his right to have access thereto and rights of rectification, erasure and to restrict the processing of personal data.

These are only a few of the very many aspects of personal data protection that are addressed in this comprehensive regulation. Interestingly, for example, Article 22(1) gives the data subject the right not to be subject to a decision based solely on automated processing which produces legal effects that significantly affect him.

5.11 EUROPEAN INTERPRETATION OF THE HUMAN RIGHT TO PRIVACY

While it is conceivable that other human rights may be relevant to the evolving IoT, attention in the context of the issues being discussed in the present section naturally centres on the right to a private life. As was noted earlier, Article 8(1) of the European Convention on Human Rights provides that 'everyone has the right to respect for his private and family life, his home and his correspondence'. Paragraph 2 of that Article then limits the circumstances in which a public authority can legitimately interfere with that right. It is therefore clear that there is an interplay between Article 8 rights and data protection, with notions of physical and psychological integrity and personal autonomy being recognised by the court as part of, or underlying, private life.[143] Clearly a house, flat or other dwelling where an individual, whether as part of a family or otherwise, resides would come within the idea of 'home'. 'Correspondence' will include developments in modern communications technology, such as email, Twitter, Facebook, Snapchat and other social media.

Delving a little deeper, a State has a positive obligation to put in place a legislative and administrative framework to ensure the full realisation of Article 8 rights and this may involve adopting measures to secure respect for private life even in the relations between individuals.[144] So this potentially translates into an obligation on the State to ensure individuals comply with the law and to put in place an appropriate framework.

[143] D. J. Harris, M. O'Boyle, E. P. Bates and C. M. Buckley, *Law of the European Convention on Human Rights* 3rd edn (Oxford: Oxford University Press, 2014), page 526.

[144] Ibid., page 533 citing *X and Y* v. *Netherlands*, A91(1985); 8 EHRR 235, para 23.

However, it should be borne in mind that human rights law is all about the relationship between the individual and the State or related authorities. That is why, on the assumption that misuse of IOT data is attributable to a private undertaking as opposed to a State or its agencies, human rights issues are only likely to arise if the State has failed to adopt appropriate measures to secure respect for the individual's private life. Accordingly, from a human rights law perspective a claim seems only likely to arise if there has been a failure by the State to take those appropriate measures and if that failure can be shown to have had adverse consequences for the individual's private life. Again, it must be appreciated that this is a heavily abbreviated summary of the most apparently relevant parts of the law.

On 20 October 2016 the European Data Protection Supervisor discussed the 'personal information management system' concept in which the current provider-centric system is transformed into a system centred on individuals able to manage and control their online identity. 'In principle, individuals should be able to decide whether and with whom to share their personal information, for what purposes, for how long, and to keep track of them and decide to take them back when so wished.'[145] The Opinion comments 'keeping control of personal data in the era of the Internet of Things and Big Data cannot succeed without an automated and reliable, yet controlled, enforcement of data protection rules. We believe this is one of the critical areas where research and investment effort should focus.'[146]

It is evident from this discussion of the IOT that there is the potential for data protection and human rights issues to arise from the processing of data associated with the use by private citizens of IOT-linked smart objects. There are reportedly security vulnerabilities associated with such IOT linkages, and it has been suggested that manufacturers have been paying too little attention to the need to ensure on-going patching of such vulnerabilities.[147] While private

[145] European Data Protection Supervisor, Opinion 9/2016, EDPS Opinion on Personal Information Management Systems: Towards more user empowerment in managing and processing personal data, 20 October 2016, pages 5 and 6, available at https://secure.edps.europa.eu/EDPSWEB/webdav/site/mySite/shared/Documents/Consultation/Opinions/2016/16-10-20_PIMS_opinion_EN.pdf.

[146] Opinion 9/2016, n.145, at page 10. Note a US legislation initiative reported at H. Stark, 'A Bipartisan Bill To Strengthen Cyber Security for the Internet of Things', 20 August 2017, available at www.forbes.com/sites/haroldstark/2017/08/20/a-bipartisan-bill-to-strengthen-cybersecurity-for-the-internet-of-things/#3f51e1a95a5f.

[147] For information as to the market in such vulnerabilities, see, e.g., P. N. Stockton and N. Golabek-Goldman, 'Curbing the Market for Cyber Weapons', 32 *Yale Law and Policy Review* 239, (2013) 248 and A. C. Emery, 'Zero-Day Responsibility: The Benefits of a Safe Harbor for Cyber Security Research', 57 *Jurimetrics* 4, Summer 2017, 483, 488–489; for a

citizens demonstrate a frequent willingness to share often intimate personal details via social media, the insecure processing of what might in time come to be regarded as personal data derived from the IOT raises the possibility of challenges by customers of manufacturers under data protection and perhaps contractual provisions, and of governments if it can be shown that State authorities have not taken adequate steps, e.g., to protect relevant human rights.

5.12 DOMESTIC LAW SYSTEMS IN RUSSIA, CHINA AND INDIA

In this penultimate section, we shall broaden the discussion by briefly referring to certain aspects of the domestic law regulation of relevant cyber activities in certain other countries, namely Russia, China and India. Available space does not permit an extensive treatment of this element of the topic.

Data protection laws in Russia bind all data operators and third parties acting under their authorisation and, subject to certain exemptions, regulate all personal data processed by such persons. Personal data is considered to be any information directly or indirectly related to an identifiable individual. All acts of data processing are covered and the legislation addresses electronic and manual personal data records. The rules seem to apply to data processing in or targeted at Russia and to collection, storage and use of personal data of Russian citizens. Roskomnadzor must be notified before personal data processing takes place giving specified information and prescribed procedures follow the making of such a notification. There are particular provisions, e.g., as to categories of personal data, the purposes and duration of processing, data subject consent and the appointment of data protection officers. Specific,

discussion of approaches to regulating the zero-day market, including criminal law, applying liability to software vendors and export control regulation, see Emery ibid., 489–496. Emery concludes that a balance needs to be struck between maximizing safety and encouraging zero-day vulnerability research and proposes ex post liability for sellers who do not perform reasonable pre-sale due diligence, ibid., 502. As to reward programmes for hackers that disclose such vulnerabilities, see for example Google's Vulnerability Reward Program Rules, available at www.google.com/about/appsecurity/reward-program/. Note, however, a report suggesting that in November 2017 China had a process for evaluating whether high threat vulnerabilities had utility in intelligence operations before publishing them to China's National Vulnerability Database. Research suggested that records of the dates when some vulnerabilities were published were altered on the database, the inference being that these were the vulnerabilities being considered by the Ministry of State Security for use in offensive cyber operations; P. Moriuchi and W. Ladd, 'China Altered Public Vulnerability Data to Conceal MSS Influence, Recorded Future', 9 March 2018, available at www.recordedfuture.com/chinese-vulnerability-data-altered/.

informed and wilful consent of the data subject is usually required before personal data can be processed.[148]

Russian laws on cyber security matters, outside areas that are heavily regulated such as state secrets and personal data, are generally considered to be sparse.[149] Personal data laws do, however, address organizational, legal, technical and physical safeguards for personal data. Reportedly the Russian privacy regulator, Roskomnadzor, issued an order on 22 August 2017 requiring companies to notify it in advance of personal data processing, including 'information of safeguards in place to prevent data breaches and whether the company intends to transfer data outside Russia' and if so, where. 'Companies must also confirm their compliance with Russia's data localization law, which requires that companies processing data of Russian citizens store that data on servers located within Russia.'[150]

The relationship between personal messaging services and State authorities in Russia has recently been the subject of a decision of the Russian Supreme Court which rejected an appeal by such a messaging service, Telegram, against a decision by Roskomnadzor. The regulator had ordered Telegram to provide the Federal Security Service access to users' messaging data.[151]

China's Cyber Security Law came into effect on 1 June 2017 and reflects the Chinese goal of having jurisdictional control of content on the Internet in turn reflecting the Chinese view that within Chinese territory the Internet is under the sovereignty of China. The Cybersecurity Law applies to network owners, managers and providers and networks are computers and related equipment that gather, store, transmit, exchange or process information. The law applies to critical sectors, such as businesses involved in communications, information services, energy, transport, water, financial and public services and government electronic services. These network operators must

[148] This para is based on the account given in S Medvedev, Data Protection in Russian Federation: Overview, Thomson Reuters, 2017, available at https://uk.practicallaw.thomsonreuters.com/2–502-2227?transitionType=Default&contextData=(sc.Default)&firstPage=true where further details are given on these matters and, e.g., in relation to special rules, the rights of individuals, security requirements, processing by third parties, data transfer and enforcement.

[149] *The Privacy, Data Protection and Cyber Security Law Review*, Edition 3, Russia dated November 2016, available at http://thelawreviews.co.uk/edition/the-privacy-data-protection-and-cybersecurity-law-review-edition-3/1140166/russia.

[150] Hunton and Williams, 'New Data Processing Requirements Take Effect in Russia', *Hunton Privacy Blog*, 29 August 2017, available at www.huntonprivacyblog.com/2017/08/29/new-data-processing-notice-requirements-take-effect-russia/.

[151] Radio Free Europe/Radio Liberty, 'Telegram App Under the Gun in Russia After Court Ruling', 20 March 2018, available at www.rferl.org/a/russia-telegram-messaging-app-under-pressure-fsb-supreme-court/29111309.html.

cooperate with Chinese crime or security investigators and must inter alia allow full access to data on request.

There are requirements as to internal security management systems and network security protections, as to measures to prevent viruses and other forms of cyber attack, as to network safety monitoring and recording, back-ups and encryption. Network operators in critical sectors must store in China data that is gathered or produced there and information and data on Chinese citizens gathered in China must be kept on domestic servers and must not be transferred abroad without permission. There are requirements for spot checks and verifications, but concerns have been expressed about the vagueness or undefined character of some of the core elements in the law.[152] Moreover, in China new rules have been issued requiring the administrators of online groups properly to administer their groups and online information, to verify the identities of group members and to take measures to protect group members' personal information. Similar requirements apply to social media accounts that release information to the public.[153]

Where India is concerned, the Information Technology Act 2000[154] makes extensive provision. After defining relevant terms in Section 2, Sections 3 and 3A address the use of digital and electronic signatures and the authentication of electronic records. The recognition and use of electronic signatures and the retention, attribution, acknowledgment and despatch of electronic records are provided for in succeeding sections. Chapter IX of the Act provides for penalties and compensation if a person without the permission of the owner or person in charge of a computer performs certain acts, including: accesses the computer; downloads, copies or extracts data; introduces a contaminant, a notion defined in the legislation, or virus into the computer; damages the computer, data or other programmes on the computer; disrupts the computer or denies access of an authorised person to the computer.[155]

[152] As to the above, and for a more detailed critique of the law and an assessment of its implications, see, e.g., J. Wagner, 'China's Cybersecurity Law: What You Need to Know', *The Diplomat*, 1 June 2017, available at http://thediplomat.com/2017/06/chinas-cybersecurity-law-what-you-need-to-know/. Note that Apple Inc will be storing Chinese users' iCloud accounts in a Chinese data centre to comply with Chinese law; S. Nelliss and C. Cadell, 'Apple Moves To Store Icloud Keys In China, Raising Human Rights Fears', Reuters, 24 February 2018, available at www.reuters.com/article/us-china-apple-icloud-insight/apple-moves-to-store-icloud-keys-in-china-raising-human-rights-fears-idUSKCN1G8060.

[153] The regulations took effect on 8 October 2017; M. Xuequan, 'China asks online group, WeChat account admins to take responsibility', Xinhuanet, 19 September 1917, available at http://news.xinhuanet.com/english/2017-09/08/c_136592568.htm.

[154] Act 21 of 2000.

[155] Section 43 of the Act.

Where data protection is concerned, Section 43A, inserted into the Act in 2009, stipulates that '[w]here a body corporate, possessing, dealing or handling any sensitive personal data or information in a computer resource which it owns, controls or operates, is negligent in implementing and maintaining reasonable security practices and procedures and thereby causes wrongful loss or wrongful gain to any person, such body corporate shall be liable to pay damages by way of compensation to the person so affected.'[156] Importantly, 'sensitive personal data or information' is defined by the Act to mean such personal information as may be prescribed by the Central Government in consultation with such professional bodies or associations as it may deem fit. It should also be noted that a Committee of Experts has been appointed to deliberate on a system of data protection for India. The committee is to identify key data protection issues and to recommend methods of addressing them.[157]

Criminal offences relating to computing activities are provided for in Chapter XI of the Act. They include tampering with computer source documents,[158] sending offensive messages through computer resources or communication devices,[159] identity theft,[160] cheating by personation[161] and[162] the violation of privacy. So, for example, the offence of identity theft is committed if a person fraudulently or dishonestly makes use of the electronic signature, password or any other unique identification feature of any other person. Cyber terrorism and the publication or transmission of obscene material in electronic form are criminalised by Sections 66 and 67 respectively. The interception, monitoring or decryption authorised by or on behalf of the Indian Government or the government of an Indian state of information generated, transmitted, received or stored in a computer device is provided for in Section 69. Prescribed procedures and safeguards are provided for in the Act and persons in charge of the computer device must provide or secure access to the relevant computer resource, must intercept, monitor or decrypt the

[156] Note the Information Technology (Reasonable security practices and procedures and sensitive personal data or information) Rules, 2011 make additional and supplementary provision.

[157] Ministry of Electronics and Information Technology Office Memorandum no. 3(6) 2017-CLES dated 31 July 2017.

[158] Section 65 of the Act.

[159] Section 66A of the Act. See the case of *S. Singhal* v. *Union of India*, Supreme Court of India Criminal Case Number 167 of 2012 in which the constitutionality of this provision was contested and in which the Supreme Court struck down the section in its entirety as being violative of Article 19(1) of the Indian Constitution.

[160] Section 66C of the Act.

[161] Section 66D of the Act.

[162] Section 66E of the Act.

information as the case may be or must provide information stored in the computer resource.[163]

5.13 CONCLUSION

So what can each of the processes that we have discussed in this chapter teach the others? The Tallinn Manuals process can certainly demonstrate that concentrated effort by experts with a similar view of the world can produce agreed and detailed provisions relatively rapidly, but these provisions will not necessarily attract global acceptance, at least in the short term. The UN GGE process can demonstrate that it is only by including experts from a representative selection of States that an inevitably less detailed, more generalised set of provisions can be achieved. It will then be those that consider that extant law goes further than the GGE experts have acknowledged that will be dissatisfied. Curiously, therefore, neither process generates unanimity of view in the short term. Arguably, it is the kind of process associated with the Tallinn Manuals that will set a more prescriptive legal agenda for the global community to, perhaps gradually, move towards.

Alternatively, or perhaps contemporaneously, voluntary codes of best practice may enable States to achieve more granular provision if they feel more comfortable with a process that is not binding at the legal level. Such approaches may also have the advantage of inherent flexibility, a quality that will likely prove important as technology in this area is likely to develop rapidly and not always in predictable directions. Mutual suspicions between major actors would, however, need to be set aside if meaningful progress is to be achieved by such a process, indeed by any collective process.

The need to reach a balance between notions of free speech, privacy and security on the Internet and the challenges associated with the Internet of Things both demonstrate the centrality of human rights considerations in the peacetime application of these technologies. Balancing conflicting, or potentially conflicting, interests involves according appropriate weight to each and reaching a sensible compromise between them. It probably, for example, requires acceptance that absolute security is not realistically achievable but that absolute privacy poses unacceptable risks for the community as a whole. Likewise in the military context, absolute certainty that there will be no civilian casualties or damage is often unachievable, yet commanders must continue to strive to achieve that goal.

[163] Section 69(3) of the Act.

Hard law at the international level in respect of cyber conflict may, certainly in the short to medium term, prove to be hard to accomplish, and may even be undesirable as involving too much inflexibility. Hard law is, however, the route that domestic jurisdictions are increasingly pursuing to deal with cyber crime and to address data protection and associated issues. Perhaps this difference rather demonstrates the core distinction between the objectives of international and domestic law in this respect. Domestic law is seeking to regulate, to demonstrate which activities are and respectively are not acceptable to a particular society. International law, reflecting as it must the consensus that States with differing approaches manage to accomplish, is always going to be more difficult to iron out.

This chapter has shown that the process of agreeing international arrangements for the regulation of cyber activity has proved anything but straightforward. Like-minded processes, whether in the form of the initiatives that led to the preparation of both Tallinn Manuals or in the form of the draft International Code of Conduct, can achieve agreed language relatively rapidly. The Tallinn Manuals approach involves the application of existing law principles and rules, adapted where necessary, to the cyber domain. Likewise, EU regulations on data protection can as we have seen be applied to this novel phenomenon. One of the major challenges, however, is to understand in which direction the evolving cyber technology will move, as only when that can be established is there any realistic hope of achieving legal arrangements that will adequately address that emerging reality.

It seems to be at the domestic law level that the need to achieve a proper balance between the security imperative to know about secretive terrorist-based activities and the citizen's rights to privacy must be established and maintained while the technology develops and while our understanding of the technology gradually becomes a little sharper. Perhaps it is because the State is liable to be challenged if it gets that balance wrong that you see detailed legislation at national level but relatively vague assertions coming out of the GGE process.

The Tallinn Manual propositions are based on logical and accurate transpositions of established international law into the cyber context, and a good number of its rules seem destined to achieve general acceptance in due course. Nevertheless Russia and China continue to voice opposition and the GGE reports may be seen by some observers as the more reliable indicator of the low level of contemporary customary law provision addressing these matters. This chapter has put forward a strong argument in favour of the customary status of many of the principles and rules reflected in the Tallinn Manual. Moreover, it seems likely that the compelling legal logic of the

Tallinn texts will, in the longer term, secure for them a more general level of acceptance. In the meantime, however, the detailed legal rules of cyber warfare will remain a matter of expert opinion while the legal rules relating to peacetime cyber activities will be governed by domestic law written in as much detail as the relevant State thinks it requires. As the scale, scope and understanding of online threats increase, society will face the increasingly complex challenge of achieving a proper balance between the right to personal privacy and the requirements of public security. In both the warfare and peacetime contexts, difficult public policy issues must be confronted if widely accepted legal arrangements for the regulation of cyber activities are to be achieved.

Before closing this chapter it would be remiss to fail to mention yet another balance between conflicting, or potentially conflicting, interests that arguably has yet to be properly worked out. We have already discussed difficult balances, paradoxes and trade-offs, e.g., between free speech and privacy and security on the Internet, between individual privacy and collective security and between profitability in the manufacture of smart appliances and the need for a long-term commitment to patch vulnerabilities. These are some of the areas in which what can perhaps best be described as friction between the Internet companies on the one hand and regulatory authorities, legislators, national Information Commissioners and governments on the other seem likely to become apparent.

However, another aspect of this cyber topic that has not been specifically discussed in these pages but that seems destined to be the subject of continuing attention is the more general need to strike an appropriate balance between the power of the major Internet companies, such as Google, Microsoft, Facebook, Amazon and others, and the authorities whose task it is to keep members of society safe, to protect our personal data and our privacy, to ensure fair taxation and to achieve a situation in which the quest for profit, for market share and even for political power is not at the expense of the legitimate rights, needs and interests of private citizens. As we have seen, the domestic law and policy of States will be the venues in which States will usually seek to achieve this balance and in which they will articulate these protections. International organisations such as the European Union may over time develop policies, directives, regulations or other instruments addressing these issues but the focus for legal and policy provision and practical enforcement is likely to rest with national authorities.

Events will determine whether hard domestic law provision is generally regarded as necessary. Indications at the time of submission of the present manuscript are that best practice arrangements or self-regulation by these companies may not achieve the level of popular assurance that is perceived as being necessary, particularly in a democratic society, and yet small States are likely to be ill equipped to take on the huge, powerful Internet companies. Initiatives of a collective nature may therefore prove to be the more attractive option. Perhaps collective action involving groups of States working with the Internet companies will provide the required outcomes; time will tell.

It is evident, however, that even powerful Internet companies have their Achilles heels. Reputational damage can erode the perceived worth of such companies. Customer bases that can rapidly expand in this novel technological field might be capable of diminishing with similar speed. If the confidence of customers, on which any company selling a service must ultimately depend, is eroded, that company's prospects are bound to suffer. So, while regulation may well be part of the solution to these issues, other approaches may be similarly, or perhaps more, effective and thus may ultimately need to be factored in.

It seems likely that events will drive public opinion, and that public opinion will influence, but may not necessarily determine, whether legislators will act and, if so, how. At some point it seems likely that society will have to ask itself at what point do we conclude that the soft touch has not worked with reference to the Internet companies, and that a selection of measures, perhaps including regulation and perhaps on a collective basis, needs to be developed and enforced?

6

Highly Automated and Autonomous Technologies

WILLIAM H. BOOTHBY

6.1 INTRODUCTION

The present chapter will follow the same analytical approach as that adopted in Chapter 5. It begins by considering what the relevant technologies consist of, then we will look at some aspects of the law that regulates the employment of automated and autonomous weapon systems. Thereafter the chapter will consider initiatives for the development of the law relating to such weapons. The chapter will go on to address the use of automated and autonomous technologies in certain domestic, peacetime environments and some of the associated issues that arise, or that seem likely to arise, from such use. The concluding section of the chapter tries to draw the discussion together and to consider some tentative conclusions.

So the opening sections of the chapter will address the application of autonomous and highly automated technologies to weapons. Chapter 2 explained that the law of armed conflict includes prohibitions and restrictions relating to mines, booby-traps and other devices. The present chapter does not address such weapon systems and references below to automated or autonomous weapon systems should be interpreted accordingly. Rather, the focus of the present discussion is, for example, on weapon platforms that employ autonomous or highly automated technologies as those descriptors are explained below. It was also noted in Chapter 2 that it is by reference to the existing weapons-law rules that the lawfulness of new technology weapons must be determined. Accordingly, this chapter will assess how those existing rules apply to highly automated and autonomous weapon systems. However, before we start to apply the law, it is important to understand the terminology and some of the technologies with which we are concerned. That is the purpose of the next section of this chapter.

6.2 THE TECHNOLOGIES AND WHAT THE TERMS MEAN

6.2.1 *Man on the Loop Systems*

A weapon system is treated in this chapter as a 'man on the loop' system if it is capable of highly automated or autonomous operation, discussed below, but is supervised by an operator who has the capability to intervene and over-ride an attack decision that the highly automated or autonomous decision-making process has made. The system will include an 'up-link' that communicates *inter alia* instructions to the platform[1] and a down-link that conveys, again *inter alia*, data from on-board sensors to the operator who is monitoring the platform's activities. The operator is not positioned, either physically or structurally, within the decision-making loop facilitated by the up- and down-links and the weapon system may be capable of undertaking its own attack decisions. The 'man on the loop' element consists of the presence of a human operator who, while not involved in the autonomous or highly automated process whereby the weapon system makes its attack decision, is nevertheless in a position to observe the decisions being made and the action being taken by the weapon system and to intervene and countermand any decision or action which seems likely to lead to an unlawful or undesirable attack. In short, an autonomous or highly automated weapon system can become a 'man on the loop' weapon system if, notwithstanding the ability of the machine to operate in an autonomous or highly automated way, a person is nevertheless so positioned as to be able to intervene and countermand the machine's decisions should the need arise.

A 'man on the loop' platform might be used to undertake a variety of military tasks, which may but will not necessarily include attack. However, if its uses are limited to, for example, the gathering of information or the undertaking of reconnaissance missions, it is unlikely to raise issues under the law of targeting. This is because obtaining timely, accurate information as to the situation in an area in which an attack is being considered, planned or undertaken is likely to promote compliance with targeting law rules, not detract from such compliance.

Any assessment of the legal acceptability of such a 'man on the loop' weapon system must start by applying the criteria that are routinely the focus of weapon reviews and that were explained in Chapter 2. However, the superfluous injury/unnecessary suffering principle and the customary and

[1] The term 'platform' is used to refer to the aircraft, vessel, sub-surface craft or land vehicle the operation of which is either highly automated or autonomous.

treaty-based environmental protection rules[2] are unlikely to be relevant to the legal acceptability of the 'man on the loop' aspect of such a weapon system that is to be used to undertake attacks. They will, however, be of relevance, of course, if the legal review also encompasses the missile, bomb or other projectile or munition that the weapon system employs. There is no ad hoc IHL rule prohibiting or limiting the use of such 'man on the loop' platform technologies. Furthermore, such systems will generally be equipped with sensors and associated systems that are specifically designed to limit attacks to lawful targets, so such a man on the loop platform system is most unlikely to be indiscriminate by nature.

The remaining important issue is whether a particular weaponised 'man on the loop' platform system is capable of being used in compliance with targeting law rules, including the distinction and discrimination principles, the proportionality rule and the rules as to precautions in attack.[3] This is not one of the standard criteria for legal review as outlined in Chapter 2. However, and as mentioned in that chapter, when the weapon system is capable of highly automated or autonomous operation, it becomes vitally important to establish that the planned manner of using the weapon system enables the principles and rules of targeting law to be complied with. Satisfying this additional weapon review criterion in the case of a 'man on the loop' system presupposes that the operator has the practical capability to intervene and countermand unsatisfactory machine-made attack decisions, that that operator has sufficient visibility of what the machine is deciding, that he has sufficient time to be able to intervene if the need arises, and that his intervention is capable of stopping an unsatisfactory attack. The workload of the relevant individual must also be such that he is able to exercise proper supervision of the man on the loop platform's attack decision-making and that he will intervene when it is appropriate to do so.

If an autonomous weapon system were to be capable of being used in such a way that the judgments, decisions and actions required by targeting law are made and taken by the weapon system itself, the law contains no over-riding requirement that a person be either controlling the platform or that he or she be 'on the loop' in the sense in which that term is used in the present discussion. Rather, the capacity of a person 'on the loop' to cancel an automated or autonomous attack decision, if it is necessary to do so, has the effect that the decisions and actions that the law of targeting

2 The superfluous injury principle and the environmental protection rules were discussed in Chapter 2.

3 API, Articles 48, 51(1), 52(1), 52(5)(b) and 57.

requires can in principle be made and taken by that person. That person, or the persons, 'on the loop' therefore become the agency for compliance with targeting law.

Such an intervention to countermand a 'machine-made' decision may become necessary for a variety of reasons. If it becomes clear that the machine-made attack decision involves directing an attack at a civilian taking no part in the hostilities, or at a civilian object, the man on the loop would be required to intervene and stop the attack.[4] Likewise, if the attack that the machine has decided to undertake would be expected to cause injury or death to civilians or damage to civilian objects, or a combination of these things, that would be excessive in relation to the concrete and direct military advantage that is anticipated to flow from the attack, the operator on the loop would be obliged to intervene to stop the attack.[5] Such intervention will also be necessary, for example, if the object of the machine's attack decision is a person or object having special protection under the law of armed conflict,[6] or if the attack that the weapon system has decided upon would be contrary to the commander's intent. It should be stressed, therefore, that the tasking of the operator 'on the loop' may need to be such as to enable the required precautions to be taken with an acceptable degree of care.[7]

Finally, it should be appreciated that this analysis is based exclusively on existing applicable law.

6.3 SWARM TECHNOLOGY

A 'swarm', for the purposes of this chapter, can be described as a coordinated group of platforms of any size that is performing, or is intended to perform, military tasks in which the individual platforms may not be acting in formation. The reference to the word 'platform' indicates that swarm technology is capable of being applied in the land, air, maritime surface and sea subsurface

4 API, Articles 51(2), 52(1) and 57(2)(b).

5 API, Articles 51(5)(b) and 57(2)(b).

6 Consider, for example, the special protection of medical personnel and chaplains under Article 24 of GCI, of cultural property under the Convention for the Protection of Cultural Property in the Event of Armed Conflict, The Hague, 14 May 1954 (Hague Cultural Property Convention), of medical units under Articles 12 and 13 of API and of certain works and installations containing dangerous forces under Article 56 of API.

7 The word 'may' is used here because there may be other elements of the weapon system or of its method of operation that do enable particular precautions to be taken. The point remains, however, that if legal compliance relies on a man on the loop and if that person is over-tasked in whatever way, compliance is put at risk.

environments.[8] The air environment will, however, be referred to for illustrative purposes.

Swarms will frequently consist of numerous aircraft. The relative sizes of the specific aircraft in the swarm will not be relevant to the characterisation of the group as a swarm. Some swarms will operate on the basis that the individual members maintain a fixed formation among themselves, while other swarms may involve dissimilar movements. Some swarms may consist exclusively of remotely piloted aircraft, of highly automated systems or, as the case may be, of autonomous systems. Other swarms may comprise a mixture of these classes of platform. Irrespective of the manner of operation of the members of the swarm, in formation or as individuals performing dissimilar movements, sufficient coordination will be required to prevent collisions and other mutual interference. Furthermore, the manner of operation may determine whether human control arrangements are required and if so the form those arrangements need to take. At any event, the arrangements under which a swarm being used in attack operations is operated must be such as to ensure that targeting law rules, including the distinction and discrimination principles and the rules as to proportionality and precautions in attack, can be complied with.

Using a swarm to undertake reconnaissance, information gathering or other tasks that are not part of an attack is unlikely to raise concerns under the law of targeting. If a swarm is being used to undertake attacks, the number of aircraft in the swarm, the number, nature and circumstances of the targets that are to be attacked, the nature, quality and reliability of the up- and down-links to each aircraft and the manner of operating the swarm will be among the factors that determine how many controllers or operators may be required. If compliance with targeting law rules requires that there be human presence in or on the loop,[9] it will again be important to ensure that the relevant personnel are so located and tasked as to be able to provide effective control and to take the precautions that the law requires.

8 As to the law applicable to unmanned maritime systems, see W. Heintschel von Heinegg, 'Unmanned Maritime Systems: Does the Increasing Use of Naval Weapon Systems Present a Challenge for IHL?' In W. Heintschel von Heinegg, R. Frau and T. Singer (eds.), *Dehumanization of Warfare* (Springer, 2018), 119 and see Chapter 11.

9 If the technology incorporated into the swarm is such that the individual weapon systems are capable of making the judgements and decisions required by targeting law, the presence of a person or persons in or on the loop of elements of the swarm may not be required. It will be a question of fact whether the swarm does indeed have that technical capability and whether the judgements and decisions required by targeting law can thereby be made with an acceptable level of reliability. The legal requirement is that the judgements and decisions prescribed by targeting law are made.

If individual aircraft within a swarm have different functions, a number of operators may be required for each swarm or some members of the swarm may have to be able to perform their activities automatically or autonomously. Highly automated or autonomous attack operations by aircraft within a swarm will raise the same legal issues as are discussed below in respect of non-swarm automated or autonomous targeting. The mixing of autonomous, automated and remotely piloted aircraft in a swarm is also likely to raise complex airspace management issues, some but not necessarily all of which may be satisfactorily addressed by sophisticated 'see and avoid' technologies.

6.4 DEFINING HIGHLY AUTOMATED AND AUTONOMOUS WEAPONS

While the automation of some elements of attack decision-making is a feature of some existing weapon systems, significant contemporary research seeks to achieve completely autonomous decision-making in attack.[10] The terms that are central to this discussion, namely 'highly automated' and 'autonomous', are unfortunately not the subject of internationally recognised legal definitions. Interpretations of both terms will therefore be offered for the purposes of this chapter.

Starting with automation, a 'highly automated system is constrained by algorithms that determine its responses by imposing rules of engagement and setting mission parameters which limit its ability to act independently'.[11] Such a system is not remotely controlled and functions in a self-contained and independent manner once it has been deployed. A highly automated weapon system independently detects and then verifies a predetermined type of target and then fires or detonates a weapon. Automation is not new, and has for example been employed in the past in mines and booby-traps.[12] The more that

[10] United States Department of Defense Unmanned Systems Integrated Roadmap, FY 2011–2036, 11-S-3613, at page vi; some systems with autonomous features are already in use, such as the maritime Phalanx system in service with the Royal Navy and described at www.royalnavy.mod.uk/The-Fleet/Ships/Weapons-Systems/Phalanx; the United States Navy MK 15 - Phalanx Close-In Weapons System, described at www.navy.mil/navydata/fact_display.asp?cid=2100&tid=487&ct=2; the Russian Arena-E Active Protection System; the Mutual Active Protection System; the Diehl BGT Mutual Active Protection System described at www.defense-update.com/20110112_maps.html; South Korean border security robots; and the Israel Aircraft Industries Harpy autonomous anti-radar SEAD system referred to at www.iai.co.il/2013/16143-16153-en/IAI.aspx.

[11] See UK Development, Doctrine and Concepts Centre, UK Air and Space Doctrine JDP 0–30, July 2013, para 215.

[12] DoD *Law of War Manual*, June 2015, para 6.5.9.1. Consider Distributed Denial of Service cyber operations and certain computer worms as an example of automated, as opposed to

the scope for such decision-making is constrained by the way in which the weapon system is set up, i.e. the more prescriptive that the pre-set parameters are, the more likely it is that the system will be categorised as highly automated rather than autonomous in nature.[13]

A weapon system may have diverse automated functions, e.g., platform or munition navigation, the coordination or fusion of sensor data, functions associated with the fusing, the locking on by an air to air missile to a target aircraft and so on. The 'highly automated' descriptor as employed here shows that weapon systems so designated apply automation specifically to both target selection and engagement. Certain weapon systems already in use incorporate these highly automated features but do not come within the definition of autonomous systems (as to which see below).

Autonomous systems have been described in terms of understanding higher-level intent and direction. 'From this understanding and its perception of its environment, such a system is able to [take] appropriate action to bring about a desired state'.[14] Whether notions of understanding can properly be ascribed to what will essentially remain mechanical processes is highly debatable. Some important characteristics of autonomous weapon systems are that they independently identify and decide to engage targets of their choosing but any definition must refer to the single most important defining characteristic of an autonomous weapon, namely its application of artificial reasoning as part of the decision-making process. So an appropriate definition for the purposes of the present discussion is a weapon that is programmed to apply reasoning[15] to determine whether an object or person is a target and whether it should be attacked. They are not programmed to target a specified object or

autonomous, operations; S. Stadlmeier, 'CPU and Keyboard: Weapons of Mass Disruption?', in W. Heintschel von Heinegg, et al. (eds.), *Dehumanization of Warfare*,147, 150 and 158.

[13] Note G. Heppner and R. Gillmann, 'Autonomy of Mobile Robots', in W. Heintschel von Heinegg, et al., (eds.), *Dehumanization of Warfare*, 77, 80–81.

[14] JDP 0–30, n 11, para Lexicon-5; the UK view is that autonomous systems are self-governing and set their own rules, that this is neither welcome nor useful in the military context and the UK is committed to maintaining human oversight over weapons release decisions; ibid., para 215. Note the suggestion by Lyria Moses that 'even when machines undertake highly complex tasks, if machines produce the reasoning employed in reaching their outcome. . .[h]uman decision makers can then scrutinize the logic of the reasoning rather than the outcome itself'; Comments attributed to L. B. Moses, 'Report of the Symposium on Ethical, Legal and Social Implications of Emerging Military Technologies, 29–30 May 2017', *Melbourne Law School*, August 2017, 5.

[15] An alternative way of expressing this thought is that the system is programmed to apply evaluative analysis to make the relevant determinations. The references hereafter to 'reasoning' should be construed as including this idea of evaluative analysis.

person. It is the software that reasons and then decides which target to prosecute, how and when.

The programming aspect of the definition shows that the system's software must have been so engineered as to enable, perhaps require, the system to apply its reasoning to the facts it detects or that are communicated to it. While the reasoning process will likely imitate that of a human, the logical processes may not necessarily be identical.[16]

The reasoning will be applied by the weapon system to make two classes of decision. The first involves determining whether the object or person that the weapon system is observing is a lawful target, i.e. whether it is a combatant,[17] a civilian taking a direct part in the hostilities[18] or an object that is a military objective.[19] If that is determined to be the case, the second decision addresses whether it would be both lawful and appropriate to attack the observed object or person. These are distinct and probably sequential decisions. Both involve the application by the autonomous system of the targeting law rules[20] and, in the latter case, of broader, strategic and political considerations.

Certain issues arising from heritage weapon systems, e.g., as to manning requirements, liability to cyber attack, jamming or other kinds of enemy interference and the growing requirement for extremely rapid decision-making may be expected to render the human component in such traditional systems the greatest point of vulnerability in the modern, high-speed battlespace.[21] Indeed, these and other similar factors may be expected increasingly to drive the development and fielding of autonomous systems. Moreover, in the high-speed battle, it is the degraded ability of the human decision-making process that needs to be weighed against the performance of the highly automated or autonomous system, although the latter will only be legally acceptable if the required discriminating, precautionary and other tasks are performed to a satisfactory standard.

[16] Georg Heppner and Ruediger Dillmann suggest as an essential feature of autonomous systems the ability to learn new things; Heppner and Dillmann, 'Autonomy of Mobile Robots' 13, 86.

[17] For States party to API, see Articles 43 and 44. For States that are not party to API, consider Regulations Respecting the Laws and Customs of War on Land, Annexed to Hague Convention IV, The Hague, 18 October 1907, Articles 1–3.

[18] API, Article 51(3) and Common Article 3 to the Geneva Conventions, 12 August 1949.

[19] API, Article 52(2).

[20] Consider, for example, API, Articles 48 to 67 inclusive, Hague Cultural Property Convention, and the rules as to specific and special protection set forth in the AMW Manual.

[21] DoD *Law of War Manual*, June 2015, para 6.5.9.1.

Autonomous offensive attack decision-making[22] is for the time being unlikely to be capable of applying the customary distinction and discrimination principles and the rules as to proportionality and precautions in attack.[23] It therefore follows that the employment of such autonomous weapon systems in a 'man out of the loop' mode, i.e. permitting the machine autonomously to make its own attack decisions without human supervision, should usually and for the foreseeable future be rejected during a weapon review. The employment of autonomous weapon systems would, however, be potentially lawful if: (1) a person will be placed 'on the loop' in the manner discussed earlier in this chapter; or (2), and as noted below, if the search by the autonomous system for targets can be limited in time and space to locations for which assessments as to proportionality and other required targeting law decisions can be made at the mission planning stage that will remain valid throughout the area and period of search. The next section of this chapter explains how existing law applies to these technologies.

6.5 APPLYING EXISTING LAW TO HIGHLY AUTOMATED AND AUTONOMOUS WEAPONS

When undertaking a weapon review, under API, Article 36 or under the implied rule, the first matter to consider is whether the weapon system complies with the customary principles and treaty rules of weapons law, it being recalled that notwithstanding the evident novelty of these weapon technologies, it is existing legal principles and rules that provide the yardstick as to a weapon's acceptability or otherwise.

The superfluous injury and unnecessary suffering principle and the environmental protection rules, both discussed in Chapter 2, are unlikely to be of any relevance to the automated or autonomous decision-making aspect of the

[22] By 'offensive attack decision-making' is meant the process whereby a weapon platform searches for its own target and determines that it is an object or person that it has been programmed to attack. This is to be distinguished from, e.g., point defence systems that are designed to defend a location or object, such as a warship, by engaging in-bound weapon systems such as torpedoes or rockets.

[23] Yoram Dinstein identifies that the 'distinguishing feature of an autonomous weapon system is that, upon activation, it can strike targets without further intervention by a human operator'. Such systems make 'their own judgment calls in the face of unforeseen events' free from human intervention, 'exercising discretion when encountering complex scenarios'. After referring to the proportionality rule, Professor Dinstein wonders: 'How can it be envisaged that AI robots will be up to the task?'; Y. Dinstein, 'Autonomous Weapons and International Humanitarian Law', in W. Heintschel von Heinegg, et al. (eds.), *Dehumanization in Warfare*, 15, 17–20.

weapon system. There are no ad hoc rules of weapons law that specifically address the highly automated or autonomous aspects of such platforms.[24] Where the indiscriminate weapons rule is concerned, the performance of the autonomous or automated target recognition technology, whether in tests or under battlefield conditions, should determine whether the system can properly be regarded as indiscriminate by nature. The intended military purpose for the weapon system is likely to require that it reliably recognises and engages military objects or lawful human targets and practical testing, computer modeling and other available empirical data, including that arising from any use of the technology or system by other States, will determine whether that is so. So, for example, highly automated attack technology may be installed in an aircraft system and may be designed to recognise the particular features of, say, a tank. If the recognition software performs satisfactorily in tests which reasonably reflect the sorts of operational circumstance in which it is intended to use the weapon system, the weapon is unlikely to be found to breach the indiscriminate weapons rule.[25] The rule may, however, be broken in the case of a similar weapon system whose technology is found, during tests, for example to be incapable of differentiating between civilian and military objects with the result that it would attack either of them without distinguishing between them.[26]

However, a weapon review of a weapon system employing highly automated or autonomous attack technologies should also assess whether the particular weapon system is capable of applying the law of targeting.[27] This is because the human operator, who is the agency through whom targeting

[24] 'S. S. "Lotus " (*Fr.* v. *Turk.*)', 1927 P.C.I.J. (ser. A) No. 10, (Sept. 7) at page 18 suggests that restrictions upon the independence of States cannot be presumed. Applying this principle, the absence of specific reference to such technologies implies that they are not the subject of a prohibition, and note US DoD, *Law of War Manual*, June 2015, para 6.5.9.2.

[25] This is a reference to the rule in Article 51(4)(b) and (c) of API which has customary law status and thus binds all States.

[26] The word 'may' is used here because the weapon system would not be indiscriminate by nature if its intended circumstances of use were to be limited to areas where, and times during which, relevant civilian objects are known to be absent.

[27] This view is explicitly rejected by the USA. A Working Paper submitted to a CCW Group of Governmental Experts meeting considering Lethal Autonomous Weapon Systems reasserted the US position that the law of war does not require machines to make legal determinations and that persons authorising or directing the use of, or operating, autonomous and semi-autonomous weapon systems must, among other things, do so in accordance with the law of war; US Working Paper, 'Autonomy in Weapon Systems', submitted to the CCW GGE Meeting on Lethal Autonomous Weapon Systems, Geneva, 13–17 November 2017 (text in the possession of the author), paras 10–15. At para 12, this is stated to be the US position 'even if the weapon (e.g., through computers, software, and sensors) may be characterized as capable of taking some form of action or decision in a given moment in the absence of direction by a

law compliance is normally achieved in the case of traditional weapon systems, has been removed from the attack decision-making process. Absent human involvement in the attack decision-making process, the legal question therefore becomes whether the highly automated or autonomous weapon system itself is capable of applying targeting law principles and rules.

Targeting law was summarised in Chapter 3, and it is the precautions that Article 57 of API obliges an attacker to take that are likely to require most careful consideration in a weapon review of such a weapon system. The Article 57(1) duty to take constant care to spare the civilian population, civilians and civilian objects provides the context in which the remaining rules of the Article should be interpreted.

It is those who 'plan or decide upon an attack' that are charged with undertaking many of the prescribed precautions. The characteristics of the autonomous or highly automated weapon system and the arrangements that are made for its employment must be such that decisions as to attack are made. Technology may be developed sufficiently in the future to enable all of the required precautionary decisions to be made by the system itself. In the nearer future, personnel controlling the systems, supervising them or planning the mission will have to be able to take sufficient precautionary steps in advance of, or during, the mission if, as is being assumed, the weapon system itself is not capable of applying the prescribed precautions either at all or in full. However, we should consider how some of the detailed precautionary rules apply in the present context.

There is a caveat of feasibility associated with the obligations in Article 57(2)(a)(i) and (ii) of API. It follows from this that if one of these obligations can be fulfilled using a manned mission but cannot be fulfilled using an autonomous or highly automated mission, the manned mission should be employed. Such

human being, such as whether to fire the weapon or to select and engage a target.' The author rejects this view on the basis that if the attack decision is being made by the weapon system with no human involvement, it is only the weapon system that is in a position to apply targeting law. If the weapon system is incapable of doing so, it should not be used if the result would be that the decisions required by targeting law are not taken. If the autonomous weapon system does not make the decisions required by the law of targeting and if no human being is involved when such decisions are being taken, it follows that targeting law will not be applied. In the author's view, the obligation of an attacker is to ensure that targeting law rules including required precautions are applied and taken, whether by a person or by a machine. Consider in this regard the comments of Jeffrey Thurnher at J. S. Thurnher, 'Feasible Precautions in Attack and Autonomous Weapons', in W. Heintschel von Heinegg, et al. (eds.), *Dehumanization of Warfare*, 99, 108–114, in which the precautionary rules are in turn analysed, essentially by reference to the ability, or otherwise, of the autonomous weapon system to perform the relevant task.

an autonomous or highly automated system must therefore be so programmed as to recognise when its employment would preclude the taking of sub-paragraph (i) or (ii) precautions that could be taken if some other weapon system, such as a manned one, were to be employed. An inability of the autonomous or automated system to fulfill the obligations in such circumstances does not render them non-feasible. Accordingly, if an autonomous or automated system is not capable of such a determination, it fails to meet the required precautions obligations and should not be employed in the relevant circumstances.

As was noted in Chapter 3, Article 57(2)(a)(i) of API requires that planners and decision-makers do everything feasible to verify that the object of the attack is a military objective and that it is not entitled to special protection. The algorithm-based[28] target recognition technology will have been designed specifically to enable an autonomous or highly automated system satisfactorily to differentiate between the objects it is programmed to identify and those it is not, and thus between military objects and civilian objects. So this technology will be the means whereby such a system will comply with this precautionary rule.[29] If an autonomous or highly automated system were to be designed to identify, and attack, human targets, the challenge under the same provision would be to show that the target identification software or mechanism satisfactorily differentiates between lawful targets and those that must be protected, i.e. between able-bodied combatants[30] or civilians directly participating in the hostilities and combatants or directly participating civilians who are *hors de combat*, persons who are non-combatants and civilians who do not directly participate. So far as is known, currently available systems are not capable of satisfying all of these requirements and it is not clear when such technology will become available.[31]

Article 57(2)(a)(i) of API also obliges attackers to do everything feasible to verify that it is not prohibited by the Protocol to attack the intended targets. Attacks may, for example, be prohibited by the Protocol because the attack

[28] An algorithm is 'a process or set of rules to be followed in calculations or other problem solving operations, especially by a computer'; *Concise Oxford English Dictionary*, 11th edn (2006), 32.

[29] The reference here to 'military objects', as opposed to military objectives, is intentional; the technology is likely to be configured so as to distinguish, for example, between an artillery piece or a tank on the one hand and a civilian vehicle of comparable size on the other.

[30] Note, for example, that if, as seems possible, nanotechnology enables the development of military firearms with reduced metal content, an autonomous or highly automated system that would differentiate between combatants and civilians only on the basis of their metallic signature may prove problematic.

[31] Consider, however, the discussion of biometric technologies in Chapter 14.

would breach Article 51(4) (discrimination principle), 51(5)(a) (separate and distinct military objectives treated as one), 51(5)(b) (proportionality), 53 (cultural objects), 54 (objects indispensable to the survival of the civilian population), 35(3) and 55 (protection of the natural environment), 56 (works and installations containing dangerous forces and military objectives in their vicinity), 41 (safeguarding of persons hors de combat), 12 and 15 (protection of medical units and personnel) and 21 to 28 (protection of medical transports). As noted, these are simply examples of the persons and objects that are protected in API from attack. Where these rules simply prohibit attacks directed at the relevant object or person, the autonomous or highly automated system must be capable of detecting that a person or object comes within one of these protected categories and must accordingly refrain from attacking that person or thing. Where the rules relating to such objects or persons require, e.g., complex warnings to be given before any attack may take place, even more difficult issues will obviously arise.

However, evaluative decisions are required by Article 51(5)(a), by the proportionality assessment referred to in Articles 51(5)(b) and 57, and by Article 57 (2)(a)(ii) and 57(3) of API. For an autonomous or highly automated system, as opposed to a human being involved in the operation of such a system, to comply with Article 57(2)(a)(ii), for example, it would need to be capable of deciding whether an attack should be undertaken using a remotely piloted aircraft, by an automated or autonomous platform, or indeed by some other means, in order to minimise incidental civilian injury and damage.

If the technology, and the circumstances of a particular mission, enable a controller or operator to make the necessary evaluations, whether before or during the mission, this is likely to overcome the difficulty.[32] If that is not the case, the system may only be employed if it can do so itself.

Currently available autonomous and highly automated technology does not enable the evaluative precautionary decisions that have been discussed in this Section to be made by the weapon system itself. To take another example, it is not possible using currently available technology to determine the military advantage to be anticipated from attacking a specified military objective in particular circumstances, nor to determine the civilian injuries and damage to be expected from the particular kind of attack and still less to compare the two

[32] It may, for example, be realistic for autonomous attack operations to take place in remote desert or remote maritime areas if those persons planning and authorising the operation can limit the area and time period of search such that proportionality and other targeting law judgements can properly be made in advance of the mission and will continue to be valid throughout the area and time of the search by the autonomous platform.

in order to determine whether the latter may be expected to be excessive in relation to the former. Accordingly, applying the proportionality rule would appear currently to be beyond the capabilities of a highly automated or autonomous weapon system operating without human supervision. Care must therefore be taken to ensure that such systems are only used in such a way that required precautions can be taken by personnel, probably before the deployment or launch of the platform. While it is likely that there will with the passage of time be some improvement in the sophistication of these technologies, the foregoing analysis has shown that for currently available highly automated or autonomous technology to be used lawfully for offensive[33] attack operations a person will need to remain sufficiently engaged and suitably located in order to be able to observe what is taking place and, if necessary, to over-ride mechanically reached attack decisions. The foreseeable workloads of such personnel must be limited so as to ensure that proper decisions are made, taking all available information into account.

6.6 ARTIFICIAL LEARNING INTELLIGENCE (ALI)

ALI seems likely to be the essential ingredient in the evolution of autonomous weapon systems. In the future, the 'learning' aspect might involve the machine developing its own criteria for recognition based on observations that it may make in the battle-space, or that it may have made in the past. So, for example, such a machine may observe the pattern of life in the area of interest and may then adjust a target list fed into it in advance of the mission to take account of the observations that it has made. Such developments may, in the future, enable an autonomous system to react appropriately, for example, to the absence from, or presence of, particular kinds of target at a specific location scheduled for an attack by refraining from the planned attack or by adjusting the way in which the attack is to be prosecuted, e.g., by selecting a particular weapon to be employed or, perhaps, by altering the timing of the

[33] The intended distinction here is between platforms that seek out their targets in locations that are not necessarily known in advance, termed here 'offensive' attack systems, and point defence weapon systems which activate autonomously to counter inbound threats such as missiles, rockets or mortars, or in the maritime context such as torpedoes and missiles, that have been fired towards the location that the autonomous system has been deployed to defend. The objects that the autonomous point defence system is designed to engage are by definition military objectives and before deploying the system, proportionality and other targeting law considerations can be taken into account by those planning or deciding on the deployment. Consider, for example, in the maritime domain the Phalanx naval system and in the land domain the Iron Dome system and note the US view in *US Working Paper*, 'Autonomy in Weapon Systems', para 19.

attack. ALI might, however, also enable the machine to detect that a planned attack would no longer comply with the discrimination rule, if for example it were able to detect the arrival of hostages in such numbers that the resulting civilian loss of life may be expected to be excessive. Perhaps less ambitiously, a somewhat more basic autonomous technology might be programmed to detect whether there has been any material change in the circumstances pertaining in the area of search compared to that suggested by the pattern of life data that was fed into the mission control software in advance of the mission; in that event, the autonomous search for targets would, and should, be called off. Later refinements to the technology might then enable the software to distinguish between changes in the circumstances pertaining in the target area that are legally material and those which are not.

Appropriate limits must be placed on what the learning part of such an ALI system is permitted to learn. These limits must be designed to ensure that the whole ALI process remains properly controlled by personnel and that the learning process remains within the limits that have been set. If learning can be effectively limited to enabling the machine to ensure with improving reliability that it only undertakes lawful attacks, this will be a positive aspect when the evaluation of the technology is undertaken by a State contemplating its acquisition. An ALI system that permits the loosening of constraints on decision-making that are reflective of targeting law rules is most unlikely to be legally acceptable.

It should be noted that some commentators consider that elements of ALI will incrementally be employed in armed conflict and that there is a need to regulate or control such activity to ensure it complies with established norms. Other commentators propose that autonomous technology should be banned as such in warfare. Initiatives for the prohibition or regulation of certain highly automated and autonomous technologies will be discussed in section 6.9. We should first, however, recognise the extent of human involvement in the procurement and use of such weapons systems and the responsibilities of such involved persons for the proper performance of their duties.

6.7 HUMAN INVOLVEMENT IN AUTONOMOUS OPERATIONS

At first glance, a powerful argument can be made that human beings should make decisions that involve the taking of human life and that therefore the authority to undertake an attack should not be delegated to an autonomous weapon system because autonomous systems cannot be held accountable for their actions. As the US DoD Law of War Manual points out, the law of war obligations in relation to attacks impose obligations on persons.[34] While it is correct, as that Manual goes on to note, that the law of war does not require a weapon to determine that the object of its attack is a military objective, the law does require that objects which are not military objectives should not be made the object of attack, and if an autonomous system is so configured that no human being is making that determination, it would only be lawful to use it if the system itself does so.

Human beings will, however, inevitably be closely involved in the acquisition and employment of autonomous weapons. States, and the procurement officials they employ, are responsible for the decision to procure the technology and for the testing and evaluation that must take place before the decision to acquire new weapon systems is made. The engineers and scientists that conduct those tests are responsible for the proper conduct of their duties in accordance with established standards. The engineers, computer programmers, technicians and operators will be responsible for the management and operation of the autonomous and highly automated systems in the same way that they are responsible for managing existing weapon systems that are often complex and that often include automated features.

Lawyers and others conducting legal reviews of autonomous and highly automated weapon systems will be responsible for ensuring that comprehensive legal reviews are prepared identifying whether the system under review is prohibited, or whether its lawful circumstances of use are restricted, by the international law applicable to the relevant State. The responsibility rests with planners and commanders to so acquaint themselves with the characteristics of the weapon system, and with the constraints imposed by the applicable law, to ensure that it is in fact used in a lawful manner.[35] The responsibility will rest with procurement staffs only to procure weapon systems that, following

[34] US DoD *Law of War Manual*, June 2015, para 6.5.9.3. Consider, for example, a project to use computer vision, an aspect of computer learning and of deep learning, autonomously to extract objects of interest from still or moving imagery; C. Mellerin, 'Project Maven to Deploy Computer Algorithms to War Zone by Year's End', *US DoD News*, 21 July 2017. While the computer technology may extract the objects of interest, note that 'what AI will do is complement the human operator'; ibid.

[35] Note the similar US view, i.e. that it is for commanders to authorise the use of lethal force in accordance with ROE and IHL on the basis of the commander's understanding of the tactical situation, on the basis of the weapon system's performance informed by testing and operational experience, and on the basis of the employment of tactics, techniques and procedures for that weapon. So, in considering lethal autonomous weapon systems, the USA notes that the commander will remain accountable, considers that States will not field weapons they cannot control as such weapons are not militarily useful, and focuses therefore on 'appropriate levels of human judgment over the use of force', a concept that apparently reflects US DoD policy on autonomous weapons. Accordingly the USA rejects the suggestion of 'a new requirement that

appropriate testing,[36] are found to comply with the existing law applicable to the State in question and to ensure that limitations on use identified in the weapon review are properly communicated to those who need to know and to implement them. Planners will be responsible for carefully preparing such autonomous missions.

Commanders will be responsible for ensuring that the orders they give as to the undertaking of such missions, e.g., as to the timing, location, purpose and conduct of the mission, will not give rise to attacks that breach targeting law. This will require commanders to obtain and take properly into account advice from suitably qualified specialists as to the capabilities of the autonomous system and its vulnerabilities and limitations so that the commander's decision is properly informed. Commanders will also likely require legal advice before approving such operations.

Technicians and other personnel will be responsible for the accurate preparation of the mission- and target-specific data that is fed into the mission control software, for correctly refueling, servicing and loading ordnance onto the aircraft and for ensuring that lessons learned from previous sorties inform the taking of precautionary steps before subsequent sorties are authorised. The negligent performance of any of the tasks discussed in this section will attract such sanctions against service personnel as their service code may provide and will attract such disciplinary consequences in the case of civilians as their contracts of employment may stipulate. Moreover, criminal conduct undertaken using autonomous or highly automated weapon technologies will not be deprived of its criminal nature by the nature of the technology that is being used.

6.8 INTERFERENCE, DECEPTION AND HUMAN RESPONSIBILITY

If, as seems foreseeable, future conflict is characterised by widespread, difficult to detect cyber deception operations that corrupt, or deny, the data that

machines make legal determinations' and considers that different weapon systems and operational contexts can require different levels of human judgment with some functions being better done by computers and others being better done by human beings; CCW Group of Governmental Experts on Lethal Autonomous Weapons Systems, US Intervention on Appropriate Levels of Human Judgment over the Use of Force, Geneva, 15 November 2016. Georg Heppner and Ruediger Dillmann suggest, perhaps realistically, that 'in safety critical applications the human will always have to make the final decision;' Heppner and Dillmann, 'Autonomy of Mobile Robots', 96.

36 For the US view that individuals will have responsibilities in accordance with their roles, see US Working Paper, 'Autonomy in Weapon Systems', paras 28 and 29.

supports the adversary's targeting decision-making, this would seem to have legal implications for the conduct of remote attack operations involving remotely controlled weapon platforms as well as those using autonomous and highly automated systems. The relevant law focuses on perfidy, on treacherous killing, wounding and capture of the enemy,[37] which are narrowly defined notions, and on misuse of certain indicia.[38] If those limited rules are not breached, there is no legal rule that, in terms, prohibits cyber activities that so corrupt the opponent's understanding of the battle-space or so degrade the operation of his essential systems that either he can no longer prosecute effective attacks or that his ability to comply with distinction, discrimination, proportionality and precautions rules is impaired.

It is foreseeable that remotely controlled, highly automated and autonomous attack systems will be the subject of cyber interference. Arguably, the obligations to take 'constant care' and to do 'everything feasible to verify' imply, in relation to unmanned operations, that once the prospect of cyber interference with, for example, computer navigation links, weapons control and guidance systems, target identification software and other systems becomes evident, everything feasible will need to be done to seek to ensure those systems remain robust against the kinds of known cyber interference that would render the use of such weapon systems indiscriminate. Good military reasons for doing this include ensuring that the intended targets are in fact engaged. Humanitarian concerns will also be served by taking such precautions, as successful interference may cause loss of control of the weapon and consequent danger to civilians or civilian objects. It might be the case that by diminishing or extinguishing the links between a controller and the weapon platform, and thus by essentially making it partly or wholly autonomous, the possibility of such cyber interference may be reduced and dangers to civilians and civilian objects arising from such interference may as a result be limited or extinguished.[39]

This would suggest that it would be desirable, but not necessarily legally required, for remotely controlled weapon platforms and highly automated and autonomous attack systems to be designed so that they detect, and then disclose to the operator or other person who is supervising the mission, when cyber interference occurs or when for any other reason there is an adverse effect on their performance or reliability. Such systems might also usefully be so designed that if cyber interference that would adversely affect the proper

37 API, Article 37(1) and Hague Regulations 1907, Article 23(b).

38 API, Articles 38 and 39 and Hague Regulations 1907, Article 23(f).

39 J. S. Thurnher, 'Feasible Precautions in Attack and Autonomous Weapons', 106.

performance of the attack decision-making technology occurs, that the mission is aborted. Sophisticated cyber interference may, however, be designed so as to conceal the intrusion or interference. Complex legal issues are likely to arise when, e.g., the operational picture of a commander who is seeking to comply with distinction, discrimination, proportionality and precautionary rules is distorted with the direct result that civilians and/or civilian objects are erroneously made the object of attack.

In order to try to assess the practical implications of what is being discussed here, let us consider situations in which a hacker is able to cause the transmission of false target coordinates, to cause a distorted sensor feed to be delivered to the operator of a remotely controlled weapon system or directly to the control mechanism of a highly automated or autonomous weapon system, or to cause stored targeting data to be corrupted. In each case the false information could deceive the highly automated or autonomous system or indeed the operator who is 'on the loop' of such a system, or both, into targeting an object that the system and/or the operator are led to believe is a military objective but which may turn out to be a civilian object.

Foreseeably, evidence may be forthcoming to link the attack to the State whose weapon system was hacked. Detection of the cyber intrusion may be hard to achieve either at all or in a timely way. Nevertheless, if detection of some kinds of cyber intrusion becomes technically feasible, satisfying Article 57's 'all feasible precautions' obligation might increasingly be interpreted as requiring the acquisition and use of appropriate detection equipment, particularly if such intrusion and interference becomes a likelihood. Detection and counteracting of such intrusion and interference operations may well become essential to ensuring that the mission complies with targeting law and, indeed, achieves its desired military purpose.

However, it must always be appreciated that acts of cyber interference that erode the utility of the enemy's targeting system are lawful operations against military objectives. Cyber deception operations that distort the enemy's picture of the battle-space do not as such invite his confidence as to protection under the law and do not therefore amount to perfidy. Rather, they seek to erode his confidence as to, say, the location of the military objectives that he would like to attack,[40] which is a classic example of a lawful ruse.[41] This rather leads one to deduce that current law would only prohibit such a cyber deception operation if its explicit purpose were to be to cause the deceived

[40] The notion of perfidy under API, Article 37(1) requires, inter alia, the deception of the enemy as to protected status under the law of armed conflict.

[41] Lawful ruses of war are defined in API, Article 37(2).

party to the conflict to attack civilians, civilian objects or persons or objects entitled to specific protection and if it directly caused that result. Cyber deception operations the purpose of which is merely to erode the confidence of the adversary as to his own targeting processes, or to cause those targeting processes to malfunction, or to cause the resulting attacks to become less reliable or accurate, would not breach current law.

Design requirements for remotely controlled, highly automated and autonomous weapon systems should therefore, it is suggested, prioritise:

a. robustness against intrusion or interference;
b. the detection and communication to the operator or to those commanding the mission of any cyber intrusion or interference that does take place;
c. communication to the operator or to those commanding the mission of the effect of any cyber intrusion or interference on control of the system or on the accuracy of the operational picture that forms the basis of attack decisions or on the integrity of the performance of autonomous or automated functions and
d. making the operator or those commanding the mission aware of the effect, if any, of such intrusion or interference on the reliability of any relevant related system.

Additional design priorities for autonomous and highly automated weapon systems should, it is suggested, be:

a. for those commanding the mission to be able if necessary to terminate the mission, or
b. for them to be able to divert the platform elsewhere and/or to re-task it mid-mission, or
c. for them to be able to alter the platform's method of operation so that any further platform or weapon action requires operator control, or
d. for the system to be so configured that the platform itself will take such action, or
e. for the system to have a suitable combination of such facilities.

If a cyber intrusion or interference operation causes the hacker to take control of a remotely controlled weapon system, of an automated or autonomous weapon system or of the munitions such systems carry, legal responsibility for subsequent wrongful use of the system and of its munitions will transfer to the persons who have taken and retain control and to a State if those taking control are organs of that State or if they are acting under direction and control exercised by that State with knowledge of the circumstances.[42] It follows from this that it will be important, as the case may be, for the controller of the weapon system or for the commander of the autonomous mission to be aware of, and to be able to demonstrate, when such a loss of control occurs.

A cyber hacker who achieves control of such a weapon system or of its munition has legal responsibility for any subsequent use by him of the system or munition. It follows from this that it is the hacker's responsibility to ensure that his subsequent use of the weapon system against the adverse party to the conflict complies with distinction, discrimination and proportionality and that required precautions in attack are taken. If he does not achieve control of the platform and its weapons but interferes with the way in which the platform and weapons are operated by the enemy the hacker cannot be regarded as the attacker but remains bound by the obligations under Article 58 of API.

If only sufficient cyber action is taken to prevent the enemy's weapon from attacking its originally intended target, such 'parrying' operations should not, it is suggested, cause the cyber hacker to be regarded as the attacker[43] with the result that he should not thereby be deemed to have breached Article 51(2) or 52(1) of API, for example, if the effect of the 'parrying' operation is that the missile impacts on civilians or civilian objects. This is because, while the notion of 'attack' in Article 49 of API encompasses any use of violence against the adversary, a parrying action of the sort being discussed involves a redirection of the violence of the other party away from its intended object, rather than directing that violence positively towards a person or object that the cyber hacker has decided should be attacked.

These fine distinctions are based on the intent of the cyber hacker, an intent which it is likely to be difficult to establish.

If a cyber hacker intrudes, for example into an RPA's precision guidance system and reduces its reliability so that it will foreseeably cause the weapon to attack military objectives or civilians or civilian objects without distinction, it would be difficult to reconcile such action with the obligation to take constant care to spare civilians and civilian objects.[44] For the cyber

[42] ILC, Draft Articles on Responsibility of States for Internationally Wrongful Acts, 2001, Article 17.

[43] Such an event would not at any event amount to an attack if, from the perspective of the hacker, it does not amount to an act of violence against the adverse party.

[44] If the civilians or civilian objects that are imperiled by such operations are within the control of the party employing the cyber deception operations, such operations may also breach the API, Article 58(c) obligation to 'take the other necessary precautions to protect the civilian population, individual civilians and civilian objects under their control against the dangers resulting from military operations.'

hacker actually to be regarded as the attacker in such circumstances it would be necessary for him to become the initiator of the operation as it ultimately transpires and for the operation to amount to an act of violence against the adverse party.

The preceding discussion suggests the following:

(1) If a cyber hacker exercises such a degree of control of a remotely controlled, autonomous or highly automated weapon system such as to enable him knowingly or intentionally to direct its munition, or knowingly or intentionally to cause the autonomous system to direct it, at a target or category of targets of his choice, he becomes responsible for the consequences of the employment of the weapon.
(2) This paragraph applies: (a) if the degree of control referred to in paragraph (1) is not achieved, but the cyber hacker nevertheless adversely affects the reliability of the munition firing or guidance system or of the target recognition software of the targeted weapon system; (b) if the cyber hacker does this with the expectation and effect, by doing so, of causing the munition to attack civilians, civilian objects, persons or objects entitled to specific protection or to undertake indiscriminate attacks; and (c) if the cyber hacker can properly be regarded as having caused the munition to be directed towards objects or persons of his choosing. If this paragraph applies, the cyber hacker should be held responsible for the consequences of the use of the weapon.
(3) If in the circumstances described in paragraph 2 the cyber operation foreseeably causes the adverse party's attack(s) to become indiscriminate the cyber operation may, depending on the circumstances, conflict with obligations under Articles 57(1) and 58(c) of API.
(4) If the cyber hacker interferes with the system's guidance mechanism intending merely to deflect the weapon from its intended target, he should not be seen as responsible for the injury or damage that the weapon causes because he is not taking over directional control of the munition with a view to using it as his weapon. His involvement cannot be equated with the 'act of violence' stipulated in Article 49(1) of API. It should however be considered whether the requirements of Article 58 of API were complied with by each party to the conflict in such a case.

Having considered how existing international law rules apply to the weapon systems that are the subject of the present chapter, attention now moves to the proposals that have been made for the prohibition or legal regulation of autonomous and certain highly automated weapon systems.

6.9 NEW LAW FOR AUTONOMOUS AND CERTAIN HIGHLY AUTOMATED WEAPON SYSTEMS

Suggestions have been put forward for a treaty and for national laws aimed at prohibiting the development, production and use of fully autonomous weapons.[45] One proposal refers to 'robots that are capable of selecting targets and delivering force without any human input or interaction' or that, although they operate under the oversight of a human operator who can override the robot's actions, are subject to such limited supervision that there is no effective human input or interaction.[46] This description is likely to include weapons technologies that would be regarded as highly automated, as opposed to autonomous, as those terms have been used in the present chapter. Some observers may therefore conclude that the scope of this Human Rights Watch proposal is over-inclusive.

A UNIDIR discussion paper makes the sensible point, alluded to above, that there might be a difference in the acceptability of an autonomous but static system that is a 'last line of defence' to counter an incoming attack versus a system that employs superhuman decision-making speed to carry out an attack.[47] Autonomous weapon systems that are the 'last line of defence' and that are so programmed that they can effectively distinguish between lawful targets and persons or objects that are entitled to protection are not prohibited under current international law.

As is evident from the terms of the preceding discussion, autonomous technology has yet to achieve maturity, so to ban all such technology now on the basis that in its current less than mature state it cannot meet applicable legal criteria would fail to acknowledge the possibility that with further development, probably over a considerable period, technology may be achieved to meet or even exceed the stated requirements. Furthermore, the effective countering of, for example, rapidly materialising mass attack threats may necessitate decision-making processes that operate at a computer's, as opposed to at human, speed[48] and it should be recalled that the decisions of

45 C. Haynes, 'Report to the UN General Assembly of the Special Rapporteur on Extrajudicial, Summary or Arbitrary Executions', A/HRC/23/47, dated 9 April 2013, available at www.ohchr.org/Documents/HRBodies/HRCouncil/RegularSession/Session23/A-HRC-23-47_en.pdf.

46 Human Rights Watch, 'Losing Humanity: The Case Against Killer Robots', (2012), available at www.hrw.org/sites/default/files/reports/arms1112ForUpload_0_0.pdf, page 2.

47 UNIDIR, 'Framing Discussions on the Weaponisation of Increasingly Autonomous Technologies', (2014), available at www.unidir.org/files/publications/pdfs/framing-discussions-on-the-weaponization-of-increasingly-autonomous-technologies-en-606.pdf, 6.

48 J. S. Thurnher, 'Feasible Precautions in Attack and Autonomous Weapons', 106.

robotic technologies will not be distorted by fear, anger, vengeance, amnesia, panic, tiredness or other peculiarly human fallibilities. Autonomous systems that are limited to ensuring with improving reliability that only lawful attacks are undertaken would of course positively contribute to legal compliance, so calls for a blanket ban on autonomous, and certain kinds of automated, weapon systems should be approached with some caution.[49]

These and other factors may be expected to inform discussion of proposals for legal provision on the matter. In that regard, it should be noted that discussions are currently being undertaken through the auspices of the Conventional Weapons Convention on what are described as 'Lethal Autonomous Weapon Systems' (LAWS).[50] The United States has proposed 'as an interim step, that States could begin work on a document setting forth best practices for conducting a comprehensive weapons review with respect to LAWS.'[51] The Report from the Chair of the 2015 CCW meeting noted the importance of rigorous legal weapons reviews to ensure that LAWS could be used consistently with IHL.[52]

[49] It should be recalled that the operative provision in the St Petersburg Declaration 1868 was overtaken by subsequent developments; note for example Hague Draft Rules of Aerial Warfare, 1923, Article 18(2). It should also be recalled that Declaration IV, 1 of 1899, as updated by Declaration XIV of 1907, prohibited the discharge of explosives and projectiles from balloons or by other new methods of a similar nature, a prohibition that was also overtaken by later technical developments, such as the use of airships and aircraft to prosecute bombing missions during World War I.

[50] The text of the discussion mandate and the documents associated with the informal meeting held at the UN in Geneva from 13 to 17 April 2015 are available at www.unog.ch/80256EE600585943/(httpPages)/6CE049BE22EC75A2C1257C8D00513E26?OpenDocument.

[51] M. W. Meier, 'Lethal Autonomous Weapon Systems (LAWS): Conducting a Comprehensive Weapons Review', 30 *Temple Int'l & Comp L J* 1 (2016) 119, 120. In a Working Paper submitted to the 2017 CCW GGE LAWS meeting, the US has also argued that it is unnecessary for the GGE to adopt a specific working definition of LAWS and that instead the focus should be on gaining a general understanding of their characteristics, noting that legal definitions are usually developed by reference to specific legal rules, not in the abstract. Curiously, in later paragraphs of the same Working Paper, definitions that the US DoD has used for domestic policy purposes are then set forth. Also in the same Working Paper, the US argues: that definitions based on specific technological assumptions should be avoided as they may become obsolete; and that articulation of specific levels of autonomy or types of machine reasoning should be avoided. For the US the focus should be on how human beings are using the weapon and what they expect it to do; US Working Paper, CCW GGE, 'Characteristics of Lethal Autonomous Weapon Systems', submitted to the CCW GGE LAWS Meeting on 10 November 2017 (text in the possession of the author), paras 2–9.

[52] Paragraph 51(b)(v) of Chair's Report of 2015 CCW LAWS Meeting, M W Meier n.51, page 122. The US proposed that work should begin on an outline document detailing a weapon review process that would apply to the development or acquisition of LAWS. For a US statement on the importance of weapon reviews in the context of LAWS, see Opening Statement by C. Trumbull, Geneva, 15 November 2016 (text in the possession of the author) and for a

From 11 to 15 April 2016 a further such CCW meeting took place in Geneva again under a discussion mandate. The Chair's report records what were clearly wide-ranging and intensive discussions, noting that no fully autonomous weapon systems currently exist,[53] and that there has been discussion of 'meaningful human control' and of 'appropriate human judgment' as possible requirements and of what such requirements might imply. In this regard, the various kinds of human involvement in the operation of autonomous and highly automated weapon systems, summarised in Section 6.7, should be borne in mind. Whether such involvement amounts to 'meaningful human control' will, of necessity, depend on the precise meaning that is given to that term were it to be included in any treaty or other document that emerges from the CCW discussions and/or from ensuing processes. Key in that regard will be the extent to which the human involvement as a whole has the effect of constraining the range of possible decisions and actions that the weapon system is permitted to choose from. 'Man on the loop' technology would clearly satisfy likely interpretations of 'meaningful human control' but it is not likely to be the only method of doing so and decisions made in advance of the mission, coupled with technical limitations imposed on the kinds of lesson that the technology is permitted to learn and other constraints may, collectively, constitute sufficient human control for these purposes. However, Thilo Marauhn, having charted the emergence of the term as a political notion in the course of the CCW discussions from 2013 to 2016, argues persuasively that the focus in the discussion of LAWS should be on the established rules of IHL and, perhaps, of arms control.[54] He acknowledges that '[t]he concept of 'meaningful human control' facilitates implementation of the law as it stands but 'should not become part of the law',[55] a view with which the present author profoundly agrees.

description of how the US applies its weapon review criteria and weapon testing practices to weapons that employ autonomous target selection and engagement functions, see US Working Paper, n. 27, paras 3–9.

53 Report of the 2016 Informal Meeting of Experts on Lethal Autonomous Weapon Systems, submitted by the Chairperson, and available at www.unog.ch/80256EDD006B8954/(httpAssets)/DDC13B243BA863E6C1257FDB00380A88/$file/ReportLAWS_2016_AdvancedVersion.pdf (2016 LAWS Report), para13.

54 T. Marauhn, 'Meaningful Human Control – and the Politics of International Law', in W Heintschel von Heinegg, et al. (eds.), *Dehumanization of Warfare*, 207, 210–212.

55 Marauhn ibid., 217. Note, however, the important ethical concerns over the erosion of human agency and intent in decisions to kill, injure and destroy, the related concerns as to loss of responsibility and of human dignity and the resulting vital ethical requirement for human control, powerfully set forth in N. Davison, 'Autonomous weapon systems: An ethical basis for human control?', ICRC Humanitarian Law and Policy blog, 3 April 2018. However, for a sceptical, somewhat critical philosophical analysis of the linkage often made between notions

Most interestingly for the purposes of this book, the Chair's report observes: 'The dual-use character of autonomously operating technology and their benefits for civilian applications was raised by a number of delegations. There was a widely shared view that legitimate developments in the civilian sphere should not be hampered by regulatory measures taken with regard to LAWS. In this context, Protocol IV of the CCW was cited as an example of banning a future weapon category without curtailing research and development in the civilian sphere.'[56] The Report shows that a broadly accepted understanding of autonomy has yet to emerge[57] but that there is a generally held view that 'machines are simply incapable of executing legal judgments as required by IHL'. The adequacy of the current rules of IHL to address all issues arising from LAWS seems to be a matter of dispute but even if LAWS could be used in compliance with IHL, there is a recognised ethical dilemma over delegating decisions on human life to machines.[58] Under the auspices of the CCW a meeting of a Group of Governmental Experts took place in Geneva from 3 to 17 November 2017.[59]

Having discussed the legal implications of applying autonomous technology to weapon systems, the chapter will now look at some examples of the employment of autonomy and automation in the peacetime environment[60]

of human dignity and concerns arising from the development of autonomous weapon systems, see A. Pop, 'Autonomous Weapon Systems: A Threat to Human Dignity', ICRC Humanitarian Law and Policy Blog, 10 April 2018.

56 2016 LAWS Report, para 22.

57 2016 LAWS Report, paras 35 to 42.

58 2016 LAWS Report, paras 46 and 47. There is also the argument, articulated, e.g., by the US and certain commentators, that the law of armed conflict applies to humans not machines, that a person launching an autonomous weapon or ordering it to be launched has a responsibility under IHL to ensure that the attack will be lawful, that this obligation cannot be delegated to a machine and that this means that some necessary human involvement is required to make sure that the principle of distinction and rules as to proportionality and precautions are adhered to; P. Scharre, 'Human Judgment and Lethal Decision-Making in War', ICRC Humanitarian Law and Policy Blog, 11 April 2018.

59 For some discussion of aspects of that meeting and an explanation of a 'compliance-based' approach to the issues posed by lethal autonomous weapon systems, see N Srürchler and M. Sigriest, 'A "Compliance-Based" Approach to Autonomous Weapon Systems', Ejil Talk, 1 December 2017, available at www.ejiltalk.org/a-compliance-based-approach-to-autonomous-weapon-systems/.

60 States are likely to issue national guidance on these matters. In the UK, BSI, the business standards company, has published BS8611, Robots and Robotic Devices: Guide to the Ethical Design and Application of Robots and Robotic Systems (2016). The guidance is intended for robot and robotics device designers and managers as well as for the general public. It identifies potential ethical harms associated with such technologies, provides risk reduction guidelines such as safe design, protective measures and information for the design and application of robots and addresses, e.g., industrial and cybernetics applications, safety measures and devices, hazards, control systems and instructions for use.

and will consider how law, regulations and guidance seek to deal with the issues involved.

6.10 REGULATION OF PEACETIME APPLICATIONS OF AUTONOMY

There are numerous applications of artificial intelligence in the peacetime environment[61] and in this section three particular fields have been chosen for specific consideration. The choice of these three fields should not blind the reader to other applications the regulation arrangements for which may be of particular relevance to the broader discussion in this book. Similarly, the introduction of automation and autonomy technologies in the commercial sector seems likely to have far reaching implications for employment, but those aspects lie outside the intended scope of the present chapter.[62] Thus, for example, AI is capable of being used by Internet providers to identify and exclude unacceptable or unlawful material from the Internet.[63] Nevertheless, we shall start by considering the application of these technologies in one aspect of civil aviation.

6.10.1 *Automation and Autonomy in Civil Aviation*

Significant effort is currently being devoted to the development of automated and autonomous civil aviation systems.[64] The following paragraphs do not claim to constitute an exhaustive explanation of all of the rules and guidance applicable to such civil automated and autonomous aircraft operations.[65] That

[61] Consider, for example, the integration of autonomous systems with the Internet of Things in the context of 'smart cities' as foreseen by Siemens and reported in A. Greenfield, *Radical Technologies* (New York: Verso, 2017), 52, and for a sceptical assessment of the 'smart city' concept, ibid., 53–59.

[62] For a discussion of these aspects, see A .Greenfield, *Radical Technologies*, 183–207.

[63] See, e.g., M. White, 'New Tech Helps Facebook Remove 99% of IS Posts Before Users Flag It Up', Sky News, 29 November 2017, available at https://news.sky.com/story/new-tech-helps-facebook-remove-99-of-is-posts-before-users-flag-it-up-11148386.

[64] Consider, for example, the CityAirbus project which is initially intended to operate on a remotely piloted basis but with a view to eventual autonomous operation; for a description of the project see www.airbus.com/newsroom/press-releases/en/2017/10/cityairbus-demonstrator-passes-major-propulsion-testing-mileston.html (accessed on 9 February 2018) and see J. Ostrower, 'Airbus "Flying Car" Takes to The Skies for the First Time', CNN Tech, 1 February 2018, available at http://money.cnn.com/2018/02/01/technology/airbus-vahana-first-flights/index.html.

[65] This section of the Chapter does not, for example, discuss the proposals issued by the European Aviation Safety Agency for the introduction of a regulatory framework for the operation of all drones regardless of their maximum take-off mass; see European Aviation

would clearly lie well outside the intended scope of the present chapter. This section does, however, cite some relevant rules, guidance and treaty provision in order to seek to illustrate how the relevant authorities seek to address the challenges that are posed by autonomous civil flying operations.[66] Certain rules and guidance issued by the UK Civil Aviation Authority (CAA) and by the US Federal Aviation Administration (FAA) form the basis of the discussion.

Operation of small unmanned aircraft must be in accordance with Article 94 of the UK's Air Navigation Order 2016 (ANO).[67] This requires, inter alia, that the person in charge of the aircraft must maintain 'direct, unaided visual contact with the aircraft sufficient to monitor its flight path in relation to other aircraft, persons, vehicles, vessels and structures for the purpose of avoiding collisions'.[68] It also requires that small unmanned aircraft having a mass excluding fuel of more than 7 kg may not be flown either in certain classes of airspace or in other specified airspace without air traffic control permission, and may not otherwise be flown at an altitude greater than 400 feet. Article 95 of the ANO prescribes additional flying restrictions in relation to small unmanned surveillance aircraft. The order does not make specific provision in relation to autonomous flight operations. Where other unmanned aircraft are concerned, part 2 of Schedule 4 of the Order suggests they will be classified as would any other aircraft in accordance with paragraph 1 of the Schedule, will be subject to the Air Navigation Order rules applying to such aircraft and will be 'further classified' as unmanned. This suggests that Articles 66 and 67 as to the carriage of crew on the aircraft apply and that the pilot in command must, inter alia, discharge the responsibilities set forth in the ensuing Articles.

The Safety and Airspace Regulation Group of the United Kingdom Civil Aviation Authority addresses UK Flight Information Services under a document CAP 774 dated November 2013 where the possibility of autonomous

Safety Agency document NPA 2017–05 (A) (2017) where reference is made to autonomous flight.

66 Note that BAe Systems has been testing a modified full size aircraft, a Jetstream 31, to undertake autonomous flights from its base at Warton to Inverness. The aircraft uses cameras to detect clouds and other aircraft and to take appropriate action. 'A ground control station, commanded by a qualified pilot, supervises the autonomous phase of each flight'; G. Corfield, 'BAe Systems' Autonomous Research Aircraft Flies Itself To Scotland', *The Register*, 5 December 2016, available at www.theregister.co.uk/2016/12/05/bae_systems_autonomous_flying/.

67 2016 SI 763. The Order defines small unmanned aircraft as 'any unmanned aircraft other than a balloon or a kite having a mass of not more than 20 kg...'.

68 2016 SI 763, Article 94(3).

flight in Class G airspace[69] is noted, but not regulated.[70] The UK CAA policy and guidance on unmanned air systems is to be found in CAP 722.[71]

The first thing to understand about CAP 722 is the most obvious point taken from its title, i.e. it is guidance. Clearly the purpose of the guidance is to seek to achieve safety in the employment of Unmanned Aircraft Systems (UAS) in UK airspace. This emphasis on safety is something of a recurring theme in the discussion of the three peacetime applications of automation and autonomy discussed in the present section. The expression 'unmanned aircraft' is defined as '[a]n aircraft which is intended to operate with no human pilot on board... [and]...is capable of sustained flight by aerodynamic means; is remotely piloted and/or capable of degrees of automated or autonomous operation; is reusable; and is not classified as a guided weapon or similar one-shot device designed for the delivery of munitions.'[72]

Chapter 3 of CAP 722 sets forth the CAA's policy on operation of autonomous systems in UK airspace. Having noted that currently no UAS system meets the CAA definition of autonomy, the document defines an 'autonomous aircraft' as '[a]n unmanned aircraft that does not allow pilot intervention in the management of the flight' and the term 'autonomous operation' is interpreted accordingly.[73] Systems are 'highly automated' for the purposes of CAP 722 if they 'still require inputs from a human operator (e.g., confirmation of a proposed action) but [...] can implement the action without further human interaction once the initial input has been provided.'[74] This interpretation would appear to be potentially consistent with the earlier discussion of 'man on the loop systems', in the sense that the 'initial input' referred to in the CAA definition can be understood as the input at the commencement of the mission prescribing, e.g., the task that the platform is to undertake. The capability of the platform to 'implement the action without further human interaction' is consistent with the suggestion that a man on the loop system will be capable of automated or autonomous operation, but in the man on the loop case, of course, the operator retains the power to intervene and

69 In the UK Class G airspace is uncontrolled. Accordingly there are no restrictions on which aircraft can enter such airspace, on what equipment they must carry or on the routes that they must take. In the UK all other airspace is controlled and aircraft are directed by air traffic controllers; see CAA, 'How is UK Airspace Structured?', available at www.caa.co.uk/Consumers/Guide-to-aviation/Airspace/How-is-UK-airspace-structured-/.

70 Note to para 5.1 on page 38.

71 UK Civil Aviation Authority, Unmanned Aircraft System Operations in UK Airspace – Guidance, CAP 722, sixth edn, dated 31 March 2015.

72 CAP 722, Glossary at pages 19–20.

73 CAP 722, Glossary on page 17.

74 CAP 722, Glossary at page 18.

countermand machine-decided action. The CAA policy then describes an autonomous system as one that will do everything for itself using high authority automated systems. 'It will be able to follow the planned route, communicate with Aircraft Controllers and other airspace users, detect, diagnose and recover from faults and operate at least as safely as a system with continuous human involvement.'[75] Paragraph 3.8 requires High Authority Automatic Systems to be deterministic, i.e. always to react in the same way to the same set of data, noting that the actions related to a dataset must be appropriate and must not hazard the aircraft or any third parties in the same area. CAP 722 notes that there may be varying degrees of system authority including full authority in which the systems are capable of operating without human control or oversight,[76] equating broadly with the notion of autonomy developed earlier in this chapter.

CAP 722 notes that intervention by a human to avert dangers may not be practical in particular circumstances, e.g., due to signal latency, so the UAS would need to be able to mitigate that risk. Paragraph 3.14 addresses learning/self-modifying systems, noting that they have the potential to be used in UAS but that the overall safety requirements would still apply such that 'it may not be possible to use these systems to their full potential'. Paragraph 3.16 of CAP 722 seems to be contemplating the possible future use of what it describes as 'non-deterministic' systems, broadly equating with fully autonomous aircraft systems, to improve system flexibility and performance. However, '[w]hilst there are no regulations that specifically prohibit this, the use of non-deterministic systems will drive a number of system and operational safety assessment issues that will need to be addressed before the use of this type of technology could be accepted for use in aviation.' So there are parallels here with the position in relation to autonomous weapons. The possibility of autonomous functions in aviation is recognised, but there are appreciated concerns that need to be addressed first and those concerns relate to the systems themselves and to operational safety issues.

It is hardly surprising that highly automated and autonomous technologies, particularly the latter, raise specific issues in the peacetime air navigation field. The understandable, indeed comforting, focus is on the maintenance of air safety, including safety of navigation, for all air users. This reflects a core function of the CAA as reflected in the Civil Aviation Act 1982[77] and in numerous provisions of the Air Navigation Order.[78] Full autonomy is

[75] CAP 722, para 3.4 on page 63.
[76] CAP 722, para 3.11 on page 64.
[77] See Section 3(c) of the Civil Aviation Act 1982 (1982 c. 16).
[78] Consider for example Articles 113, 184 and 213 of the Air Navigation Order, 2016.

recognised as raising difficult issues and is not therefore currently 'accepted for use in aviation'. That does not mean of course that it will not at some future point become accepted, nor that automation of particular systems will be unacceptable. Indeed numerous systems are already automated in civil aviation. It is interesting that the CAA guidance document specifically addresses learning systems, and that it distinguishes between the technologies on similar lines to those identified in Section 6.6. It can be safely concluded that obtaining approval from the relevant national aviation authority or authorities both as to the operation of autonomous systems at all and as to the arrangements for their lawful operation in particular airspace will be matters requiring careful attention before such systems can properly be introduced.

So, as a Parliamentary Postnote observes, '[f]or the foreseeable future, direct responsibility via human pilots will still be required for unmanned flights, especially those operating in close proximity to other air users and populations ... It is expected that fully autonomous systems would need full 'situational awareness' being able to follow or plan a route, sense, avoid and communicate with other air users, and find and rectify faults.'[79]

The Federal Aviation Authority of the United States has issued guidance and rules for the operation of small drones. Small unmanned aircraft systems are regulated under Title 14 of the Code of Federal Regulations (Part 107). Small unmanned aircraft systems are for these purposes unmanned aircraft of less than 55 pounds weight. The operator of the unmanned system must register the aircraft and must hold a FAA remote pilot certificate. The applicable rules include requirements that the aircraft must be flown within visual line of sight at or below 100 miles per hour, must yield right of way to manned aircraft, must not be flown from a moving vehicle (unless in a sparsely populated area) and must not be flown near other aircraft or directly over people. The drone must not be flown in controlled airspace near airports without FAA permission and it may only be flown during daylight or civil twilight at or below 400 feet. Certain of the rules are subject to waiver.[80] The rules are clearly centred on operator-controlled platforms. Part 107 does, however, make it clear that autonomous operations of such drones are permissible, but this is subject to the important proviso that 'the remote pilot in command must retain the ability to direct the small unmanned aircraft to

[79] Parliamentary Office of Science and Technology, Postnote, Number 479, October 2014, 3.

[80] Federal Aviation Administration, Fly Under the Small UAS Rule, 15 December 2017, available at www.faa.gov/uas/getting_started/ and for detailed guidance on conducting small UAS operations in accordance with Part 107, see FAA Advisory Circular 21–2 dated 21 June 2016, available at www.faa.gov/uas/media/AC_107-2_AFS-1_Signed.pdf.

ensure compliance with the requirements of part 107.'[81] That is an arrangement which, again, seems to amount to what we described earlier as 'man on the loop' operation and would not seem to constitute full autonomy, or even proper automation. It is worth making the point that while Part 107 and equivalents elsewhere, such as the small UAS regime in the UK, are often primarily intended to address leisure and recreational use, the US regime says nothing about the purpose of the flight and therefore commercial operations may take place in the US without further approval provided that the requirements of Part 107 are met. These requirements appear to be inconsistent with plans for, e.g., beyond line of sight autonomous drone delivery systems, such as have been discussed for example by Amazon in the US.

As something of a postscript, it should be recalled that Article 8 of the Chicago Convention 1944 refers to pilotless aircraft. The Article provides: 'No aircraft capable of being flown without a pilot shall be flown without a pilot over the territory of a contracting State without special authorisation by that State and in accordance with the terms of such authorisation. Each contracting State undertakes to insure that the flight of such aircraft without a pilot in regions open to civil aircraft shall be so controlled as to obviate danger to civil aircraft.'[82] Currently, this is most likely to apply to remotely piloted aircraft. It would, of course, be capable of application to highly automated or autonomous aircraft, and would have the effect that such aircraft can only be operated in the national airspace of another State if that State has specifically authorised such operations. Any State contemplating granting such authorisation would need to stipulate in advance the detailed restrictions on the times, places, altitudes etc. where such highly automated or autonomous flight operations can take place with a view to ensuring de-confliction with other civil aircraft movements and would be well advised to stipulate the technical and operational safeguards that must be complied with by those undertaking automated or autonomous flight operations in order to ensure the safety of all aircraft flying within the airspace for which it is responsible.

Perhaps the most obvious, yet sensible, theme that seems to run through all of the rules, guidance and treaty provisions that have been cited in this part of the discussion is the common focus on ensuring the safety of operation of civil aviation. While pilotless aircraft can indeed be operated, national schemes of

[81] See 81 FR 42063, Operation and Certification of Small Unmanned Aircraft Systems, FAA, 28 June 2016, Section VII Automation; see also E. 'Ackerman, FAA Unveils Drone Rules: Autonomy is In, Drone Delivery is Out', *IEEE Spectrum*, 16 February 2015, available at http://spectrum.ieee.org/automaton/robotics/drones/faa-proposed-commercial-drone-rules.

[82] Convention on International Civil Aviation, Chicago, 7 December 1944, Article 8.

regulation must 'obviate danger to civil aircraft', and highly automated and autonomous aircraft operation must be limited sufficiently to achieve that core requirement. Where CAP 722 is concerned, while the text is presented as guidance and thus does not appear to have directly regulatory status, it can perhaps best be seen as guidance as to how the CAA will interpret and discharge its responsibilities and functions and thus effectively sets the basis on which aviation activities, operations etc. will be permitted and, e.g., whether authority to fly will be granted.

6.10.2 *Automation and Autonomy in Surgical Practice*

The use of robots in surgery procedures has attracted a certain amount of publicity in recent years.[83] Advantages in the use of such equipment include the increased precision that can be achieved in certain kinds of surgery, and reduced intrusiveness, reportedly resulting in improved post-surgery recovery times. There is also the possibility that such technologies can assist in dealing with matters such as complex joint fractures. It is evident, however, that the equipment continues to be a tool that is employed by the surgeon, as opposed to a device that takes over and itself becomes the surgeon. As a recent discussion of the topic identified, the technological challenges involved in giving the robot the skill and judgment of a surgeon, the risk of legal challenge and the reassurance of having an expert on hand are reasons limiting the introduction of automation into surgery.[84] Accordingly, it appears that surgical systems such as the da Vinci system are used under the supervision of, or are employed to assist, a surgeon and as a result yield, e.g., the precision benefits already mentioned earlier.[85] In 2016 it was, however, reported that the Smart Tissue Autonomous Robot (STAR) successfully completed open bowel surgery in pigs. The STAR system 'is entirely autonomous ... and able to react to the uncertain, dynamic landscapes of soft tissue surgery without direction ... STAR reacts to a changing environment, similar to how self-driving cars are

[83] A. Eisenberg, 'When Robotic Surgery Leaves Just a Scratch', *New York Times*, 17 November 2012.

[84] M. Piesing, 'Medical Robotics: Would you Trust a Robot with a Scalpel?' *The Guardian*, 10 October 2014.

[85] P. Ganapati, 'Surgical Robots Operate With Precision', *Wired*, 9 November 2009, available at www.wired.com/2009/09/surgical-robots/. Note the emphasis in the Surgical Robot Challenge 2017 on 'shared autonomy between the surgeon and the robot'; The Hamlyn Symposium 2017, Surgical Robot Challenge 2017, available at http://hamlyn.doc.ic.ac.uk/hsmr/events/surgical-robot-challenge-2017. For a practical example, see M. Simon, 'A Robot Ear Surgeon Drills into the Future of Medicine', *Wired*, 15 March 2017, available at www.wired.com/2017/03/robot-ear-surgeon-drills-future-medicine/.

programmed to not only drive on the highway, but also to react to someone else making a mistake and getting in your lane.' The robot also has a highly sophisticated, three-dimensional visual system and reportedly out-performed manual procedures in several open-bowel surgeries.[86]

In the United States it is the Food and Drug Administration that is responsible for the clearance of, inter alia, Robotically Assisted Surgical (RAS) Devices. 'The FDA has cleared RAS devices for use by trained physicians in an operating room environment for laparoscopic surgical procedures in general surgery, cardiac, colorectal, gynecologic, head and neck, thoracic and urologic surgical procedures. Some common procedures that may involve RAS devices are gall-bladder removal, hysterectomy and prostatectomy (removal of the prostate).'[87] The critically important word here is 'assisted' which suggests that the use of the device is to assist the surgeon who must be taken to retain prime responsibility for the procedure being undertaken. This implies that the robot remains a tool employed by the surgeon, as opposed to becoming the surgeon itself. It would therefore probably equate with a degree of automation, but could not properly be described as autonomous in the sense that that word is being used in the present chapter.

Some commentators have proposed a framework of six levels of autonomy for medical robotics. Level 0 corresponds to no autonomy, and includes, e.g., prosthetic devices in which the device responds to the user's commands. At level 1, robot assistance, the robot provides mechanical guidance or assistance during a task, but the human retains continuous control, e.g., lower limb devices with balance control. Level 2, task autonomy, involves, e.g., the robot undertaking suturing in places prescribed by the human but with the human monitoring the robot ready to intervene as necessary. Conditional autonomy, i.e. level 3, consists of a human selecting from strategies developed by the machine or approving an autonomously selected strategy. High autonomy at level 4 involves the robot making medical decisions under the supervision of a qualified doctor while level 5, full autonomy, consists of a robotic surgeon that performs an entire surgery.[88] This framework seems to make sense in that it

[86] L. Bushak, 'Robot Surgeon Succeeds Without Help From Human Doctors', *Medical Daily*, 13 May 2016, available at www.medicaldaily.com/robot-surgeon-smart-tissue-autonomous-386079.

[87] US Food and Drug Administration, Computer-Assisted Surgical Systems, last updated 29 July 2015, available at www.fda.gov/MedicalDevices/ProductsandMedicalProcedures/SurgeryandLifeSupport/ComputerAssistedSurgicalSystems/default.htm.

[88] G.-Z. Yang, et al., 'Medical robotics: Regulatory, Ethical And Legal Considerations for Increasing Levels of Autonomy', 2 *Science Robotics* 4, 15 March 2017, available at http://robotics.sciencemag.org/content/2/4/eaam8638.full.

breaks down the degree of autonomy that the robot is permitted and the degree and nature of required human supervision and control in a way that facilitates discussion of regulatory and related issues. As the level of autonomy in these devices increases so the stringency, costs and time involved in obtaining approval also increase. The important point is that at the higher levels of autonomy, level 5 and possibly also level 4, these factors may be further increased by the fact that the device is 'practising medicine' for which approval is required from the relevant professional body. Critical is the recognition that these developments involve risk that must be appropriately managed.[89]

A couple of points emerge from this part of the discussion. The difficulty in equipping a robot with the skill and judgement of a surgeon would seem to have parallels in the inability, identified earlier, of currently available autonomous weapon technologies to undertake the evaluative decisions required by targeting law. The human surgeon will inevitably be confronted with the repeated requirement to make judgements, comparisons, evaluative assessments and to apply experience to make decisions that do not boil down to arithmetic calculations, and it is the inability of currently available technology to do this that renders the human surgeon as irreplaceable as the weapon operator.

The risk of legal challenge has also been cited as inhibiting the adoption of autonomous or highly automated surgical procedures. Perhaps this is a time-limited objection. After all, if, say, an automated surgical procedure were to be developed that achieves levels of reliability and surgical success that human surgeons can no longer match, once that technical superiority has been proved, it is imaginable that it would be the allocation of a human surgeon in a particular case that might precipitate legal challenge.

The reassurance of having an expert on hand will remain as a powerful reason for the continued requirement for surgeons even in cases in which automated or even autonomous procedures have become the norm. The foreseeable challenge then would be to ensure that the surgeons concerned retain sufficient experience and, thus, expertise for their continued presence actually to be a proper source of reassurance.

6.10.3 *Driverless Cars*

The testing and proposed introduction of driverless cars offers another, important example of the application of autonomous technologies in the non-

89 G.-Z. Yang, et al., 'Medical Robotics'.

military context. The first thing to note is that autonomous vehicles 'do not drastically alter the regulatory theory perspective of road behavior.... Road rules, the police and courts, development standards, and infrastructure are no less influential on road behavior when vehicles operate autonomously.'[90] Secondly, it is becoming increasingly evident that this is a topic that is developing very rapidly. Numerous motor car manufacturers are devoting substantial resources to the development of the technology. States are similarly actively engaged in working out policies, and in some cases legislation, to address the testing, and potentially the use, of autonomous technology on the roads.[91] The following discussion will address some of the initiatives but cannot hope to chart all of the relevant developments worldwide.

Legal rules are being developed to regulate autonomous vehicle operation in California. Both at federal and California state levels, the documents that are discussed below contain extensive and detailed provisions. It should be appreciated that in this section only the provisions of greatest evident relevance to the themes of the chapter and of the book are discussed.

At the US federal level, the National Highway Traffic Safety Administration (NHTSA) has issued its Federal Automated Vehicles Policy.[92] The Policy notes that 35,092 people died on US roadways in 2015 and that 92 per cent of all crashes can be tied to human choice or error. 'An important promise of highly automated vehicles is to address and mitigate that overwhelming majority of crashes'.[93] Accordingly, road safety is placed firmly at the centre of the policy justification for the legalisation of driverless vehicles. The federal policy incorporates vehicle performance guidance, which outlines best

[90] Comments attributed to A/Prof C Dent, Report of the Symposium on Ethical, Legal and Social Implications of Emerging Military Technologies, 29–30 May 2017, Melbourne Law School, August 2017, 4.

[91] The Automated and Electric Vehicles Bill is, at the time of writing, before the UK Parliament. In related to autonomous vehicles, however, the legislation only addresses compulsory motor vehicle insurance arrangements. The Bill would include provisions: as to the maintenance of a list of automated vehicles considered to be capable of safely driving themselves; as to the liability of insurers when an accident is caused by an automated vehicle; as to contributory negligence; as to accidents caused by unauthorised alterations to, or failure to update, software; and as to the right of an insurer to claim against the person responsible for an accident. Interestingly, a vehicle would be 'driving itself' for the purposes of the proposed legislation if 'it is operating in a mode in which it is not being controlled, and does not need to be monitored, by an individual'; Automated and Electric Vehicles Bill (HC Bill 112), available at https://publications.parliament.uk/pa/bills/cbill/2017–2019/0112/cbill_2017–20190112_en_2.htm#pt1-l1g1.

[92] The policy was released on 20 September 2016 and is available at https://one.nhtsa.gov/nhtsa/av/av-policy.html.

[93] NHTSA Federal Automated Vehicle Policy, Executive Summary.

practice for the design, development and testing of automated vehicles, and model policies for states that have the evident purpose of achieving a harmonised approach across states within the US. The policy document explores the current regulatory framework and identifies new tools, authorities and regulatory structures to facilitate these technological developments. It is, however, clear that it is the State of California that has been taking the lead in exploring the issues associated with the introduction to the public roads of driverless vehicles. So having recognised the federal framework, the discussion will now shift to look at how the regulation of testing arrangements applicable in California has evolved.

On 10 March 2017 the US State of California Department of Motor Vehicles published proposed regulations to establish a path for the testing and deployment of fully autonomous vehicles in California. Already twenty-three manufacturers had been testing autonomous vehicles with test drivers behind the wheel on the roads of the state. In the following paragraphs, material will be considered that is taken from the proposal that was published.[94]

Senate Bill 1298 of 2012 enacted vehicle code 38750, which requires the Department for Motor Vehicles (DMV) to adopt the regulations that are needed to secure the safe operation of autonomous vehicles on the public roads whether or not a driver is inside the vehicle. It is therefore made evident that the initiative for the introduction of driverless vehicles is motivated by road safety considerations and that the testing of such technology must be undertaken in a way that meets any safety concerns.

The regulations of 2014 required the presence of a driver but the Senate Bill contemplated the development of technology that would not necessarily require the presence of a driver in the vehicle and manufacturers have now developed systems capable of operating without a driver. The proposal was therefore to amend the relevant provision, called Article 3.7, to permit the safe testing on public roads of such vehicles. The Article sets forth the requirements for a Manufacturer's Testing Permit and the intention was to promote the development of autonomous technology with the potential to enhance mobility and increase safety. So the requirement placed on the DMV was to adopt regulations giving the requirements associated with applications by autonomous vehicle manufacturers for the operation, referred to as deployment, of autonomous vehicles by the public on public roads. Article 3.8 would

[94] State of California, Department of Motor Vehicles, Title 13, Division 1, Chapter 1, Initial Statement of Reasons – Article 3.7 and Article 3.8.

be added to specify the requirements to deploy autonomous vehicles including those not requiring the presence of a driver in the vehicle.

Under a federal statute of 1966,[95] no motor vehicle may be sold for use on US public roads unless the manufacturer certifies that the vehicle meets specified performance requirements. So federal safety requirements must be reflected in the regulations, including a requirement for manufacturers to certify that their autonomous vehicles satisfy the specified performance requirements. Where testing without a driver of all levels of autonomous vehicles is concerned, the proposed regulations required the manufacturer to submit a fifteen-point safety assessment letter pursuant to the 'Vehicle Performance Guidance for Automated Vehicles' contained in the Federal Automated Vehicles Policy to which reference was made earlier.[96] So, the purpose of the regulations is to allow manufacturers of autonomous vehicles to develop and test autonomous vehicles on public roadways and to deploy the vehicle in a way that addresses public safety concerns.

It is informative to note the options that were considered and rejected. These included simply prohibiting the testing of such vehicles, but 'the rapid advancement in the development of autonomous technology' and 'mobility options' are cited as reasons for permitting such testing.

Another possibility was also considered, namely to require manufacturers to have a vehicle demonstration test conducted by an independent third party to assess the vehicle's ability to perform driving tasks and then to require the submission of a report certifying that the vehicle performed in such a way as would be necessary for safe operation on public streets. It was, however, concluded that the inability to prescribe pass/fail criteria might result in anomalous results.

Accordingly, under the proposed regulations, autonomous operation would involve a combination of hardware and software, both remote and on-board, that performs the dynamic driving task with or without a natural person actively monitoring the driving environment.[97] This would not include systems limited to providing driver assistance or that do not perform the dynamic driving task on a sustained basis or without the constant active monitoring or control of a natural person.[98] The 'dynamic driving task' would mean all real time functions required to operate a vehicle in on-road traffic. The term would not include the selection of intermediate and final

95 National Traffic and Motor Vehicle Safety Act 1966.

96 See NHTSA Policy, 20 September 2016.

97 Article 3.7, para 227.02 Definitions.

98 See levels 3 to 5 inclusive of the taxonomy of automation in the NHTSA Federal Automated Vehicle Policy.

destinations, but would incorporate the following: object and event detection, recognition and classification; object and event response; manoeuvre planning; steering, turning, lane-keeping, and lane changing including providing the appropriate signal for the lane change or turn manoeuvre; and acceleration and deceleration.[99]

There were certain other stipulations in the proposed state regulations. The autonomous vehicle would be required to have a mechanism to engage and disengage the autonomous technology that is easily accessible to the operator; to have a communication link to the remote operator to provide information on its location and status; and to provide continuous two-way communication between the remote operator and any passengers in the vehicle should failures take place that endanger safety or that stop the vehicle from functioning as intended.

On 26 February 2018, the State of California Department of Motor Vehicles and Office of Administrative Law adopted regulations to deal with the testing of autonomous vehicles on the state's public roads.[100] The definition of 'autonomous operation' that appeared in the draft regulation is reflected in broadly similar terms in the definition of 'autonomous mode' in the 2018 Regulation.[101] Crucially, the definition of 'autonomous test vehicle 'requires a human test driver or a remote operator to continuously supervise the vehicle's performance of the dynamic driving task'.[102] To meet the requirements of an 'autonomous vehicle test driver', the person must be licensed to drive the relevant class of vehicle and must be 'capable of taking over active physical control of the vehicle at any time'. The definition of 'dynamic driving task' seen in the draft regulation is repeated in the substantive regulation. The definitional requirements for a person to be a 'remote operator' in accordance with the regulation address the safety concerns highlighted in the language used in the draft regulations. So, a remote operator must be a natural person who is appropriately licensed for the class of vehicle, who is not in the driver's seat of the vehicle and who 'engages and monitors the autonomous vehicle; [and] is able to communicate with occupants in the vehicle through a communication link.'[103] The remote operator may also be able to perform the dynamic driving task for the vehicle or cause it to achieve a 'minimal risk condition'.[104] The Regulation specifies requirements for the issue of a

99 Article 3.7, para 227.02, sub-division (g).

100 Order to Adopt, Title 13, Division 1, Chapter 1, Article 3.7 – Testing of Autonomous Vehicles, dated 26 February 2017.

101 Ibid., Definitions, para (a).

102 Order to Adopt, n. 100, Definitions, para (b).

103 Order to Adopt, n. 100, Definitions, para (n).

104 Ibid.

manufacturer's testing permit, requiring, inter alia, that the manufacturer conduct the testing. Comprehensive provisions dealing with all relevant aspects of vehicle testing, specifying the requirements for autonomous vehicle test drivers and dealing with numerous other relevant aspects are set forth in these lengthy and apparently comprehensive regulations.

Interestingly, there have been a number of US federal-level initiatives which also illustrate evolving thinking on the matter. The Self Drive Act has been passed by the US House of Representatives but, given the role of the Senate, it is at this stage only an indication of the form federal legislation might take on the matter. Nevertheless it is worth noting that the legislation would set forth a basic federal framework for autonomous vehicle regulation by giving the National Highway Traffic Safety Administration (NHTSA) (as opposed to individual states) power to regulate the design, construction, and performance of such vehicles. It would be for the NHTSA to specify what manufacturers must show in relation to safety, what features require performance standards, how to demonstrate the safety of a vehicle's sensor configuration, its cyber-security, etc. Vehicle manufacturers would be required to have privacy plans for the protection of passengers' personal data relating to their use of the vehicle and their action associated with it.[105] The US Department of Transportation is also engaged in a broad programme of consultations in relation to 'the safe development, testing and deployment of automated vehicle technology.'[106]

In May 2017, the upper house of the German parliament approved a law under which self-driving cars can be used on German roads. Road testing is permitted of such cars if a driver is behind the wheel at all times and ready to take back control if so prompted by the autonomous vehicle. A 'black box' must record the journey and the driver will retain responsibility unless 'the self-driving system is in charge and a system failure is to blame' in which case 'the manufacturer will be responsible' for an accident.[107] Data protection issues have yet to be addressed, but may feature when the legislation is reviewed in 2019.

[105] A. Marshall, 'Congress Unites (Gasp) to Spread Self-Driving Cars Across America', *Wired* 6 September 2017, available at www.wired.com/story/congress-self-driving-car-law-bill/.

[106] See USDOT Automated Vehicles Activities, US Department of Transportation, 23 February 2018, available at www.transportation.gov/AV.

[107] 'Germany Adopts Self-Driving Vehicles Law', Reuters, 12 May 2017, available at www.reuters.com/article/us-germany-autos-self-driving/germany-adopts-self-driving-vehicles-law-idUSKBN1881HY. See also N. Lomas, 'UK Kicks Off Driverless Car Law Review to Get Tech On The Road by 2021', *TechCrunch*, 7 March 2018, available at https://techcrunch.com/2018/03/07/uk-kicks-off-driverless-car-law-review-to-get-tech-on-the-road-by-2021/.

Returning to the US approach, however, the interesting aspects are the formulation of a policy at the national level, the generation of state-level proposals, consultation in relation to those proposals and all with a view to the creation and ultimate implementation of regulations, i.e. hard domestic law, governing a testing regime the purpose of which is to evaluate the introduction of such vehicles onto the state's public roads. This is clearly a cautious, step-by-step process designed to identify what the issues are and then to formulate regulations that will address those issues. Experience gained from the testing process will obviously inform the content of the regulations that in due course will apply when the testing phase is completed and public use of automated vehicles is introduced. One can detect that there is a public relations dimension at work here. The prospect of vehicles on the public roads without a driver in control will raise obvious concerns in the minds of many citizens. On the other hand, there is the implicit appreciation that the technology may be expected to significantly out-perform human drivers in safety terms. So intuitive assumptions seem likely to be challenged by empirical experience, but only when the safety, indeed the superiority, of the new technology has been proven will public acceptance of it, maybe even demands for it, be likely to be generally forthcoming. The resulting cautious approach to policy development, consultation, testing and eventual implementation is, however, the way to avoid things going wrong. Where public support might easily be lost if say a series of unfortunate accidents were to occur, a sensible, logical and cautious approach will be necessary to ensure that the eventual arrangements that are made match the popular view.

6.11 CONCLUSIONS

We should start this concluding section by seeking to answer the question that constitutes the theme for the book as a whole, namely whether there are aspects of the arrangements for the peacetime regulation of autonomous technologies that could benefit the law as to their military application or vice versa. When considering the employment of autonomous technology in civil aviation, we concluded that parallels do indeed exist with the position in relation to autonomous weapons. We observed that the possibility of autonomous functions in aviation is recognised, but there are understood concerns that need to be addressed first and those concerns relate to the systems themselves and to operational safety issues. There is no explicit recognition in the law of armed conflict of autonomous technology and, more particularly, there are no rules that make specific provision for autonomous weapons of the kind currently under discussion. The generally applicable law of armed conflict rules as to precautions in attack seem to be the nearest equivalent of the civil aviation focus on operational safety. This is particularly reflected, for example, in Article 8 of the Chicago Convention to which reference was made earlier.

In relation to automated and autonomous surgery, we noted that the human surgeon is needed to make judgments, comparisons and evaluative assessments and to apply experience on a regular basis. It is the inability of machines to make decisions that do not boil down to mathematical calculations that renders the human surgeon irreplaceable. This broadly equates to the problem we noted earlier, namely that the inability of currently available autonomous and certain highly automated weapon systems to undertake evaluative decisions will restrict their acceptability when they are assessed in a weapon review. The reassurance of having an expert surgeon available when autonomous or automated surgery is taking place has its equivalents in insistence under current arrangements that a driver should be in a position to take back control of a driverless car and that a person must, until technology improves radically, remain 'on the loop' or 'in the loop' of autonomous and highly automated weapon systems that are engaging in offensive attack operations as that notion has been explained in the current chapter.

The noted challenge of ensuring that such monitoring surgeons retain sufficient experience and, thus, expertise for their continued presence actually to be a proper source of reassurance also seems to apply in the context of driverless cars and in the context of persons on or in the loop of autonomously operating weapon systems. This, it seems, is a challenge in all applications of autonomous technologies and breaks down into the need to ensure that the observing human has sufficient skill and experience while also ensuring that the observing human remains sufficiently aware of what is going on, which in turn will be a function of technical connectivity and of human concentration.

The possibility remains that autonomous technologies may perform relevant tasks in both the military and civil contexts more efficiently, more accurately and more reliably than human operators. If and when it is demonstrated conclusively that this is so, it will be for society at large to decide, in the various contexts that have been discussed in this chapter and for that matter in all other contexts in which autonomous technology is to be employed, whether continued human involvement can be justified. This, of course, would represent something of a reversal of the current basis of debate in which the status quo seems to be seen as inherently preferable to future technological possibilities.

It must be appreciated that the uses of robot technology in the peacetime environment that this chapter has addressed are merely examples of the potential applications of such technology. Another example would consist of the application of highly automated processes to the laying of bricks for, e.g., house construction.[108] There can certainly be little doubt that high degrees of automation, and even autonomous systems, will have an increasingly important role to play in many aspects of future peacetime life and will have implications for employment that, as previously explained, lie outside the intended scope of the current chapter.

It is difficult to reach hard and fast general conclusions about the acceptability of autonomy and high levels of automation in the various applications where it seems likely to feature. Perhaps, as such technologies are increasingly seen in particular fields of activity, they will be regarded as somewhat more acceptable in other fields. So for example, employment of automation in surgery, in commercial aviation and on the public roads may well render its employment in warfare, and in particular in relation to weapon systems, potentially less controversial than is the case at the moment. Certainly, if autonomy on the roads is seen as a safer option than driver control as appears currently to be the case, it would be but a short step to apply the same logic to the use of weapon systems. But that assumes that cold logic will inform public attitudes, and it is perhaps just as likely that while applying such technologies in peacetime may be seen as a sensible, even as an advantageous, step, vesting machines with the power of deciding who is to live and who is to die will likely be seen by many as unacceptable in ethical/moral terms.

Developing the point made earlier in this concluding section, there will be some who will wonder what will happen if the machine-made decisions over wartime targeting are shown to be consistently superior to and more reliable than the decisions made by human operators. Imagine a situation in which it can be shown that civilian casualties and damage arising from machine-made targeting decisions are consistently fewer than those arising from human controlled targeting. In such a circumstance, the difficult question would be how much unnecessary civilian death, injury and damage will be the acceptable price to pay, as it were, for an ethical preference for man-made targeting decision-making.

It will be interesting to see whether California's experiment with driverless cars throws up concerns or issues.[109] The suspicion is that if the driverless car

[108] A .Tovey, 'Rise of The Machines: How Robot Brickies Could Be On Site Soon', *Daily Telegraph*, 27 June 2017.

[109] Consider, for example, J. C. Wong, 'Rage against The Machine: Self-Driving Cars Attacked by Angry Californians', *The Guardian*, 6 March 2018, available at www.theguardian.com/technology/2018/mar/06/california-self-driving-cars-attacked.

process involves for the most part, or even exclusively, automated processes in which the machine is programmed to react in a predetermined way in specified circumstances, the challenges may well prove to be manageable if sufficiently prescriptive rules are developed to ensure user and third party safety. Clearly, it is when the decision-making goes from the automated to the autonomous that the legal, regulatory and other challenges become rather acute. It is the very unpredictability that is core to autonomy that at the same time poses the challenge.

This all suggests that the introduction of artificial learning intelligence (ALI) into driverless vehicles will present legislators, testers, administrators and potentially lawyers with some complex issues. Maybe the solution will involve the placing of specified constraints on what the ALI system is permitted to learn, such that it may only learn lessons designed to improve passenger and third party safety. Indeed, constraining the lessons that may be learned and applied by autonomous systems, and constraining the discretion that is granted to such systems is likely to be a pre-condition to their acceptable employment in a variety of contexts.

The legal consequences if highly automated or autonomous technologies malfunction or function correctly but with undesirable consequences differ as between the armed conflict and domestic peacetime situations. The law of armed conflict does not contemplate individual rights to claim compensation,[110] and the consequences of either unlawful or otherwise undesirable attacks decided upon by autonomous weapon systems are unlikely to give rise to such compensation. States will however be anxious to avoid such incidents for other reasons, including the erosion of public confidence and support at home that such events may cause, and the international condemnation that would arise if such wrongful or mistaken attacks became frequent. Failures of autonomous domestic systems in peacetime may well, by contrast, be expected to give rise to damages claims. Contract law, tort and product liability are among the foreseeable legal fields in which such claims may be attempted, and it may be imagined that the reputational damage associated with such claims and of course the potential financial implications of successful claims will be powerful incentives for manufacturers, testers, those drawing up the regulations and those having the responsibility for implementing them to discharge their duties with suitable care. It is in this respect instructive to see the careful approach being undertaken in the State of California, with public consultation taking place over draft regulations designed to address the testing

[110] Hague Convention IV, 1907, Article 3 and API, Article 91 provide for State to State claims for compensation only 'if the case demands' such payment.

of such technologies. These regulations in their final form will have been improved by the consultation process and the experience derived from the testing of the vehicles on the roads will inform further refinements of the regulations. That said, all involved in the process will have their responsibilities the proper discharge of which will be critically important.

So how are we to view the calls for a ban in the weapons context? Does it make sense that a technology that in the context of motor vehicle use is seen as likely to promote safety, in the weapons context is seen as ethically unacceptable? In a sense the California approach is to examine what the technology can do and then to determine whether it is acceptable for use on the public roads. This sort of process could it is suggested be also sensibly applied to the weapons domain. So, rather than prohibit such technology in relation to weapons in ignorance of what it might be capable of offering, a more rational approach would seem to be to develop and to explore the potential for such technologies in weapon platforms with a view to making a more balanced assessment of its acceptability in the light of established facts.

This analysis is, however, based on a somewhat cold application of rather legalistic logic and is unlikely to satisfy those who have a fundamental objection to the development of machines that decide on which human beings are to be targeted and which are to be spared. It is for society at the national and international levels to decide whether the absolutely valid ethical objections should trump the cold legalistic logic or vice versa.

7

Military Human Enhancement

IOANA MARIA PUSCAS

7.1 INTRODUCTION

Human enhancement for military purposes has long been a subject of fascination for the Sci-Fi genre. Increasingly, breakthroughs across biotechnologies and neurotechnologies are making the prospect of a physically and cognitively enhanced soldier real thus bringing forth attendant ethical, operational and legal concerns. This chapter will look into the legal and ethical implications of enhancing human soldiers in operations. It will also compare the operational implications and possibilities for regulation in both the military and civilian realms.

A declassified report from the United States Defense Advanced Research Projects Agency (DARPA) dating from 2002 observed that, in the light of the revolutionary advances in military technology and weaponry, the human element was 'becoming the weakest link' in defence systems.[1] Indeed, by many measurable standards – and as alluded to in DARPA documents – the human soldier's persistent frailty was an encumbrance in war; bound by physical and cognitive limitations, soldiers can make brash and counterproductive decisions in a moment of exhaustion or when anxiety impairs their judgement; unlike a computer, they cannot transfer their knowledge from one to another, and finally, they can easily die.

At the same time, humans are indispensable to war, notwithstanding that autonomous robots and unmanned vehicles increasingly accompany humans

[1] See statement by Dr Tony Tether, submitted to the Subcommittee on Terrorism, Unconventional Threats and Capabilities, House Armed Services Committee, United States House of Representatives, 27 March 2003, 12, www.darpa.mil/attachments/TestimonyArchived(March%2027%202003).pdf. The statement mentions DARPA's Strategic Plan submitted to the Congress in February 2003 and lists 'Bio-Revolution' as one of the eight strategic research thrusts of the Agency (p. 2).

on the battlefield. One solution to this predicament is to enhance soldiers themselves with the help of technology. Human enhancement methodologies and technologies have been given different labels by the armed forces, such as 'performance optimisation', 'peak performance'[2] or 'protecting human assets'.[3]

The idea of enhancing soldiers is not new in the armed forces. For example, during the Vietnam War, the armed forces would prescribe a moderate dosage of 20 mg of dextroamphetamine for 48 hours of combat – a prescription that was hardly followed, as amphetamines became available, in the words of a veteran pilot, 'like candies'.[4] In a span of only three years, between 1966 and 1969 (as revealed in a Report by the House Select Committee on Crime), the armed forces used 225 million tablets of stimulants, mostly Dexedrine, an amphetamine derivative.[5] Despite the problem of overdose and rampant addiction, the effect of the drugs was still limited. They allowed soldiers to be more energetic, fighting their fears and boredom, yet soldiers remained limited in their capacity to fight and continued to be essentially human: no significantly enhanced physical or mental capacities resulted from the use of drugs. The new generation of enhancement procedures and technologies is more radical and potent, and their continuous development promises to give human soldiers literal upgrades to their bodies, allowing them to survive in extreme conditions, repress negative emotions such as fear and panic, be better decision makers and stay focused for many hours. Customised procedures could provide on-demand reactions, such as stimulating certain parts of the brain for better memory and vigilance in conditions of extreme sleep deprivation.

The idea of enhancing human soldiers carries practical and operational opportunities. Three rationales stand out. The first is straightforwardly about performance: an enhanced soldier is more efficient and productive. The second concerns costs: fewer enhanced soldiers could compensate for, and achieve as much as, larger units; this also implies decreased costs for veteran affairs in the long run. It is worth recalling that in 2014, each US service member deployed in Afghanistan cost the United States around $2.1 million annually, and this excluded long-term veteran support costs. This cost surged

2 Statement by Robert Leheny, submitted to the Subcommittee on Terrorism, Unconventional Threats and Capabilities, House Armed Services Committee, United States House of Representatives, 20 May 2009, 24, www.darpa.mil/attachments/TestimonyArchived%20(May%2020%202009).pdf.

3 Ibid., 27.

4 Ł. Kamieński, *Shooting Up: A History of Drugs and War* (New York: Oxford University Press, 2016), 189.

5 Ibid.

from roughly $1.3 million per deployed soldier between 2008 and 2013.[6] The third rationale concerns human cost: with steady resistance to boots-on-the-ground strategies, fewer enhanced soldiers sent into the battle would also mean less human presence and, in theory, a reduced human cost.[7]

Why is enhancement then problematic at all in the armed forces? As a rule, armed forces seek a comparative advantage and therefore prepare personnel continually to be more efficient during combat. If the same technologies were to be applied in society, in the workplace and in schools, they would raise immediate concerns of misuse, of abuse and breach of ethics or meritocracy, but warfare constitutes a uniquely different environment to which 'civilian' ethical concerns are not always readily transferable and where operational success is a key priority, even the overriding consideration.

Nevertheless, soldier enhancement creates profound predicaments in the context of military ethics, challenges the application of the laws of armed conflict and, last but not least, has serious implications for the conduct of operations. However, to appreciate the implications and consequences of enhancement technologies fully, it is imperative to clarify what enhancement is and why and in what ways it is different from the previous generation of tools, procedures and weapons employed to boost a soldier's performance in war.

7.2 DEFINITION AND CONTROVERSIES

Generally speaking, countless human activities are examples of enhancement? because they improve our lives in one way or another: from drinking coffee, to securing a shelter, to education or having vaccinations. From that perspective, 'the story of humanity is the story of enhancement'.[8] Human enhancement then becomes meaningless as a concept because it is too broad and inclusive. In the context of new technologies, which allow radical transformations to the human body, human enhancement must be defined much more narrowly.

Bioethicist Eric Juengst already clarified this back in the late 1990s at a time when enhancement technologies were more rudimentary than today. He defined human enhancement as the suite of medical and technological interventions to enhance human capabilities beyond what is necessary to

[6] T. Harrison, 'Chaos and Uncertainty: The FYI 2014 Defense Budget and Beyond', *Centre for Strategic and Budgetary Assessments*, October 2013, 11–12.

[7] I. Puscas, 'La Quête du Soldat Augmenté', *Le Monde Diplomatique*, September 2017, 3.

[8] H. T. Greely, 'Regulating Human Biological Enhancements: Questionable Justifications and International Complications', *Santa Clara Journal of International Law* 4:2 (2006), 88.

achieve or sustain health. Enhancement is therefore different from treatment or therapy, which merely aim to heal or repair damaged functions and bring the human body back to a 'normal', 'healthy' state. In other words, enhancements are interventions that 'do not respond to legitimate medical needs'.[9]

This definition is not without its controversies and the medical field has often had to grapple with the complexity of 'normal' in assessing parameters of standard or 'typical' human functions. Notwithstanding the inherent complexity of the medical concept of 'normality', there are countless cases where the distinction between treatment and enhancement is unambiguously clear. For example, if we take the standard definition of normal human vision (calculated on the 20/20 chart), interventions to correct vision that is below this standard is a case of 'treatment', whereas vision that is beyond this level is an example of enhancement. In other words, corrective eye surgery is categorically different from retinal implants that would allow night vision: the former is treatment, the latter is an enhancement.

The distinction between enhancement and treatment must always be contextualised. Bostrom and Sandberg provide a relevant example by looking at a spectrum of memory parameters. Boosting the memory of an otherwise healthy individual, with average or above-average memory skills, is a case of enhancement. The distinction becomes blurrier when we consider the case of a person with *naturally poor* memory, who even after receiving treatment to address that deficiency is still left with a memory below the level of a person who suffered from an identifiable pathology (e.g., early stage Alzheimer's). The person with naturally poor memory received an enhancement, whereas the Alzheimer's patient benefited from a treatment. In such instances, enhancements do not necessarily lead to extraordinarily high capabilities – in other words, they do not always create 'super humans'. In this particular case, enhancement is merely a form of improving a subsystem in some way.[10] In the military context, this ambiguity is probably less present because the rigours underlying the recruitment of personnel would have likely singled out individuals with naturally poor memory. The question then arises as to whether recruitment should change in the era of enhancements, allowing even individuals with suboptimal physical capabilities to join, in the expectation that

9 E. Juengst, 'The Meaning of Enhancement', in E. Parens (ed.), *Enhancing Human Traits: Ethical and Social Implications* (Washington, DC: Georgetown University Press, 1998), 29–47. See also: E. Juengst, 'Can Enhancement Be Distinguished from Prevention in Genetic Medicine?', *The Journal of Medicine and Philosophy* 22 (1997), 125–142.

10 N. Bostrom and A. Sandberg, 'Cognitive Enhancement: Methods, Ethics, Regulatory Challenges', *Science and Engineering Ethics* 15:3 (September 2009), 312.

they would soon be enhanced anyway. This remains one of the open ethical questions about military human enhancement.

Returning to some of the difficulties in conceptualisation, a few other caveats are worth mentioning.

A common assumption in defining enhancement is that tools such as exoskeletons or climbing aids are soldier enhancements. This is, however, a conceptual overstretch. An exoskeleton worn on the outside certainly improves the endurance of a soldier but does not directly alter the soldier's biology. Accordingly, in addition to the treatment/enhancement aspect, the inside/outside distinction is also useful in clarifying the concept of enhancement.

A device or weapon that remains outside of the body is easily removable and would not therefore, in accordance with our developing concept, be a form of human enhancement. However, a device such as a brain–computer interface or a chip implanted under the skin or in the brain is most likely an enhancement because it provides an 'always-on' access and connectivity.[11] Here too, the distinction between inside/outside the body has its limitations and needs contextual scrutiny. Similarly, it is important to clarify that the mere insertion of a device into the human body need not be giving the 'enhanced' individual an extra capability that is incrementally greater than what an external device could provide. We are, however, more likely to define an object inserted into the human body as an enhancement, as opposed to an external device even if that external device has a substantially similar functionality and effect. This is revealing of the moral judgement that we are inclined to make when it comes to merging humans with technology, as well as the difficulty in accepting the idea that we have a 'right' to customise and enhance our bodies through scientific means.[12]

Furthermore, the emotional reactions to certain procedures might vary depending on the effects of the procedure at different times, not based on the procedure itself. For example, a pill ingested to reduce fatigue and increase stamina in sleep-deprived individuals will more likely be seen as less

[11] P. Lin, M. J. Mehlman and K. Abney, 'Enhanced Warfighters: Risk, Ethics and Policy', *Report Prepared for the Greenwall Foundation*, 13 January 2013, http://ethics.calpoly.edu/greenwall_report.pdf.

[12] H. T. Greely, 'Regulating Human Biological Enhancements', additionally explains the religious and cultural factors that make modern human enhancement very controversial. He considers that the poor reputation associated with such enhancement traces back to religion and to the sorts of theme explored in the Book of Genesis, where mankind is punished for overreaching. Then, there is a long history of literature that exposed the dark side of enhancement, for example: Goethe's *Faust*, or Huxley's *Brave New World*.

intrusive, and less problematic from an ethical point of view than a pill that erases negative memories and cures post-traumatic stress disorder (PTSD), or further, than a pill that blocks certain protein synthesis in the brain and is conducive to ruthless, guilt-free behaviour.[13]

The concept of human enhancement can present ambiguities but individual, contextual assessments will help clarify it. Separating enhancement from treatment can be a fairly straightforward intellectual and legal exercise if the fundamental premise of the concept guides the case-by-case analysis: enhancement is about giving the mind or the body abilities beyond that individual's natural competencies in a normal, healthy state. Those enhanced abilities are also often above the level that the said individual could achieve by rigorous training alone. The 'individual' matters a lot in this classification because, in addition to species-specific parameters, the physiology of any individual is unique to that individual.

7.3 ORIGINS OF PRESENT-DAY ENHANCEMENT

Twenty-first century human military enhancement is unique in scale and scope due to unprecedented understandings of human biology. This knowledge was not available before the advances made in neuroscience and neurochemistry, and a suite of other technologies that helped advance our insights into the human brain. Increasingly, we are able to understand the functions of specific nerves and specific neurochemicals in shaping social behaviour, and implicitly the way in which manipulating those same chemicals will alter moral judgement. For the first time in history, scientists are able to look into how the brain produces moral attitudes and their corresponding neurochemicals. An increasing body of neuroscientific research, for example, presents evidence about the role of serotonin in moral judgement and behaviour. Serotonin is a neurotransmitter shown to modulate human concerns for harm and fairness.[14]

However, it is important to understand modern ideas of human enhancement in a historical context: while interventions to the human body can now be more invasive, the idea behind enhancement in the armed forces, i.e. to

[13] See Lin, et al., 'Enhanced Warfighters'. The authors also explore these nuances; they mention that interventions such as, for instance, taking anabolic steroids can prompt different reactions if taken by patients with muscular dystrophy or by a healthy athlete, depending on the perceived 'moral justification' of the interventions.

[14] J. Z. Siegel and M. Crockett, 'How Serotonin Shapes Moral Judgment and Behavior', *Annals of the New York Academy of Sciences*, Issue: Sociability, Responsibility and Criminality: From Lab to Law (2013), 44.

boost soldiers' endurance and efficiency, is as old as the history of warfare. Historically, state combatants and fighters in non-state groups have always been eager to experiment with and employ mind-altering or mood-enhancing substances.

During the Napoleonic Wars, two French infantry divisions charged with seizing the Pratzen Heights during the Battle of Austerlitz were distributed triple rations of brandy, leading them to 'burst with eagerness and enthusiasm'.[15] Rum consumption was rampant among the British soldiers and sailors, and it is estimated that in the second half of the nineteenth century, the 36,000 men of the British Army drank 550,000 gallons of rum annually.[16] It was believed that giving rum to soldiers made them better fighters – indeed, the courage acquired after drinking alcohol became colloquially known as 'Dutch courage', in reference to the English soldiers who had fought the Dutch in the seventeenth century's wars and who had gained their courage from sipping Dutch gin.[17] During World War I, the role of alcohol in boosting soldiers' morale was recognised further. Doctors writing in the *British Medical Journal* recommended it for the treatment of soldiers' mental disorders and a medical officer giving testimony to a parliamentary committee on the issue of 'shell shock' among troops (PTSD in psychiatric classification today) stated that: 'Had it not been for the rum ration, I do not think we should have won this war'.[18]

Non-alcoholic enhancers have also been used widely in war: Zulu fighters fighting the British enhanced their courage and stamina with a variety of cannabis prepared by their shamans. This led them to fight with fanaticism. The twentieth century decisively consecrated the role of drugs in warfare. Methamphetamine – a stimulant of the central nervous system, which mimics endogenous neurotransmitters such as serotonin and dopamine[19] – was widely employed in combating fatigue, and to keep troops alert and awake. Nazi and Japanese soldiers during World War II relied heavily on psychostimulants like methamphetamines. Amphetamines were available in industrial quantities to US soldiers fighting in the Vietnam War, with far-reaching implications. Moreover, amphetamines are highly addictive and can make consumers jittery and often aggressive. The search for safer alternatives has led to the discovery of new drugs of choice, such as modafinil (frequently sold under the brand name Provigil) and Ritalin.

15 Kamieński, *Shooting Up*, 8.
16 Ibid., 9.
17 Ibid., 9.
18 P. Fussell, *The Great War and Modern Memory*, in Kamieński, *Shooting Up*, 19.
19 B. K. Logan, 'Methamphetamine – Effect on Human Performance and Behaviour', *Forensic Science Review* 14:133 (2002), 136.

Modafinil is usually prescribed for narcolepsy and sleep disorders but in healthy individuals it functions as an alert-enhancing drug, with fewer side effects compared to amphetamines. Taken in low doses of 100 mg at regular intervals, modafinil can attenuate the effects of fatigue and improve the mood of fighters for up to thirty-seven hours of continuous wakefulness.[20] Recognising its potential, the UK Ministry of Defence ordered 5,000 tablets of Provigil in 2001, the same year when the Allied forces began their offensive in Afghanistan and 4,000 tablets in 2002, before the invasion of Iraq.[21]

This historical snapshot confirms that certain kinds of enhancer for soldiers have been employed for centuries in warfare. It does not, however, seem to have affected how the law of armed conflict has developed to date. Indeed, when rum was called a 'weapon of war',[22] the meaning of 'weapon' was certainly meant as a metaphor (perhaps, at times, humorously), not in the context of how weapons and weapons systems are defined under international law.

The raison d'être of international humanitarian law is to strike a delicate balance between military necessity and the principles of humanity.[23] This limits suffering and cruelty and any destruction that is not necessary for attaining the military purpose. These limitations to how war is fought have followed at least some advances in science and in weapons technology. For example, the St Petersburg Declaration of 1868 drew attention to the need to consider the legal implications of new weapons. In 1863, a new bullet was invented and introduced in the arsenal of the Imperial Russian Army: a bullet that exploded on contact with hard surfaces. The bullet was upgraded in 1867 so it was also able to explode on soft surfaces, including the human body.[24] The St Petersburg Declaration addressed the need for legal reviews in stating that:

> The Contracting or Acceding Parties reserve to themselves to come hereafter to an understanding whenever a precise proposition shall be drawn up in view of future improvements which science may effect in the armament of

[20] J. Caldwell, et al., 'Modafinil's Effects on Simulator Performance and Mood in Pilots during 37 Hours without Sleep', *Aviation, Space and Environmental Medicine* 75:9 (September 2004), 777–784.

[21] I. Sample and R. Evans, 'MoD Bought Thousands of Stay Awake Pills in Advance of War in Iraq', *The Guardian*, 29 July 2004.

[22] See I. Williams, *Rum: A Social and Sociable History* (New York: Nation Books, 2005).

[23] See M. N. Schmitt, 'Military Necessity and Humanity in International Humanitarian Law: Preserving the Delicate Balance', *Virginia Journal of International Law* 50:4 (2010), 795–839.

[24] J. Kelly, *Is Medical Ethics in Armed Conflict Identical to Medical Ethics in Times of Peace?* (Newcastle: Cambridge Scholars Publishing, 2013), 145.

troops, in order to maintain the principles which they have established, and to conciliate the necessities of war with the laws of humanity.[25]

A battalion consuming alcohol or any kind of stimulant was hardly a matter of legal concern. This does not, of course, imply that those soldiers would be permitted to commit atrocities but, as long as their behaviour remained within the parameters acceptable in the norms of warfare of the time, the focus was firstly on how they used their weapons. Because soldiers remained essentially human, no difficulty was envisaged in classifying them as *distinct* from a weapon system, means or method of warfare.

In the light of new advances in enhancement technologies, this distinction will be at times become more problematic and many legal scholars and ethicists have started to debate this.[26] A host of other legal instruments are challenged as well. Military ethics and operational standards are affected too in unique ways.

7.4 PHYSICAL AND COGNITIVE ENHANCEMENT TODAY

The United States, through its DARPA, is at the forefront of developing and sponsoring human enhancement technologies today. The USA is also among the very few countries that provide open-source publications and releases on such projects; hence DARPA's work on enhancement technologies has been the subject of closer scrutiny. The questions related to ethics and law that arise from enhancements are nevertheless relevant globally and they signal the novel challenges generated by these technologies.

DARPA's mission was from its creation in 1958 a hefty one: to prevent strategic 'technological surprise' and ensure the United States would be at the forefront of innovation. The Agency known as the technological engine for the Department of Defense soon realised that to prevent technological surprise was to *create* technological surprise for the adversaries of the United States. Indeed, some of DARPA's innovations truly changed warfare and the

[25] 'Declaration Renouncing the Use, in Time of War, of Explosive Projectiles under 400 Grammes Weight'. Saint Petersburg, 29 December/11 December 1868, https://ihl-databases.icrc.org/ihl/a/568842c2b90f4a29c12563cd0051547c?opendocument.

[26] See, for instance, P. Lin, 'Could Human Enhancement Turn Soldiers into Weapons That Violate International Law? Yes', *The Atlantic*, 4 January 2013, www.theatlantic.com/technology/archive/2013/01/could-human-enhancement-turn-soldiers-into-weapons-that-violate-international-law-yes/266732/. For a thorough legal analysis exploring several legal dimensions related to human enhancement in the military: H. Dinniss and J. K. Kleffner, 'Soldier 2.0: Military Human Enhancement and International Law', *International Law Studies* 92:432 (2016).

course of history: it contributed to the creation of the Internet (initially called Arpanet), stealth aircraft, the GPS technology and much more. What counted as innovation in defence has changed over the decades and now includes innovative development to the bodies of soldiers themselves.

Already in the 1990s, a paradigmatic shift took place at DARPA as the Agency started to be more interested in biology. In 1999, DARPA created the Defense Sciences Office and named a biologist and venture capitalist, Michael Goldblatt, as its director. For Goldblatt, defence sciences needed to demonstrate that 'the next frontier was inside of our own selves' and this conviction emboldened his efforts to establish the creation of a supersoldier as an imperative for twenty-first century warfare.[27]

In 2004 the Agency created the Biological Technologies Office with a mission to bring together scientists working in biology and physical sciences. It formally stated that 'starting today, biology takes its place among core sciences that represent the future of defense technology'. The human soldier is central to these endeavours, and a core function of the Office is to develop technologies to 'maintain peak warfighter abilities'.[28]

These efforts include cognitive and physical enhancements although this distinction can be crude as the two are often interconnected. The brief technological survey below lists a few of the initiatives to enhance the human body and mind with the help of technology. While not all of these initiatives are referred to below, the discussion indicates the ways in which emerging technologies will impact warfare, the sorts of capabilities that will be enhanced, the envisaged ways of doing so and, quite simply, what it is that enhancements do to warfighters. Moving forward, any discussion of ethics and law must be well grounded in such an understanding, in order to avoid unrealistic, hyperbolic and ambiguous misrepresentations. This is also important for a thorough understanding of how the applicable law and ethical concerns impact these emerging technologies.

Therefore, while the 'end product' of an enhancement procedure is an enhanced soldier, the nature of the enhancement and the physical procedure (e.g., electrical neurostimulation) or device (e.g., implantable chip) used to achieve that enhancement are critical elements in themselves. This is because they have associated legal and ethical consequences. In some cases, an enhancement could be a separate entity such as a distinct device inserted into the brain. In other cases, it can be indistinguishable from the soldier, and

[27] A. Jacobsen, 'Engineering Humans for War', *The Atlantic*, 23 September 2015.

[28] DARPA Launches Biological Technologies Office, 1 April 2014, www.darpa.mil/news-events/2014-04-01.

indeed, impossible to neutralise without affecting the soldier as well. An example of this would be brain stimulation with electricity, which involves a *procedure* applied before actual entry into combat or flight operations, but which is not present as a separate physical device that the soldier carries. So, the brief discussion that follows looks at a sample of enhancement technologies and procedures.

7.4.1 *Physical Enhancement*

7.4.1.1 Surviving Blood Loss

Better digestion, improved metabolism, mitigating the effects of sleep deprivation, preventing muscle fatigue, and surviving blood loss are some of the highlights of DARPA's endeavours in military physical enhancement.

DARPA's 'Surviving Blood Loss' programme aims to develop new strategies to extend the time that injured warfighters can survive despite critical blood loss. The priorities of the programme are ambitious. It aims to identify the mechanisms that preserve cellular function even in conditions of low oxygen delivery, to explore ways of controlling the metabolic state on demand, to achieve the induction of a hibernation-like state, and to develop therapies which reduce the tissue's demand for oxygen and metabolites.[29]

A DARPA-funded project at the University of Wisconsin studied hibernation by analyzing how squirrels preserve their energy and regulate their body temperature as they hibernate. Understanding this process is of interest to the armed forces because if a similar process could be replicated in soldiers, it would have serious life-saving potential, allowing wounded soldiers to enter a hibernation-like state, stop the bleeding and wait until help arrived.[30]

Blood loss is of considerable concern to the armed forces. The US Department of Defense estimates that around 80 per cent of preventable deaths in war are a direct result of blood loss, and timing is critical: usually, soldiers need to receive blood product resuscitation within an hour in order to survive. Of the total number of deaths caused by haemorrhage, it is estimated that in about 33 per cent of the cases, the use of tourniquets would be enough

[29] DARPA Program Information Document, 'Surviving Blood Loss', available at www.darpa.mil/program/surviving-blood-loss.

[30] N. Shachtman, 'Squirrel=Super Soldier', *WIRED*, 3 December 2007, available at www.wired.com/2007/03/soldiers-if-you/.

for haemorrhage control, but about 67 per cent are 'non-compressible/non-tourniquetable'.[31] There have been important advances in haemostatic agents in recent decades. For example, the British armed forces have experimented with zeolites, which are 'microporous minerals derived from volcanic rocks', and 'chitostan, a complex carbohydrate derived from crushed shellfish'.[32] In the USA, the Department of Defense has taken on a more ambitious task with its 'Haemorrhage and Resuscitation Research and Development Program', which was established with the purpose of conducting research on innovative methods, drugs, interventions or devices that could mitigate the consequences of blood loss.

An important breakthrough was announced in 2015 at the University of Alabama. After years of trial-and-error research exploring the mechanisms of prolonged survival in spite of severe blood loss, researchers discovered that a dose of the estrogen 17β-estradiol (E2) could protect both males and females against septicaemia. The E2 was then tested for blood loss, and the results showed that it improved heart performance, heart output and liver function.[33] DARPA had a critical role through its aforementioned 'Surviving Blood Loss' programme. Somewhat later, a pharmaceutical company began to synthesise the estrogen in a variant called EE-3-SO4.[34] The drug does not stop a patient from bleeding per se but mobilises other physiological, cellular and biochemical mechanisms that help the body cope with the haemorrhage. EE-3-SO4 helps the heart beat more efficiently, lowers resistance to blood flow to vital organs, increases blood pressure, and finally, it recruits fluid into the blood stream from surrounding tissues thus increasing blood volume and

31 A. E. Pusateri, 'US Department of Defence Haemorrhage and Resuscitation Research and Development Program', presented at the Remote Damage Control Resuscitation Conference, Bergen, Norway, 19 June 2002, available at http://rdcr.org/wp-content/uploads/symp2012pdf/US%20Department%20of%20Defense%20Hemorrhage%20and%20Resuscitation%20Research%20and%20Development%20Program%20-%20Anthony%20Pusateri.pdf.

32 S. Marcus, cited in William Rowell, 'Haemorrhage Control: Lessons from the Military', *Emergency Nurse* 21:7 (November 2013), 22.

33 Jeff Hansen, 'Female Sex Hormone May Save Injured Soldiers on the Battlefield', University of Alabama News Portal, 21 October 2015: www.uab.edu/medicine/news/latest/item/874-female-sex-hormone-may-save-injured-soldiers-on-the-battlefield. In 2005, DARPA launched a competition inviting applicants to produce a treatment that would enable soldiers to survive as long as three hours after significant blood loss. If successful in this phase, the winner would pass to Phase II, a challenge of six hours. Not only was DARPA determined to help extend the one-hour timeframe of intervention to six but it also tasked researchers to produce a deliverable material that would be small and fit conveniently in a battlefield pack. The team that discovered E2 proved that a single injection of their formula allowed survival for 3 hours without fluid resuscitation and was able to diminish the physiological effects resulting from the lower delivery of oxygen that results from low blood pressure.

34 Ibid.

compensating for the blood loss. This immediately caught the attention of the US Department of Defense, which provided an additional $10 million in funding through the Combat Casualty Care Research Program to begin testing on humans.[35]

'Metabolic flexibility' would mean that soldiers could slow or adapt their metabolism until conditions in the environment changed. Hydrogen sulphide could reduce the metabolic rate of a soldier and induce a state akin to suspended animation until help arrived. Exposure to low levels of this chemical has been shown to reduce cellular oxygen consumption and induce a hibernation-like state in some mammals.[36] Another of DARPA's initiatives, the 'Wound Stasis System' addresses rapid and uncontrolled blood loss and aims to keep injured warfighters alive until they can be provided surgical intervention. A way to achieve this is by developing an agent to achieve rapid haemostasis that could be encapsulated in a portable delivery system.[37]

7.4.1.2 Engineered Digestion and Metabolism

Cocktails of vitamins can lead to improved endurance but interventions at the genetic and cellular level can go further in boosting stamina and maintaining physical performance despite a lack of access to food.

At the University of Iowa, agricultural experts obtained a grant to study fibro-biotic bacteria. These are bacteria which allow soldiers to obtain nutritional value from substances not normally digestible by humans. The researchers studied how pigs digest cellulose, and the possibility of humans mimicking this process. This 'Crystalline Cellulose Conversion to Glucose Program' studied ways to enable soldiers to eat inedible plants such as grass.[38] The undertaking was part of the DARPA Intestinal Fortitude Program.[39]

7.4.1.3 Preventing Muscle Fatigue

Researchers working within the Peak Soldier Performance programme studied a new drug that could drastically limit muscle fatigue. The drug is called

[35] B. Shepard, 'UAB to Start First Human Trials of Drug to Prevent Death Due to Severe Blood Loss', 21 October 2015, https://uab.edu/news/innovation/item/6624-uab-to-start-first-human-trials-of-drug-to-prevent-death-due-to-severe-blood-loss.

[36] Ibid.

[37] Col. M. Hepburn, 'Wound Stasis System', www.darpa.mil/program/wound-stasis-system.

[38] DARPA study, cited in P. Lin, et al., 'Enhanced Warfighters' (2013), 25.

[39] N. Shachtman, 'Be more. . .', The synopsis of the study is also presented on the webpage of the US Department of Agriculture: https://portal.nifa.usda.gov/web/crisprojectpages/0411287-improved-dietary-fiber-digestion-in-the-large-intestine-through-selection-of-fibr-biotics.html.

S107 and was demonstrated to prevent calcium leakage (calcium leaking into muscle cells is a key cause of muscle fatigue and inflammation) in mice.[40] The drug's full effects on humans have, however, not been tested. S107 could increase endurance much more than existing, similar drugs, some of which block calcium channels without necessarily increasing athletic performance.

7.4.2 *Enhancing Cognitive Performance*

The notion of 'mind fitness' in the army is not a twenty-first century invention. Military training was always as much about the mind as about the body. The markedly different aspect of cognitive enhancement in the twenty-first century relates to the extent to which the limitations of human cognitive skills can be loosened following strides made in drugs research, neuroscience and the neurotechnologies of brain stimulation.

In general, enhancements of the brain comprise improvement of motor systems, cognition, and mood.[41] For example, modulating neural plasticity may enhance learning skills; pharmacological interventions (e.g., Modafinil) can improve attention despite prolonged sleep deprivation, or can decrease the intensity of very emotional memories, such as the drug propranolol,[42] which was shown to dampen post-traumatic stress disorder symptoms when given to patients who had recently undergone a traumatic event. Other neuroenhancements are more invasive and far more potent in effect, to the extent that they directly act in the nervous system and alter its functions in order to boost the performance of very specific tasks.

These varied methods can be generally divided into two main approaches: pharmacological (relying on drugs), and technological (employing technology and various devices) – although the distinction is not always clear-cut.

7.4.2.1 Pharmacological Approaches

Pills have a long history of military use. The more 'archaic' methods for sustaining performance, such as sleep management programmes (manipulating the duration of sleep) or hypnotics do not always guarantee that they can

[40] Statement by Dr R. Leheny, Acting Director of DARPA, 20 May 2009, www.darpa.mil/attachments/TestimonyArchived%20(May%2020%202009).pdf.

[41] A. Chatterjee, 'The ethics of neuroenhancement', *Handbook of Clinical Neurology*, 118 (2013), 323.

[42] See R. K. Pitman, K. M. Sanders, R. M. Zusman RM, et al., 'Pilot Study of Secondary Prevention of Posttraumatic Stress Disorder with Propranolol', *Biological Psychiatry* 51:2 (2002), 189–192.

even prolong wakefulness, let alone allow humans to maintain 'normal' performance levels. Stimulants, on the other hand, have been demonstrated to sustain alertness in situations of prolonged and intense combat, and they could be the critical enabler for operational success. Amphetamines were used in the armed forces for a long time to enhance alertness and their use subsequently gave rise to many controversies. Today, a more common drug is the Dexedrine (dextroamphetamine), which is prescribed in certain operational contexts, such as single-seat and dual-seat flight missions.[43] Dexedrine has been available since the 1960s and its stimulant effects occur through dopaminergic action.[44] With a single dose of 20 mg of dextroamphetamine, cognitive performance returns to baseline levels and is maintained even after 48 hours of sleep deprivation. Multiple 10 mg doses can sustain performance of pilots for as long as 64 hours.[45] Another drug used for cognitive enhancement is the methylphenidate (sold under the name Ritalin). Ritalin is very often used for ADHD because it blocks the re-uptake of dopamine and norepinephrine, thus increasing the levels or prolonging the availability of these neurotransmitters.[46] In healthy adults, low doses of methylphenidate lead to an increase in attention, but higher doses can cause hyperactivity.

In searching for a safer alternative for military use that has a reduced potential for causing addiction and abuse, attention has turned to Modafinil in recent years (although recent studies showed that the risk of abuse is not as low as was previously thought). Modafinil (sold under the brand name Provigil in the USA since 1998) is a psychostimulant first created in the 1970s. It was recognised for its outstanding capacity to promote wakefulness; it continues to be prescribed medically for various sleep disorders. In healthy individuals, it has been proven to enhance pattern recognition, reaction time and spatial planning.[47] Armed forces have been keen to harness the potential of this drug. Modafinil has been studied specifically for its uses in aviation and is believed to work neurochemically by counteracting the dopamine reuptake transporter. This effect leads to an increase in the extracellular levels of dopamine.

[43] Lin et a.l, 'Enhanced warfighters', 6.

[44] Dopaminergic refers to the action and effects which are activated or transmitted by dopamine.

[45] A. Estrada, 'Modafinil as a Replacement for Dextroamphetamine for Sustaining Alertness in Military Helicopter Pilots', *Aviation, Space and Environmental Medicine* 83:6 (June 2012), 557.

[46] K. R. Urban, 'Performance Enhancement at the Cost of Potential Brain Plasticity: Neural Ramifications of Nootropic Drugs in the Healthy Developing Brain', *Frontiers in Systems Neuroscience* 8:38 (May 2014), 2.

[47] The Royal Society, 'Neuroscience, Conflict and Security', February 2012, https://royalsociety.org/~/media/Royal_Society_Content/policy/projects/brain-waves/2012-02-06-BW3.pdf 35–36.

Consumption of Modafinil was formally authorised by the US Air Force in 2003 for certain missions, such as dual-pilot bomber missions lasting over 12 hours and F-15E missions lasting over 8 hours. In 2006, Modafinil was authorised (usually in doses of 200mg) for single-seat fighter operations.[48]

Interest in Modafinil grew noticeably following a 2002 incident in Afghanistan. In April of that year, two US pilots mistakenly dropped a bomb on a Canadian infantry unit, killing four soldiers and wounding eight others. The accident is known as the 'Tarnak Farm incident'. During the subsequent investigation, the pilots claimed they had been pressured to take amphetamines, which had left them in a state of agitation.[49] Their defence lawyers further argued their misjudgements were due to the amphetamines that the Air Force authorised for long missions.[50]

Such incidents, in addition to the general controversy surrounding the use of amphetamines, led to a search for safer alternatives like Modafinil/Provigil. Pilots taking the pill could stay awake for as long as 40 hours, with no major adverse effects. Modafinil—and stimulants generally—increasingly seem to be an indispensible tool for Western armed forces. Data released by the UK Defence Medical Supplies Agency revealed that the UK MoD had purchased over 24,000 tablets of Provigil since 1998, with a peak of orders coming in 2001, the year when allied forces entered Afghanistan (the second largest order was in 2002). In France, the drug had been used as early as the Gulf war in 1991.[51] In the United States, an investigation into the costs and scales of stimulation medication showed that it increased more than fivefold: from \$7.5 million in 2001, to \$39 million in 2010.[52]

7.4.2.2 Technological Approaches to Neuroenhancements And 'Techno-Integration'

Therapies with electricity are already in use for certain mental disorders, but the armed forces envisage using it on healthy individuals to augment some

[48] Estrada, 'Modafinil', 557.

[49] O. Burkeman, R. Norton-Taylor, 'US Pilots Blame Drug for Friendly Fire Deaths', *The Guardian* 4 January 2003, www.theguardian.com/world/2003/jan/04/afghanistan.richardnortontaylor.

[50] D. Stout, 'US Pilots Avoid Prosecution for Bombing Canadian Troops', *New York Times*, 19 June 2003, www.nytimes.com/2003/06/19/national/us-pilots-avoid-prosecution-for-bombing-canadian-troops.html.

[51] I. Sample, 'Wired Awake', *The Guardian*, 29 July 2014, www.theguardian.com/education/2004/jul/29/research.highereducation.

[52] R. A. Friedman, 'Why Are We Drugging Our Soldiers?', *New York Times*, 21 April 2012, www.nytimes.com/2012/04/22/opinion/sunday/why-are-we-drugging-our-soldiers.html?_r=0.

brain functions. In 2014, the Pentagon made public that it was considering stimulating troops' brains with electricity, in order to boost attention and vigilance. Low levels of electricity applied to certain areas of the brain can lead to an increase in attention span and attentiveness in sleep-deprived soldiers. Preliminary tests were conducted at the Air Force Research Laboratory in Ohio. Results showed a marked increase in both alertness and acuity in subjects.[53]

In recent years, it has been argued that brain stimulation is a more powerful and viable route to performance enhancement. The aforementioned example of the Pentagon exploring electrical stimulation in soldiers' brains is only one of the many examples of human enhancement technologies that have been tested. More insidious methods for cognitive enhancements, rapid skills learning, and memory restoration or alertness have been studied simultaneously.

Human-computer interfaces, chips implanted in the brain, and other devices integrated in the human body could result in a convergence between human soldiers and technological components. *Techno-integration,* which is defined as 'the symbiotic coupling of humans with technology' is a paradigm that diversifies the possibilities of enhancement by merging humans with machines or smart devices.[54] This refers to all possibilities which enable communication between man and machine (not counting extremities). It covers both invasive and non-invasive methods, from magnetic stimulation of parts of the brain to implants in the brain (such as microelectrodes). The information exchange between man and machine can therefore be established either invasively by directly contacting or acting on nerves in the brain and body, or non-invasively, by using indirect signals.

As would be expected, most of these technologies are used both in health treatment and for military purposes. Braingate, for instance, is a small square chip with 100 microelectrodes, and has already been tested on paralyzed patients. The chip allows them to control the movement of a cursor by

[53] Crucial to these approaches has been our understanding of fatigue – both its biological and neurophysiological characteristics. Some evidence suggests that the central nervous system may experience fatigue due to lack of glucose (which is the main source of energy for the brain), or due to an imbalance of neurotransmitters/neuromodulators, such as dopamine, serotonin or adenosine. See B. Bender, 'Pentagon Considers Using Electricity to Stimulate Troops' Brains', *Boston Globe*, 18 February 2014, www.bostonglobe.com/news/nation/2014/02/18/air-force-stimulates-brain-waves-improve-performance-drone-operators/V8ZG5DEYze4lCoGlloq14H/story.html.

[54] S. Reschke, J. B. F. van Erp, A. M. Brouwer and M. Grootjen, 'Neural and Biological Soldier Enhancement: From SciFi to Deployment', Paper Presented at the NATO RTO Human Factors and Medicine Panel, Sofia, 5–7 October 2009, http://publica.fraunhofer.de/documents/N-194631.html.

imagining the cursor. It is implanted in the brain intracortically, where the cortical surface picks up and subsequently decodes electric signals. In the military context, such neural implants could act as sensory enhancers. They would allow the brain to process information such as images or targets at much higher speeds than normal, and could accelerate a soldier's ability to recognise danger.[55]

Many recent brain-enhancement technologies are similar in scope to psychostimulant drugs, as they aim to enhance cognition, decision-making skills, and the maintenance of high functionality even in conditions of sleep deprivation. Others, however, go much further. These latter allow for communication and control of the brain beyond one's body. Below is a listing of some of the more far-reaching and futuristic technologies.

7.4.2.3 Implantable Neural Interfaces

DARPA's Neural Engineering System Design (NESD) Program works 'to develop an implantable neural interface able to provide advanced signal resolution and data transfer bandwidth between the brain and electronics'.[56] Building on interdisciplinary breakthroughs in the fields of neuroscience, synthetic biology, low-power electronics and systems engineering, NESD seeks to eventually achieve a 'biocompatible device no longer than one cubic centimeter in size'. This interface would translate the electrochemical language of neurons to the 'zeros and ones'-based language of digital communication.[57] Such a device would have far more capabilities than existing brain-computer interfaces, as it would increase the accuracy and quality of visual or auditory information into the brain. A similar project is still in the research phase. DARPA's Neural Dust programme is looking at the possibility of implanting millimetre-scale wireless devices (a commercial prototype measures 0.8 mm × 3 mm × 1 mm) through ingestion or needle injection (either in muscles or the peripheral nervous system). These tiny devices, called 'neural dust', are based on ultrasound technology; they are made of a special crystal which records activity in the body continuously, detects electrical activity of nerves, and then converts the ultrasound waves from outside the body into electricity.[58] Neural dust does not need batteries and would continue to

55 The Royal Society, 'Neuroscience, Conflict and Security', 39–40.

56 Dr P. Alvelda, 'Neural Engineering System Design' (NESD), www.darpa.mil/program/neural-engineering-system-design.

57 Ibid.

58 'Implantable "Neural Dust' Enables Precise Wireless Recording of Nerve Activity', 3 August 2016, www.darpa.mil/news-events/2016-08-03.

function for up to 10 years in the human body. Like most of these technologies, the applications of neural dust are wide-ranging: from stimulating nerves in patients with epilepsy to improving the immune system by prompting it into action.[59]

The RAM (Restoring Active Memory) Program is a DARPA project under the umbrella of the BRAIN initiative. It aims to develop a wireless implantable neural interface device which would enable military service members to better retrieve memories formed prior to an injury. The programme, which is also known as 'RAM Replay', would benefit both military veterans and civilians who have suffered traumatic brain injuries, which affect millions of people each year in the USA.[60] The device is designed to perform a targeted neural stimulation aimed at memory function restoration. Militarily speaking, better memory is a critical asset in operations, allowing soldiers to sieve through a wealth of data and information, which in turn plays an essential role in decision-making. The DoD is especially interested in the outcomes of the project because new knowledge about memory formation could help combatants' struggle with rehabilitation and recovery. Proven ability to mitigate PTSD would likely assist recruitment, and faster recovery times would accelerate a soldier's return to the battlefield. The Director of the programme makes this justification explicit:

> "Military personnel carry a growing responsibility to recount, report, and act upon knowledge gleaned from previous experiences, and how well these experiences are recalled can make a difference in how well these individuals perform in combat and other challenging situations".[61]

7.4.2.4 Neurostimulation

The notion that deep brain stimulation through prosthetic devices can improve cognition or treat depression has been around for some time.[62] Neurostimulation has been studied in the context of pathological and

[59] I. Johnston, 'Tiny Implant Could Connect Humans and Machines Like Never Before', 3 August 2016, www.independent.co.uk/news/science/neural-dust-implant-sensor-brain-nerves-humans-machines-prosthetics-berkeley-a7170251.html.

[60] Dr J. Sanchez, 'Restoring Active Memory (RAM)', www.darpa.mil/program/restoring-active-memory.

[61] 'DARPA Aims to Accelerate Memory Function for Skill Learning', 27 April 2015, www.darpa.mil/news-events/2015-04-27-2.

[62] See National Research Council, *Opportunities in Neuroscience for Future Military Applications* (Washington: National Academies Press, 2009), 86.

neurological conditions such as Parkinson's and Alzheimer's diseases. Military interest now centres on experimentation with neurostimulation in an effort to optimize soldiers' brains for specific tasks.[63]

In March 2016, DARPA announced a comprehensive new Targeted Neuroplasticity Training Program (TNT). The programme explores peripheral nerve stimulation (by boosting synaptic plasticity) as a way to accelerate learning. While the program aims to enhance soldiers' performance, an underlying objective is to reduce the cost of the DoD training and learning regimen.

The peripheral nervous system has become a central focus of military research because its manipulability renders the soldier a more capable and enduring resource. TNT seeks to advance the possibility of cognitive skills learning through precise activation of peripheral nerves. Once activated, they would consolidate neuronal connections and accelerate learning. Furthermore, the stimulation of peripheral nerves could also trigger neurochemical reactions that support this process by reorganising neural connections. Soldiers would learn more effectively and at greater speeds, reducing the time needed for learning a foreign language, conducting intelligence analysis, or studying other aspects related to an operation.

The enhancement component of this programme is explicit: 'unlike many of DARPA's previous neuroscience and neurotechnology endeavors, it will aim not just to restore lost function but to advance capabilities beyond normal levels.'[64] One can imagine a soldier deployed to urban combat in a foreign country, for instance, quickly mastering vast tangles of information concerning the geography, topography, and infrastructure of the urban setting, or acquiring conversational knowledge of the local language in a short period of time. It is easy to see the advantages of rapid knowledge acquisition in an operational setting.

[63] The most common methods of brain stimulation are transcranial magnetic stimulation (TMS) and transcranial direct current stimulation (tDCS). TMS is a neurostimulation and neuromodulation technique that relies on electromagnetism (more specifically, Faraday's Law of electromagnetism). TMS works by delivering brief current pulses through a coil of wire placed on the scalp. The magnetic field that is formed penetrates the scalp and induces effects which outlast the initial moment of stimulation. Transcranial electrical stimulation functions differently: it works by applying electrodes on the scalp, and a weak current flows through the brain between the anode and the cathode electrodes. Full description in: Jean Levasseur-Moreau, et al., 'Non-Invasive Brain Stimulation Can Induce Paradoxical Facilitation. Are These Neuroenhancements Transferable and Meaningful to Security Services?', *Frontiers in Neuroscience* 7:449 (August 2013), 2.

[64] 'Boosting Synaptic Plasticity to Accelerate Learning', 16 March 2016, www.darpa.mil/news-events/2016-03-16.

Technologies of neurostimulation draw on – and contribute to – our knowledge concerning brain plasticity and the correlation between brain regions and behaviour.[65] For example, electric stimulation of the prefrontal cortex can either enhance or inhibit the tendency to lie, or it can encourage individuals to accept and respect social norms. The technique allows for increasingly on-demand interventions.[66] In addition to methods that involve direct interaction of TMS with neural activity, another mechanism has been the so-called 'addition-by-subtraction', through which the TMS disrupts or inhibits the part of the brain which competes with or distracts from the essential task at hand. In effect, this produces 'temporary network reorganisation' in the brain.[67] The effects of brain stimulation can go even further in modulating responses and reactions that are typically part of an individual's distinct personality. Some experiments in neuroenhancement have looked at modalities to reduce impulsive behaviour through a process called 'Theta Burst Stimulation.' At the end of these procedures, the subjects showed a change in preference for larger and delayed rewards as opposed to immediate rewards.[68] Neurostimulation could also alter decision-making mechanisms, risk-taking, planning skills or deceptive capabilities. For example, one session of tDCS with both electrodes placed over the dorsolateral prefrontal cortex indicated a more risk-averse response.[69]

Importantly, these techniques not only stimulate but can also suppress neural activity. The effect can last for up to several hours, or even days. They can also lead to changes in neurochemistry. For example, repetitive TMS can severely disturb neurotransmitters, and stimulation of the frontal lobe at 20 Hz was shown to lead to an increase of dopamine in the hippocampus. Other after-effects have included seizures, endocrine responses and increases in heart rate, to name a few. What is certain is that the full picture of the side- and after-effects is not yet known.[70]

65 When non-invasive brain stimulation induces a deficit in a neural activity, it is called 'virtual lesion' and when it leads to a functional enhancement, the phenomenon is called 'paradoxical facilitation'.

66 C. Chambers, 'Neuro-Enhancement in the Military: Far-Fetched or an Inevitable Future', *The Guardian*, 7 October 2016, www.theguardian.com/science/head-quarters/2013/oct/07/neuroscience-psychology.

67 B. Luber and S. H. Lisanby, 'Enhancement of Human Cognitive Performance Using Transcranial Magnetic Stimulation', *Neuroimage* 85: 03 (15 January 2014), 6–7, www.ncbi.nlm.nih.gov/pmc/articles/PMC4083569/pdf/nihms-494160.pdf.

68 Ibid., 5.

69 Ibid., 5–6.

70 S. Rossi, et al., 'Safety, Ethical Considerations, and Application Guidelines for the Use of Transcranial Magnetic Stimulation in Clinical Practice and Research', *Clinical Neurophysiology* 120 (2009), 2018–2024.

7.5 LEGAL CONUNDRUMS

A general question that may be asked from a legal point of view is: what gets to be regulated, the enhanced individual or the enhancement? No easy or clear-cut answer can be provided because human enhancement is not one technology, with one distinct effect. Rather and as has been shown in the preceding pages, it comprises several kinds of interventions including the insertion of physical components into the body, the ingestion of pills and changes to the neurochemical pathways. These tend to be methods applied before the soldier enters into combat, such as TNT mentioned above, so their effects do not generally endure beyond the period of combat. In most cases, indeed, enhancements will alter the body of the soldier but they cannot be classified as distinct weapons (see Chapter 2). However, there remains a difficulty where an enhancement integrates an offensive capability, e.g., if a brain–computer interface can operate a weapon system such as a drone. Such an enhancement becomes a legitimate target for the enemy and, in attempting to neutralise the enhancement, the enemy would simultaneously make the soldier a target, as the soldier hosts the technology within his or her body.[71]

Chapter 8 presents a comprehensive study of the implications of the law of armed conflict and of human rights law for military human enhancement. The present legal discussion complements that chapter and more specifically, tackles those legal aspects that are also closely connected to the ethical challenges posed by these technologies.

7.5.1 *Transparency and IHL*

A critical point here concerns transparency. The following fictional scenario is revealing. Consider the example of a soldier who has been enhanced with a combination of the following procedures: a genetic therapy has modified digestion and allows the soldier to survive on minimal nutrients; a nano-scale medical robot has been inserted into the body, which in case of blood loss, starts releasing a dose of the E2 oestrogen hormone; the soldier's brain has been subjected to neurostimulation before entry into combat by stimulating his peripheral nerves for accelerated learning; and two neural devices have been implanted into the soldier's brain. The first device allows him to transfer information to the military base in real time, or simply enables the suppression of his appetite (one possible application of the 'neural dust' on the peripheral

[71] Dinniss and Kleffner, 'Soldier 2.0', 438–439.

nervous system)[72] thus enabling prolonged fighting. The second neural device provides a brain–computer link that enables the soldier's brain to exert direct control of a weapon system. Despite these enhancements, the soldier looks perfectly normal to an outside observer.

a) It is likely that the last of these enhancement technologies would be classified as a part of a means of warfare and both neural implants might be classed as methods of warfare. It is questionable whether the enhancements would immediately become apparent to the enemy, given the fact that the enhanced soldiers would not be visibly distinguishable. This non-detectability is not the case with weapons typically in use today. The CAR-15 (Colt Automatic Rifle-15), used widely by the United States during the Vietnam War, or the M-4 carbine in wide use today, are unmistakably recognisable as weapons. In the case of enhanced soldiers, their potential status as part of a weapon system is less identifiable. That said, there is no legal obligation to inform the enemy of improvements in methods and means of warfare, and States will frequently go to considerable lengths to keep such improvements secret. Indeed, it is perfectly lawful to allow the enemy falsely to believe that it is in a strategically stronger military position than is actually the case.

The ethical arguments may be more poignant, however. It is unlikely that most enhancement procedures that will be in use in the next few decades will qualify as distinct weapons, means or methods of warfare. Given that a decision not to reveal the use of enhancements to the adversary would not be an illegal act, but the simple exploitation of its technological superiority, what we are really discussing here is another example of the asymmetry that has always existed in warfare. The main challenge may, therefore, not be in relation to law, but, rather, may be ethical in nature. This may have long-term strategic implications, and has parallels with the ethical conundrums prompted by the use of unmanned vehicles and is discussed below in the Section on Ethics.

b) An even more likely case involves enhancements that have not been previously classified as weapons, means or methods of warfare. When would it be incumbent on States to notify the enemy of the employment of such enhancements? Such a need to inform the enemy about soldier enhancement would only be likely to arise in particular circumstances.

Consider, for example, the situation of a captured soldier who has previously been enhanced with a substance that requires periodic ingestion such that withdrawal is painful or problematic (see Chapter 8). In such situations a

[72] J. Gallego, 'New Dust-Sized Implants May be the Key to Monitoring Your Internal Health', *Futurism*, 5 August 2016, https://futurism.com/new-dust-sized-implants-may-be-the-key-to-monitoring-your-internal-health/.

State may wish to hastily inform the detaining forces that the captured soldiers were previously enhanced. The enhancing State will likely seek to secure the early repatriation of its soldier. The Detaining Power may be reluctant to agree to such early repatriation. If the enhancing State is unwilling to disclose the details of the required treatment, it is difficult to see how the Detaining Power can be criticised for a failure to give treatment the required nature of which it cannot necessarily be expected to know. Humanity might require the release of the prisoner to the enhancing Power against a parole undertaking not to participate further in the armed conflict, but some armed forces prohibit the giving of such parole undertakings. An alternative might be a transfer to and internment by a Neutral Power under arrangements provided for in Geneva Convention III. The purpose of this part of the discussion is to show that enhancements, depending on their nature, may raise a multiplicity of complex issues under the law of armed conflict. It will of course be important for the Detaining Power to avoid any suggestion that by depriving the prisoner of necessary treatment it is practicing torture or cruel or inhuman treatment.

In general, however, the State deploying enhanced soldiers to the battlefield seems likely in most cases to be able to do so without a formal requirement of transparency.

7.5.2 *Deception in Warfare – Could Enhanced Soldiers Be a 'Ruse Of War'?*

It is then worthwhile inquiring when and under which circumstances, enhancement(s) could be linked to another important provision in IHL which addresses deception operations and under which 'ruses of war', including camouflage, are considered lawful whereas acts of perfidy giving rise to death, injury or (in the case of States that are party to API) capture are designated as unlawful.

Under the Hague Regulations of 1907: 'Ruses of war and the employment of methods necessary for obtaining information about the enemy and the country are considered permissible.'[73]

Article 37(2) of the 1977 Additional Protocol I develops the law by providing that:

> Ruses of war are not prohibited. Such ruses are acts which are intended to mislead an adversary or to induce him to act recklessly but which infringe no rule of international law applicable in armed conflict and which are not

[73] Annex to Hague Convention IV, 1907, Regulations Respecting the Laws and Customs of War on Land, The Hague, 18 October 1907, Article 24.

> perfidious because they do not invite the confidence of an adversary with respect to protection under the law. The following are examples of such ruses: the use of camouflage, decoys, mock operations and misinformation.

The Lieber Code of 1863 includes several provisions on deception. Article 15 states that:

> Military necessity ... allows ... of such deception as does not involve the breaking of good faith either positively pledged, regarding agreements entered into during the war, or supposed by the modern law of war to exist.

Article 16 concludes that 'Military necessity ... admits of deception, but disclaims acts of perfidy...'

The provisions of 1907 are widely accepted as reflecting customary international law and, thus, bind all States. Deception operations, as such, are not regarded as inherently unlawful but some examples given below reveal some nuanced interpretations of the meaning of deception and perfidy.

Australia's LOAC Manual of 2006 covers deception in the following way:

> Ruses of war and the employment of measures necessary for obtaining information about the enemy and the enemy country are permissible. Ruses of war are used to obtain an advantage by misleading the enemy. They are permissible provided they are free from any suspicion of treachery or perfidy. Legitimate ruses include surprises, ambushes, camouflage, decoys, mock operations and misinformation. *Psychological operations are also permitted.*[74] [emphasis added]

Canada's LOAC Manual of 2001 states that:

> Ruses of war are more formally defined as acts, which are intended to mislead an adversary or to induce that adversary to act recklessly. Ruses must not infringe any rule of the LOAC. Ruses are lawful *if they are not treacherous* [emphasis added], perfidious and do not violate any express or tacit agreement.[75]

Israel's Manual on the Rules of Warfare of 2006 more explicitly links the notion of deceit with transparency, in stating that:

> *Surprise, deception and dishonesty are among the most basic principles of war, gaining a tactical and even a strategic advantage for the army.* The ban in the annals of the rules of engagement is not designed to prevent armies from

[74] IHL Database, Practice Relating to Rule 57. Ruses of War, https://ihl-databases.icrc.org/customary-ihl/eng/docs/v2_rul_rule57.

[75] Ibid.

> using surprise tactics and *does not require each side to be 'transparent' in the face of its enemy* [emphasis added].[76]

New Zealand's Military Manual of 1992 states that:

> Ruses of war are measures taken to gain advantage over the enemy by mystifying or misleading him.[77]

The United States Field Manual of 1956 states that:

> Ruses of war and the employment of measures necessary for obtaining information about the enemy and the country are considered permissible. [...] Ruses of war are legitimate so long as they do not involve treachery or perfidy on the part of the belligerent resorting to them. They are, however, forbidden if they contravene any generally accepted rule. The line of demarcation between legitimate ruses and forbidden acts of perfidy is sometimes indistinct.

The US Law of War Manual of 2015 also re-states the legality of ruses of war as long as they 'do not infringe upon any rule of international law applicable to armed conflict'.[78] The Manual includes a long list of examples of deceptions that are not prohibited, but the list is not exhaustive, which means that other deception activities may, in the US DoD view, amount to ruses, depending on the circumstances.

What does this mean for military human enhancement?

The way in which transparency, or the lack of it, is described in the selected excerpts demonstrates that the use of enhancements – provided they are not *a priori* classified as illegal – does not incur a moral or legal obligation on the part of the State to inform the enemy about their use. States may even use these legal provisions on deceptions to justify the deployment of enhanced soldiers.

This can be a strategic choice especially in urban warfare, which confronts both militarily weak and strong countries with awkward choices. In those instances, enhanced soldiers could indeed 'mystify' the enemy. In a Report to US Congress detailing the situation during the Gulf War, it was mentioned that: military operations in urban settings represent a 'form of fighting that is costly to attacker, defender, innocent civilians and civilian objects'.[79] Urban warfare presents unique challenges because it compels stronger, militarily

76 Ibid.
77 Ibid.
78 *Law of War Manual*, US Department of Defense (2015), 302–307.
79 US Department of Defence, Report to Congress on the Conduct of the Persian Gulf War (1992), 721.

mightier nations to adapt to an unconventional setting, and slows down their operations. At the same time, it brings an opportunity for the weak to obtain strategic advantages, despite glaring technological asymmetries.[80] Indeed, asymmetric urban warfare has often exposed the 'militarily superior side's political vulnerabilities'.[81] Staying fully compliant with IHL 'lessens [the asymmetrically superior]'s ability to quickly and decisively achieve [a] strategic objective' and then, in trying to achieve its military objectives, navigating through densely populated urban areas 'makes it difficult to stay within lawful parameters'. This serves the weaker side, which will try to exploit politically any attacks on civilians, knowing that the stronger side faces pressure to achieve a quick victory with minimum casualties.[82]

The deployment of enhanced soldiers could resolve this paradox. Let us imagine that in this urban war theatre, the technologically weaker warring party removes or changes street signs to confuse the adversary. However, that adversary is an enhanced soldier who can fight in extreme conditions, does not get tired after thirty hours in combat, has enhanced vision which allows him to see even at nighttime and, in addition, has a neural implant through which he is able to receive accurate directions from his military base. The opponent is confused and exhausted, expecting the morale of the other to drop, but the enhanced soldier shows no sign of fatigue.

For the enemy, enhanced soldiers could be seen as a dishonest way of fighting, but one which is legally permissible under IHL (see, for example, the Israeli definition above), or as contributing to a 'psychological operation', which is also permissible (see, for example, the Australian definition). So in most circumstances at least, consideration of the law relating to deception suggests that the deployment of enhanced soldiers may not only satisfy military necessity but also constitute a lawful ruse of war.

80 I. Arreguin-Toft, 'How the Weak Win Wars. A theory of Asymmetric Conflict', *International Security* 26:1, 93–128.

81 M. John-Hopkins , 'Regulating the Conduct of Urban Warfare: Lessons from Contemporary Asymmetric Armed Conflicts', *International Review of the Red Cross* 92:878 (June 2010), 471. Political theory has explained this by comparing the survival objectives of the weak and the strong: strong actors have a comparatively lower interest in winning because their survival is not at stake, weaker actors have a higher interest in winning because it is only victory that can ensure their survival. See A. Mack, 'Why Big Nations Lose Small Wars: The Politics of Asymmetric Conflict', *World Politics* 27:2 (January 1975), available at: https://web.stanford.edu/class/polisci211z/2.2/Arreguin-Toft%20IS%202001.pdf. This framework has been partially refuted by other theorists (e.g., Arreguin-Toft), observing that for a country such as the United States, which brands itself 'the leader of the free world', a defeat in a distant part of the world can be materially insignificant, but it carries deeper military consequences and reputation costs.

82 John-Hopkins, 'Regulating the Conduct', 471.

However, ultimately, with the sort of enhancement technologies currently envisaged, the greatest challenges will most likely arise not in the domain of IHL, but domestic law and military ethics. The next sections will elaborate on these.

7.6 ETHICS AND OPERATIONAL CHALLENGES

Military human enhancement will bring to the fore issues related to transparency, legitimacy, military necessity, and will also challenge military ethics and army values.

7.6.1 *Ethical Risks and Operational Difficulties*

7.6.1.1 Military Values

Human enhancement technologies will challenge some of the fundamental values shared among servicemen and women. The Australian Army, for instance, lists four core values: courage, initiative, respect, and teamwork; most soldiers internalise these values to the point that 'their profession becomes a large part of their identity'.[83] The impact of enhancement technologies may prove to be profoundly disruptive because they may be expected to cause the army to redefine the meaning of some virtues that create loyalties and solidarity, and which make military ethics so uniquely distinct from civilian ethics.

7.6.1.1.1 HONOUR The United States Law of War Manual of June 2015 reaffirms the three interdependent principles – military necessity, humanity, and honour – that provide the foundation for other principles in the law of war, such as proportionality and distinction.[84] Honour, also called 'chivalry', requires 'a certain amount of fairness in offense and defense', as well as 'certain mutual respect between opposing forces'.[85] This point is elaborated as follows: 'opposing forces should respect one another outside of the fighting because they share a profession, and they fight one another on behalf of their respective States and not out of personal hostility.'[86] Especially important is the section on 'Honor and the Class of Combatants', which states that: 'in

[83] M. Beard, J. Galliott and S. Lynch, 'Soldier Enhancement: Ethical Risks and Opportunities', *Australian Army Journal* XII:1 (Winter 2016), 13.
[84] *Law of War Manual*, United States Department of Defense (June 2015), 50.
[85] Ibid., 66.
[86] Ibid., 67–68.

demanding a certain mutual respect between the opposing forces, honor also reflects the premise that combatants are a common class of professionals who have undertaken to comport themselves honorably.'[87]

A war that opposes enhanced and non-enhanced soldiers certainly challenges the application of this principle because it diminishes the sense of respect for the enhanced soldiers, whose courage and endurance may now be a function of their enhancement. The notion of 'common class of professionals' remains technically true, but the artificially generated difference in capabilities weakens the legitimacy of the principle and may, depending on the technology and the circumstances, call into question whether the conflict is being conducted honourably. Whether an activity that is found not to be consistent with notions of honour would necessarily also be considered treacherous is an open question. For what it is worth, the present author suggests that breaches of honour would not necessarily amount to treacherous conduct, and that the mere enhancement of a soldier's operational capabilities without more does not amount to treachery. However, it cannot be excluded that unlawful behaviour may be committed by enhanced soldiers and that the enhancement may have contributed directly to its commission.

There is, furthermore, a risk that the resentment felt by non-enhanced soldiers for their enhanced opponents could lead to more cruelty. The use of drones in warfare signals the moral consequence of sharp technological asymmetry. Retired general Stanley McCrystal cautioned in a 2013 interview about the effects of using drones in Afghanistan. He stated that drones created resentment, a perception of arrogance and visceral hatred against the USA.[88] The deployment of enhanced soldiers could lead to similar perceptions (and an incentive for more vicious use of force against them).

7.6.1.1.2 COHESIVENESS AND OPERATIONAL CHALLENGES The principle of honour was developed to guide relations with enemy forces, not among soldiers *within* the same army and the same country. For centuries if not millennia, the understanding of mutual respect, teamwork and solidarity within the army of the same country was taken for granted for a simple reason: those soldiers identified themselves as part of a community, serving the same nation. While differences in rank have existed for centuries, they have been and continue to be an essential part of an established institutional order.

[87] Ibid., 68.

[88] 'Retired General cautions against overuse of "hated" drones', *Reuters*, 7 January 2013, www.reuters.com/article/us-usa-afghanistan-mcchrystal/retired-general-cautions-against-overuse-of-hated-drones-idUSBRE90608O20130107.

Soldier enhancement risks generating an unprecedented challenge to established notions of unit cohesion and may eventually involve a revised interpretation of honour.

An interesting question to consider is whether unit morale and cohesion could be maintained as faithfully if enhancements were to be used selectively, or only for some units, or if units were to have a mixed composition of enhanced and non-enhanced soldiers? The precise manner in which enhancements will be rolled out is of critical importance and will have far-reaching consequences. Some questions commanders will need to grapple with are: should enhancements be restricted to special or elite forces? Which types of enhancements should be used where and why? Additionally, should enhancements be removed or should that be optional?[89] How would remuneration, reward and promotion be equitably distributed in such a system? Furthermore, how will recruitment be affected? In particular, should less meritorious individuals be recruited, who despite their shortage of skills and/or other competencies, would be eligible to receive the necessary enhancements such as to put them on a par with others?

In addition, given the higher costs invested in enhanced soldiers, as well as the fact that, by definition, they are more physically able, should such soldiers be spared the menial tasks and only be sent to fight? This may be expected to increase their sense of entitlement and to further alienate them from their non-enhanced peers. The commander/subordinate relationship may also suffer severely, particularly if the commander is not enhanced.[90]

The resulting resentments could take various forms and may affect morale, the command structure and, ultimately, operational success. It is important to recall that in 2013, when (now retired) Lieutenant General David Morrison, Chief of the Australian Army, added the value of 'respect' to the existing list of values to be upheld by the Australian Army (courage, initiative and teamwork), his initiative was partly due to the number of allegations of abuse, bullying and assault that had plagued the Australian Defence Force.[91] The introduction of enhancements could lead to acrimonious and hostile interpersonal relations and may be detrimental to the socialisation within the armed forces of values of respect – both vertically and horizontally.

89 P. Lin, 'Ethical Blowback from Emerging Technologies', *Journal of Military Ethics* 9:4 (2010), 322.

90 Lin, et al., 'Enhanced Warfighters', 39–40.

91 P. Collins, 'The Value of Respect: What Does it Mean for an Army?', *Journal of Military Ethics* 16:1–2, 2.

That is why, without careful implementation, military enhancements could end up being counter-productive.

7.6.1.1.3 COURAGE AND MERIT In a rescue mission in November 1965, during the Vietnam War, US Capt. Ed W Freeman risked his own life by flying an unarmed helicopter through enemy fire; 14 rescue mission trips later, he saved about 30 wounded soldiers. In 2001, he was awarded the Congressional Medal of Honor for his extraordinary perseverance, and intrepidity that was 'above and beyond the call of duty or mission'.[92] Courage is one of the most valued virtues in army life, and it involves a strong sense of duty and risk-taking abilities, which in extraordinary circumstances can manifest in heroic acts and, subsequently, rewards according to the merit system established in each country. Human enhancements could bring unprecedented challenges to this fundamental value.

Is it conceivable at all that an enhanced soldier could receive a Medal of Honor for courage in combat and for taking risks 'above and beyond the call of duty'?[93] Could a soldier be rewarded with a Bronze Star Medal, normally awarded in recognition of 'heroic or meritorious achievement or service'? Hypothetically, the enhanced soldier's endurance may have been the result of a therapy to annihilate pain, and their risk-taking ability the result of neurostimulation of the relevant brain parts.

Therefore, how can courage be measured in an enhanced soldier? A similar question has been asked in the context of pilots of unmanned aerial vehicles who are physically distanced from danger. Such warfighters, whose own risks of injury or death are either severely restricted or eradicated altogether, have been called 'post-heroic soldiers'.[94] This would, of course, be an inaccurate description for enhanced soldiers who, despite their advantage in combat, would continue to encounter dangers as they would be physically present on the battlefield. But the issue of courage still presents some ethical conundrums. Not all enhancements will reflect in a heightened ability to take risks and show courage; indeed, some may target only specific physiological functions. Even in this case, however, it is important to note that the added value of a physical enhancement may well have outputs in terms of the amount of courage displayed: by prolonging physical endurance, a soldier may feel more empowered and emboldened on the battlefield.

[92] 'Bush presents Congressional Medal of Honor', *CNN*, 16 July 2001, http://edition.cnn.com/TRANSCRIPTS/0107/16/se.02.html.

[93] Description of Medals, http://valor.defense.gov/Description-of-Awards/.

[94] C. Enemark, cited in Beard, et al., 'Soldier Enhancement', 13.

In other cases, a form of enhancement may lead to an increase in risk-taking abilities, however, if that soldier remains fully aware of the risks and of the implications of his or her actions (including risk of loss of life), the element of courage is still arguably present. The question for military ethics then is whether the courage displayed while enhanced has the same intrinsic value as the courage demonstrated in an entirely non-enhanced state.

Some ethicists suggested that the predicament of merit could be resolved if the army presented and justified the enhancement of soldiers not as a way of gaining an advantage over un-enhanced enemies but in order to be in a more balanced contest against enhanced opponents. (This would, of course, work only if the enemies are also enhanced.) The second suggestion was to re-evaluate what courage means, and promote a view that courage means the ability to do what is right, and not necessarily physical courage.[95] These theoretical solutions are more likely to be rebuffed than adopted because it is practically difficult to change the meaning of a value such as courage only from top-down, even in an institution that is premised on conformity. If enhancements are perceived as unfair, they will risk demoralising many in the military and do irreparable damage to the sense of community.

7.6.1.1.4 SACRIFICE REDEFINED? Human enhancement may diminish, or adjust, the meaning of courage but it could paradoxically also lead to expanding the understanding of sacrifice. Enhancement procedures should not be simplistically branded only as a conveniently quick fix for improving performance; they can have serious health implications, and may compromise the soldier's health and even identity. From this point of view, enhancements could lead to a new understanding of personal sacrifice. For example, if a form of enhancement has been demonstrated to be very effective but to have long-term adverse side effects (e.g., cognitive decline), the soldier volunteering to undergo that enhancement procedure, while fully aware of the risks, is subscribing to a new understanding of heroism. That soldier will be sacrificing his or her chances of a potentially healthy and normal life after deployment in order to a) make a vital contribution to the success of the mission, and b) possibly, to spare his fellows and reduce the human cost of the operation (assuming that fewer enhanced soldiers can achieve as much as larger numbers of unenhanced personnel). Conversely, this scenario may create an ethos for volunteering for enhancements (possibly entailing competition to undergo the riskiest procedures), and such a tendency may be expected to

95 Beard, et al., 'Enhancement', 14.

have wide-ranging implications when those soldiers return from the battle and need to reintegrate to civilian life.

7.6.1.2 Informed Consent and Autonomy

Not all soldiers may volunteer for enhancement; in other cases, the decision will come as an order from the military superior. This will face the military, an institution which already curtails individual autonomy, with new questions about informed consent, privacy, confidentiality, and liberty to choose.

The right to bodily integrity (see Chapter 8) refers to the right of any human being, including children, to have autonomy over their bodies, and simply put, for the individual to determine what is to be done to his or her body. This collection of rights faces limitations in the military.[96] As Michael Gross points out: while bioethics is focused on the patient, military ethics takes the interests of three distinct actors into account: combatants, non-combatants, and the State. In the armed forces, and particularly during periods of armed conflict, Gross notes, the State abridges individual autonomy, while it establishes 'benign paternalism'. Ironically, he contends, as combatants lose their right to life, they gain the right to kill.[97]

The notion of 'informed consent' to enhancements is complicated by several factors.

For example, in the case of the United States, a soldier refusing vaccination can be punished by a commander under Article 15 of the Uniform Code of Military Justice.[98] Article 90 of the Uniform Code of Military Justice prohibits willfully disobeying a superior officer and Article 92 prohibits failing to obey an order.

In a number of countries, soldiers have an obligation to accept any intervention that is considered 'standard procedure' and a soldier is expected to

[96] There remain, however, absolute rights in this category (for example, the prohibition against slavery or torture), which cannot be violated by States in any circumstance. At the same time, the freedoms of soldiers are restricted. So, for example, they may, depending on the domestic and human rights law applying in the State in question, not have the right to refuse medical care or treatment (e.g., vaccinations), and refusal may, again depending on the applicable domestic law, constitute disobedience of a lawful order, and, thus, may amount to a disciplinary offence.

[97] M. L. Gross, 'Bioethics and Armed Conflict: Mapping the Moral Dimensions of Medicine and War', *Hastings Centre Report* 34:6 (2004), 23.

[98] M. J. Davidson, cited in L. R. Robbins, 'Refusing to be All That You Can Be: Regulating against Forced Cognitive Enhancement in the Military', in M. L. Gross and D. Carrick (eds.), *Military Medical Ethics for the 21st Century* (Surrey: Ashgate, 2013), 128.

submit to medical care, for example in preparation for active duty.[99] However, what is seen as amounting to 'standard procedure' is not without controversy.

Two main issues may arise in the context of military enhancements.

1. There is a risk that, as the armed forces develop these technologies, they become classified as standard medical care, which implies that they may be seen as compulsory for soldiers. The immediate response should be strict guidelines and regulations to frame the boundaries between treatment and enhancement.[100] A difficulty lies, however, with the language itself because of the military tendency to shy away from using the term human enhancement. The US military, for instance, has usually preferred terms such as 'performance optimisation' or 'peak performance'. In their report, 'Enhanced Warfighters: Risk, Ethics, and Policy', the authors consider this a public relations choice, which helps to avoid the usual ethical questions associated with human enhancement.[101] This choice of language can have implications for the way legal debates proceed. Framing as 'optimisation' what is effectively a form of enhancement can have the effect of changing the procedure's connotation and making it appear more benign. The only way for law-makers to distinguish between what enhances the body beyond its limitations, and what is merely therapeutic is to sieve through bureaucratic jargon and grasp the full effects and implications of each form of biomedical intervention.
2. The meaning of 'informed' in the term 'informed consent' is not absolute. A soldier may have doubts as to whether the procedure or medicine that he or she is being required to take is safe and/or as to whether its side-effects are understood. In 2000, the case of Major Bates from the US Air Force garnered national attention because he refused to be vaccinated against anthrax, and thus refused the direct order. While it was known beyond doubt that Anthrax presents a deadly threat, Major Bates exposed the medical controversy relating to the vaccine, which had been shown to be effective for animals, but not for humans. The legal controversy associated with this case may well indicate the kind of controversies that enhancements will raise. Ultimately, military judges find it difficult to accept that an order to submit to vaccinations is unlawful; nevertheless, this case and others demonstrated how a lack of trust can erode the willingness of military personnel to comply with

99 Ibid.
100 Ibid.
101 Patrick Lin, et al., 'Enhanced Warfighters', 16.

> certain superior orders. Other studies showed that the fear of the anthrax vaccines caused a significant number of service members to leave the armed forces because they did not trust the government and ultimately feared for their health.[102]

In general, it is difficult to build a case in favour of the application of enhancements as a standard procedure. Such a policy could only be justified if those interventions were proven beyond doubt to be safe. An especially concerning prospect is that of soldiers losing their 'cognitive liberty', essentially their right to think independently and not have one's state of mind manipulated by neurological enhancements.[103] If a soldier has been subjected to a procedure that has this kind of effect that soldier may be rendered incapable thereby of giving an informed and free consent to any further enhancement procedure or, indeed, to any other action for which consent is required.

7.6.1.3 More Wars or More Violent Wars?

The emergence of military robots and autonomous weapons has led to heated debates about the future and ethics of war in general. Many have expressed concerns that decreased human presence on the battlefield, coupled with diminished risks of loss of life (assuming robot 'casualties' could outnumber human casualties), will lower the barriers to start a war. Similar arguments have been attempted for the development of human enhancement technologies in the armed forces: will it become easier to start a war when the human cost is expected to be low and when victory seems likely to be achieved more quickly? Arguably, the intent to reduce human casualties in war might have the effect of rendering war less horrific, but more likely.

On the other hand, as Patrick Lin wonders, would this mean that only by raising the barriers to go to war could war be avoided? This implies making war as brutal and savage as possible, which also appears counter-intuitive and inherently undesirable.[104]

Additionally, the assumption that war casualties will be low is incorrect. As highlighted in the previous section, the overwhelming majority of war casualties in twenty-first century armed conflicts occur in urban warfare, where

[102] R. D. Katz, 'Friendly Fire: The Mandatory Military Anthrax Vaccination Program, *Duke Law Journal* 50 (2001), 1835–1865.

[103] L. R. Robbins, n. 98; Richard G. Boire, 'Civil Liberties for the Mind: At the Crossroads of Civil Liberties and the Law', *Mind Matters (Centre for Cognitive Liberty & Ethics)* 1:2 (2004).

[104] P. Lin, 'Ethical Blowback from Emerging Technologies', *Journal of Military Ethics* 9:4 (2010), 324–325.

civilians suffer the most. It therefore seems possible that soldier enhancements may reduce overall soldier casualties but not death rates among peaceable civilians. Moreover, human enhancement policies are unlikely to make the idea of war any more attractive or justifiable to public opinion, a thought which rather calls into question the idea that more frequent wars would be the result of enhancement technologies.

7.6.2 *Ethical Opportunities*

Generally, the debate on human enhancement has been very polarising and very often, the prospect of enhancing human soldiers appears frightening, if not sinister. However, the use of all military enhancements cannot summarily be discarded as being immoral, illegal or unethical. Quite the opposite, some of their contributions could well prove to be ethically desirable.

Some arguments for resisting the adoption of military enhancements, seemingly based on moral grounds, mirror the rationales against the introduction of robots on the battlefield. This narrative stresses that only humans can behave virtuously, have compassion or empathy. Similarly in the case of military human enhancement, the complete rejection of enhancements would reinforce a critical misunderstanding: the view that non-enhanced soldiers are by default more moral, humane or compassionate, and that enhancements would render them more violent, robot-like, or would turn them into human killing machines. There are forms of enhancement that can help soldiers 'do the right thing'. Indeed, in some instances, an intervention to boost cognitive functions, for example, can make the difference between a legally and operationally positive outcome, and a catastrophe.

7.6.2.1 Solution for Combat Stressors

It is estimated that between 80 and 85 per cent of accidents involving the armed forces are caused directly by human error attributed to failures of cognitive performance. These failures are the result of biological stressors[105]

[105] Field research with US military units identified the following six stressors in military operations: isolation (the sense of being estranged from family in a foreign location, often with unreliable communication), ambiguity (unclear norms, confusion), powerlessness (imposed restrictions and separation from one's culture), boredom and alienation (long periods of repetitive work, long missions), danger and threat (stress related to real risks of injury or death) and workload (long duration of deployments and long working days). See . Bartone, 'Forging Stress Resilience: Building Psychological Hardiness', in R. Armstrong, et al. (eds.), *Bio-Inspired Innovation and National Security* (Washington: National Defence University Press, 2010), 245.

such as: fatigue or inadequate sleep, calorific deficiencies, extreme heat or cold, and cognitive overload.[106]

There are military situations – rarely applicable in the civilian context – where the use of enhancements is advisable. Drugs to cope with sleep deprivation, to fight anxiety and maintain vigilance could eventually improve the chances of survival of the individual and the group.[107] At the same time, benefits can quickly turn into risks, and it is important to understand the full spectrum of possible side-effects of any type of intervention on the brain. The aforementioned case at the Tarnak Farm in Afghanistan in 2002 is a reminder about the possible deleterious implications of enhancers.

7.6.2.2 Trauma and PTSD

Stress and trauma in combat have immediate detrimental effects on judgement and attention-spans. For example, empirical studies on Vietnam veterans who suffered from PTSD showed they had smaller hippocampi – the brain region that controls conscious memory. Prolonged exposure to high levels of cortisol (also known as the 'stress hormone') and trauma lead to deficits in cognitive functioning: recent studies revealed that US soldiers who served in Iraq showed continuing symptoms of memory lapses and an inability to focus – for over two months following their return from combat.[108] In the same study, soldiers who saw combat from April 2003 to May 2005 were compared to soldiers who had not been deployed. The former performed significantly worse than the latter in tasks which measured spatial memory, verbal ability and focus. Deployment, in short, is associated with visible risks of 'neuropsychological compromise'. However, the soldiers who had been in operations demonstrated better abilities in terms of quick reactions to simple tasks. This implied that their brains had managed to adapt to the speed of combat conditions but – critically – at the expense of other mental capacities.[109]

[106] M. L. Thomas and M. B. Russo, 'Neurocognitive Monitors: Toward the Prevention of Cognitive Performance Decrements and Catastrophic Failures in the Operational Environment', *Aviation Space and Environmental Medicine* 78:5 (May 2007), 144–145.

[107] A. Chatterjee, 'Ethics of Neuroenhancement', 329.

[108] E. Stanley, 'Neuroplasticity, Mind Fitness, and Military Effectiveness', in R. Armstrong (ed.), *Bio-Inspired Innovation*, 259–269.

[109] J. J. Vasterling, et al., 'Neuropsychological Outcomes of Army Personnel Following Deployment to the Iraq War', *Journal of the American Medical Association* 296:5 (August 2006), 526–528.

Brain stimulation or drugs that can selectively target and erase memories could have beneficial uses for soldiers during and after combat. They could help individuals live more meaningful lives, without the paralysing fear, emotional numbness, nightmares or intrusive thoughts that result from trauma. A method used for many years, Critical Incident Stress Debriefing (CISD), popular with several organisations in the USA including the US Department of Defense and with the Israeli Army, was based on the principle that people experiencing traumatic events should speak about them sooner thereafter so that the memory does not get 'sealed over'. The method had, however, a low success rate, and was in many cases counter-productive, leading to more depression and PTSD.[110] Increasingly, it appears the most effective methods are the more invasive ones.

Experiments with different modalities of brain stimulation found that repeated stimulation of the right and left dorsolateral prefrontal cortex led to significant improvements in PTSD symptoms.[111]

A 'forgetting pill' has also been researched for a similar purpose, making memory optional, by targeting the range of neurochemicals associated with memory formation. Memory is formed through a process called long-term potentation, involving a cascade of gene activations and protein synthesis. At New York University, one team discovered that by delaying the creation of new proteins during the act of remembering, the memory would cease to exist.[112] Over time, a molecule in particular was identified as particularly relevant to the regulation of long-term potentation, namely protein kinase C. After over a decade of research, it was discovered that a form of this protein, PKMzeta, was specifically linked to how stable recollections form or disappear. Once PKMzeta inhibitors are used widely, they could make people forget not only memories, but even pain – a dangerous prospect indeed, as pain is often educational and has a survival value.[113]

Technologies of this sort that are developed for therapeutic purposes are always, of course, at risk of being adapted for offensive purposes. Were technologies of the sort discussed in the previous paragraph to be developed and weaponised, for example, so as to cause targeted enemy personnel to forget their orders or to forget operationally important information or skills, such a technology would require legal review in accordance with Article 36 of

[110] J. Lehrer, 'The Forgetting Pill: How a New Drug can Erase Your Worst Memories', *WIRED*, 17 February 2012, www.wired.com/2012/02/ff_forgettingpill/.

[111] Vladan Novakovic, et al., 'Brain Stimulation in Posttraumatic Stress Disorder', *European Journal of Psychotraumatology* 2 (2011), www.ncbi.nlm.nih.gov/pmc/articles/PMC3402102/.

[112] Lehrer, 'The Forgetting Pill'.

[113] Ibid.

API and in accordance with the implied rule that binds States not party to API. A State contemplating the introduction of such a weapon would *inter alia* have to consider whether it breaches the principle that prohibits weapons, projectiles and material and methods of warfare that are of a nature to cause superfluous injury or unnecessary suffering (discussed in Chapter 2).

Any proposal to employ a therapeutic treatment based on PKMzeta technology for the benefit of prisoners of war or protected persons, e.g., to treat conditions caused by memories associated with combat, would be potentially lawful provided that recognised rules, including those reflected in GCIII, GCIV and API are complied with. Applying treatment that is not indicated by the state of health of the patient and which is not consistent with generally accepted medical standards would be unlawful.[114]

7.7 ENHANCEMENTS IN SOCIETY

This section draws a parallel between the military and civilian facets of human enhancement technologies. The contrast between the two approaches explains some of the persistent differences in ethical standards and regulatory issues.

7.7.1 *Sports*

Doping in sports is the main area of performance enhancement that is strictly regulated globally, as well as morally rejected. The stigma attached to doping can ostracise an athlete for the rest of their career. In contrast, the armed forces have a less consistent approach.

Over two decades ago, a group of military professionals defended the use of psychostimulants in war and considered that the elimination of amphetamine use put aircrews at increased risk. They conceded that:

> Using drugs to enhance performance in sports may be 'immoral', but war is not a sporting event. Success in combat is not a question of fairness but of power (…). We do *not* seek to equalize each side's chance of success prior to initiating contact (as we do in sports), but we *do* seek to obtain every advantage for our forces. However, this does not mean that we should rely upon amphetamine indiscriminately to create performance edge on every

[114] API, Article 11(1). Note also the prohibition of medical or scientific experiments in Article 11(2)(b). Persons in the power of an adverse party to the conflict also have the right to refuse a surgical operation. Note also Article 11(4).

day of combat operations. As with most things in life, we should consider costs and benefits prior to taking specific actions in various situations.[115]

Notwithstanding the cynical tone of these remarks, they capture the moral contrast between enhancements in sports and warfare. In sports, the guiding principle for imposing bans on enhancers is fairness; in warfare, it should be moderation and a cost-benefit analysis.

Another obvious contrast is that military human enhancement will always be a mainly national issue so long as the implications of the enhancements do not challenge the application of international law. This is because such enhancements generally raise issues under domestic and human rights law rather than under international law.

This is a resoundingly different approach from doping in competitive sports, where in addition to national anti-doping agencies (such as the UK Anti-doping Agency), there is a World Anti-doping Agency (Wada), which was established in 1999 and functions as an independent international agency. Wada is also guardian of the World Anti-doping Code that harmonises anti-doping policies across countries, and monitors implementation and compliance with its standards. Its regulatory framework includes a complete list of prohibited substances and methods.[116]

Among the substances and methods banned from competitive sports are also many which have been mentioned in this Chapter in the context of military human enhancement: alcohol, stimulants (including Amphetamine, Modafinil, Epinephrine among others), anabolic agents, beta-blockers; hormone and metabolic modulators. Notably, the January 2017 Prohibited List also lists the following banned methods: manipulation of blood and blood components; chemical and physical manipulation, and gene doping (adding or modifying genes in order to increase performance).[117]

7.7.2 *Workplace and Society*

Human enhancement has increasingly been studied for its potential to impact and change the workplace.

[115] Lt Col R. Cornum, Dr J. Caldwell and Lt Col K. Cornum, 'Stimulant Use in Extended Flight Operations', *Airpower Journal* (Spring 1997), 57, available online: https://pdfs.semanticscholar.org/8128/64ff145b4a3916b824311ed1dbb554384178.pdf.

[116] See 'World Anti-Doping Code' (2015), *World Anti-Doping Agency*, www.wada-ama.org/sites/default/files/resources/files/wada-2015-world-anti-doping-code.pdf.

[117] World Anti-doping Agency, *Prohibited List*, January 2017, www.wada-ama.org/sites/default/files/resources/files/2016-09-29_-_wada_prohibited_list_2017_eng_final.pdf.

In November 2012, the Royal Society, the Academy of Medical Sciences, the British Academy and the Royal Academy of Engineering issued a joint report entitled 'Human enhancement and the future of work'.[118] The study analysed the practical challenges and opportunities arising from the employment of human enhancement technologies in a workplace environment. Confronted with demographic shifts and an ageing work force (the report referred to the UK, but this applies to Europe, largely), the possibility of enhancing healthy individuals in order to make them more productive presents new opportunities for employees and employers.[119] The report, however, concluded that the implications will be 'complex, unpredictable or divisive'.[120]

Cognitive enhancing drugs or brain stimulation can improve memory and concentration, which can have returns in terms of productivity. Visual enhancement, such as retinal implants or cell replacement procedures, can help to counter degenerative conditions and can also be used by healthy individuals to extend the range of normal human vision – even allowing night vision. This last facility could, for instance, be applied to security guards or safety inspectors.[121]

Concerns and caveats similar to those that have been highlighted for the armed forces will also be applicable to the workplace. Inevitably, tensions will arise because enhancements will not be equally relevant across all sectors of activity. They will also raise questions concerning equal access, affordability, regulation, and personal freedoms (the possibility to choose to abstain from certain enhancements). In very practical terms, enhancements will also add additional costs and pressure on government institutions, especially if the line becomes blurred between medical need and 'enhancement of the healthy'. Enhancement technologies will also radically change social norms and values, such as those relating to the meaning and worth of some occupations and as to how disability is viewed.[122]

Hypothetically, a reassuring contrast between enhancements in the armed forces and in the workplace is that the development and implementation of relevant regulations is more likely to be achieved sooner in the social context.

[118] 'Human enhancement and the future of work', Report from a joint workshop hosted by the Academy of Medical Sciences, the British Academy, the Royal Academy of Engineering and the Royal Society, November 2012, available online: www.acmedsci.ac.uk/viewFile/publicationDownloads/135228646747.pdf.

[119] The report also analysed the technologies that could facilitate the access of disabled people – about 50 per cent of which are of working age in the UK – to work.

[120] 'Human enhancement and the future of work', 37.

[121] Ibid., 25–26.

[122] Ibid., 46–47.

Regulations would be driven by the fact that society – unlike the armed forces – is generally a more open environment, where concerns about health impacts and changes in values will foster dialogue and eventually lead to regulation. However, beyond this hopeful hypothesis, there are countless practical difficulties.

The regulation of enhancement technologies is a very complex matter because it involves a wide spectrum of technologies and multiple types of actors, including employers, governments, employees and suppliers of enhancement products. There is no 'one size fits all' approach that can work. Regulation could take many forms: a bottom-up approach in some cases (governmental), or self-regulation (rules determined by employers), or a hybrid approach (top-down on some aspects, complemented by self-regulation).[123] Irrespective of the approach, further complications arise from the fact that regulation would need a certain level of harmonisation in order to avoid competitive disadvantages. For example, if enhancements that allow a certain group of professionals to work for longer hours are accepted in one country but not in others, the latter are – theoretically – facing a position of disadvantage which may or may not conflict with trading arrangements and regulations that may apply between the respective States.

Finally, aside from these theoretical scenarios, the enhancements that have already entered into use in society – at home, on university campuses or in the workplace – have inspired little regulation provision.

In July 2017, a technology company in Wisconsin offered its employees the opportunity to implant microchips the size of a grain of rice between the thumb and index finger. The functionalities of the chip are apparently harmless, and comparable to what a normal badge enables: entering the office or paying at the cafeteria. The initiative was very popular with employees, who volunteered in large numbers. However, once inserted, there is no way of protecting against other possible uses, such as it becoming an instrument of surveillance or monitoring.[124] The use of chips is not necessarily new, nor is it *a priori* illegal. In fact, chips worn for medical reasons were approved by the Food and Drug Administration in the USA in

[123] R. Brownsword and M. Goodwin, *Law and the Technologies of the Twenty-First Century: Text and Materials* (Cambridge: Cambridge University Press, 2012), 24–45, cited in 'Human enhancement and the future of work', 50.

[124] M. Astor, 'Microchip Implants for Employees? One Company Says Yes', *New York Times*, 25 July 2017, www.nytimes.com/2017/07/25/technology/microchips-wisconsin-company-employees.html.

2014. However, there are ample concerns about the future of the technology as it gains a foothold and opens the way for a myriad of new applications.[125]

Cognitive enhancement devices have entered commercial use in recent years and can be purchased online. The most commonly marketed product for brain stimulation involves transcranial direct current stimulation and it is employed to help fight anxiety, migraine, and in healthy individuals it has been promoted for its promise to improve memory and focus abilities in mathematics and languages. Its effects are usually considered to be short-term but other studies demonstrated effects that can last up to six months. The procedure is painless and involves weak electrical currents applied on the head via electrodes for up to 20 minutes. The unintended consequences, especially if the device is used incorrectly, can be drastic, including long-lasting changes in neurobiology, and adverse effects such as impaired automaticity, or atypical brain development when used on a developing brain.[126] However, despite the fact that such brain stimulation devices present safety risks comparable to medical devices, they are not regulated accordingly. For instance, they are not covered by the EU Medical Devices Directive.[127]

The definition of a medical device under this EU Directive specifies that a medical device is intended and used for diagnostic and/or treatment purposes, which is not the case where cognitive enhancement devices are concerned. Such devices are only regulated in the sense that any product is covered by regulations for general product safety, i.e. by the General Product Safety Directive.[128] While product safety criteria include risks to health and safety, they do not explicitly refer to requirements for rigorous clinical assessment before commercialisation.[129]

In the United States, there is a regulatory framework for neuroprosthetic devices, such as Cochlear implants (approved by the FDA since 1995), Argus II, which is a retinal implant approved in 2013, and deep brain stimulation

[125] B. J. Feder and T. Zeller Jr., 'Identity Badge Worn Under Skin Approved for Use in Health Care', 14 October 2014, www.nytimes.com/2004/10/14/technology/identity-badge-worn-under-skin-approved-for-use-in-health-care.html?_r=0.

[126] H. Maslen, T. Douglas, R. Cohen Kadosh, N. Levy and J. Savulescu, 'Mind Machines. The Regulation of Cognitive Enhancement Devices: Extending the Medical Model", *Oxford Martin Policy Paper*, 2014, www.oxfordmartin.ox.ac.uk/downloads/briefings/Mind_Machines.pdf, 9–10.

[127] Ibid., 5

[128] Directive 2001/95/EC of the European Parliament and of the Council of 3 December 2001 on general product safety, Full text available here: http://eur-lex.europa.eu/legal-content/EN/TXT/PDF/?uri=CELEX:32001L0095&from=EN.

[129] Maslen, et al., 'The Regulation of Cognitive Devices', 73–74.

(used to treat Parkinson's disease since 1997). Most of the FDA guidelines for such devices have been put forward to treat diseases like Parkinson's, and therefore they were studied in the context of their medical use.[130] Several working groups in the European Union have made some disparate efforts to present proposals on how to regulate enhancement devices. These efforts have, however, been limited in scope (focusing mostly on pharmaceutical enhancements, and enhancement in the workplace) and impact.

More concrete and workable proposals for the regulation of cognitive enhancement devices, generally, were suggested in recent years by a group of researchers from the UK and Australia. The prohibition of cognitive enhancement devices was not considered a realistic or defensible option. Indeed, once the benefits of the technology are known, banning it from the market is seen as likely to encourage the private construction of rudimentary capabilities using unsafe or off-the-shelf components. Devising appropriate regulatory options would seem to be a better approach to the problem.

The authors of the study suggested the following options as possible road-maps:

I. New regulation, specifically for cognitive enhancement devices, should be established. Under such a process, devices for brain stimulation should be held to a higher/lower/the same standard compared to medical devices;
II. Cognitive enhancement devices should be included in the same legislation as medical devices, to a higher/lower/the same regulatory standard as medical devices.[131]

The strongest argument that could support Option I, namely the establishment of a new regulatory process, would be that enhancement devices are very different from medical devices. Nevertheless, this option would encounter practical difficulties, and possibly require the creation of new governmental agencies, while presenting difficulties for manufacturers who would find themselves producing the same or similar products under two regulatory regimes. Additionally, enhancement devices are not fundamentally different from medical devices; they can actually be used for both treatment and

[130] 'Neurological Devices', www.fda.gov/MedicalDevices/ProductsandMedicalProcedures/NeurologicalDevices/default.htm and 'Standards and Guidances for Neurological Devices', www.fda.gov/MedicalDevices/ProductsandMedicalProcedures/NeurologicalDevices/ucm528789.htm.
[131] Ibid., 77–78.

enhancement, sometimes using similar parameters.[132] Including cognitive enhancement devices under existing medical regulatory regimes seems to be the best solution: it ensures more coherence and sets a precedent for how non-therapeutic devices could be regulated as medical devices.[133] The prescriptive model would also involve: classifying such devices in terms of the risks they pose; prohibiting those devices with a high-risk profile; engaging with manufacturers; asking manufacturers to provide evidence-based information to consumers; and criminal sanctions to protect the vulnerable or non-competent third parties (e.g., as applicable for instance to the supply of alcohol to children).[134]

Importantly, this prescriptive model remains a *model* for the time being. It has been, however, a useful starting point for stimulating debate and garnering public attention for a development that will become riskier if and when commercialised enhancement devices become more invasive. At the moment, there are regional and global forums where the regulation of medical devices is discussed. One of them is the International Medical Device Regulators Forum, which was created in February 2011, and is based on the previous work of the Global Harmonisation Task Force on Medical Devices. Global efforts for the regulation of enhancement devices could finally be moved to such forums but initial steps must begin with the national and regional parties directly concerned, and with the States and regions where enhancement devices are already available and popular.

7.8 WHAT OPTIONS FOR REGULATION?

The comparison between human enhancement in the military environment and in society points to the conclusion that the regulatory arrangements associated with such enhancements will be benchmarked against different ethical and medical questions.[135] Alternatively, even if the same clinical questions were to be asked, for example: 'does this enhancement device/technology/drug present a low, moderate or high risk for the user', the answer might still *not* lead to the same result. If, for example, the answer is 'high risk', that form of enhancement will most probably be prohibited in the civilian

[132] R. Cohen Kadosh, 'Using Transcranial Electrical Stimulation to Enhance Cognitive Functions in the Typical and Atypical Brain', 4 (March 2013), *Translational Neuroscience*, 20–33, cited in Maslen, et al., 'The regulation of cognitive devices'.

[133] Maslen, et al., 'The regulation of cognitive devices', 80–81.

[134] Ibid., 92–93.

[135] This comparison overlooks the parallel with sports because, as alluded in the main text, the arguments for the ban of performance enhancement in sports will probably never be a source of inspiration and model for regulating enhancement in the military.

domain but not necessarily for the armed forces. Certainly, if the risk involves death or irreparable physiological damage, the armed forces would not have the right to impose it on their personnel either. However, if the 'high risk' is defined in terms that do not involve an attack on the right to life, or any other non-derogable rights, the military might subject to other applicable law and still use those enhancements, or perhaps propose them to soldiers on a voluntary basis.

Questions related to fairness, justice, dignity, safety, or privacy are likely to arise in both the military and civilian contexts, but, again, probably with different outcomes. A form of enhancement may, for example, be prohibited in society because it is considered to be too socially disruptive, and that ban would apply to all individuals in society. In the armed forces, however, this could take another format. For example, if an enhancement is thought likely to lead to tensions and resentment among soldiers, the approach might be to create a different tier of enhanced units, separating them from unenhanced personnel, rather than simply banning that form of enhancement as such.[136]

The distinction between enhancement in the armed forces and civilian sectors indicates that – even when the risks are high – there will be greater scope for enhancements to be acceptable in a military setting. However, there remains a realistic prospect for the development of regulation and the fostering of responsible use of enhancement technologies.

There is a very fine and extremely dangerous line between the opportunities and risks associated with enhancements, and this will eventually trigger the need for a sound framework or set of guidelines addressing how each type of enhancement should be used and when. There are many risks of enhancements backfiring and leading to unwanted repercussions. Consider the possible employment by the armed forces of beta-blocking drugs, which generally prevent the formation of traumatic memories, and of strong emotions. While the purpose in employing such drugs may be to eliminate emotions and thus to seek to prevent 'crimes of rage' during warfare, the same drugs may have the effect of rendering 'decent soldiers in a state of emotional deadness', and thus give them a propensity for indiscriminate killing.[137] Such interventions at the neurochemical level would be reckless if pursued without full knowledge of

136 The complication would be, however, that the tension could no longer be avoided when those enhanced soldiers returned back to civilian life, in a society that otherwise prohibited that form of enhancement.

137 'Soldier Enhancement: New Technologies and the Future Battlefield', 27 May 2014, event co-organised by the International Committee of the Red Cross – Australia Mission, and the Asia Pacific Centre for Military Law, www.icrc.org/en/event/soldier-enhancement-new-technologies-and-future-battlefield.

the long-term and/or adverse implications. They will likely be damaging to the individual and may lead to battlefield acts that breach IHL.

Considerations about values and ethics will also influence whether military enhancements are used and, if so, how. The principle of 'military necessity' will not justify the use of enhancements if the result is ultimately counter-productive or if the State may be expected to bear costs that are too high. Therefore, although regulation of armed forces' use is unlikely in the framework of international law, a set of elementary rules must govern the military practice. Maxwell J. Mehlman et al. developed a policy framework underscored by nine rules. These rules reflect the following factors: 1. there must be a legitimate military purpose; 2. the enhancements must be demonstrably necessary in the context of the operation and offer the best risk-benefit ratio in the context of the laws of war; 3. the benefits must outweigh the risks (meaning net risk, i.e. risks from using the enhancement minus the risks of not using it); 4. the dignity of the warfighter must be maintained; 5. the burdens on the warfighter must be minimised; 6. there must be consent; 7. there must be transparency (although some part of the research may need to be kept secret, some part of the research should be made available); 8. there must be fair distribution of risks and benefits and 9. there must be accountability of superiors (to avoid illegal commands).[138]

Checklists such as this one can build the foundation for an 'ethical and legal benchmark of acceptability', and might over time consolidate the practice of using enhancements into formal regulations or more voluntary guidelines. The uniform applicability of such frameworks is, nevertheless, problematic from the beginning because notions related to consent, dignity and transparency in the armed forces vary across countries. Furthermore, in the societal context, notions of fairness and of what is acceptable in the sporting, employment, medical and other contexts will tend to vary as between national and social groups.

7.9 CONCLUSIONS

Military human enhancement comes with a host of challenges: it risks creating sharp polarisations between enhanced and non-enhanced soldiers; it tests the limits and applicability of law, and risks causing new sources of tension within and among military units. Only a definite conclusion that benefits

[138] M. J. Mehlman, P. Lin and K. Abney, 'Enhanced Warfighters: A Policy Framework', in M. L. Gross and D. Carrick (eds.), *Military Medical Ethics for the 21st Century*, 121–124.

outweigh risks can legitimately justify the use of such technologies, and even then, the ethical considerations may outweigh perceived net benefits.

The foregoing discussion clearly shows that, at the very least, ethical codes and procedural guidelines must accompany the development and implementation of these technologies.

The case for stricter regulation can be made based on the fundamental principle that rights are not inconsistent with regulation, and that in many cases, regulations enhance freedom. After all, it is by complying with traffic laws that we are able to enjoy freedom of movement on the roads in reasonable safety. It is those very traffic regulations that enable us to make informed driving decisions and to anticipate the actions of others.[139] This logic can be extended to enhancements in the military and in society generally. Legal and ethical regulations and guidelines may create some limitations as to how and when enhancements are used, and may prohibit the use of particular enhancements in specified circumstances. The agreement and adoption of such regulations and guidelines will be the necessary way of ensuring that the benefits of enhancement technologies are realised while their potential drawbacks are limited or excluded.

[139] P. Lin and F. Allhoff, 'Against Unrestricted Human Enhancement', 18 *Journal of Evolution and Technology* 1, May 2008, 35, 4.

8

Legal Aspects of Human Enhancement Technologies[1]

HEATHER A. HARRISON DINNISS

8.1 INTRODUCTION

At a time when there is so much ongoing activity associated with human enhancement technologies, many of which are in the early stages of research and development, a full survey across all of the relevant fields of science and technology is impossible in a work of this kind. Further, as many of the technologies are not even at the field prototype stage, it would be premature to try to give a definitive picture of how they will work when operationalised. Rather, this chapter has selected a few of the technologies that are currently being developed and that seem to pose interesting questions for the law generally and for the law of armed conflict in particular. That said, many of the questions raised by human enhancement (both legal and ethical) cannot be answered solely by reference to existing technologies. While the majority of the examples referred to in these pages either already exist or are the subject of active programmes, occasionally this chapter will refer to popular culture to provide an illustration of the legal questions raised by more futuristic technologies. The hypothetical nature of such examples will be clearly indicated to avoid confusion.

8.2 HUMANS, PROPERTY AND WEAPONS

One of the major tropes of any work of science fiction involving human enhancement is the question of determinacy of the enhanced individual. As dystopian speculative fiction author Margaret Atwood has put it, 'how far can

[1] Parts of this chapter are drawn from, and build on, a previous work by the author: H. A. Harrison Dinniss and J. K. Kleffner 'Soldier 2.0: Military Human Enhancement Technologies' (2016) 92 *International Law Studies* 432, available at http://digital-commons.usnwc.edu/ils/vol92/iss1/14/ [hereinafter 'Soldier 2.0'].

we change, before we cease to be ourselves'.[2] Far from being a question only of interest to philosophers and science fiction aficionados, it is a question with profound legal consequences when applied to human enhancement generally and military human enhancement in particular. As human beings begin to enhance themselves, is there a point on the spectrum between entirely unenhanced and completely 'upgraded' at which we are prepared to say that someone is no longer a person for the purposes of protection of the law; that they are no longer human, but rather, have become something else entirely. If that is the case, is there a point at which we need to think differently about the rights and protections that should be offered to such individuals, or should we be considering whether they need to be subject to different regulation from that applying to ordinary humans. Certainly, at the time of writing, there is no information available in the public domain about any technologies currently in development that would change the fundamental humanity of the enhanced individual; but it raises the question of what it is that we consider would achieve that effect. Clearly, the use of prostheses, whether the replacement of a missing limb or even several limbs, has not been considered as removing an individual from the protection of human rights law.[3] This is the case even when those limbs bestow above-average human abilities on the affected individual.[4] Nevertheless, how far does that protection stretch? To take an example from the realms of science fiction – do we consider, for example, that the same protections should apply to those whose entire skeleton has been replaced (or perhaps merely permanently supplemented) by a prosthetic structure? Is it the existence of a 'ghost in the shell',[5] a human mind or soul, that triggers the protection of human rights law? Further, is it that same level of enhancement that would move an individual from the protection offered to individuals and populations, i.e. people, under the law of

[2] M. Atwood ' Margaret Atwood: the Road to Ustopia', *The Guardian*, 14 October 2011, available at www.theguardian.com/books/2011/oct/14/margaret-atwood-road-to-ustopia.

[3] See, for example, the ECtHR case: *Zarzycki* v. *Poland*, App 15351/03, decision of 12 March 2013 (complainant with forearm prostheses).

[4] For example, the case of Oscar Pistorius, the 'blade runner', whose replacement prosthetic legs allowed him to train and perform at the level of an Olympic athlete. In the subsequent trial following the death of his partner, there was no suggestion that he was not entitled to a fair trial or that the conditions of his subsequent detention should not meet human rights standards accorded to those with disabilities. The example is not perfect, however, as the murder charge was only tangentially related to the fact of his disability.

[5] *Ghost in the Shell* (Paramount Pictures, 2017). There is an interesting point of distinction raised by the difference between the 2017 film (in which the human brain is transplanted into a synthetic body) and the original source material (in which it is only the ghost/soul which is transplanted). However, the author considers the latter too far removed from current science for inclusion in this chapter.

armed conflict into the realm of rules regulating the use and protections of objects, or even into those rules specifically regulating weaponry? This is a matter of immense importance for the law of armed conflict, which allows a party to an armed conflict to requisition objects and to use any weapon captured on the battlefield for one's own purposes, but makes it a war crime to force an individual to fight for the opposing side.

An alternative theory of applicability might argue that the retention of rights by individuals such as Olympic athlete Oscar Pistorius is linked to the treatment aspect of their enhancements. That is, despite the above-average abilities bestowed on a select few individuals, for the most part, prostheses are primarily designed to return people to their previous abilities. Any above-average functionality in one aspect is balanced out by deficiencies in others, not necessarily by design but by the state of the technology. For example, increased grip strength in a prosthetic hand cannot compensate for a lack of sensation or decreased load-bearing capacity. Thus those with prostheses are not truly super-human but merely 'differently abled'. As prosthetic technologies advance, however, this rationale may change.

One of the side-effects of large-scale conflicts is the increased numbers of young and otherwise healthy individuals returning from the battlefields having lost limbs and needing prostheses. In turn, this prompts a surge in the development of prosthetic technology and in funding into this research; the long wars in Afghanistan and Iraq have proved no exception. To take an example, DARPA-funded researchers have already created a prosthetic arm controlled by thought processes that enables a sense of touch, and that can perform a bicep curl of 45 lbs (20 kg).[6] Traditional socket prostheses are being replaced with osseointegrated prostheses giving increased stability as well as important ramifications for the patient's perception of their own bodily integrity.[7] This has prompted debate in legal circles about whether these prostheses should more properly be viewed as extensions of the body,

6 'Prosthetic Limbs, Controlled by Thought', *New York Times*, 20 May 2015, available at www.nytimes.com/2015/05/21/technology/a-bionic-approach-to-prosthetics-controlled-by-thought.html (accessed 9 March 2018).

7 Osseointegrated prostheses allow the attachment of a prosthetic to a titanium implant which is inserted into the bone of the remaining limb. Human bone cells graft onto the implant making the titanium implant a permanent anchor for the prosthetic. This results in increased stability for the attached limb. The fact that the patient's own body grafts with the implant leads to increased osseoperception, with the patient feeling as if the prosthetic were part of their own body. See generally www.osseointegration.eu/osseointegration/; H. Van de Meent, M. T. Hopman and J. P. Frölke, 'Walking Ability and Quality of Life in Subjects With Transfemoral Amputation: A Comparison of Osseointegration With Socket Prostheses', *Archives of Physical Medicine and Rehabilitation* 94 (2013), 2174–2178.

rather than as property, thus extending rather than curtailing the extent of the rights of these individuals.

The same acceptance can be seen with reference to those technological implants which correct for physical disabilities. Cochlear implants to assist with hearing difficulties in deaf children, pacemakers, cardioverter defibrillators, catheters and heart valves, have all become increasingly technologically based and completely accepted by society.[8] We do not question whether these people have ceased to be human by the mere fact that their physiology has been supplemented by a technological device.

Likewise, drug or pharmacological-based biological enhancements do not appear to pose such questions. Perhaps owing to the (currently) temporary nature of the enhancement or the fact that these enhancements' effects are entirely internal to the body, we do not seem to question the basic humanity of the individual affected in this way. However, such factors are certainly taken into account when discussing questions of criminal responsibility or ethical fitness.[9] So society clearly considers that these differences should be reflected and accounted for through legal regulation and appropriate consequences on the basis of an implicit recognition that those involved retain their status as human beings.

Genetic differences may prove another matter, although to date, those with naturally occurring genetic differences, such as those born with chromosomal differences like Down's Syndrome or other genetic anomalies such as cancer, have not been viewed as anything other than human by the law. While researchers, lawmakers and ethicists have struggled with the permissibility of correcting genetic anomalies in hereditary diseases, and many countries have banned the manipulation of the human germ-line, the law has not denied the humanity of any potential offspring.[10] To date the debate has centred on therapeutic use, correcting for genetic disease, but the spectre of designer babies and genome-edited individuals has appeared prominently in the debates and has been reflected in the resulting legal regulations.[11] What then

[8] See, for example, J. Pelegrín-Borondo, E. Reinares-Lara and C. Olarte-Pascual, 'Assessing the Acceptance of Technological Implants (the Cyborg): Evidences and Challenges', (2017) 70 *Computers in Human Behaviour*, 104–112, 105.

[9] See, for example, the discussion of doping in sports in Section 7.7.1 in Chapter 7.

[10] See generally Z. Corbyn 'Crispr: Is It A Good Idea to "Upgrade" Our DNA?' *The Observer (London)*, 10 May 2015, available at www.theguardian.com/science/2015/may/10/crispr-genome-editing-dna-upgrade-technology-genetic-disease (accessed 9 March 2018).

[11] See, for example, Council of Europe, *Convention for the Protection of Human Rights and of the Human Being with regard to the Application of Biology and Medicine* (Convention on Human Rights and Biomedicine, or the Oviedo Convention) (CETS n. 164), adopted in Oviedo on 4 April 1997 (entered into force on 1 December 1999), (Oviedo Convention). Article 13 of the

of those who hack their own genome or those of their (as yet unborn) children for societal or career advancement? Would the 'frankenfoods' debate around genetically modified foods extend itself to humans to the extent that society would consider such individuals as requiring different regulation, or as not being entitled to the same rights?[12]

Some authors have even suggested that were such processes to be applied to otherwise healthy members of the armed forces the effect may be to transform the individual such that they should be considered as biological weapons, and thus regulated under the Biological Weapons Convention.[13] Apart from the teleological and interpretational difficulties confronting it,[14] such an argument would fundamentally change the legal status of the individual and the protections available to them under the law of armed conflict. Reflective of the underlying principle of humanity, the law of armed conflict contains within it a fundamental difference in the ways in which we treat, respectively, people and objects. Weapons, being objects, may be captured by the enemy, commandeered by the party to the armed conflict and put to use by their armed forces. On the other hand, it is a war crime to force a person, be they a prisoner of war or other protected person to fight for their enemy.

8.3 HUMAN ENHANCEMENT IN TIMES OF ARMED CONFLICT

The prevailing *lex specialis* for parties to an armed conflict is the law of armed conflict or international humanitarian law. This section explores the impact of human enhancement technologies on this body of law and the ability of the parties to comply with existing law when using fighters (be they members of the armed forces or members of organised armed groups) whose natural capabilities have been enhanced. While every technology will be different, for the most part the enhancement technologies under development at the time of writing will have an impact on the ability to *implement* the principles of the law of armed conflict rather than on the principles themselves. This is

Convention allows for interventions to the human genome for preventive, diagnostic or therapeutic purposes and 'only if its aim is not to introduce any modification in the genome of any descendants'; Article 18 also provides protection for embryos *in vitro*.

12 One might foresee overly zealous or cautious governments requiring any genetically modified individuals to register themselves with national authorities and restricting or preventing them from procreating, in order to protect the human germline.

13 P. Lin, M. J. Mehlman and Keith Abney, *Enhanced Warfighters: Risk, Ethics, and Policy* (2013), available at http://ethics.calpoly.edu/greenwall_report.pdf.

14 For example, there is a widely accepted definition of biological agent which refers to microorganisms and does not leave room for 'secret agents'. Further, States party to the Convention had no intention of regulating this kind of agent.

because the technologies are generally used on a party's own armed forces in order to boost their capabilities, as opposed to being directly applied against an enemy force. The use of enhanced fighters may also properly be viewed as a method of warfare, if and when their use constitutes an integral part of offensive activities at the strategic and tactical levels.[15]

8.3.1 *Distinction*

The principle of distinction is one of the 'cardinal principles' of international humanitarian law. It requires that parties to an armed conflict at all times distinguish between the civilian population and combatants and between civilian objects and military objectives and that they direct their operations only against military objectives.[16] How then do the new wave of enhancement technologies currently under research and development affect the principle?

Clearly, any 'enhancement' which results in a combatant becoming an automaton-like indiscriminate killer or that induces some sort of uncontrolled, frenzied outburst of undirected violence, in either case such that they would be unable to distinguish between civilian and military targets, would violate the principle of distinction and would be prohibited as an indiscriminate method of warfare.[17] However, such a drug is unlikely to be fielded by a modern armed force. As noted in Chapter 2, the principle prohibiting indiscriminate weapons and methods of warfare would prohibit the development and fielding of any weapon or means or method of warfare that cannot be directed against a specific military objective, or the effects of which cannot be limited to military objectives.[18] While there is some debate over the customary nature of the obligation to review methods of combat, the customary international humanitarian law prohibition against indiscriminate attacks clearly also refers to methods of combat. Moreover, such an approach would be of limited military utility in modern armed conflicts where much of the fighting takes place 'amongst the people'.[19]

Not all new enhancement technologies will be so clear-cut. In an effort to relieve the intense information overload on modern armed forces, they have

[15] Ibid., 439.

[16] The current iteration of the customary international law principle in found in Articles 48, 51 and 57 of Additional Protocol I.

[17] Article 51(4)Additional Protocol I.

[18] Chapter 2, Section 2.3. Note that the customary nature of the obligation to review methods of warfare is currently debated and is likely only to be binding on those States party to API as a matter of treaty law.

[19] See Rupert Smith, *The Utility of Force* (London: Allen Lane, 2005) on war amongst the people.

been investigating ways to manage the massive increase in data that the modern fighter has to deal with. For some, the end concept is to integrate the data flow to a heads-up display. This would initially be inside the soldier's helmet but eventually the data may be communicated directly to an implant in the soldier's brain or to electrodes located on the head. A number of different technologies have been researched to relieve this problem. Some of the technologies do not pose any issue for the principle of distinction. For example, the human mind is particularly adept at recognising anomalies in data that is being shown to a person before the person is consciously aware of having detected the anomaly. Defence and intelligence research agencies have been working with scientists and industry to develop image triage systems that exploit this ability in order to scan and identify threats.[20] The utility of the technology is being investigated at two levels that are significant for the law of armed conflict. The first level relates to intelligence analysts who must review thousands of hours of drone and satellite imagery to detect rocket launches, roadside bombs and the like. The second level, at the field or operational scale, concerns soldiers on patrol or in a foxhole scanning their environment for possible threats. In both cases the technology works by the user wearing an electrode cap that detects the P300 wave in the brain, an event that signifies the early stages of recognition, the image is then flagged for further review. According to one of the scientists involved in the intelligence analyst studies, these image triage systems allow images to be processed and threats detected at three times the speed of an unenhanced observer.

At the operational level, the United States Defence Advanced Research Projects Agency (DARPA), has been working with industry to bring this type of technology to soldiers in the field. A joint venture between Boeing and General Motors has created the Cognitive Technology Threat Warning System that uses the same electrode cap for the observer and six cameras placed at 60-degree intervals around the position giving 360-degree visibility. The cap-wearer looks at a screen that is fed images in rotation from the surrounding cameras every tenth of a second; if one of these images stimulates a P300 signal in his or her brain the system alerts the observer and lets them look properly at the image in question. In tests, people using the technology spot nearly twice as many threats as those scanning their surroundings with conventional field glasses.[21]

[20] See generally, 'Know Your Enemy: How to Make Soldiers' Brains Better at Noticing Threats', *The Economist*, 27 July 2017, available at www.economist.com/news/science-and-technology/21725543-target-recognition-warfare-how-make-soldiers-brains-better-noticing (accessed 13 February 2018).

[21] Ibid.

However, where the principle of distinction is concerned, identifying targets accurately is more important than speed. Once the technology was combined with deep brain stimulation for a period of twenty minutes, the researchers reported a 13 per cent increase in the ability to detect threats accurately as compared with a control group. Other experiments have reported even higher increases in accurate threat reporting. One study, conducted at the United States Air Force's Human Effectiveness Directorate at Wright-Patterson air base in Ohio, reported that six hours after receiving approximately twenty-five minutes of brain stimulation, subjects were spotting twice as many tanks, missile launchers and other targets in images as those in control groups.[22]

Other studies are also looking to read brain waves in order to prevent friendly-fire incidents through misidentification of troops.[23]

> In conjunction with the Australian and British defence ministries, [the US] Army Research Laboratory, in Maryland, is analysing the brain's alpha, beta, delta and gamma waves. These, unlike P300, are always running. But their strengths and frequencies vary according to what the brain is up to.
>
> After deciding that he has seen an enemy, a soldier with his finger on the trigger takes between 400 and 450 milliseconds to pull it. But what if he changes his mind during that period? According to Jean Vettel, a neuroscientist who is working on the project, brainwave data can indicate that he has realised he has made a mistake in as little as 200 milliseconds. This should be enough time for a shot to be blocked by disabling the firing mechanism automatically, and for a comrade's life thus to be saved.

If this mechanism is approved for use in the field, it would not only be of potential importance for decreasing friendly-fire incidents but also for reducing civilian deaths though misidentification of civilians as combatants or otherwise lawful targets. Integrating the brain waves of the soldier into the weapon system itself could prove an interesting development in weapons technology that may improve compliance with the principle of distinction.

8.3.2 *Proportionality and Precautions*

Misidentification of individuals can also play a role where compliance with the rule of proportionality is concerned. The rule requires that parties to an

[22] Ibid.

[23] Ibid.; See also 'Simulating friendly fire to learn how to avoid it' 05 April 2016, available www.canterbury.ac.nz/news/2016/simulating-friendly-fire-to-learn-how-to-avoid-it.html (accessed 13 February 2018).

armed conflict refrain from, or cancel attacks that may be expected to cause incidental loss of civilian life, injury to civilians, or damage to civilian objects which would be excessive in relation to the concrete and direct military advantage anticipated.[24] The mechanism discussed above which integrates brain waves of the soldier into the firing mechanism certainly appears to play into the obligation to cancel attacks. However without further details on exactly how this mechanism would work in a prototype it is hard to estimate whether the higher level decision-making required for the proportionality calculation would be too complex for such a technology.

As most human enhancement technologies are applied only to one's own armed forces, other enhancement technologies do not seem to have a direct impact on the rule. However the proportionality rule and the requirement to take precautions in attack may also be affected by other techniques designed to mediate the flow of information to war-fighters. War-fighters face unprecedented levels of information and raw data in the modern era.[25] While this increase in information creates the possibility for increased compliance with the principle of distinction and with the proportionality and precautions rules, the human mind's ability to deal with data is still limited. The results of 'information overload' on the armed forces can have tragic consequences; for example, it was the underlying cause of one of the worst losses of civilian life in Afghanistan involving US armed forces, in which twenty-three civilians were killed when a convoy of Afghan villagers was attacked by American helicopters.[26] Techniques to increase the mind's ability to deal with large amounts of data have been researched, including the passive image triage techniques outlined above. Other ideas that have been mooted, albeit in the civilian context, have consisted of the removal of unwanted visual information from the user. Researchers in Germany have designed software to filter visual information passing through augmented reality lenses such as the much-derided Google Glass and its equivalents such as contact lenses.[27] Although the author is unaware of any intention to utilise this research in a military

[24] The rule of proportionality runs as a thread through many of the substantive provisions of IHL. See, e.g., Additional Protocol I, Articles 51(5)(b), 57(2)(a)(iii), 57(2)(b); CIHL, Rule 14.

[25] One estimate puts the increase in data flow across the US armed forces at 1,600 per cent since the events of 9/11. T. Shanker and M. Richtel, 'In New Military, Data Overload Can be Deadly', *New York Times*, (New York, 16 January 2011) A1, www.nytimes.com/2011/01/17/technology/17brain.html?pagewanted=all&_r=0, (accessed 17 February 2018).

[26] Shanker and Richtel, 'In New Military'.

[27] E. Morozov 'The Perils of Perfection', *New York Times* (3 March 2013) SR1, www.nytimes.com/2013/03/03/opinion/sunday/the-perils-of-perfection.html?_r =0, (accessed 17 Februrary 2018); E. Champion, 'Thirty-five Arguments against Google Glass', *Reluctant Habits* (14 March 2013), www.edrants.com/thirty-five-arguments-against-google-glass/, (accessed 17 February 2018).

context, modern armed forces do operate similar augmented reality systems. For example, the Q warrior system designed by BAE is modelled on the same principles as Google Glass.[28] However in a military setting, any technology that removes contextual information from those deciding on or conducting an attack will have to be examined carefully to ensure that continued compliance with the principle and rules is not thereby prejudiced. The same analysis will, in the future, be required if the technologies are to be applied to augmented reality systems that more directly interface with the operator, either through visual prosthetics or cybernetic implants.

8.3.3 *Superfluous Injury and Unnecessary Suffering*

One of the oldest principles of the modern law of armed conflict is the prohibition against causing superfluous injury and unnecessary suffering; its roots can be seen in the Lieber Code of 1863 and in the preamble of the St Petersburg Declaration of 1868.[29] Along with the principle of distinction, it was recognised as one of the 'cardinal principles' of the law of armed conflict by the International Court of Justice in its Nuclear Weapons Advisory opinion,[30] and applies in international and non-international armed conflicts as a matter of customary international law.[31] The principle has been restated in many treaties. The most recent versions include Article 35(2) of API, which provides: 'It is prohibited to employ weapons, projectiles and material and methods of warfare of a nature to cause superfluous injury or unnecessary suffering.' Identical wording is used in the preamble to the Convention on the Certain Conventional Weapons.[32] The present author has written elsewhere regarding the evolution and differences in the wording of the principle and as to its impact on human enhancement technologies.[33] For present purposes it suffices to note that it is the intended design and normal effect of the weapon,

[28] See generally, Q-Warrior® Helmet-mounted Display, www.baesystems.com/en/product/qwarrior-helmet-mounted-display, (accessed 17 February 2018).

[29] Instructions for the Government of Armies of the United States in the Field, General Order № 100, 24 April 1863 (Lieber Code), Article 16 contains a prohibition against 'the infliction of suffering for the sake of suffering'.

[30] *Legality of the Threat or Use of Nuclear Weapons, Advisory Opinion*, 1996 ICJ 226, ¶ 238 (July 8).

[31] CIHL, Rule 70.

[32] Convention on Prohibitions or Restrictions on the Use of Certain Conventional Weapons Which May be Deemed to be Excessively Injurious or to Have Indiscriminate Effects, Oct. 10, 1980, 1342 UNTS 137 [hereinafter 'CCW Convention'].

[33] H. A. Harrison Dinniss and J. K. Kleffner 'Soldier 2.0: Military Human Enhancement and International Law', 92:432 *International Law Studies* (2016), 439–440.

projectile, material or method which should be considered when applying the principle to such technologies.[34]

As human enhancement technologies are generally only directly applied to one's own armed forces, the legal question is less about the technology itself, and is more concerned with whether the *use* of enhanced soldiers would in some way contravene the prohibition against causing superfluous injury and unnecessary suffering. Thus the application of the principle leads us back to the integration of the human body and mind with technology and the question of how to classify the individuals involved. As noted in the previous section, the human enhancement technologies currently being researched or developed, according to information that is available in the public domain, are not such that they would transform an individual from a person into a weapon. However the use of enhanced soldiers, whether enhanced through pharmaceutical, cybernetic, prosthetic, or any other means, may be considered a method of warfare. Furthermore, certain enhancements will themselves be considered to be weapons, as distinct from the individual or a part of a weapons system incorporating the individual, and will thus be considered a means of warfare. An example of the former would be a prosthetic arm incorporating a firing mechanism, or other offensive capabilities. The latter would include, for example, armed drone technologies flown through a neural interface or, from the realms of science fiction at this stage, some form of avatar technology that integrates a human mind as an integral part of a robotic soldier. In each case, the normal use of that method, means or weapon and the injury and suffering caused by it will need to be compared against the legitimate military aim to be achieved and the other options available to the force in order to determine the necessity of any suffering caused. It is a calculation that is entirely dependent on the enhancement technology involved and one that cannot be made in the abstract.

A calculation must also be made in relation to any weapons designed specifically to counter enhanced soldiers. This is particularly the case where the type of enhancement is such that it is not clear whether the individual is enhanced or not and there is a real risk of the weapon or counter-measure being used against enhanced and unenhanced soldiers alike.

In addition, there are also some injuries which society has a particular aversion to – it has led to the banning of particular weapons such as blinding laser weapons. A similar aversion may well be found for any weapon which causes lasting or permanent brain damage, such as that which could be caused

[34] For a brief discussion of the principle, see Chapter 2, Section 2.3.

by overloading a neural interface in order to disable it. The same reasoning, that it is unnecessary to blind an individual permanently, may well apply to a weapon designed to permanently degrade an individual's ability to function normally.

8.3.4 *Principle of Protection*

The principle of protection requires that those who do not, or who no longer, take part in hostilities be treated humanely, without any adverse distinction based on race, nationality, religious belief or political opinions, and without any other adverse distinction founded on similar criteria. With respect to the use of human enhancement technologies in armed conflicts this will relate primarily to detention situations, whether civilian or prisoner of war, rather than the conduct of hostilities.

Chapter 7 details several technologies that are under development for the treatment of injuries in the field: for example, haemostatic agents to arrest severe blood loss. While many of these will amount to treatment mechanisms and thus fall outside the scope of this work, a number may also be considered enhancement technologies by virtue of either their prophylactic use, or the level to which they return the treated individual. Thus any technology which is used for the treatment of protected persons under the care of an adverse party to the armed conflict must be carefully assessed so as to be in compliance with the law of armed conflict. Human enhancement technologies may impact the principle of protection in a number of ways. These primarily consist of issues raised by the detention of enhanced individuals, and issues associated with the use of enhancement technologies against non-enhanced individuals.

First, a State may find itself in the position of detaining previously enhanced individuals. This may result in a number of difficulties for that State. In particular, if a detainee requires regular doses of a substance in order to maintain their enhancement and thereby to maintain their good health, the Detaining Power may find itself accused of failing properly to look after its detainees as required by the Conventions, e.g., as the physiological or mental condition of prisoners is seen to deteriorate as they suffer the effects of withdrawal from their enhancements. As prisoners of war are not obliged to provide any details other than identifying information,[35] medical staff may find themselves with no knowledge of the existence of the enhancement, no

[35] Convention (III) Relative to the Treatment of Prisoners of War, Aug. 12, 1949, 6 U.S.T. 3316, 75 U.N.T.S. 135 [hereinafter GC III], Article 17 provides that 'Every prisoner of war, when

knowledge as to how to differentiate between 'normal' withdrawal and a seriously incapacitated patient at risk of permanent damage, and no knowledge as to what the appropriate course of treatment may be. The Third Geneva Convention relating to prisoners of war requires States to repatriate all seriously wounded or seriously sick prisoners of war to their own country.[36] This applies in particular to those whose mental or physical fitness seems to have been gravely diminished; those who are incurable; those who have recovered, but whose mental or physical fitness seems to have been gravely or permanently diminished; and those who require treatment and are unlikely to recover within a year.[37] The possibility that a prisoner may be enhanced will require medical personnel to form good faith assessments of individual prisoners based on available intelligence and on generalisations as to what comprises normal human functioning. Those prisoners who are merely suffering withdrawal, and who are returning to a normal, unenhanced but still otherwise fit state (their individual baseline) would not qualify for repatriation. However in the event that a prisoner's inability to maintain the required doses would result in grave and permanent damage to the prisoner, and where the Detaining Power is unable to provide the appropriate treatment in its detention facilities, the law requires that the prisoner be transferred to a military or civilian medical facility where the appropriate treatment can be provided.[38] Other than the difficulties that this lack of information poses with respect to medical treatment of enhanced detainees, it should be noted, as a general matter, that the fact that a prisoner is enhanced has no bearing on the general obligations of the Detaining Power in relation to the conditions of their detention.

The second situation in which human enhancement techniques may impact the principle of protection concerns the use of enhancement technologies to affect or treat otherwise non-enhanced individuals. This might arise in one of two ways - either in an attempt to make prisoners more willing to cooperate, or by medical staff attempting to treat or prevent medical conditions. With regard to the former, Article 17 of the Third Geneva Convention categorically prohibits *any* form of coercion of prisoners of war. A prohibition for protected persons (i.e. civilian detainees) also exists under the Fourth Geneva Convention.[39] Thus any attempt to coerce a prisoner to

questioned on the subject, is bound to give only his surname, first names and rank, date of birth, and army, regimental, personal or serial number, or failing this, equivalent information.'

[36] GCIII, Article 109.

[37] GCIII, Article 110.

[38] GCIII, Article 30.

[39] Convention (IV) Relative to the Protection of Civilian Persons in Time of War, 12 August 1949, 6 UST 3516, 75 UNTS 287 [hereinafter 'GC IV'], Article 31. The prohibition is not

cooperate through the administration of biochemicals or other techniques would be prohibited.

The lawfulness of using enhancement techniques to treat prisoners' medical complaints will depend in large part on whether such treatments are considered to accord with generally accepted medical standards or are viewed as experimental treatments. Both treaty and customary international humanitarian law prohibit any medical or scientific experiments or any other medical procedure not indicated by the state of health of the person concerned and not consistent with generally accepted medical standards.[40] For those States party to Additional Protocol I, Article 11(2) makes clear that the prohibition applies irrespective of the consent of the person concerned. Given the strength of the prohibition, and assuming that an enhancement is being provided in the best interests of the patient, much will therefore turn on whether a particular technique is considered experimental. While there may be some variations in clinical practice between States, where a particular treatment has been approved by a State's standards authority, it should not be viewed as experimental.

The second limb requires that treatment conform with generally accepted medical practice. Chapter 7 outlined a number of enhancement techniques currently being developed for the use of States' own forces, which one can envision being used to treat wounded or sick enemy personnel to block pain, promote healing etc. While many of these constitute treatment rather than enhancement and will thus fall into the treatment paradigm, the effects can be long-lived and the enhanced ability may in some cases be used prophylactically potentially bringing it within the enhancement paradigm. As noted, customary international humanitarian law and Article 11 of API protect any person who is in the power of an adverse party, or who is interned, detained or otherwise deprived of liberty as a result of an armed conflict. It is prohibited to subject such a person:

> to any medical procedure which is not indicated by the state of health of the person concerned and which is not consistent with generally accepted medical standards which would be applied under similar medical circumstances to persons who are nationals of the party conducting the procedure and who are in no way deprived of their liberty.

absolute – Some express exceptions are provided for in the Convention, e.g., Articles 79–135 in relation to internees.

40 CIHL, Rule 92. For specific treaty prohibitions, see AP I, Article 11; GC III, Article 13; GC IV, Article 32; Additional Protocol II, Article 5(2)(e).

What constitutes generally accepted medical practice will be determined by medical ethics and the standards adopted by professional medical bodies, for example, the World Medical Association and the World Health Organisation. In the absence of any international standards having been determined for more innovative enhancement techniques, a State may choose to rely on the fact that the technique has been approved by a national authority as a generally accepted practice for treatment of its own troops or citizens.

The author has noted previously that some prophylactic enhancement techniques may even be preferred.[41] Article 31, GC III requires that POWs undergo medical inspections at least once a month, specifically requiring that the most efficient methods available be employed for the detection of contagious diseases, as well as for the supervision of the general health, nutrition and cleanliness of prisoners. With technology costs inevitably decreasing over time, techniques such as diagnostic nanoplatforms may well become the most efficient way of monitoring the health of detainees.

8.4 CONTEMPORARY HUMAN RIGHTS LAW

Human enhancement technologies pose a number of challenges for contemporary human rights law, although there should not be any genuine doubt that that body of law continues to apply to enhanced individuals (see Section 8.2). The question of whether one has the right to modify one's own body and mind without limitation has become a topic of increasing interest as a growing number of DIY body 'hackers' experiment with emerging technologies. Similarly the boundaries of the permissible right to modify others' bodies, the role of informed consent (particularly in a strongly hierarchical system like the armed forces) and the possibility that some enhancement techniques could become compulsory for some professions, both within the military and in civilian life, have all raised questions about the appropriate legal regulation of these methods. This section reviews some of the major technologies under development and highlights the human rights issues associated with them both for the general population and for serving members of the armed forces.

In respect of the human rights of the armed forces, it should be noted at the outset that both European human rights bodies and the Inter-American system of human rights have treated serving members of the armed forces, whether volunteer or conscript, as citizens in uniform.[42] In particular, the European

[41] Harrison Dinniss and Kleffner, 'Soldier 2.0', 451.

[42] See, e.g., *J.S.C.H. & M.G.S., Case* 12.689, Inter-Am. Comm'n on HR, Report No. 80/15, OEA/Ser.L./V/II.156, doc. 33 (2015).

Court of Human Rights (ECtHR) has been a major contributor to the jurisprudence and collective understanding of the rights of members of the armed forces. As far back as 1976, the Court acknowledged the application of the law to members of the armed forces but also recognised that the 'particular characteristics of military life' and their 'effects on the situation of the individual members of the armed forces' must be taken into account when applying the law.[43] It has continued to develop the law in subsequent cases observing that the extent of protection given to members of the armed forces must take into account the characteristics of military life, the nature of the activities they are required to perform and the risk arising from those activities.[44] This recognition of the special qualities associated with military life and activities may well mean that many of the human rights issues raised by human enhancement technologies differ for serving members of the armed forces, particularly when they are on active service in war or when dealing with a public emergency threatening the life of the nation.

In addition to the possible derogation from certain rights in times of war and public emergency, many rights contained in the relevant treaties also allow for limitation by law in certain circumstances. These qualified rights often contain clauses specifically providing for limitations and restrictions for reasons of national security and/or public safety, both of which are likely to affect members of the armed forces more than the general population by virtue of their role in society.

With regard to the underlying right to modify one's own body, the European Court of Human Rights has noted that the notion of personal autonomy is an important principle underlying the right to a private life (contained in Article 8 of the ECHR).[45] The concept of a private life is broad and covers, *inter alia*, the physical and psychological integrity of a person; the Court further commented that it 'may also include the opportunity to pursue activities perceived to be of a physically or morally harmful or dangerous nature for the individual concerned'.[46] The Court went on to note that even where a particular conduct poses a danger to life or health, State interference by imposing 'compulsory or criminal measures' is regarded as impinging on the right and requires justification.[47] It is thus a qualified right as noted above.

[43] *Engel and Others* v. *Netherlands* (1976) Series A no. 22.

[44] See, e.g., *Grigoriades* v. *Greece*, 1997-VII ECHR.

[45] *Pretty* v. *United Kingdom* [2002] ECHR 427, para 61.

[46] Ibid.

[47] Ibid., para 62.

Any restriction must be in accordance with the law, and be necessary in a democratic society and for a legitimate aim.[48] Thus States may well prohibit the ability to modify the human genome or germline for reasons of protection of health or the rights of others. They would have less margin of appreciation for example in prohibiting implantation of radio-frequency identification (RFID) chips except in cases where such implantation poses a threat to national security. Clearly each enhancement technique would need to be assessed on its own merits with any restriction being judged by its proportionality in the whole context.

8.4.1 *Biochemical Enhancement*

While biochemical enhancement covers a wide range of possible enhancement technologies for both civilian and military applications, in respect of international human rights law the primary questions it raises are the same. The right to bodily integrity, including the right either to refuse or to consent to the modification of one's physical systems or attributes, whether it be the ability to focus the mind on a particular task for prolonged periods of time, to learn at an accelerated pace or to function efficiently with little sleep, is fundamental to the idea that the human body should be inviolable and under the dominion and control of its owner. In fact, the right to bodily integrity consists of a bundle of different rights and is expressed slightly differently in different treaty regimes. It incorporates the right to life, security of the person,[49] the right to humane treatment,[50] the prohibition against slavery,[51] the prohibition against torture and inhuman or degrading treatment,[52] as well as the prohibition against medical and scientific experimentation without freely given consent.[53]

As noted in Chapter 7, the issue of free and informed consent in the armed forces has been a vexed one in relation to the use of biochemical enhancements and other preventative measures such as vaccines.[54] By way of example, pressure on pilots to use or at least carry amphetamines while flying combat

[48] ECHR, Article 8(2). Legitimate aims are the interests of national security, public safety or the economic well-being of the country, the prevention of disorder or crime, the protection of health or morals and the protection of the rights and freedoms of others.

[49] UDHR, Article 3.

[50] ACHR, Article 5.

[51] UDHR, Article 4; ICCPR, Article 8; ECHR, Article 4; ACHR, Article 6.

[52] ICCPR, Article 7; ECHR, Article 3; ACHR, Article 5(2).

[53] ICCPR, Article 7.

[54] See Chapter 7.6.2 .

missions, and field pressure from officers to take experimental anthrax vaccines have added to the US armed forces' troubled relationship with biochemical enhancement.[55] The United States is certainly not the only State whose military forces have this issue. To a certain extent, the right to refuse an enhancement within the armed forces will be dependent on the nature and status of the particular drug. Biochemical enhancements that have been formally approved by the appropriate drug approval authority for use within the relevant State may well become part of the standard pre-deployment medical regime in the same manner as conventional vaccines. Refusal to comply with such approved medical requirements may result in the soldier being declared unfit for deployment to a particular region as his failure to comply with such approved medical requirements may be seen as rendering the service person a risk to his or her unit's military efficiency or unsuitable for a particular posting. Where the drug is more experimental, the soldier will be more likely to have the right to refuse the enhancement. Many of these issues played out in 2003 in relation to the use of unapproved anthrax vaccines during the deployment of armed forces from a coalition of States to Iraq. Different States took different approaches to the refusal of some armed forces personnel to receive the inoculation: some merely sent the troops concerned back to their home State, others instigated disciplinary action against them.[56] Other authors have also noted with concern that in some cases the right to refuse vaccination has not been communicated to troops in the field, thereby negating any truly informed consent.[57] Outside of the military context, the European Court of Human Rights has held that the testing of a new drug on a patient without his or her consent was unacceptable, in the light of international standards, and that it amounted to inhuman and degrading treatment in breach of Article 3 of the European Convention on Human Rights.[58]

Similar considerations must be applied to biochemical enhancement technologies with regard to the right to privacy. In some jurisdictions it is this right that has been the basis for decisions regarding the right to refuse medical

55 See Harrison Dinniss and Kleffner, 'Soldier 2.0', 460–461.

56 'Anthrax Jab Side-Effects Withheld', *BBC News* (21 February 2004), http://news.bbc.co.uk/2/hi/asia-pacific/3509037.stm.; 'Court-Martial for Refusing Anthrax Shot', *New York Times* (5 April 2003), www.nytimes.com/2003/04/05/nyregion/court-martial-for-refusing-anthrax-shot.html.

57 See J. Bird and G. Bird, 'Human Rights and the Military: The "Chemical Soldier"', *Alternative Law Journal* 30:81 (2005), 81–85; G. J. Annas, 'Protecting Soldiers from Friendly Fire: The Consent Requirement for Using Investigational Drugs and Vaccines in Combat', (1998) 24 *American Journal of Law & Medicine* 245, at 253, n.42.

58 *Bataliny* v. *Russia, no. 10060/07, Judgment of 23 July 2015* (in respect of a patient detained involuntarily in a psychiatric institution and given an experimental drug without consent).

treatment.[59] The ECtHR has noted the broad concept of the right to a private life covering the physical and psychological integrity of the person. Unlike the prohibitions against torture and inhuman and degrading treatment, which are absolute, the right to privacy is a limited right and may be restricted by the State under certain conditions. For example, under the ECHR restrictions may be imposed where the restriction is in accordance with the law, pursues a legitimate aim (for example, national security) and is necessary in a democratic society.[60]

Other human rights issues may come to the fore when States encounter other States' biochemically enhanced soldiers as a result of detention. Although some new generation pharmaceuticals which may be used for enhancement (such as Modafinil) highlight their apparently non-addictive qualities,[61] others remain highly addictive, for example the amphetamines used by air force pilots discussed above. Capture and detention of personnel who are in some way dependent on an enhancement drug will implicate the detaining State in providing for their medical assessment and treatment in the same manner as when other drug-dependent prisoners are incarcerated. In addition to the requirements under IHL, failure to provide drug-addicted detainees with the requisite medical care has resulted in the ECtHR concluding that the prison authorities' treatment of a detainee had violated the prohibition against inhuman and degrading treatment.[62]

Other issues concern the permissible treatment of detainees. For example, sleep deprivation has been denounced by multiple UN human rights bodies and has been recognised as a form of inhuman and degrading treatment by the European Court of Human Rights. However, in its landmark judgement *Ireland* v. *United Kingdom* the Court noted that the minimum level of severity required to fall within the prohibition was relative and dependent on all the circumstances of the case.[63] The Court went on to note that these circumstances include the 'duration of the treatment, its physical or mental effects and, in some cases, the sex, age and state of health of the victim, etc'. In an age of enhanced humans, these factors may well also include the fact that the detainee is under the influence of an enhancement technique that means that

59 See for example, *Pretty* v. *United Kingdom* [2002] ECHR 427, para 61–3; *in re Quinlan* (1976) 70 NJ 10, 40, 355 A.2d 647, 633.

60 ECHR, Article 8(2).

61 Although it should be noted that the non-addictive qualities of Modafinal have been questioned in recent years.

62 *McGlinchey and Others* v. *the United Kingdom* (application no. 50390/99) 29 March 2003, http://hudoc.echr.coe.int/eng-press?i=003–741378–753326 (accessed 4 April 2018).

63 *Ireland* v. *United Kingdom* (1978) 25 ECHR (ser. A) 58, para 162.

he or she requires less sleep than a 'normal' human does, or that counteracts the harmful effects of sleep deprivation. The risk is that without knowing the parameters of these enhancement techniques, States may be tempted to assume that detainees are more resilient than is in fact the case, with the result that the detainees are subjected to treatment widely recognised to be unlawful.

Any purported use of biochemical enhancement techniques against detainees may also implicate the fair trial rights of those detained. In particular, the subjection of detained persons to oxytocin or other substances intended to make them more cooperative is likely to infringe the right to silence and the privilege against self-incrimination.[64] Such a practice is specifically banned in many jurisdictions[65]; however, the 2013 Court approval of the use of a so-called 'truth serum' in the United States highlights the need for clarity on the issue.[66]

Other biochemical enhancements may affect fair trial rights because of their effects on memory recall. One of the techniques being researched has far-reaching implications in both the civilian and military worlds. Scientists have been investigating the use of pharmaceuticals and other techniques to reduce the incidence of post-traumatic stress disorder (PTSD) following traumatic events such as rape, terror attacks, or combat tours by armed forces personnel. One such study has looked at the use of beta-blockers to affect the formation of traumatic memories[67] by attempting to reduce the emotional impact of an event before it is encoded in the memory and before the risk that the patient goes on to develop PTSD arises. This enhanced ability to remain unaffected by traumatic events would be of obvious benefit to armed forces that are highly dependent on troops conducting multiple tours of duty in extended conflicts, to civilian populations living under threat of terrorist attacks and to law enforcement forces that routinely have to deal with the kinds of situation that are liable to give rise to PTSD responses. However any enhancement technique that affects memory formation, recall or otherwise

64 ICCPR 14(3)(g), ACHR 8(2)(g), *Heaney and McGuinness* v. *Ireland* 2000-XII ECHR 419, para 40.

65 See for example, the Swedish Instrument of Government: Reringsformen 2:5 (prohibiting 'any medical influence aimed at extorting or suppressing statements') and at the international level, the broad wording of the Rome Statute of the ICC, Article 55(1)(b) (providing that a person shall not be subjected to *any form* of coercion, duress . . .' emphasis added).

66 Ed Pilkington, ' Judge Approves Use of "Truth Serum" on Accused Aurora Shooter James Holmes', *The Guardian* (12 March 2013), www.theguardian.com/world/2013/mar/12/judge-approves-truth-serum-james-holmes (accessed 5 April 2018).

67 J. Giles, ' Beta-blockers Tackle Memories of Horror', *Nature* (28 July 2005), 436, 448–449.

manipulates memory in any way risks impacting the fair trial rights of those affected, who may of course include those with whom the enhanced person has been dealing. Such memory manipulation will create doubt as their veracity and reliability as witnesses, either in their own defence or against another in any court action.

8.4.2 *Cybernetic Implants and Brain Machine Interfaces*

As with biochemical enhancements, the human rights issues raised by cybernetic implants and other brain machine interfaces will be dependent on the particular technologies involved, however some generalisations may be made. Similar issues with regard to consent and bodily integrity are raised by implant technology as discussed in the previous section and will not be repeated here. In addition to these issues, of particular relevance will be the right to privacy and that of freedom of thought, conscience and expression. These are all limited rights, however, and may be infringed where such action is prescribed by law, is necessary in a democratic society and is undertaken for a legitimate aim.

Human enhancement technologies which affect the flow of information between the outside world and the individual, such as cybernetic implants and other brain machine interfaces, raise challenges in this regard. The problem is particularly acute in a military setting where it is the State that will determine the parameters of the information made available to the individual user concerned. In the civilian setting, this determination may be made by the State, by a private company (such as Google, or Neuralink) or by a research conglomerate (such as Braingate). While these technologies are still in their infancy, the human-rights implications need to be considered now so that they can be factored into the design process as the technology advances. Some of the technologies that have been under research and development are designed to filter information provided to the user. For example, one DARPA-initiated programme (now ended) was conducted with the object of preventing information overload to soldiers operating under stress.[68] At present these are wearable devices such as head-up displays. However, they integrate EEG readings (taken from a sensor cap) as well as other physiological indicators. The system would then filter and prioritise data depending on the individual's cognitive state. The obvious danger in such a device being used

[68] See, e.g., the DARPA-initiated augmented cognition programme, now ended. Committee on Opportunities in Neuroscience for Future Army Applications, *Opportunities in Neuroscience for Future Army Applications* (2009), 117, www.ncbi.nlm.nih.gov/books/NBK207983/.

by the military is that operationally relevant information will be filtered out by software that cannot comprehend complex operational environments, thus potentially interfering with the ability of the soldier to make decisions and judgements required under the law of armed conflict (such as those relating to distinction, proportionality and precautionary measures – see earlier sections). From a human-rights perspective, it is a violation of the right to freedom of expression, which includes 'the right to receive and impart information and ideas without interference by public authority and regardless of frontiers'.[69] Needless to say, in respect of its application to the armed forces, this right is often qualified for reasons of national security. It might be wondered why civilians would require this kind of intervention, however it should be noted that the military is not the only entity experimenting with information filtering technology. While the civilian version does not include the cybernetic aspect, reports surfaced in 2013 that German developers had devised software that enabled the deletion of visual information from smart glasses (e.g., Google Glass or Cool Glass One) or their contact lens equivalents.[70] As one futurist was quoted as saying, '[s]o if you decide you don't like homeless people in your city, and you use this software and implant it in your contact lenses, then you won't see them at all.'[71] While this may be morally questionable from a human-rights perspective, it differs from the military situation in that the user has downloaded the software and the filter appears to be controlled by the user. Unless a public authority is controlling the information filter, it is unlikely to raise the same issues. It should, however, be noted that other features of these forms of augmented reality technology (whether implanted or wearable) may involve other human rights issues; for example, the Chinese civilian version incorporates facial recognition technology, which utilises a vast database established by the Chinese State in 2017.[72] This would clearly raise issues in relation to the privacy rights of those being observed.

[69] ECHR, Article 10. Other international human rights instruments contain similar or broader wording.

[70] E. Morozov, 'The Perils of Perfection', *New York Times*, 3 March 2013, at SR1, www.nytimes.com/2013/03/03/opinion/sunday/the-perils-of-perfection.html?_r=0; Edward Champion, 'Thirty-five Arguments against Google Glass', *Reluctant Habits* (14 March 2013), www.edrants.com/thirty-five-arguments-against-google-glass/.

[71] P. Khanna and A. Khanna, *The Pleasure and Danger of Augmented Reality*, BIG THINK, http://bigthink.com/hybrid-reality/the-pleasure-and-danger-of-augmented-reality (accessed 13 September 2018).

[72] Deena Zaidi, ' Inside China's Growing Use of Facial Recognition Technology' *Venture Beat*, 15 February 2018, https://venturebeat.com/2018/02/15/inside-chinas-growing-use-of-facial-recognition-technology/ (accessed 6 April 2018).

To return to the present topic of cybernetic implants and brain machine interfaces, it is the privacy implications for those who are using the technology which are the primary focus. Brain computer interfaces (BCIs) translate brain activity into computer code to control external devices and prostheses. To mitigate against the risks of unintended BCI-related control signals (so-called brain-hacking) some scientists have suggested that all biosignals and control commands should be stored in a black box (a similar concept to that currently employed in aircraft) for a certain period of time.[73] This storage raises questions about the right to privacy, how the information should be stored, and who will have the right to access it. Similarly strict policies will need to be developed in relation to ocular implants or any other implantable monitoring device which is connected to a system with a data recording capability or which provides for the upload of data to a network or to the cloud. Unlike the wearable devices mentioned above, implant technology cannot be taken off and left at the front door or put in a cupboard, so the design of the technology will need to factor in the privacy rights of the wearer and those of other persons with whom they come in contact.[74] So, for example, the non-consensual recording and outside observation of private family moments would be a clear breach of the right to a private life.

Fair trial rights may also be implicated when brain machine interfaces or implants connected to recording devices impact on the privilege against self-incrimination. On a basic level, if an ocular implant is recording and transmitting everything that the wearer sees, a question arises over whether this infringes against the right not to incriminate oneself in any subsequent court proceedings in which the footage may be used. Much will depend on the nature of the device. Pure video footage, such as that collected by helmet or body cameras, is classified as objective evidence and can be compelled for a legitimate purpose. If an ocular implant is merely acting in this capacity, it is unlikely that any meaningful distinction will be drawn between footage obtained in this manner and that of a helmet camera. However the nature of the implant technology may be to provide some degree of processing of the images by the brain (as opposed to a pure optical prosthetic, such as a glass eye). This being the case, it is arguable that it may be sufficient to transform

[73] Jens Clausen, et al., 'Help, Hope, and Hype: Ethical Dimensions of Neuroprosthetics', *Science*, 356:6345 (30 Jun 2017), 1338–1339, DOI:10.1126/science.aam7731.

[74] Note that one of the main reasons for the failure of public adoption of the Google Glass product was the privacy concerns of individuals with whom users came in contact. There are anecdotal reports of early adopters being ejected from public spaces such as bars and cafes, and in some cases even physically assaulted because of other patrons strenuously enforcing their expectation of privacy.

any data retrieved post-processing into evidence that is subject to the right. For example, any monitoring and/or recording of synaptic responses to the imagery may not be able to be compelled. A more complex problem is posed by the responses processed by brain machine interfaces. Far from being an objective recording of observable fact, because our ability to retrieve and reliably use brain activity governing complex thought processes is still very basic, we tend only to capture and decode some of the billions of neurons that may be implicated in a response. Those that we do so capture in fact represent an insight (albeit an incomplete one) into our intentions and emotional responses and should therefore fall within the privilege. If, as discussed above, amplified brain signals are stored in a so called 'black box' in order to determine responsibility for actions, would that evidence be subject to the privilege against self-incrimination? The clearest method of ensuring that individuals remain in control of their own information is to encrypt all the data recorded and place the key solely under the individual's control. Assuming the key remains something that the person alone knows (information similar to a password) current law would consider the key to be subject to the right.

8.4.3 *Prosthetic Technologies*

Prosthetic technology has undergone a revolution in recent years. Advances in integrating prostheses with cybernetic input have resulted in thought-controlled limbs, limbs that can provide sensory feedback (e.g., a sense of touch) and most recently, limbs that promote a sense of kinaesthesia (knowing intuitively where your body is in space).[75] As noted in Section 8.2 above these advances in technology remove these high-end prosthetic devices from the realm of 'ordinary' prosthetics that might be equated to other wearable technology, to a truly integrated part of the individual such that it falls within the realm of human enhancement (where its functionality is sufficiently advanced). The underlying assumption in this section is that any fitting of prosthetic devices is done for therapeutic reasons (rather than to replace otherwise healthy limbs), and that the enhancement is something that is in excess of returning the individual to their previous state.

[75] See, for example, S. Fan, 'New Bionic Arm Blurs Line Between Self and Machine for Wearers', *Singularity Hub*, 4 April 2018, available at https://singularityhub.com/2018/04/04/new-bionic-arm-blurs-line-between-self-and-machine-for-wearers/#sm.0000aeon81dp1ecg115s6ehz1nqub (accessed 6 April 2018).

For military users the primary human-rights issue will come, not while they are on active service but at the end of service and as a matter of reintegration into civilian life. At present, highly functioning prostheses are very expensive and only available to a select few. The most advanced are still in research and development and only exist as prototypes, meaning even those who test them and are providing their experiences to help bring the technology forward cannot yet take them home at the end of the experiment. One can foresee a situation in which an active duty soldier is provided with a high-level prosthesis as part of their military service, but is required to relinquish it on their exit from the armed forces to be replaced by a version that is more appropriate for civilian life. Certainly, some researchers have raised the question whether enhanced military personnel should be required to have *all* enhancements removed before their integration back into civilian society so as not to give an unfair advantage in the workplace.[76] This suggestion not only raises complex issues of property law (prosthetic devices are considered to be property, rather than a part of the individual, in many jurisdictions). It also implicates the right to bodily integrity, both physical and psychological, under human rights law.

The debate surrounding bodily integrity and prostheses has tended to view the current prosthetics as something external to the body and therefore outside the realm of physical bodily integrity. The new and emerging prostheses may challenge that view. Their deeper connection to the body via osseointegration, the effects of the integration of the prosthesis using brain computer interfaces and the sensory feedback are all features that provide a new landscape to which human rights law must adapt and all such features have relevance to both civilian and military users. There are legitimate areas of debate as to whether an assault the effect of which is limited to such a prosthetic would be viewed as an affront to the wearer's physical bodily integrity or as property damage. An even more complex question arises when one considers what the legal implications of a cyber-attack against the neural interface software that helps control the prosthetic would be. There are no clear legal answers to any of these questions, and the debate remains a lively one.

The right to bodily integrity, whether expressed as a function of the right to a private life or otherwise, also includes the right to psychological integrity. With advances in prosthetic technologies, particularly the development of the most recent devices that can restore kinaesthesia, the line between self and machine in the individual user's mind has become increasingly blurred.[77] In a study published in March 2018, patients reported that a prosthetic limb that

[76] Lin, et al., 'Enhanced Warfighters', 71; see also section 7.7.2 in the previous chapter.
[77] Fan, 'New Bionic Arm'.

created the kinaesthetic effect felt like theirs, giving them a greater sense of agency over the limb.[78] Replacing such a limb with a basic model might well affect the right to bodily integrity as well as having all sorts of potential psychological consequences for the affected individual.

The issue relates not only to replacement limbs but also to the use of visual prostheses to provide augmented vision to those with normal vision and to restore vision to those who have lost it.[79] Optical prosthetics which in some way mediate information between the outside world and the brain will also be subject to the same concerns regarding the right to receive information as discussed above in relation to brain machine interfaces.

8.5 NEW HUMAN RIGHTS FOR AN AGE OF ENHANCEMENT

Despite the human rights issues detailed above and referenced in Chapter 7, there is a sense that the current body of human rights law does not fit easily with the massive advances in human enhancement technologies. While law inevitably lags behind technological development, the exponential pace of development particularly in the field of neuroscience and brain machine interfaces has raised particular challenges to the relationship between the law and the mind and mental processes. A growing body of researchers is suggesting that new human rights should be developed to protect individuals' freedom of the mind and expressly to extend to the mind the protections currently offered to the body. In one such paper, Marcello Ienca and Roberto Andorno have suggested that four new human rights should be considered: the right to cognitive liberty, the right to mental privacy, the right to mental integrity, and the right to psychological continuity.[80] Ienca and Andorno are concerned with the advances in neurotechnology and in particular the risks posed by brain machine interfaces. They argue that the current law is not sufficient to cope with emerging neurotechnology scenarios – for example, in relation to the right to privacy, in particular because they relate to an

[78] P. D. Marasco, et al., 'Illusory Movement Perception Improves Motor Control for Prosthetic Hands', *Science Translational Medicine* 10:432 (2018), available at http://stm.sciencemag.org/content/10/432/eaao6990.

[79] See, e.g., P. Sherwell, ' Blind Man Describes Joy at Seeing Wife for First Time in Decade Thanks to 'Bionic Eye', *The Telegraph* (London, 24 February 2015), available www.telegraph.co.uk/news/worldnews/northamerica/usa/11433149/Blind-man-describes-joy-at-seeing-wife-for-first-time-in-decade-thanks-to-bionic-eye.html.

[80] M. Ienca and R. Andorno, 'Towards New Human Rights in the Age of Neuroscience and Neurotechnology', *Life Sciences, Society and Policy* 13:5 (2017), available at https://lsspjournal.springeropen.com/articles/10.1186/s40504-017-0050-1.

individual's ' inner life and personhood'.[81] While undoubtedly the advances in new technologies will pose direct challenges to human rights law, that law has proved remarkably adaptive. Where inner life and personhood are concerned, the European Court of Human Rights has found that a person's sexual orientation, gender identity and health status all fall within the scope of the right. These are all aspects of a person's inner life (although clearly each may also have outward expression).[82]

The right to cognitive liberty is proposed as an 'underlying neurocognitive substrate' to all other rights such as freedom of thought, religion, expression etc. Ienca and Andorno describe the right as a complex right comprising the liberty to make choices about one's own mental domain, the right to mental integrity, and the possibility to take control of one's mental life.[83] Their argument that cognitive liberty cannot be covered by existing rights fails to comprehend the breadth of the underlying international rights laws – particularly so in the European context. As discussed above, the right to bodily integrity incorporates the right to determine (to consent to, or refuse) what may be done to one's own body. It is, in effect the underlying substrate which Ienca and Adorno seek to provide. The right to a private life is explicitly recognised as also applying to psychological integrity of the person. Freedom of thought contains an inner facet which is expressed as an absolute right. Thus, many of the concerns raised by the authors (and others) can be solved by looking outside purely national rights regimes to the international sphere and by incorporating the rich body of law and jurisprudence available in those regimes into the domestic context.[84]

8.6 CONCLUDING REMARKS

Throughout this chapter and in Chapter 7, the linkages between the use of human enhancement technologies, their legal and ethical regulation, and the underlying philosophical challenges of determinacy and humanity have been raised and explored. A chapter of this size cannot hope to examine all of the legal issues involved, particularly when many of the technologies discussed, while having firm bases in on-going research and development programmes,

[81] Ibid., 14.

[82] See generally, European Court on Human Rights, *Guide on Article 8 of The European Convention on Human Rights: Right to Respect for Private and Family Life* (30 April 2017), available at www.echr.coe.int/Documents/Guide_Art_8_ENG.pdf.

[83] Ienca and Adorno, 'Towards New Human Rights', 11.

[84] See also Chapter 13, discussing the human rights issues posed by brain-machine interfaces in the context of the United States.

have yet to be field-tested as prototypes. This contribution represents a look at the main legal issues to be raised in the future by both military and civilian uses of certain human enhancement technologies under the law of armed conflict and human rights law. With respect to the former, it is clear that for the most part, human enhancement technologies will affect the ability to implement the law rather than the content of any particular rule. This is primarily a function of the fact that the technology will generally be applied to a party's own fighters rather than against an opposing party. This remains true whether the enhanced individual is to be a civilian or member of the armed forces.

Human rights such as the right to bodily integrity, privacy, freedom of thought, conscience and expression also apply irrespective of the military or civilian status of the individual, subject to limitations imposed under that body of law. Although there may be a need for new rights to be recognised in the light of advances in these technologies, the law has proved remarkably adaptive to new technologies to date. A thorough survey and examination of the potential scope of the current rights should be prioritised over assumptions that new technologies require new rights. While many of the technologies discussed in these chapters are ground-breaking in their sophistication and extent, those considering their introduction have the legal analysis associated with their more low-tech and rudimentary historical antecedents to draw upon.

9

Human Degradation Technologies and International Law

HARRY AITKEN AND HITOSHI NASU

9.1 INTRODUCTION

The use of human degradation technology as a non-lethal means of violence has been prevalent from the early periods of human history. Early accounts of warfare – in the times of Ancient Greece and the Roman Empire – are replete with examples of the hostile use of poisons, hallucinogens and other early biological and chemical weapons to overcome the limitations of conventional weapons or the dangers of close combat. The industrial and technological advances of the modern age have contributed to the mass production of a variety of new means to degrade the human body, such as incapacitating chemical agents, blinding lasers, tasers, acoustic devices and directed energy weapons. While non-lethal in design purpose, these human degradation technologies have deleterious effects on the sensory-motor function of the human body, with a potentially lethal result depending on pre-existing medical conditions or other factors such as age and gender.

Many commentators have discussed and criticised the terminology and discourse of 'non-lethal weapons' as a way of circumventing the existing legal regulation governing the use of lethal force despite their potentially lethal consequences.[1] The same challenge applies to human degradation technologies; perhaps it applies even more acutely given that they employ subtle and

[1] See generally, A. Dymond-Bass and N. Corney, 'The Use of "Less-lethal" Weapons in Law Enforcement', in S. Casey-Maslen (ed.), *Weapons Under International Human Rights Law* (New York: Cambridge University Press, 2014) 32; S. Casey-Maslen and S. Connolly, *Police Use of Force under International Law* (New York: Cambridge University Press, 2017) 100. D. A. Koplow, *Non-Lethal Weapons: The Law and Policy of Revolutionary Technologies for the Military and Law Enforcement* (New York: Cambridge University Press, 2009), 9–10; D. P. Fidler, 'The International Legal Implications of "Non-Lethal" Weapons', *Michigan Journal of International Law* 21:51 (1999), 63.

less understood forms of violence. In the absence of any comprehensive treaty prohibiting or restricting the use of human degradation technologies as such in peacetime or in armed conflict, the regulation in this area remains sparse. The critical questions that have been set aside in the literature are: which threshold criteria apply to the use of particular weapons under the different legal regimes governing peacetime law enforcement and armed conflict; and what do the inconsistent approaches adopted under these regimes mean for the overall regulation of the use of human degradation technologies across the spectrum of situations?

This chapter will begin by providing the first comprehensive definition of human degradation technology, drawing from literature on non-lethal weapons, contemporary military and police practice and other sources. In doing so, it will lay down parameters that can be applied to current degrading weapons and to emerging technologies that may be weaponised in the future. The chapter will then critically examine the bodies of international law relevant to the prohibition or restriction of the use of human degradation technologies, with a view to identifying regulatory gaps or inconsistencies. The conclusion that will be drawn from this enquiry is that, while human degradation technology may fall within the scope of the considered bodies of law, its use will be prohibited or constrained only in limited cases. The novel means by which human degradation technology interferes with its target (including its variable effects) and, the fact that to date it has been principally used in the context of peacetime law enforcement, results in the application of the law in a patchwork manner. This, it is submitted, reflects a system of laws developed with lethal weapons in mind, which is ill-equipped to deal with the permutations of technologies designed to cause anything from short-term debilitation to serious mental or physical damage. The chapter concludes by suggesting that the prohibition on inhuman and degrading treatment is one rule that has the potential to address the complex circumstances that can be associated with the use of human degradation technology. While no jurisprudence on this question has yet emerged, the prohibition's ability to focus on both the physical and psychological effects of human degradation technology, rather than on such a weapon's design purpose, points the way to a more comprehensive and coherent regulation of the field.

9.2 'HUMAN DEGRADATION' AS AN ACT OF VIOLENCE

9.2.1 *Defining 'Human Degradation'*

The term 'degradation' is defined variably as '[t]o reduce from a higher to a lower rank, to depose from a position of honour or estimation'; '[t]o lower in estimation; to bring into dishonour or contempt'; '[t]o lower in character or quality; to debase'; and '[t]o descend to a lower grade or type; to exhibit a degradation of type or structure; to degenerate'.[2] As Sir Gerald Fitzmaurice observed in his separate opinion in *Ireland* v. *United Kingdom*, 'degrading' is used 'in every day speech…very loosely and figuratively. On such a basis, essentially anything that is personally unpleasant or disagreeable can be regarded as degrading by those affected'.[3]

For present purposes, in the context of weapons technologies, we adopt a human performance-based understanding of 'degradation'. A weapon will be degrading where it interferes with one or more of the human sensory systems with a view to reducing or eliminating the ability of an individual to respond to a situation as he or she ordinarily would by producing an automatic motor response (in the immediate term) or an involuntary behaviour (over the long term). This definition is consistent with, if not more developed than, that adopted in the limited policy discourse on the degrading effects of weapons. The US National Research Council and National Academy of Engineering, for example, have observed that new military technologies such as chemical non-lethal weapons 'offer possible opportunities to inhibit or reduce the performance of adversaries'.[4] The UK Royal Society has also referred to the military application of neuroscience technologies with the goal of 'diminishing the performance of one's enemy'.[5]

There is significant overlap between human degradation technology and non-lethal weapons, and many of the same legal issues confront both categories of weapon. Attempts at defining non-lethal weapons in military doctrine have focused on their ability to incapacitate targets while reducing fatalities. NATO defines non-lethal weapons as those that 'are explicitly designed and developed to incapacitate or repel personnel, with a low probability of fatality or permanent injury…'.[6] The policy of the US Department of Defense (DoD) on non-lethal weapons provides a more expansive understanding of their purpose, which is to achieve 'immediate target response and … provide

2 *Oxford English Dictionary*, 2nd edn (Oxford: Clarendon Press, 1991).

3 ECtHR, *Ireland* v. *United Kingdom* (19 January 1978) Application No. 5310/71, Separate Opinion of Judge Fitzmaurice, para 22.

4 J.-L. Chameau, et al., *Emerging and Readily Available Technologies and National Security: A Framework for Addressing Ethical, Legal, and Societal Issues* (Washington: National Academies Press, 2014) 72.

5 The Royal Society, 'Brain Wave Module 3: Neuroscience, Conflict and Security' (February 2012), p. 1 https://royalsociety.org/~/media/Royal_Society…/brain-waves/2012–02–06-BW3.pdf.

6 NATO Policy on Non-lethal Weapons (13 October 1999, updated 6 November 2008) para 3 www.nato.int/cps/ic/natohq/official_texts_27417.htm?selectedLocale=en.

predictable and intended reversible effects, allowing the affected target to return to pre-engagement functionality.' Consistent with NATO's definition, however, the DoD provides that non-lethal weapons are 'developed and used with the intent to minimize the probability of producing fatalities, significant or permanent injuries' but they 'are not intended to, eliminate risk of those actions entirely'.[7]

On the one hand, human degradation technology can be viewed as a particular sub-class of non-lethal weapons. Unlike a number of non-lethal weapons, such as plastic bullets, bean bag rounds, water canons and batons which achieve incapacitation through kinetic means, human degradation technology does so by engaging with different sensory systems in different ways, including somatic sensation (touch), vision, hearing and olfaction (smell). Through this engagement, the technology interferes with parts of the nervous system, including the brain, in an attempt to produce an involuntary motor response in the target that serves the strategic interests of the user. In this respect, like non-lethal weapons, human degradation technology is defined by its *intended outcome* on the target; but, unlike non-lethal weapons, it is also characterised by its *method of interference* (sensory interference) and its *intended effect* (a motor response).

On the other hand, certain forms of human degradation technology may also fall outside the definition of non-lethal weapons. Not all technology aims to produce 'immediate' or 'reversible' effects as required by the DoD definition. Certain weapons, by their engagement with sensory functions, produce deleterious (and sometimes irreversible) effects that cause a more gradual change in behaviour. This is a second kind of *intended effect* of human degradation technology.

Finally, defining non-lethal weapons and human degradation technologies by their intended outcome – incapacitating or repelling a target – does not imply that they are not lethal. This is reflected in the DoD and NATO

7 Department of Defense Directive 3000.03E (25 April 2013 as revised 27 September 2017), www.esd.whs.mil/Portals/54/Documents/DD/issuances/dodd/300003p.pdf?ver=2017-09-27-125836-647. Stuart Casey-Maslen and Sean Connolly define less-lethal weapons as 'those that have a low risk of killing or causing serious injury as a result of ordinary or reasonably predictable use. The term does not encompass firearms charged with metal-jacketed rounds or cyber weapons'; Casey-Maslen and Connolly, *Police Use of Force*, 147. The Human Effects Advisory Panel established by the US Joint Non-Lethal Weapons Directorate has proposed a quantitative definition, under which a weapon would be classified as non-lethal if it incapacitates 98 per cent of the people it is used against, while killing no more than 0.5 per cent, permanently injuring no more than 0.5 per cent, and having no effect on 1 per cent (as referred to in D. Koplow, 'Tangled up in Khaki and Blue: Lethal and Non-Lethal Weapons in Recent Confrontations', *Georgetown Journal of International Law* 36 (2005), 703, 708 fn 3).

definitions. In fact, such weapons may result in more fatalities than conventional weapons. It has been observed that 'if a weapon is one tenth as lethal as a hand gun but is used ten times more often, then there is no net improvement'.[8] Moreover, the technological means to degrade the physical functions of an individual are capable of being used – whether lawfully or unlawfully – as force enablers for conventional weapons by incapacitating targets and allowing for a safer and more effective kill.[9]

9.2.2 *Applications of Human Degradation Technology*

Alongside advances in traditional weapons (from spears, to cannons, to bombs) there have emerged alternative means of destruction for those occasions when conventional weapons were ineffective or insufficient. Greek mythology depicts the hero Hercules using poison-tipped arrows from the venom of Hydra, illustrating a very early awareness of the idea of biological weapons.[10] There are records of the Greeks using bio-weapons with hallucinogenic properties in the colonisation of Ionia in about 1000 BC to cause temporary insanity in the enemy.[11] During the Roman siege of Syracuse in 212 BC, philosopher/mathematician Archimedes was said to have used a 'heat ray' to burn Roman ships by creating a huge reflective surface of polished bronze shields.[12] Sound was deployed as a weapon of war in the biblical Battle of Jericho, in which Joshua's army marched around the city blowing rams' horns,[13] and it has been said that bagpipes were deployed by the Scottish to instil fear in the enemy.[14] In modern times, potent weapons based on similar ideas have been deployed to devastating effect, including the use of

8 Dymond-Bass and Corney, 'The Use of "Less-lethal" Weapons in Law Enforcement', 32, fn 2, citing S. Sweetman, 'Report on the Attorney General's Conference on Less than Lethal Weapons', National Institute of Justice, Washington DC, 1987, 26.

9 See, e.g., Report of the International Commission of Inquiry mandated to establish the facts and circumstances of the events of 28th September 2009 in Guinea, UN doc. S/2009/693, 18 December 2009, §198, recounting how Guinean forces shot tear gas into an opposition rally at a football stadium and then fired directly into the confused and disoriented crowd.

10 A. Mayor, *Greek Fire, Poison Arrows, And Scorpion Bombs: Biological and Chemical Weapons in the Ancient World* (New York: Overlook Press, 2009) 44.

11 Ibid., 165–167.

12 Ibid., 217. Archimedes' method was successfully tested in a research project at MIT in 2005: J. Schwartz, 'Recreating an Ancient Death Ray', *New York Times*, (18 October 2005), www.nytimes.com/2005/10/18/science/recreating-an-ancient-death-ray.html.

13 Joshua 6:1–27.

14 K. Ravilious, 'The Secrets of Sonic Weapons', *The Guardian*, (8 November 2005), www.theguardian.com/science/2005/nov/08/g2.weaponstechnology.

asphyxiating gas in World War I and, a century later, the use of chemical weapons against civilian populations in Iraq and Syria.

As a result of modern technological advances, the range of possible means to interfere with the sensory-motor functions of an individual has considerably expanded, with the most common forms deployed in law enforcement through conducted energy weapons such as tasers and through chemical agents such as riot control agents ('RCAs'), including OC (pepper spray) and CS gas (tear gas).[15] Tasers use compressed air to fire two darts that trail an electric cable back to the handset.[16] On impact with the target, the darts transmit a 50,000 volt charge which degrades the body by overpowering its normal electrical signals, temporarily confusing the nervous system, and making muscles rigid without affecting the heart.[17] RCAs are defined in the Chemical Weapons Convention as:

> Any chemical not listed in a Schedule, which can produce rapidly in humans sensory irritation or disabling physical effects which disappear within a short time following termination of exposure.[18]

RCAs reduce the physical performance of a human body by interfering with the human sensory system through the eyes and nose in order to disable or repel targets. The important distinguishing features of an RCA are the rapidity with which it produces that sensory or disabling effect and the shortness of the time that is needed following termination of exposure for the irritation or disabling physical effects to disappear.

Another type of chemical agent is the malodorant, which works by targeting both the olfactory and trigeminal nerves not only to create an unbearable odour but also to generate strong feelings of alarm.[19] Malodorants have found

[15] S. Casey-Maslen 'Crowd Management, Crowd Control, and Riot Control', in Casey-Maslen (ed.), *Weapons under International Human Rights Law*, 32, 70–71.

[16] Taser (now rebranded as Axon) claim that their non-lethal weapons have been used 3,805,660 times in the field: https://uk.axon.com/how-safe-are-taser-weapons.

[17] Amnesty International, 'Less Than Lethal? The Use of Stun Weapons in US Law Enforcement' (18 December 2008) AMR 51/010/2008 p. 6, www.amnesty.org/en/documents/AMR51/010/2008/en/; 'How the Stun Gun Works' (1 August 2001), *BBC*, http://news.bbc.co.uk/1/hi/uk/1468188.stm; 'Taser stun guns and their use in the UK', *The Guardian* (24 August 2001), www.theguardian.com/world/2011/aug/24/police-tasers-research-safety.

[18] Article II(7). See, also, M. Crowley, 'The Use of Riot Control Agents in Law Enforcement', in Casey-Maslen (ed.), *Weapons under International Human Rights Law*, 334, 335. The ECtHR has discussed the use of tear gas and pepper spray in several cases, including: *Guler and Ongel* v. *Turkey* (4 October 2011) App. Nos. 29612/05 and 30668/05; *Ali Güneş* v. *Turkey* (10 April 2012) App. No. 9829/07; *Abdullah Yaşa and Others* v. *Turkey*, (16 July 2013) App. No 44827/08.

[19] C. 'Sid' Heal, 'The Quest for the "Magic Bullet"', in M. Dando (ed.), *Non-lethal Weapons: Technological and Operations Prospects* (London: Jane's Information Group, 2000), 32, 35–36.

a range of applications, with a notable example in the counter-piracy context, where water defences on ships can be used in combination with malodorants that interfere with the sensory function to disable normal neural responses, causing panic and confusion.[20]

The 'holy grail'[21] in the human degradation space is considered to be the incapacitant, which attempts to produce a disabling condition (disorientation, sedation, loss of consciousness, etc.) by affecting the higher regulatory activity of the central nervous system.[22] The International Committee of the Red Cross (ICRC) has stated that despite the 'ongoing attraction' of this technology it is not clear to what extent it has progressed from academic and industrial uses to law enforcement and military uses.[23] The principal problem with using incapacitants as weapons is that the same dosage can have radically different effects on a person depending on their sex, age, health and other characteristics. The same dose might sedate one person but kill another. Thus the British Medical Association has concluded that 'it is and will continue to be almost impossible to deliver the right agent to the right people in the right dose'.[24]

Despite this, chemical incapacitants have been used in practice. The best known example is the Russian government's response to a hostage crisis at a theatre in Moscow in 2002.[25] On 23 October 2002, more than forty terrorists belonging to the Chechen separatist movement took more than 900 people hostage in the Dubrovka theatre (also known as the 'Nord-Ost').[26] The theatre was booby-trapped and suicide bombers carrying explosives were positioned among the hostages. The resulting siege lasted for three days, during which time the Russian authorities and the terrorists did not achieve a negotiated settlement of the situation. On the morning of 26 October 2002, Russian

[20] A. Priddy, 'The Use of Weapons in Counterpiracy Operations', in Casey-Maslen (ed.), *Weapons under International Human Rights Law*, 122–160, 154.

[21] D. A. Koplow, *Non-Lethal Weapons: The Law and Policy of Revolutionary Technologies for the Military and Law Enforcement* (New York: Cambridge University Press, 2009) 21.

[22] M. Crowley, 'The Use of Incapacitants in Law Enforcement', in Casey-Maslen (ed.), *Weapons Under International Human Rights Law*, 357– 381, 358.

[23] ICRC, Expert Meeting: Incapacitating Chemical Agents, Implications for International Law, Montreux, Switzerland, 24–26 March 2010, p. 3.

[24] British Medical Association (BMA), The Use of Drugs as Weapons: the Concerns and Responsibilities of Healthcare Professionals, BMA, London, 2007, p. 1.

[25] For other instances in which chemical non-lethal weapons have been used or considered in hostage situations, see A. Dolnik and R. Pilch, 'The Moscow Theater Hostage Crisis: The Perpetrators, their Tactics and the Russian Response', *International Negotiation* 8 (2003), 577, 600–601.

[26] Reports on the numbers of hostage-takers, hostages and other facts differ significantly. We rely on the numbers set out by the ECtHR in *Finogenov and Others* v. *Russia* (20 December 2011) Application Nos. 18299/03 and 27311/03 ('*Finogenov* v. *Russia*').

security forces pumped an incapacitating gas into the theatre auditorium through the air conditioning system prior to storming the building. The gas rendered unconscious most of the hostages and some of the terrorists. When Russian special forces entered the theatre, they shot and killed most of the terrorists (including those who were unconscious) and released the majority of the hostages. However, 129 of the hostages died during and shortly after the operation, mostly due to the effects of the gas.

The Russian government was widely criticised for the high casualty rate and its response to the crisis was the subject of litigation both domestically and before the European Court of Human Rights (ECtHR). The ECtHR decided that, while Russia's use of the incapacitating gas was not in breach of the European Convention on Human Rights (ECHR), its failure to adequately prepare for and deliver appropriate medical treatment in response to the use of the gas breached victims' right to life under Article 2 of the ECHR.[27] Relevant to the finding was the government's failure to notify medical authorities of the character of the chemical/s which were used in the gas. It was disclosed several days after the operation that the gas was a derivative of fentanyl – a synthetic opioid analgesic – but the precise composition of the mixture has never been revealed by the Russian government.[28] It has, however, been argued in academic literature that the gas was a mixture of two types of fentanyl: carfentanil (used in veterinary medicine to tranquilise animals) and remifentanil (an analgesic anaesthetic used on humans).[29]

Since the Moscow theatre attack, there has been one further reported use of chemical incapacitants by Russian security forces. In 2005, Russian NTV reported that Russian forces had deployed a 'knockout gas' against Chechen separatists who launched an attack in October of that year in the Russian town of Nalchik.[30] At the time of writing, no further details about the character of the incapacitant have emerged.

[27] The Court held that a further breach of Article 2 had been occasioned by Russia's failure to properly investigate the conduct of its authorities in responding to the crisis.

[28] 'Russia Names Gas Used to End Siege', *The Guardian* (30 October 2002), www.theguardian.com/world/2002/oct/30/russia; *Finogenov* v. *Russia*, para 101.

[29] The Royal Society, 'Brain Wave Module 3: Neuroscience, Conflict and Security' (February 2012), p. 11, https://royalsociety.org/~/media/Royal_Society. . ./brain-waves/2012–02–06-BW3.pdf; J. R. Riches, et al., 'Analysis of Clothing and Urine from Moscow Theatre Siege Casualties Reveals Carfentanil and Remifentanil Use', *Journal of Analytical Toxicology* 36:9 (2012), 647–656.

[30] House of Commons Foreign Affairs Committee, 'Global Security: Non-Proliferation', Fourth Report of Session 2008–09 (3 June 2009), p. 230 https://publications.parliament.uk/pa/cm200809/cmselect/cmfaff/222/222.pdf.

A third type of human degradation technology uses acoustic waves, transmitted at high or low frequencies, to interfere with the nervous system through the hearing or other sensory systems. This interference may degrade the physical performance of adversaries by disabling or disorientating them, thereby preventing or curtailing their ability to attack a target. Long Range Acoustic Devices (LRADs) were developed as a response to the bombing of the USS *Cole* in order to provide a better means for battleships to communicate with and deter potentially hostile targets without resorting to deadly force.[31] The manufacturer, LRAD Corporation, claims that its devices can project deterrent tones up to 3 km and focus audio in a 15–30° beam.[32] The US military is developing a 'Distributed Sound and Light Array' non-lethal weapon which 'uses a combined laser, non-coherent light, and acoustics to produce a synergistic engagement system . . . capable of optical and auditory impairment and/or intelligible audible tones.'[33]

It is worth noting two cases in which acoustic technologies have been deployed – though not strictly capable of being defined as 'weapons'[34] in the context of the law of armed conflict – in order to achieve longer-term behavioural changes rather than immediate or automatic motor responses. These examples shed light on the threshold of harm that needs to be satisfied in order for a technology to be considered capable of degrading the physical function of a human body. The first case concerns the US strategy to capture Panamanian dictator, Manuel Noriega, who took refuge in the Apostolic Nunciature of the Holy See in Panama City following the US invasion of Panama in 1989. In an effort to force Noriega's exit without physically entering the Nunciature, US Navy Seals executed 'Operation Nifty Package', which involved blaring deafening rock music and other noises at the embassy from the top of Humvees over several days. It was originally claimed that the purpose was to

[31] 'The Secrets of Sonic Weapons', *The Guardian* (8 November 2005), www.theguardian.com/science/2005/nov/08/g2.weaponstechnology.

[32] LRAD Corporation, 'Extended Range Communication & Scalable Escalation of Force', www.lradx.com/application/defense/.

[33] US Department of Defense, Non-Lethal Weapons Program, 'Distributed Sound and Light Array', http://jnlwp.defense.gov/Developing-Non-Lethal-Weapons/Distributed-Sound-and-Light-Array/.

[34] In the context of the law of armed conflict, 'weapon' is authoritatively defined as 'a means of warfare used in combat operations. . . that is capable of causing either (i) injury to, or death of, persons; or (ii) damage to, or destruction of, objects': Program on Humanitarian Policy and Conflict Research at Harvard University, *HPCR Manual on International Law Applicable to Air and Missile Warfare* (Cambridge: Cambridge University Press, 2013) rule 1(ff). See also, M. N. Schmitt (ed.), *Tallinn Manual 2.0 on the International Law Applicable to Cyber Operations* (Cambridge: Cambridge University Press, 2017), 452 ('*Tallinn Manual 2.0*').

create a 'music barrier' to prevent eavesdropping by journalists on sensitive negotiations between US military leaders and the Papal envoy at the Nunciature, but it was also acknowledged that the operation was a psychological tool designed to harass Noriega and encourage his surrender.[35] The extent to which this acoustic weapon caused Noriega to surrender is unclear, but the use of technological means to merely cause annoyance falls short of being classified as a human degradation technology.[36]

The second example involves sonic attacks against US diplomats stationed in Cuba. Since late 2016, an acoustic technology has been used against more than twenty diplomats, who have suffered variously from concussion, permanent hearing loss, brain swelling, dizziness, nausea, severe headaches, balance problems and tinnitus.[37] An expert speaking on the attacks observed that '[i]t is possible to create a complex sound of certain frequencies and acoustic patterns that could affect our body function especially brain function', including the parts involved in memory, cognition and emotion.[38] The University of Pennsylvania's Center for Brain Injury and Repair ran tests on the diplomats and concluded that through their 'exposure to auditory and sensory phenomena' they had suffered 'sustained injury to widespread brain networks.'[39] The US government expelled fifteen Cuban diplomats in October 2017 on the grounds that Cuba had failed to protect the American diplomats.[40] At the time of writing, the nature of the weapon and the identity of the attacker had not been confirmed, but by contrast with 'Operation Nifty Package', the acoustic

35 R. H. Cole, *Operation Just Cause: The Planning and Execution of Joint Operations in Panama, February 1988–January 1990* (Joint History Office, Office of the Chairman of the Joint Chiefs of Staff, Washington 1995), 59–60.

36 'Music Torture: How Heavy Metal Broke Manuel Noriega', *BBC* (30 May 2017), www.bbc.co.uk/news/world-latin-america-40090809. See, also, L. Rohter, 'The Noriega Case: Panama City; Papal Envoy Asserts Psychology, Not Ultimatum, Swayed Noriega', *New York Times*, (6 January 1990), www.nytimes.com/1990/01/06/world/noriega-case-panama-city-papal-envoy-asserts-psychology-not-ultimatum-swayed.html.

37 'Mystery of Sonic Weapon Attacks at US Embassy in Cuba Deepens', *The Guardian* (15 September 2017), www.theguardian.com/world/2017/sep/14/mystery-of-sonic-weapon-attacks-at-us-embassy-in-cuba-deepens.

38 Dr Jeffrey Kim, Georgetown University Otolaryngology Department, cited in Zachary Cohen, 'New Audio Adds to Mystery of Attacks on US Diplomats', *CNN* (14 October 2017), http://edition.cnn.com/2017/10/13/politics/cuba-us-diplomats-acoustic-weapons/index.html.

39 R L. Swanson II, et al., 'Neurological Manifestations Among US Government Personnel Reporting Directional Audible and Sensory Phenomena in Havana, Cuba', *Journal of the American Medical Association Preliminary Communication*, 15 February 2018.

40 G. Harris, J. Hirshfeld Davis, E. London, 'U.S. Expels 15 Cuban Diplomats, in Latest Sign Détente May Be Ending', *New York Times*, (3 October 2017), www.nytimes.com/2017/10/03/world/americas/us-cuba-diplomats.html.

device used against US diplomats appears to be equipped with a technological capability to degrade the physical functioning of a human body.

As will be discussed below, the infliction of observable harm such as that suffered by the US diplomats would suffice to cause such an operation to be classed either as an attack, if IHL applies to the relevant circumstances (see further Section 9.4.1), or as a criminal assault under applicable domestic law (see further Section 9.4.2). There is an interesting potential contrast to be observed here between the US diplomats case and operations of the sort applied to Manual Noriega; if loud rock music is used to harass the person being subjected to it and to cause that individual to surrender, this would not only (as already noted) fall short of our definition of human degradation as discussed in the present chapter, but might also not constitute either an attack for the purposes of IHL or an assault under most criminal law provisions owing to the absence of consequent harm.

In the same way that acoustic technologies rely on sound waves, directed energy technology, such as 'Active Denial Technology', deploys high power microwave or millimetre waves to degrade the sensory-motor function of the target,[41] by producing a powerful sensation of heat without in fact burning the skin or inflicting any injury.[42] Active Denial Technology was available to US forces in Afghanistan, ostensibly for crowd control and mob dispersal purposes.[43] Lasers and other optical disorientation methods using bright light are also being adapted for military purposes. These include the Distributed Sound and Light Array, and rifle-mounted 'dazzlers' that have been used by US soldiers in Iraq to temporarily incapacitate individuals.[44] As these examples show, technology can still be degrading even when it generates only short-term or temporary effects because the target's performance is reduced, to the strategic benefit of the user, during the period of exposure to the technology.

This non-exhaustive discussion of certain technologies gives some colour and texture to the definition of human degradation set out in Section 9.2.1. While each technology operates in its own way, they share the common goal

[41] US Department of Defense, Non-Lethal Weapons Program, 'Active Denial Technology', http://jnlwp.defense.gov/Future-Non-Lethal-Weapons/Active-Denial-Technology/.

[42] Koplow, *Non-Lethal Weapons*, 20–1.

[43] E. Massingham, 'Conflict Without Casualties . . . A Note of Caution: Non-Lethal Weapons and International Humanitarian Law', *International Review of the Red Cross* 886 (2012), 673, 678. The technology was never used and was apparently recalled without public explanation of the reasons for doing so.

[44] W. Pincus, 'Senate Boosts Funding for Laser Weapons', *Washington Post* (22 September 2008), www.washingtonpost.com/wp-dyn/content/article/2008/09/21/AR2008092102432.html; B. Resnick, 'A Brief History of Militarized Lasers', *The Atlantic* (12 December 2004), www.theatlantic.com/politics/archive/2014/12/a-brief-history-of-militarized-lasers/453453/.

of interfering with the sensory system in order to produce an automatic or immediate motor response or to result in psychological deterioration that produces a longer-term behavioural outcome. As outlined above, they may be collectively classified by their method of interference and their intended effects and outcomes. When referring to these individual elements collectively in this chapter, we will use the term 'design purpose'.

This section also illustrates that human degradation technology is used in a range of situations, from law enforcement to armed conflict and complex operations that cut across the 'grey zone' between peace and war. These include counterinsurgency or counterterrorism operations in circumstances which increasingly approach the threshold of armed conflict, or the use of human degradation technology for law enforcement by an occupying power during an armed conflict. It should also be observed that while a type of technology may have been designed to be used in one circumstance (usually law enforcement) it may also be used exceptionally or with increasing frequency in another domain.

Different branches of law apply in these contexts; and each body of law brings with it the possibility of regulating the use of human degradation technology. However, the ability to do so will require the scope of the rules in question actually to engage with the characteristics of human degradation technology – their design purpose and nature – and to apply in the circumstances in which they are being used. To the extent the application of these rules is determined by different thresholds, or different criteria, or that there are inconsistencies between them, there might arise gaps in law which States might be tempted to exploit by the use of human degradation technology. The remainder of this chapter will examine these questions in relation to what we consider to be the bodies of international law with the most promising regulatory potential: the law of weapons, the law of targeting and international human rights law.[45]

9.3 PROHIBITED MEANS OF HUMAN DEGRADATION

Throughout the ages, human inventiveness has been devoted to the development of weapon technologies. Ethical concerns with certain kinds of weapon

[45] There are other bodies of municipal and international law that may also regulate the use of human degradation technology, including domestic criminal law, domestic human rights law, police laws and regulations, diplomatic and consular law, international environmental law and international criminal law. Due to space constraints, these bodies of law fall outside the scope of this chapter.

have persisted since their invention and, over time, norms have developed which have curtailed their use. Beginning with the turn of the twentieth century, international law has developed progressively to minimise the potential for biological or chemical agents and toxins to be used as weapons to cause not only death but also temporary incapacitation or permanent harm to humans.[46] In the late twentieth century, conventional weapons designed to have a particular kind of degrading effect or with certain characteristics, such as those whose primary purpose is the dispersal of fragments that are not detectable by X-rays and lasers designed as a combat function to cause permanent blindness to unenhanced vision, were added to the list of prohibited weapons.[47]

However, this does not mean that all of these technologies have fallen into disuse in the weapons space. To the contrary, there has been a renewed focus in recent decades on how non-lethal weapons can be used as a safer and more humane alternative to lethal weapons.[48] This has been made possible due to the way in which these weapons have been regulated under the existing regime of weapons law with reference to their intended application. For example, the Biological Weapons Convention does not prohibit the development, production, stockpiling or acquisition of microbial or other biological agents, or toxins, where the types and quantities can be justified for prophylactic, protective or other peaceful purposes.[49] Similarly, the Chemical Weapons Convention provides an exception where the use, development, production, acquisition, stockpiling or transfer of toxic chemicals is intended, inter alia, for peaceful, protective and law enforcement purposes and the types and quantities of toxic chemicals are consistent with such purposes.[50]

46 Chemical Weapons Convention, Article II(2); Convention on the Prohibition of the Development, Production and Stockpiling of Bacteriological (Biological) and Toxin Weapons and on their Destruction, opened for signature 10 April 1972, 1015 UNTS 163 (entered into force 26 March 1975) (hereinafter Biological Weapons Convention) Article I.

47 Convention on Prohibitions or Restrictions on the Use of Certain Conventional Weapons Which May be Deemed to be Excessively Injurious or to Have Indiscriminate Effects, opened for signature 10 April 1981, 1342 UNTS 137 (entered into force 2 December 1983): Protocol I on Non-Detectable Fragments, opened for signature 10 April 1981, 1342 UNTS 137 (entered into force 2 December 1983); Protocol IV on Blinding Laser Weapons, opened for signature 13 October 1995, 1380 UNTS 370 (entered into force 30 July 1998).

48 Note that this justification was used by some authors in relation to gas in World War I. See A. Marshall, 'War by Poison', *Nature* (1934), 952–953.

49 Biological Weapons Convention, Article I.

50 Chemical Weapons Convention, Article II(9).

Certain human degradation technologies fall *prima facie* within the scope of the Chemical Weapons Convention. This is because there is an overlap between the definition of 'chemical weapons' and the methods of interference used by some degradation technologies, such as RCAs, incapacitants and malodorants.[51] But the carve out in the Convention for technologies that would otherwise be chemical weapons by reference to their intended application – particularly their use in law enforcement – creates a potentially significant limitation. This is due, in part, to the expansive nature of modern military operations, particularly in the context of counter-terrorism and counter-insurgency where a law enforcement-type situation intersects with a non-international armed conflict at its lower end of intensity.[52]

The employment of a chemical agent by Russian *Spetnaz* Forces in the rescue operation during the Moscow Theatre hostage crisis was treated as a law enforcement operation by the European Court of Human Rights,[53] although it could well have been considered as part of the wider context of the non-international armed conflict between Russian forces and Chechen separatists.[54] On the other hand, the employment of chemical weapons in Syria was subject to an investigation by a fact-finding mission of the Human Rights Council, which identified a pattern of Government forces using chemical weapons against civilians in opposition-held areas resulting in fatalities as constituting war crimes.[55]

[51] Art II defines 'chemical weapons' as: '(a) Toxic chemicals and their precursors, except where intended for purposes not prohibited under this Convention, as long as the types and quantities are consistent with such purposes; (b) Munitions and devices, specifically designed to cause death or other harm through the toxic properties of those toxic chemicals specified in subpara (a), which would be released as a result of the employment of such munitions and devices; (c) Any equipment specifically designed for use directly in connection with the employment of munitions and devices specified in sub-para (b).' The term 'toxic chemical' is then defined in Article II(2) as: 'Any chemical which through its chemical action on life processes can cause death, temporary incapacitation or permanent harm to humans or animals. This includes all such chemicals, regardless of their origin or of their method of production, and regardless of whether they are produced in facilities, in munitions or elsewhere.'

[52] See K. Watkin, 'Chemical Agents and "Expanding" Bullets: Limited Law Enforcement Exceptions and Unwarranted Handcuffs?', in A. M. Helm (ed.), *The Law of War in the 21st Century: Weaponry and the Use of Force* (US Naval War College International Law Studies, Vol. 82, 2006), 193, 196–197.

[53] See Section 9.4.2.

[54] For detailed analysis of the event, see, D. P. Fidler, 'The Meaning of Moscow: "Non-lethal" Weapons and International Law in the 21st Century', *International Review of the Red Cross* 859 (2005), 525.

[55] Report of the Independent International Commission of Inquiry on the Syrian Arab Republic (Advance Edited Version) (8 August 2017) UN doc. A/HRC/36/55, paras 67–77. It should be noted that Syria was not a party to the Chemical Weapons Convention and therefore was not

In the Syrian case, where the weapon's design purpose was widespread lethality or permanent harm, it is clear that it was not being manufactured and used for law enforcement purposes and so it is evident that the exception under the Chemical Weapons Convention does not apply. But in cases such as the Moscow Theatre hostage crisis, where the chemical agents were arguably intended and used for law enforcement purposes, the Convention's reach is more limited and its force is less clear-cut. As the exceptions in the Convention attach to intended use, the possession or use of a toxin will not breach the Convention if its design and the quantity held are consistent with one of the specified purposes.[56] Yet, the toxin could then be used in a way that stretches the meaning of one of those purposes and falls foul of the prohibition. As these exceptions in the Convention focus on intended rather than actual use, the Convention is not well equipped to prevent the use of some human degradation technologies deployed with a lethal intent or effect in a particular context, particularly if their design and the quantities held would have been consistent with employment for non-prohibited purposes.

Human degradation technologies are also subject to the general principles of weapons law, most relevantly, the prohibitions of weapons of a nature to cause superfluous injury or unnecessary suffering and of those that are indiscriminate by nature.[57] The superfluous injury or unnecessary suffering principle regulates the use of weapons by reference to their harmful effects as balanced against the military necessity or purpose underlying the design intent of the weapon.[58] The principle must be considered by States when they assess the legality of new weapons.[59] The principle presents an opportunity to regulate entire groups of weaponised human degradation technology because the assessment of legality takes place by reference to a weapon's design purpose – how it is intended to behave in its normal circumstances of application.[60] Moreover, the kinds of harmful effect that can be regarded as

in violation of any treaty obligation by developing, producing, acquiring or stockpiling chemical weapons, although its use of such weapons was arguably in breach of customary international law.

56 It remains the subject of academic debate whether the law enforcement exception extends beyond riot control agents to incapacitating chemical agents: W. H. Boothby, *Weapons and the Law of Armed Conflict*, 2nd edn (Oxford: Oxford University Press, 2016) 127.

57 Protocol Additional to the Geneva Conventions of 12 August 1949, and relating to the Protection of Victims of International Armed Conflicts (Protocol I), opened for signature 8 June 1977, 1125 UNTS 3 (entered into force 7 December 1978) (API), Article 35(2).

58 The prohibition of indiscriminate weapons, on the other hand, is not balanced against the military necessity underlying the design intent of the weapon, but derives from the prohibition of indiscriminate attacks as discussed below in Section 9.4.1: API, Article 51(4).

59 API, Article 36.

60 Boothby, *Weapons and the Law of Armed Conflict*, 259.

giving rise to injury or suffering for the purposes of the application of the principle can be broadly interpreted. Indeed, the ICRC suggests that in reviewing the legality of new weapons in terms of health effects, it may be relevant to consider 'whether there is any predictable or expected long term or permanent alteration to the victims' psychology'.[61]

However, there remain several potential limitations, each of which is enhanced by the balancing requirement which gives discretion to States and military commanders.[62] First, the perceived non-lethal character of degrading weapons may result in their being approved and used more readily in armed conflict. As Fidler writes, new weapons technologies may mean that:

> the relationship between international law and armed conflict is characterized less by the tension between military necessity and humanitarian principle created by lethal weaponry than by an alliance of humanitarian necessity and military principle.[63]

Second, as the assessment of a new weapon's legality attaches to design purpose rather than actual use, there is the danger that a degrading weapon may be deployed for use in a particular way in a particular context and then be used in a different way in a different context. For instance, LRADs may be deployed at checkpoints to enable military officers to communicate over a long distance with approaching vehicles to ascertain their intent. However, they could also be used in a conflict situation as an anti-personnel weapon to cause acute ear pain and permanent deafness.[64] While the original use intended for LRADs would not violate the unnecessary suffering rule, they are nevertheless capable of being used in a way that would increase their harmful effects where those effects might be considered to cause additional injury or suffering for which there is no corresponding military necessity. Other devices, including lawful weapons, are capable of being misused in ways that can be expected to occasion superfluous injury or unnecessary

61 ICRC, *A Guide to the Legal Review of New Weapons, Means and Methods of Warfare* (January 2006) Part 1.3.3, p. 19. Nevertheless, the relevant victim(s) must have suffered an injury or other harm, and inconvenience, annoyance, irritation and other effects falling short of harm are insufficient to give rise to the application of the principle.

62 For details, see Boothby, *Weapons and the Law of Armed Conflict*, 259–260; H. Nasu and T. Faunce, 'Nanotechnology and the International Law of Weaponry: Towards International Regulation of Nano-Weapons', *Journal of Law, Information and Science* 20 (2009), 21, 34–43.

63 D. P. Fidler, '"Non-lethal" Weapons and International Law: Three Perspectives on the Future', *Medicine, Conflict and Survival* 17:3 (2007), 194, 200–201.

64 See the discussion of the LRAD in D. Wallace and S. Reeves, 'Modern Weapons and the Law of Armed Conflict', in G. S. Corn, R. E. VanLandingham and S. R. Reeves, *U.S. Military Operations: Law, Policy, and Practice* (New York: Oxford University Press, 2016), 41, 58–59.

suffering. That possibility of misuse does not, per se, cause the weapon to breach the principle.[65]

9.4 REGULATION OF HUMAN DEGRADATION AS AN ACT OF VIOLENCE

9.4.1 *Applicability of Targeting Law*

The law of armed conflict regulates the categories of individual and the sorts of object that, respectively, may and must not be made the object of an attack in international or non-international armed conflict. While the laws on targeting present a set of principles and rules to guide the use of human degradation technologies as a means of warfare, their application to these types of weapons requires careful examination. If the law of targeting does not apply, for example, Active Denial Technology can lawfully be employed across an entire village to flush out insurgents or other trouble-makers without distinguishing between such insurgents/trouble-makers and members of the civilian population in general, although such action may well be subject to any relevant rules of domestic law and, if applicable, human rights law as will be discussed in Section 9.4.2.

Contrary to the assumption that some commentators appear to have made,[66] there is a threshold question as to whether the law of targeting applies to the use of human degradation technologies. This question is distinct from, and precedes, the question of how the law of targeting governs the use of the technologies. The threshold arises by virtue of the fact that the law of targeting only applies in respect of 'attacks' or, in more limited cases, 'military operations'. The use of human degradation technologies must therefore constitute one of these situations in order for the relevant law to apply. As will be seen, this question directly engages with the definitional challenges associated with the legal regulation of human degradation technologies discussed above.

The starting point is the 'basic rule' of Part IV of Additional Protocol I (API), which requires belligerent parties to distinguish military objectives from the civilian population and objects, and direct their military operations only

65 It should be noted that the LRAD is specifically designed by its manufacturer with a dual purpose of communicating and deterring, even if it is only deployed for the former use.

66 Koplow, for example, writes that the laws of armed conflict 'apply to all weapons, lethal and non-lethal alike, even those (such as the acoustic, electric, netting, and blunt trauma projectiles noted above) that have not yet been subjected to any dedicated treaty regime': Koplow, *Non-Lethal Weapons*, 44.

against the former.[67] The position in customary international law, at least as it is framed in the ICRC's study on customary international humanitarian law, appears to be narrower as it only applies in the context of attacks:

> The parties to the conflict must at all times distinguish between civilians and combatants. Attacks may only be directed against combatants. Attacks must not be directed against civilians.[68]

A second general protection is the rule in Article 51(1) of API that civilians 'shall enjoy general protection against dangers arising from military operations'. A third is the requirement to take constant care to spare civilians and civilian objects.[69] However, the operative rules in relation to civilian protection and precautions apply only to 'attacks' and not to other aspects of military operations.[70] Thus, it is prohibited to make civilians the object of an attack,[71] and to engage in indiscriminate attacks,[72] and/or attacks that would breach the proportionality rule.[73] It is unlawful to engage in acts or threats of violence that are primarily intended to spread terror among the civilian population.[74] Finally, there are obligations to take specific precautions when engaging in an attack, including the requirement to do everything feasible to verify that targets are not civilians or civilian objects.[75]

The Commentary to API defines 'military operation', to which the rules in Articles 48, 51(1) and 57(1) of API apply, broadly as being 'all the movements and activities carried out by armed forces related to hostilities'.[76] According to this definition, the use of human degradation technologies arguably constitutes a 'military operation' provided it is employed by armed forces in connection with hostilities. But the question of whether the use of a human degradation technology can be considered an 'attack', and thereby engage

67 API, Article 48. Note that in the text of the Article, only the word 'operations' is used. However, it is suggested in the ICRC Commentary of 1987 to API that this refers to military operations: Y. Sandoz, et al. (eds.), *Commentary on the Additional Protocols of 8 June 1977 to the Geneva Conventions of 12 August 1949* (Geneva: ICRC, 1987) ('ICRC Commentary of 1987') para 1875.

68 J.-M. Henckaerts and L. Doswald-Beck, *Customary International Humanitarian Law* (Cambridge: Cambridge University Press, 2005) ('ICRC Study'), Rule 1, p. 3.

69 API, Article 57(1); ICRC Study, Rule 15, p. 51.

70 While this is true of treaty rules, there may be differences at customary law. For a comparison of the rules of custom and treaty in relation to precautions, see W. H. Boothby, *The Law of Targeting* (Oxford: Oxford University Press, 2012) 72–74.

71 API, Article 51(2); ICRC Study, Rule 1, p. 3.

72 API, Article 51(4); ICRC Study, Rules 11 and 12, pp. 37–43.

73 API, Article 51(5)(b); ICRC Study, Rule 14, p. 46.

74 API, Article 51(2); ICRC Study, Rule 2, p. 8.

75 API, Article 57(2)(a)(i); ICRC Study, Rule 16, p. 55.

76 ICRC Commentary of 1987 to Article 51 of API, para 1936 (see, also ICRC Commentary of 1987 to Article 48 of API, para 1875).

the operative protections in targeting law, is more problematic. It should, however, be borne in mind that the use of a human degradation technology may, for example, be a preliminary part of a more general attack. In such circumstances, all targeting law rules will apply to the attack considered as a whole.

The term 'attacks' is defined in API as 'acts of violence against the adversary, whether in offence or defence'.[77] The notion of 'violence', it will be recalled, is also contained in the prohibition of acts or threats of violence intended to spread terror among civilians.[78] As explained in the Tallinn Manual 2.0, 'it is the use of violence against a target that distinguishes attacks from other military operations'.[79] In the Tallinn Manual 2.0, the understanding given to the notion of 'attack' in cyber operations is provided in Rule 92: '[a] cyber attack is a cyber operation, whether offensive or defensive, that is reasonably expected to cause injury or death to persons or damage or destruction to objects'.[80] It is well established that the use of harmful chemical or biological substances can constitute an 'attack'.[81] It has also been accepted that other technologies, such as directed energy and lasers, may fall within this definition.[82] There is thus no closed list of weapons technologies that can be used to perpetrate an 'attack'. Consequently, the properties of a particular type of human degradation technology – whether it uses microwave technology or chemical substances – and its method of interference does not per se exempt its use from the law of targeting.

Rather, the key question lies in the type of effect that the technology is intended to have on the target. As Boothby observes:

> The absence of a kinetic impact does not deprive such capabilities of the character of weapons nor the characterization of their use as attacks. Rather, it is the injurious or damaging effect that they are designed or intended to have on the target of the attack, and thus on the adverse party to the conflict, that marks out their use as an attack. [83]

The Tallinn Manual 2.0 also adopts a 'consequential harm approach' to understanding the meaning of 'attack'.[84] It provides that the violence

77 API, Article 49(1). The authors of the ICRC Study also rely on this definition of attack in relation to customary international law: p. 4.

78 API, Article 51(2); ICRC Study, Rule 2, p. 8.

79 *Tallinn Manual* 2.0, 415.

80 Ibid.

81 See, e.g., the discussion of the use of chemical weapons and the law of armed conflict in *Prosecutor* v. *Tadic*, 'Decision on Defence Motion for Interlocutory Appeal', 2 October 1995 *International Legal Materials* 35 (1996), 89 paras 120–122.

82 Boothby, *The Law of Targeting*, 384.

83 Ibid.

84 *Tallinn Manual* 2.0, 415–416.

threshold would not be met by *de minimis* damage or destruction, but would be satisfied by injury, including serious illness and severe mental suffering.[85] The extension to psychological injury warrants attention. It is justified in the Tallinn Manual 2.0, in part, on the basis of the prohibition of acts or threats of violence designed to spread terror among the civilian population.[86] However, that rule does not necessarily equate violence with the psychological condition of terror. Rather, the rule's inclusion may imply that violence relates to a physical consequence, and that terror is something logically additional or distinct from that act as a side effect. As such, a means or method of warfare that causes fear, but does not cause any physical harm, may fall outside the definition of 'attack'.

As outlined above, the intended effect of weaponised forms of human degradation technology is to reduce the performance of the target by creating an involuntary motor response or other behavioural shift. However, lethal consequences are not ruled out, and so it may be that a human degradation technology is deployed with the knowledge that fatalities or injuries will or are likely to occur under attendant circumstances. In that case, it is clear that the use of a human degradation technology would constitute an 'attack' and therefore attract the application of the law of targeting. On the other hand, if the use of a human degradation technology resulted in fatalities or caused injury contrary to the intention or expectation of the user due to, for example, a pre-existing medical condition of the target, it would be less clear whether the use of a human degradation technology constituted an 'attack'.[87]

Based on the examples of human degradation technology examined above, however, it can be expected that their use in armed conflict would ordinarily interfere with the sensory-motor system without causing permanent injury or death. It must therefore be considered whether such interference can reach the threshold of injury required. Even accepting that there is a minimum (*de minimis*) degree of injury that must be occasioned in order for the act to be violent, there exists no criterion and no commonly accepted position on where this line is to be drawn. For instance, several of the human degradation technologies discussed above generate acute short-term pain. On one view, the intensity of the discomfort might constitute violence. On another view, the lack of any health effects on the target might indicate there was no injury. The complexities are apparent in the following examples provided by Jensen:

[85] Ibid., 416.
[86] Ibid., 417.
[87] C. Enemark, '"Non-Lethal" Weapons and the Occupation of Iraq: Technology, Ethics and Law', *Cambridge Review of International Affairs* 21:2 (2008), 199, 201.

> the use of a nanobot to infiltrate an individual's body and collect data and then transmit that data to an adversary may seem more like espionage than an attack, despite its invasive nature. Similarly, the spreading of a gene that creates an allergic sensitivity to pollen may have significant effect on a fighting force, but might not be termed an act of violence.[88]

For some types of human degradation technology, such as the sonic device that has produced permanent hearing loss and other serious and long-lasting conditions in American diplomats in Cuba (although not here used in armed conflict), it will be clear that injury above the relevant threshold has been occasioned. For other types of human degradation technology, such as dazzling lasers that do not cause any permanent reduction in vision, it is probably the case that they do not represent an act of violence, although they may well be used as part of a wider military operation that does amount to an attack. But for the majority of types of human degradation technology, which seek to avoid permanent injury as part of their design, but which are nevertheless painful and invasive, the position is unclear.

Like the laws relating to specific classes of weapon, the law of targeting has the potential to cover the use of human degradation technology. The meaning of 'attack' does not necessarily exclude weapons based on their method of interference and so each type of technology must be considered by reference to the factors that have been discussed above. The key element of 'attack' is its violent effect on the target, which is one of the constitutive elements of some of the kinds of human degradation technology that we have discussed, forming part of their design purpose. The limitation, however, is that certain types of human degradation technology, while causing harm, may not be designed to have the necessary injurious or damaging effect to meet the definition of 'attack'. If a particular human degradation technology does not involve the infliction of harm such as would be required for its use to constitute 'an attack', this may mean that military forces would be attracted to the greater use of such human degradation technologies as the specific rules on attacks would not apply. This possibility may arise in circumstances where the presence of a legitimate military target is uncertain or where the application of the law otherwise prevents an attack against the target.

88 E. T. Jensen, 'The Future of the Law of Armed Conflict: Ostriches, Butterflies and Nanobots', *Michigan Journal of International Law* 35 (2014), 253, 312.

9.4.2 *Applicability of Human Rights Law*

The use of a technological means to degrade the physical functions of an individual in peacetime is primarily regulated according to applicable rules of domestic law, which will vary from State to State.[89] For instance, Section 3(1) of the Criminal Law Act 1967 (UK) provides that: 'A person may use such force as is reasonable in the circumstances in the prevention of crime, or in effecting or assisting in the lawful arrest of offenders or suspected offenders or of persons unlawfully at large'. However, the domestic regulation of the use of force in law enforcement has increasingly been subjected to international law obligations to respect and to ensure respect for human rights such as the right to life and the freedom from cruel, inhuman and degrading treatment. It follows that the use of human degradation technology in a law enforcement operation may involve a violation of human rights, even though there is no specific prohibition on the use of such technology in this context under international human rights law.

The use of lethal force in law enforcement might engage the right to life under relevant human rights instruments. For example, Article 6(1) of the International Covenant on Civil and Political Rights provides: 'Every human being has the inherent right to life. This right shall be protected by law. No one shall be arbitrarily deprived of his life'.[90] The use of 'lethal force' is thus regulated as a potential arbitrary deprivation of a person's life. There are various legal requirements attached to the use of lethal force such as necessity, proportionality and the duty to report and investigate injury or death.[91] In this respect, the right to life applies to those cases where the limited intended outcome of using human degradation technology – simply to incapacitate or repel a target for strategic purposes – is not met because the target is killed.[92]

89 See generally, Casey-Maslen and Connolly, *Police Use of Force*, ch 5.

90 International Covenant on Civil and Political Rights, opened for signature 19 December 1966, 999 UNTS 171 (entered into force 23 March 1976) ('ICCPR'), Article 6(1). See also, Convention for the Protection of Human Rights and Fundamental Freedoms, adopted 4 November 1950, 213 UNTS 221 (entered into force 3 September 1953) as last amended by Protocol No 14, CETS No 194 (entered into force 1 June 2010) ('ECHR'), Article 2; American Convention on Human Rights, adopted 22 November 1969, 1144 UNTS 123 (entered into force 18 July 1978) ('ACHR'), Article 4(1); African Charter on Human and People's Rights, adopted 27 June 1981, 1520 UNTS 217 (entered into force 21 October 1986) ('ACHPR'), Article 4; ASEAN Human Rights Declaration, adopted 18 November 2012 ('AHRD'), Article 11.

91 See, e.g., C. Heyns, 'Report of the Special Rapporteur on Extrajudicial, Summary or Arbitrary Executions' (1 April 2014) UN doc. A/HRC/26/36 paras 55–85.

92 Ibid., paras 101–106. Indeed, from the perspective of respecting and ensuring respect for the right of life, the development of non-lethal incapacitating weapons for use in appropriate situations has been encouraged: Basic Principles on the Use of Force and Firearms by Law

But even when the use of human degradation causes death, it may prove difficult to find a violation of the right to life. This was exemplified in *Finogenov* v. *Russia*, in which the use of unknown narcotic gas in the rescue operation during the Moscow Theatre hostage crisis was alleged to have violated the right to life under Article 2 of the European Convention.[93] The ECtHR was in a difficult predicament as it was presented with diverging submissions as to the lethality of the chemical agent and did not have before it 'the exact formula of the gas'.[94] Ultimately it avoided definitively ruling on the question of the lethality of the gas, observing that it was sufficient for the purposes of Article 2 of the ECHR that it was a 'dangerous substance'.[95]

However the Court did accept, parenthetically, that 'the gas was probably not intended to kill the terrorists or hostages', classifying the unknown narcotic gas as something closer to non-lethal incapacitating weapons than to firearms.[96] This classification appears to have played a critical role in the Court's assessment of the compatibility of its use with the standard of care required in an operation involving the use of force. In *Isayeva*, the Court held, in the absence of a derogation, that the massive use of indiscriminate weapons in a civilian populated area did not meet the standard of care required in a police operation aimed at protecting civilian lives from unlawful violence.[97] In *Finogenov*, even though such use of the unknown narcotic gas was considered to be indiscriminate in nature, the Court distinguished it from *Isayeva* on the grounds that 'the gas used by the Russian security forces, while dangerous, was not supposed to kill, in contrast, for example, to bombs or air missiles'.[98] The Court further elaborated on the basis for this distinction, observing that:

> Although the gas in the present case was used against a group consisting of hostages and hostage-takers, and although the gas was dangerous and even potentially lethal, it was not used 'indiscriminately' as it left the hostages a high chance of survival, which depended on the efficiency of the authorities' rescue effort.[99]

Enforcement Officials, adopted at the Eighth United Nations Congress on the Prevention of Crime and the Treatment of Offenders, Havana, Cuba, 27 August–7 September 1990.

93 *Finogenov* v. *Russia*, para 165 *et seq.*
94 Ibid., para 200.
95 Ibid., para 203.
96 Ibid., para 202.
97 ECtHR, *Isayeva* v. *Russia*, Judgment (24 February 2005) Application No 57950/00, para 191.
98 *Finogenov* v. *Russia*, para 232.
99 Ibid.

While condemning Russian authorities for various flaws in the planning and execution of the rescue operation,[100] the Court found that the use of unknown narcotic gas in itself was not a disproportionate response due to the high chance of survival that it left the hostages.

The principal problem with the Court's reasoning is that its characterisation of the chemical agent was based on what it understood to be the intended effect and outcome of using an incapacitant in law enforcement operations in general (i.e. its design purpose), and not on the intended effect of the particular incapacitant in question (by reference to its unique properties) or on the outcome intended by the Russian special forces in the circumstances. The Court did not have before it the evidence to make the second type of assessment. As the ICRC pointed out, 'the Court was not provided information about the specific toxic chemicals used and thus was in a difficult position to judge whether the adverse effects of their use should have been foreseen'.[101] The Court was also not furnished with complete information about the quantity of gas used and what its potency would have been in light of the size of the room, the ventilation system and other relevant circumstances. Consequently, its decision that the gas left the hostages a high chance of survival was made solely on the basis of an incomplete and rudimentary understanding of the character of the gas and did not consider its intended or actual use in the operation. For instance, Russian special forces could have used the incapacitant with the intention to kill the terrorists, while accepting the collateral loss of life of the hostages that this would entail, by deliberately pumping a lethal dose into the theatre.[102] Indeed, as Enemark has observed, the use of chlorine, phosgene and mustard with deadly intent in World War I achieved a lower mortality rate than the use of the incapacitant in the Moscow theatre.[103]

In principle there should be nothing to preclude the right to life applying when human degradation technologies are used with fatal results. But *Finogenov* illustrates how the assessment framework in which an analysis of the right to life takes place might be challenged by the use of a weapon that is designed to degrade rather than kill. In a complex operation, when intervention is necessary, even with risk to civilians' lives, the use of a non-lethal

[100] Ibid., para 266.

[101] ICRC, 'Toxic Chemicals as Weapons for Law Enforcement: A Threat to Life and International Law?' (September 2012) 3, www.icrc.org/en/document/toxic-chemicals-weapons-law-enforcement-threat-life-and-international-law.

[102] If that were indeed the case, this operation would only have been lawful if it met the various requirements attached to the right to life, as discussed above.

[103] Enemark, '"Non-lethal" Weapons', 199, 209 fn 4.

weapon as opposed to lethal force might be taken to imply that the police and security forces have complied with their obligation to take all necessary precautions to protect human life. In this respect, the characterisation of a human degradation technology by reference to its design purpose might lead a court astray; i.e. it might cause a court to consider that the weapon is less indiscriminate, more proportionate and constitutes responsible force escalation by the law enforcement agencies. However, the contrary may be true. As outlined above, the definition of human degradation technology does not preclude lethal effects. An incapacitant could be lethal in a number of circumstances: a standard dosage may have more harmful effects on people with infirmities; the dosage level may be intentionally or recklessly lethal; the intended circumstances of use may render the calculation of the appropriate dose (i.e. a dose that disables the hostage takers with minimal effect on the hostages) challenging; or the incapacitant may be used as a force enabler with conventional weapons. Indeed, each of these factors may have been present in the Moscow siege case. This means that for the right to life to effectively regulate the use of human degradation technology, the weapon in question will have to be understood not just by reference to its design purpose but also to its intended use in the circumstances in question. The shortcomings of doing otherwise were demonstrated in the reasoning of the ECtHR, which made juxtapositions between bombs and gas when it might have been more relevant to consider whether a different dosage, another type of incapacitating agent, another weapon altogether or indeed another method of addressing the siege might have provided a greater chance of survival for those involved, including the hostages.

This does not mean that no other rights will be infringed by the use of human degradation technologies. Where the use of force does not result in death, but nevertheless causes harm, it might engage the prohibition against cruel, inhuman or degrading treatment. For instance, under Article 16 of the Convention Against Torture, States undertake to prevent: 'other acts of cruel, inhuman or degrading treatment or punishment which do not amount to torture as defined in Article I, when such acts are committed by or at the instigation of or with the consent or acquiescence of a public official or other person acting in an official capacity'.[104] Indeed, in *Ali Güneş v. Turkey*, the ECtHR held that the unwarranted use of capsicum spray by police constituted

[104] Convention against Torture and Other Cruel, Inhuman or Degrading Treatment or Punishment, opened for signature 10 December 1984, 1465 UNTS 162 (entered into force 26 June 1987) Article 16; see also ICCPR, Article 7; ECHR, Article 3; ACHR, Article 5; ACHPR, Article 5; AHRD, Article 14.

inhuman and degrading treatment because it subjected the applicant 'to intense physical and mental suffering and was such as to arouse in him feelings of fear, anguish and inferiority capable of humiliating and debasing him'.[105] This reasoning is reflective of the general approach in human rights jurisprudence, which is to equate degrading treatment with conduct that humiliates or debases an individual and thereby diminishes his or her human dignity.[106] However, there is little focus in the existing jurisprudence on interference with a victim's physical and mental integrity. In respect of other practices, such as interrogation, the prohibition has been engaged by conduct that has 'caused, if not actual bodily injury, at least intense physical and mental suffering to the persons subjected thereto'.[107] The further jurisprudential development on the prohibition of cruel, inhuman and degrading treatment may generate more effective regulation of human degradation technology with the adoption of an effects-based approach, free from constraint by the requirement that there be a causal link to fatality. It remains to be seen whether a conceptual alignment between degrading weapons and inhuman or degrading treatment will be established in future case law.

9.5 CONCLUDING OBSERVATIONS

With further scientific and technological advances, there will emerge a greater variety of novel means of degrading the physical functions of the human body, which may find effective use both in law enforcement and armed conflict. This chapter has reviewed three different legal regimes that apply to the use of human degradation technology as a non-lethal means of violence – the law of weapons governing specific classes of weapons, the law of targeting applicable both in international and non-international armed conflict, and human rights law applicable to law enforcement operations. The review has identified inconsistent approaches adopted to the threshold question for the applicability of each body of law – either by reference to intended use, design purpose or harmful effects – which, together with the varying degrees of uncertainty associated with the lethal effect of these weapons, present an inherent limit to the regulation of human degradation technology as a means of violence.

105 ECtHR, *Ali Güneş* v. *Turkey* (10 April 2012) App. No. 9829/07, para 43. For other cases considering the use of human degradation technology in law enforcement, see, eg, ECtHR, *Abdullah Yaşa and Others* v. *Turkey*, (16 July 2013) App. No 44827/08 paras 42–43; ECtHR, *Guler and Ongel* v. *Turkey* (4 October 2011) App. Nos. 29612/05 and 30668/05 paras 29–31.

106 See, P. G. Carozza, 'Human Dignity', in D. Shelton (ed.), *The Oxford Handbook of International Human Rights Law* (Oxford: Oxford University Press, 2013), 345, 354.

107 ECtHR, *Ireland* v. *United Kingdom* (19 January 1978) Application No. 5310/71, para 167.

This regulatory gap is unlikely to be filled by a new treaty on weapons, at least in the near future, because of the lack of political will and technical difficulties in specifying the harmful effects to be prohibited. The law of targeting may evolve to accommodate a greater variety of harmful effects in the definition of 'attacks' or through the application of general obligations imposed upon 'military operations' more broadly,[108] but it will only do so gradually, as state practice develops.

The prohibition on cruel, inhuman and degrading treatment may, as the jurisprudence develops, present a more promising opportunity to regulate human degradation technology. Unlike the right to life regime, it is not constrained by the requirement that there be a causal link to fatality; instead it adopts an effects-based approach. As discussed above, existing jurisprudence on the principle in the context of non-lethal weapons is restricted to the consideration of the effect on human dignity, and has not yet grappled with the unique physical and psychological impacts on the human body caused by degrading weapons. But were this to happen, it may have widespread consequences, as the prohibition applies both in peacetime and in armed conflict situations as a rule of customary international law.[109] The focus on the actual physical and psychological condition of a human being subjected to the use of human degradation technology, rather than its intended effect or design purpose, represents a potential way forward to develop a more coherent, effects-based approach to the regulation of human degradation technology.

[108] What is being contemplated here is a development of the interpretation of the somewhat vague notion of 'military operations' that may bring some kinds of human degradation within the ambit of Articles 48, 51(1) and 57(1) of API, which may thus require that they be used in compliance with the principle of distinction, that constant care be taken to protect civilians from them and that civilians should be generally protected from the dangers arising from their use.

[109] See, ICRC Study, Rule 90.

10

Nanomaterials: A Tale of Two Applications

KOBI LEINS AND DIANA M. BOWMAN

Helena: When, for example, the workmen struck against the Robots and smashed them up, and when the people gave the Robots firearms against the rebels and the Robots killed so many people. And then when the Governments turned the Robots into soldiers and there were so many wars and all that.

Domain: We foresaw that, Helena. You see, these were only passing troubles which are bound to happen before the new conditions are established.

***Rossum's Universal Robots*, Karel Čapek (Oxford University Press, 1920)**

10.1 INTRODUCTION: GOVERNANCE OF ENGINEERED NANOPARTICLES

'Nano' is a prefix meaning 'a billionth' (a factor of 10^{-9} of a metre) derived from the Greek νᾶνος, meaning 'dwarf'. By way of illustration, a nano-sized object is to an apple, what an apple is to the size of the Earth. Or to give another example, one nanometre particle could fit approximately 80,000 times across a human hair.[1] In the last two decades, the ability to deliberately create, manipulate, or modify existing matter at the nanoscale has resulted in rapid innovation in, and across, many fields, due to properties different from matter at the regular scale, such as increased strength, chemical resistance properties and heightened conductivity of matter at the nanoscale.

Nanoparticles are not new. They have occurred naturally for thousands, if not millions, of years in volcanic ash, sea spray and as mineral deposits, to name just a few examples.[2] These naturally occurring forms of nanoparticles, for the most part, pose few challenges to human health, as human bodies have

[1] Jennifer Kahn, 'Nano's Big Future' (2006) 209(6) *National Geographic* 98.

[2] V. Kumar Thakur and M. Kumari Thakur, *Handbook of Sustainable Polymers: Structure and Chemistry* (Abingdon: Taylor and Francis, 2016), 688.

adapted to deal with these materials over time.[3] It is not these types of nanomaterial that in the early 2000s attracted concern from consumer groups, non-governmental organisations (NGOs) and academics in relation to potential harms and great uncertainties. Rather, it was the increase in creation and/or use of purposefully engineered nanoparticles (ENPs) that generated discussion and debate across a number of different disciplines and jurisdictions.[4] Commercially, ENPs are being used in the manufacture of scratchproof eyeglasses, crack- resistant paints, anti-graffiti coatings for walls, transparent sunscreens, stain-repellent fabrics, self-cleaning windows and ceramic coatings for solar cells, to name just a few examples.[5] Armed forces are harnessing the potential of ENPs to create more efficient energy storage,[6] advanced photovoltaics,[7] and improved military protective equipment.[8,9]

Promoters of nanotechnologies, including products that incorporate ENPs, have argued that the technology has the potential to improve the performance of existing products, to offer new capabilities – especially through convergence

3 C. Buzea, Ivan P. and K. Robbie, 'Nanomaterials and Nanoparticles: Sources and Toxicity', *Biointerphases*, 2(4) (2007), 17–71 https://link.springer.com/content/pdf/10.1116/1.2815690.pdf.

4 See, for example, P. H. Hoet, I. Brüske-Hohlfeld and O. V. Salata, 'Nanoparticles – Known and Unknown Health Risks', *Journal of Nanobiotechnology*, 2(1) (2004), 12; A. D. Maynard, et al., 'Safe Handling of Nanotechnology', *Nature*, 444(7117) (2006), 267–269; P. J. A. Borm, et al., 'The Potential Risks of Nanomaterials: A Review Carried out for ECETOC', *Particle and Fibre Toxicology*, 3(1) (2006), 11; S. Foss Hansen, et al., 'Late Lessons from Early Warnings for Nanotechnology', *Nature Nanotechnology*, 3(8) (2008), 444–447; J. M. Balbus, et al., 'Meeting Report: Hazard Assessment for Nanoparticles – Report from an Interdisciplinary Workshop', *Environmental Health Perspectives*, 115(11) (2007), 1654.

5 The Scientific Committee on Emerging and Newly Identified Health Risks, *Nanomaterials* (2006) http://ec.europa.eu/health/scientific_committees/opinions_layman/en/nanotechnologies/l-3/5-nanoparticles-consumer-products.htm#opo.

6 E. S. Cho, et al., 'Graphene Oxide/Metal Nanocrystal Multilaminates as the Atomic Limit for Safe and Selective Hydrogen Storage', *Nature Communications*, (2016) www.nature.com/articles/ncomms10804.pdf.

7 Photovoltaics refer to the method of converting solar energy into electricity using semiconducting materials, usually solar panels, which convert the sun to usable power. The importance of improvement of efficiency in this function has direct and widespread application for any means of warfare that requires an energy source.

8 Much research is being undertaken by militaries to incorporate nano-elements to enhance the function of existing military applications, including gas filters, masks and materials for tanks and protective clothing for soldiers. See Symposium, 'Future Technologies for Personal Protection' (7–8 November 2013) www.defencescienceinstitute.com/2012/09/10/dsi-dsto-symposium-future-technologies-personal-protection/.

9 J. Wang and P. Dortmans, 'Nanotechnology: Selected Topics and Potential Military Application' (Paper presented at Land Warfare Conference, Melbourne, Australia, 27–30 September 2004) http://dspace.dsto.defence.gov.au/dspace/handle/1947/2559 69–81.

with other emerging technologies – and to be of fundamental assistance in addressing global problems such as climate change and food security.[10] It is therefore not surprising that a number of national governments have heavily invested in a nanotechnology future.[11]

As public- and private-sector investment has increased in nanotechnology-related research and development (R&D) activities, so too has the number of products on the market that incorporate ENPs. Alongside this innovation have been calls for more sophisticated governance frameworks that effectively regulate the technology and its by-products. As observed by Levi-Faur and Comaneshter, such corresponding debates around emerging technologies are without precedent. In their words:

> unlike other cases where the discussion of associated risks has followed the development of new technologies, the discussions of the proper regulatory frameworks for the governance of nanotechnology risks is accompanying the development of the technology and the associated products themselves.[12]

Parallel discussions in, and around, innovation and appropriate regulatory responses between policy-makers, safety regulators, industry and NGOs have resulted in a myriad of different activities and instruments to reduce the potential risks to individuals and the environment. Importantly, a number of these initiatives, such as risk-management frameworks, codes of conduct and reporting mechanisms, have crossed disciplines, sectors and geographical boundaries,[13] changing the way governance is considered and applied.

[10] C. Som, et al., 'The Importance of Life Cycle Concepts for the Development of Safe Nanoproducts', *Toxicology*, 269 (2010), 160–169.

[11] M. C. Roco, 'International Perspective on Government Nanotechnology Funding in 2005', *Journal of Nanoparticle Research*, 7(6) (2005), 707–712.

[12] D. Levi-Faur and H. Comaneshter, 'The Risks of Regulation and the Regulation of Risks: The Governance of Nanotechnology', in G. Hodge, D. Bowman and K. Ludlow (eds.), *New Global Frontiers in Regulation: The Age of Nanotechnology* (Cheltenham: Edward Elgar, 2007), 149–165, 150.

[13] See, for example, M. Hull, 'Nanotechnology Risk Management and Small Business: A Case Study of the Nanosafe Framework', in M. Hull and D. M. Bowman (eds.), *Nanotechnology Environmental Health and Safety: Risks, Regulation and Management* (New York: Elsevier, 2010), 247–295; C. Meili and M. Widmer, 'Voluntary Measures in Nanotechnology Risk Governance: The Difficulty of Holding the Wolf by the Ears' in G. Hodge, D. M. Bowman and A. Maynard (eds.), *International Handbook on Regulating Nanotechnologies* (Cheltenham: Edward Elgar, 2010), 446–461; T. Widler, et al., 'Organisational Risk Management of Nanomaterials Using SUNDS: The Contribution of CENARIOS®', in F. Murphy, E. M. McAlea and M. Mullins (eds.), *Managing Risk in Nanotechnology* (Springer International, 2016), 219–235; L. Krabbenborg, 'DuPont and Environmental Defense Fund Co-Constructing a Risk Framework for Nanoscale Materials: An Occasion to Reflect on Interaction Processes in a Joint Inquiry', *NanoEthics*, 7(1) (2013), 45–54.

In contrast, military developments in the form of both basic research and acquisition programmes have often gone unreported in some countries, or have been cloaked in secrecy, which, it has been argued, is justified by national defence interests. Although this is not a new issue for developments by, or for, the armed forces, in the case of ENPs there is a strong argument for greater transparency in the ways in which ENPs are being used, and as to whether, and to what extent, military agencies are actively considering governance in relation to the matters that are exercising their civilian counterparts – namely protection of workers, of the public and of the environment more generally. This argument is strong given the nascent knowledge about the long-term effects of novel ENPs on the human body and on the environment.[14] To the authors' knowledge, there appears to be no consistent collaboration, or structures requiring or facilitating collaboration, between commercial and military research involving ENPs. In particular, so far as is known there is no centralisation of information regarding the long-term toxicity of ENPs. It is known that ENPs are being included in military systems, fabrics and membranes, for example, with little to no understanding of what the long-term effects of these ENPs would be on combatants, civilians, or the environment.

This chapter will consider two parallel, but to date institutionally and culturally isolated, international governance frameworks applicable to ENPs: those governing commercial developments, and those governing military developments. Section 10.2 will consider the implied governance of inclusion of ENPs in commercial applications, and the relevant governance frameworks. Sections 10.3–10.7 will consider military applications of ENPs and the relevant governance frameworks. Conclusions and recommendations are set out in Section 10.8.

[14] See, for example, C. A. Poland, et al., 'Carbon Nanotubes Introduced into the Abdominal Cavity of Mice Show Asbestos-Like Pathogenicity in a Pilot Study', *Nature Nanotechnology*, 3(7) (2008), 423–428; H. Johnston, et al., 'Engineered Nanomaterial Risk. Lessons Learnt from Completed Nanotoxicology Studies: Potential Solutions to Current and Future Challenges', *Critical Reviews in Toxicology*, 43(1) (2013), 1–20; S. Hansen, et al., 'A Critical Analysis of the Environmental Dossiers from the OECD Sponsorship Programme for the Testing of Manufactured Nanomaterials', *Environmental Science: Nano*, 4(2) (2017), 282–291; D. Johnson, *Nanoparticles Found in Brains Come from External Sources* (8 September 2016) IEEE Spectrum http://spectrum.ieee.org/nanoclast/semiconductors/materials/nanoparticles-found-in-brains-comes-from-external-sources; B. Maher, et al., 'Magnetite Pollution Nanoparticles in the Human Brain', *Proceedings of the National Academy of Sciences of the United States of America*, 113(39) (2016), 10797–10801 www.pnas.org/content/113/39/10797.

10.2 GOVERNANCE FRAMEWORKS FOR COMMERCIAL USE OF NANOTECHNOLOGIES

Emerging technologies have traditionally challenged regulatory regimes. More recently it is not just the technologies themselves, but also the products that they create, that have challenged existing governance structures. The acceleration of new technologies into the market has amplified these challenges, resulting in periods of, as noted by Ludlow et al., 'under- and over-regulation'.[15] Framing it as a 'pacing problem', Marchant et al. have argued that traditional legal instruments, legislation and formal rules, are becoming increasingly cumbersome for effectively dealing with these new technologies because of the shorter cycles in which they are being developed.[16]

One could assume by the hype and hyperbole surrounding nanotechnologies in the early 2000s that the technology was 'new' or 'emerging' at that time.[17] For the purposes of regulatory and policy debate, along with the discussions regarding risk assessment, this is arguably true but only in particular contexts as explained below.[18] From an engineering/materials perspective, the purposeful manipulation of substances and materials at the nanoscale (1–100 nm) was 'old'; the information technology (IT) sector had, for example, been working at the nanoscale for more than two decades.[19] The invention of the scanning tunnel microscope by Binning and Rohrer[20] in 1981 accelerated innovation at the nanoscale as it provided scientists with 'a new method to deal

[15] K. Ludlow, D. M. Bowman and D. D. Kirk, 'Hitting the Mark or Falling Short with Nanotechnology Regulation?', *Trends in Biotechnology*, 27(11) (2009), 615–620, 615.

[16] G. E. Marchant, B. R. Allenby and J. R. Herkert (eds.), vol. 7, *The Growing Gap between Emerging Technologies and Legal-Ethical Oversight: the Pacing Problem* (New York: Springer, 2011).

[17] See, for example, D. M. Berube, *Nano-Hype: the Truth behind the Nanotechnology Buzz* (Amherst, NY: Prometheus Books, 2006); K. Kjølberg and F. Wickson, 'Social and Ethical Interactions with Nano: Mapping the Early Literature', *NanoEthics*, 1(2) (2007), 89–104; Uday Khandeparkar, 'Behind the Nano Hype', *Wall Street Journal*, (19 January 2009).

[18] Royal Society and Royal Academy of Engineering, *Nanoscience and Nanotechnologies: Opportunities and Uncertainties* (Royal Society, 2004).

[19] See, for example, D. Eigler, 'From the Bottom up: Building Things with Atoms' in G. Timp (ed.), *Nanotechnology* (Springer, 1999), 425–435; P. Smaglik, 'Building for the Future', *Nature*, 408(6812) (2000), 622; *IBM Research – Nanotechnology* (2002) IBM www.research.ibm.com/pics/nanotech/defined.shtml.

[20] G. Binnig, et al., 'Surface Studies by Scanning Tunneling Microscopy', *Physical Review Letters*, 49(1) (1982), 57. See, also, G. Binnig and H. Rohrer, 'Scanning Tunneling Microscopy', *IBM Journal of Research and Development*, 44(1/2) (2000), 279; D. Baird and A. Shew, 'Probing the History of Scanning Tunneling Microscopy', in D. Baird, A. Nordmann, and J. Schummer (eds.), *Discovering the Nanoscale* (IOS Press, 2004), 145–156.

with atoms, molecules, and nanometer-scale structures'.[21] During the 1980s and 1990s this research and development was not, however, considered to be 'nanotechnologies' as, simply put, the term had not yet been incorporated into the everyday vernacular.

With nanotechnologies being viewed as being 'new' from a policy and regulatory perspective, the entry of nano-based substances and products into the market occurred under a cloud of scientific uncertainty regarding the potential risks posed by some ENPs. While these products were not 'unregulated' *per se*, they fall under traditional regulatory regimes that are, for the most part, silent on the technological processes by which they are created,[22] so questions regarding the adequacy of existing governance arrangements quickly arose.[23]

In response to such concerns, a number of governments and/or government agencies undertook, or commissioned, reviews of their current regulatory arrangements for nanotechnologies.[24] The findings of each confirmed that

[21] G. Binnig and H. Rohrer, 'In Touch with Atoms', *Reviews of Modern Physics*, 71(2) (1999), S324–330, S324.

[22] E. Stokes, 'Nanotechnology and the Products of Inherited Regulation', *Journal of Law and Society*, 39(1) (2012), 93–112; E. Stokes and D. M. Bowman, 'Regulatory Inheritance and Emerging Technologies: The Case of Nanotechnologies and Synthetic Biology', *European Journal of Risk Regulation*, 2 (2012), 235–241.

[23] See, for example, Q. Chaudhry, et al., *Final Report: A Scoping Study to Identify Gaps in Environmental Regulation for the Products and Applications of Nanotechnologies* (2006) Safe Nano www.safenano.org/media/109893/DEFRA2005-Review_of_gaps_in_environmental_regulation_for_the%20products_and_applications_of_nanotechnologies.pdf; A. C. Lin, 'Size Matters: Regulating Nanotechnology', *Harvard Environmental Law Review*, 31 (2007), 349; R. F. Wilson, 'Nanotechnology: The Challenge of Regulating Known Unknowns', *The Journal of Law, Medicine & Ethics*, 34(4) (2006), 704–713.

[24] Health and Safety Executive, *Review of the Adequacy of Current Regulatory Regimes to Secure Effective Regulation of Nanoparticles Created by Nanotechnology: the Regulations Covered by HSE* (2006) HM Government www.hse.gov.uk/nanotechnology/regulatoryreview.pdf; Chaudhry, et al., *Final Report*, above; Food Standards Agency, *Report of the FSA regulatory review* (2008) HM Government www.food.gov.uk/sites/default/files/multimedia/pdfs/nanoregreviewreport.pdf; Committee on Toxicity of Chemicals in Food, Consumer Products and the Environment, *The Toxicology of Nanoparticles Used in Healthcare Products* (2006) HM Government https://cot.food.gov.uk/sites/default/files/cot/cotstatementnanomats200701.pdf; Environmental Protection Agency, *Nanotechnology white paper* (2007) US Government www.epa.gov/sites/production/files/2015-01/documents/nanotechnology_whitepaper.pdf; Food and Drug Administration, *Nanotechnology: A Report of the US Food and Drug Administration Nanotechnology Task Force* (2007), US Government www.fda.gov/ScienceResearch/SpecialTopics/Nanotechnology/ucm2006659.htm; K. Ludlow, D. M. Bowman and G. A. Hodge, *A Review of Possible Impacts of Nanotechnology on Australia's Regulatory Framework: Final Report* (2007) Monash University Law www.researchgate.net/publication/265997359_A_Review_of_Possible_Impacts_of_Nanotechnology_on_Australia's_Regulatory_Framework; European Commission,

despite the legislative instruments being silent on the words 'nano', 'nanotechnologies' and/or 'nano-enabled', such products were captured by the existing arrangements. While the findings of each review varied, a number of concerns were raised, including the existence of potential regulatory gaps and questions around implementation, and these concerns were identified as needing government attention.[25] The Royal Society and the Royal Academy of Engineers (RS-RAE), for example, expressed the view that the regimes that oversaw the approval (or 'listing') of chemical substances onto the market constituted the regimes that required the most urgent attention.[26] Under such regimes, chemical substances are regulated on the basis of their Chemical Abstracts Service (CAS) Number, and not on the basis of the size and/or properties exhibited by the chemical. As such, carbon-based nanoparticles, such as a carbon nanotube, would be considered to be 'graphene' or 'carbon black' for regulatory approval and oversight purposes despite the very different risk profiles and characteristics between them.[27] The characterisation of chemicals by their CAS number is reflected in the governance of the creation, transfer and use of chemicals in the Chemical Weapons Convention (CWC)

Regulatory Aspects of Nanotechnologies: SEC(2008) 2036 (17 June 2008) https://ec.europa.eu/research/industrial_technologies/pdf/policy/comm_2008_0366_en.pdf; European Commission, Accompanying document to the communication from the Commission to the European Parliament, the Council and the European Economic and Social Committee regulatory aspects of nanomaterials: summary of legislation in relation to health, safety and environment aspects of nanomaterials, regulatory research needs and related measures: Commission Staff Working Document No SEC (2008) 2036 (17 June 2008); C. Gavaghan and J Moore, *A Review of the Adequacy of New Zealand's Regulatory Systems to Manage the Possible Impacts of Manufactured Nanomaterials* (2011) www.academyconferences.com/pdf/nznanoreview.pdf; European Commission, *Second Regulatory Review on Nanomaterials (COM(2012) 572 final)* (2012) http://eur-lex.europa.eu/LexUriServ/LexUriServ.do?uri=COM:2012:0572:FIN:en:PDF; M. Saner, E. Heafey and D. M. Bowman, 'Mapping the Regulatory Environment for Nanomaterials in Canada', *Nanotechnology Law & Business*, 9(4) (2013), 343–68; D. M. Bowman, 'More than a Decade On: Mapping Today's Regulatory and Policy Landscape Following the Publication of Nanoscience and Nanotechnologies: Opportunities and Uncertainties', *NanoEthics*, 11(2) (2017), 169–186.

25 See, for example, Chaudhry, et al., *Final Report*; Ludlow, et al., *A Review of Possible Impacts*.

26 Royal Society and Royal Academy of Engineering, *Nanoscience*. See, also, D. M. Bowman and G. van Calster, 'Does REACH Go Too Far?', *Nature Nanotechnology*, 2(9) (2007), 525–526; L. L. Bergeson and J. E. Plamondon, 'TSCA and Engineered Nanoscale Substances', *Nanotechnology Law & Business*, 4 (2007), 51.

27 Royal Society and Royal Academy of Engineering, *Nanoscience*; Ludlow, et al., *A Review of Possible Impacts*, above; Bowman and van Calster, 'Does REACH Go Too Far?'; J. Monica, G. Van Calster and P. Agapi, 'A Nanotechnology Legal Framework', in M. Hull and D. M. Bowman (eds.), *Nanotechnology Environmental Health and Safety: Risks, Regulation and Management*, 2nd edn, (New York: Elsevier, 2014), pp.265–312.

(see Section 10.3 below), which runs up against the same regulatory challenges in relation to ENPs.[28]

Against this backdrop of formal reviews, questions were being raised in policy circles and the NGO community as to what, for the purposes of regulation, was nanotechnology/ies? And what is a nanoparticle and/or a nanomaterial? While seemingly somewhat pedantic in nature, such questions served to illustrate the lack of universally agreed definitions of key terms, the differences in positioning between industry, scientists and the NGO community and the inability of the scientific community to draw a clear line between those (nanoscale) particles that display unique properties when compared to their macro-scale counterparts and those that do not.

Implicit here, too, were the questions of whether or not it was possible and/or practicable to define these key terms for regulatory purposes. And as to what would be the implications, if any, of incorporating such definitions into legislative instruments.

Despite a lack of consensus within, among others, scientific and academic communities as to how a 'nanomaterial' should be defined for regulatory and policy purposes (if, indeed it should be), in 2011 the European Commission moved forward with adopting its own definition. According to the Commission Recommendation of 18 October 2011, (2011/696/EU), a 'nanomaterial':

> means a natural, incidental or manufactured material containing particles, in an unbound state or as an aggregate or as an agglomerate and where, for 50% or more of the particles in the number size distribution, one or more external dimensions is in the size range 1 nm-100 nm (Article 2).

The Commission's definition is arguably broad, yet at the same time quite narrow. It captures 'natural, incidental or manufactured material containing particles', while at the same time using 100 nm as a sharp cut off point. It also differs from the definition proposed by one of the Commission's own expert scientific bodies. It is arguably not surprising, therefore, that the adoption of the definition drew considerable attention and concern from a number of stakeholders in the field.[29] Furthermore, the consideration of the definition in relation to applications for commercial peaceful uses of the technology is far

[28] Convention on the Prohibition of the Development, Production, Stockpiling and Use of Chemical Weapons and on their Destruction, opened for signature 3 September 1992, 1974 UNTS 45 (entered into force 29 April 1997) (Chemical Weapons Convention).

[29] A. D. Maynard, 'Don't Define Nanomaterials' *Nature*, 475(7354) (2011), 31; D. R. Boverhof, et al., 'Comparative Assessment of Nanomaterial Definitions and Safety Evaluation Considerations', *Regulatory Toxicology and Pharmacology*, 73(1) (2015), 137–150. See, also, Chemical Watch, *EU Consults on Nanomaterial Definition: Revision of Recommendation*

more advanced than in relation to military applications where no definition has been provided within the international weapons governance sphere (see Section 10.3).

The definition adopted by the Commission in the Recommendation is not absolute, with the Preamble noting that some flexibility may be needed in relation to materials and scale captured by specific legislative contexts (Preamble 16). Indeed, it is worthy of note that the definition of a nanomaterial adopted by the European Parliament and European Council in Regulation (EC) No 1223/2009 of the European Parliament and of the Council of 30 November 2009 dealing with cosmetic products[30] differs, for example, from the definition that was subsequently adopted by the European Parliament and Council in Regulation (EU) No 1169/2011 of the European Parliament and of the Council of 25 October 2011 dealing with the provision of food information to consumers.[31]

As set out in the Preamble: 'The definition including descriptors should therefore be subject to a review by December 2014 to ensure that it corresponds to the needs…' (Preamble 14). Despite the express mandate provided by the Recommendation, the public consultation phase of the review was not launched until September 2017.[32] The Commission's Roadmap for the review notes a number of problems/challenges with the current approach including, for example, that '[t]here are currently difficulties in directly applying the Recommendation in legislation'.[33] With public consultation still underway at the time of writing, it is unclear how the Commission may revise the definition.

Three Years Late (19 September 2017) https://chemicalwatch.com/59016/eu-consults-on-nanomaterial-definition.

[30] Article 2(k) of Regulation (EC) No 1223/2009 defines a 'nanomaterial' as 'an insoluble or biopersistant and intentionally manufactured material with one or more external dimensions, or an internal structure, on the scale from 1 to 100 nm'.

[31] Article 2(t) of Regulation (EU) No 1169/2011 defines an 'engineered nanomaterial' to mean:

> any intentionally produced material that has one or more dimensions of the order of 100 nm or less or that is composed of discrete functional parts, either internally or at the surface, many of which have one or more dimensions of the order of 100 nm or less, including structures, agglomerates or aggregates, which may have a size above the order of 100 nm but retain properties that are characteristic of the nanoscale.
>
> Properties that are characteristic of the nanoscale include:
>
> (i) those related to the large specific surface area of the materials considered; and/or
> (ii) specific physico-chemical properties that are different from those of the non-nano form of the same material […].

[32] European Commission, Revision of Commission Recommendation 2011/698/EU on the definition of nanomaterial (Directorate-General for Environment, 2017).

[33] Ibid., 2.

The approach taken by the Commission and European Parliament differs markedly from that of other jurisdictions who have, instead, for the most part opted to maintain the status quo for regulatory and policy purposes. It is unclear at this stage what barriers this divergence may generate, and to what degree the EU may seek to amplify the divergence in the short to medium term. What is clear, however, is that the EU is paying close attention to the regulation of ENPs and will continue to keep such governance in its line of sight.

10.3 GOVERNANCE FRAMEWORKS APPLICABLE TO THE USE OF NANOMATERIALS IN ARMED CONFLICT

The focus of this section is the use of nanotechnologies by the armed forces of States. Interest in the platform technology by the military is very real.[34] For now, many military items based on, or enhanced by, nanotechnologies remain on the drawing board or in prototype phase and are still difficult to apply on a large scale, such as invisibility cloaks,[35] artificial 'disease' agents that can rest harmlessly in victims' bodies until activated by an external signal,[36] or the use of optogenetics to remotely alleviate pain or fear.[37] Traditional thermobaric weapons are already in use that incorporate nanoparticles to enhance the blast and create a bigger vacuum (and more casualties) with a smaller weapon.[38] In the future, more sophisticated applications of ENPs involving neuroscience, or other alterations of human biology, are likely to be subtle, powerful and difficult to detect.[39]

[34] Defense Advanced Research Projects Agency, 'Atoms to Product: Aiming to Make Nanoscale Benefits Life-Sized' (Media Release, 22 August 2014) www.darpa.mil/news-events/2014-08/22 ; National Nanotechnology Initiative, NNI Supplement to the President's 2017 Budget (31 March 2016) www.nano.gov/2017budgetsupplement.

[35] H. Nasu, 'Nanotechnology and Challenges to International Humanitarian Law: A Preliminary Legal Assessment', *International Review of the Red Cross*, 94 (2012), 653–672.

[36] K. Hearn, 'Future Soldiers Could Get Enhanced Minds' (19 March 2001) *United Press International* www.geocities.ws/marksrealm/project457.html. This article describes the potential use of nanotechnology to enhance soldiers. The issue of the 'bioenhanced' soldier is gaining increasing attention, but is beyond the scope of the Chapter.

[37] K. Leins, 'Shining a Regulatory Spotlight on New Lasers: Regulation of the Use of Nanolaser Technologies in Armed Conflict', *Jurimetrics*, 56 (2016), 261–278.

[38] J. Bean, 'Enhanced Blast Weapons and Forward Medical Treatment', *US Army Medical Department Journal*, (April–June 2004), 48–51.

[39] B. Viswanath and S. Kim, 'Influence of Nanotoxicity on Human Health and Environment: The Alternative Strategies', *Review of Environmental Contamination and Toxicology*, 242 (2017), 61–104.

There are two types of international law pertaining to war. The first is *jus ad bellum*, largely but not exclusively based on the UN Charter and which limits the circumstances in which Nation States may lawfully resort to the use of force. Once an armed conflict is under way, whether initiated by a lawful or unlawful act, a different body of law, namely *jus in bello*, governs how armed conflicts may lawfully be fought. Increasingly, States[40] and courts[41] are also contemplating the ongoing application of some human rights simultaneously with *jus in bello*. The weapons and methods that are used during an armed conflict are not without legal limits.[42] These restrictions date back to prohibitions on the use of poisons and poisoned weapons by ancient Hindu law, to the Greeks and the Romans,[43] to the more recent ban in 1868 on certain explosive bullets,[44] and on the prohibition of the use of 'arms, projectiles, or material calculated to cause unnecessary suffering'.[45] Humanitarian concerns have more recently prompted either prohibitions of certain specific kinds of weapon, or regulation of the ways in which other categories of weapon may

40 V. Boulanin, *Implementing Article 36 Weapon Reviews in the Light of Increasing Autonomy in Weapon Systems* (November 2015) SIPRI Insights on Peace and Security www.sipri.org/sites/default/files/files/insight/SIPRIInsight1501.pdf. For further commentary see S. Casey-Maslen, N. Corney and A. Dymond-Bass, 'The Review of Weapons under International Humanitarian Law and Human Rights Law', in S. Casey-Maslen (ed.), *Weapons under International Human Rights Law* (Cambridge: Cambridge University Press, 2014).

41 *Legal Consequences of the Construction of a Wall in the Occupied Palestinian Territory (Advisory Opinion)* [2004] ICJ Rep 136, [106]. The Court found that '[a]s regards the relationship between international humanitarian law and human rights law, there are three possible situations: some rights may be exclusively matters of international humanitarian law; others may be exclusively matters of human rights law; yet others may be matters of both these branches of international law'. Unfortunately, the Advisory Opinion did not go one step further to clarify exactly when these situations would arise.

42 *Legality of the Threat or Use of Nuclear Weapons (Advisory Opinion)* [1996] ICJ Rep 226, [74]-[87]. There are two main lessons that can be extrapolated from this decision in relation to weapons containing nanotechnology. The first lesson is that, either in the presence or absence of specific treaties governing their use, all weapons, even those that include nano-scale technology, are required to be compatible with the general principles and rules of LOAC. The Advisory Opinion noted that any weapons causing combatants unnecessary harm or uselessly aggravating their suffering would be prohibited, and that States do not have unlimited freedom of choice of means in the weapons they use. As to the general proposition that there are legal limits as to the weapons that may lawfully be used in armed conflict, note also Regulations Respecting the Laws and Customs of War on Land annexed to Hague Convention IV, The Hague, 18 October 1907, (Hague Regulations) Article 22 and API, Article 35(1).

43 J. Kelly, 'Gas Warfare in International Law', *Military Law Review*, 1 (1960), 3–5.

44 Declaration Renouncing the Use, in Time of War of Explosive Projectiles under 400 grammes Weight, opened for signature 29 November 1868, 138 CTS 297 (entered into force 11 December 1868) ('*Declaration of St Petersburg*').

45 Hague Regulations, Article 23(e). For a more modern formulation, see Article 35(2) of API where the focus is on the nature of the weapon.

lawfully be used.[46] For a more complete discussion of the law as it applies to the use of weapons in armed conflicts, the reader is referred to Chapter 2.

Prohibitions and principles require distillation into specific rules to ensure that they are respected by nations during the conflict. Esteemed disarmament specialist Goldblatt noted that the rules of conduct 'may not resist the pressure of military expedience' and that, for this reason, there is an 'intrinsic link' between the regulation of weaponry and the development of what is traditionally referred to as humanitarian law.[47] In other words, whether speaking of weapons, or the use of weapons in armed conflict, it is exactly in high-pressure military contexts that the laws and specific rules regarding specific weapons must be taken into account to ensure compliance with international law. These specific rules must embody and reflect the much older custom, general rules and legal principles.

Restrictions on the ways that wars may be fought between Nation States have consistently included specific weapons regulation by way of custom or other sources.[48] A treaty specific to a category of weapons that may have applicability, where one does exist, is the starting point for any analysis of legal regulation. For the purposes of considering the legality of weapons that include ENPs, the 1925 Geneva Gas Protocol,[49] the 1972 Biological

46 Some of these prohibitions are captured in the Convention on Prohibitions or Restrictions on the Use of Certain Conventional Weapons Which May be Deemed to be Excessively Injurious or to Have Indiscriminate Effects and Protocols I-III, opened for signature 10 October 1980, 1342 UNTS 137 (entered into force 2 December 1983). As of 7 March 2018, the Convention has 125 States Parties.

47 J. Goldblatt, *Agreements for Arms Control: A Critical Survey* (Stockholm International Peace Research Institute, 1982), 89:

> 'All laws of war suffer from one common weakness: the rules of conduct established for belligerents in time of peace may not resist the pressure of military expedience generated in the course of hostilities, and the attempts to 'humanise' war may sometimes prove futile. The danger that the weapons prohibited may, under certain circumstances, be resorted to – as has occurred on several occasions – will not disappear if these weapons remain in the arsenals of States. Hence the intrinsic link between the development of the humanitarian laws of war and progress in the field of disarmament.'

48 What constitutes international law is defined in the Vienna Convention on the Law of Treaties, opened for signature 23 May 1969, 1155 UNTS 331 (entered into force 27 January 1980), Article 38 ('VCLT'). See also J.-M. Sorel, 'Article 31 General Rule of Interpretation', in O. Corten and P. Klein (eds.), *The Vienna Conventions on the Law of Treaties: A Commentary* (Oxford: Oxford University Press, 2011), 804, 807–808.

49 Protocol for the Prohibition of the Use of Asphyxiating, Poisonous or Other Gases, and of Bacteriological Methods of Warfare, opened for signature 17 June 1925, 94 LNTS 65 (entered into force 8 February 1928).

Weapons Convention (BWC),[50] the 1993 Chemical Weapons Convention (CWC),[51] and the relevant Protocols to the CCW, including Protocol I prohibiting non-detectable fragments,[52] Protocol III regulating the use of incendiary weapons,[53] and Protocol IV prohibiting blinding laser weapons all may have potential relevance.[54] Beyond the specific treaties is a considerable body of customary international law, some of which is enshrined in existing treaties, and some of which is not. Customary law is defined as:

> State practice, including that of States whose interests are specially affected...both extensive and virtually uniform in the sense of the provision invoked; and...in such a way as to show a general recognition that a rule of law or legal obligation is involved.[55]

Needless to say, the applicable treaties and customary international law rules were formed well before the emergence of many of today's technologies including nanotechnologies. As noted previously, technology is almost always ahead of the law,[56] and the use of technology by the military is no different (as evidenced by, for example, the use of gas on the battlefields of Ypres, or the use of the nuclear bomb on Hiroshima and Nagasaki). As important as the ability of the law to respond is the ability for the law to be predictable and stable. If new laws were required for every innovation, a complete lack of

[50] Convention on the Prohibition of the Development, Production and Stockpiling of Bacteriological (Biological) and Toxin Weapons and on Their Destruction, opened for signature 10 April 1972, 1015 UNTS 163 (entered into force 26 March 1975) (Biological Weapons Convention).

[51] Chemical Weapons Convention.

[52] Protocol on Non-Detectable Fragments (Protocol I) to the Convention on Prohibition or Restrictions on the Use of Certain Conventional Weapons Which May be Deemed to be Excessively Injurious or to Have Indiscriminate Effects, 10 October 1980, 1342 UNTS 171 (entered into force 2 December 1983).

[53] Protocol on Prohibitions or Restrictions on the Use of Incendiary Weapons (Protocol III) to the Convention on Prohibition or Restrictions on the Use of Certain Conventional Weapons Which May be Deemed to be Excessively Injurious or to Have Indiscriminate Effects, opened for signature 10 October 1980, 1342 UNTS 171 (entered into force 2 December 1983).

[54] Protocol (IV) on Blinding Laser Weapons to the Convention on Prohibitions or Restrictions on the Use of Certain Conventional Weapons Which May be Deemed to be Excessively Injurious or to Have Indiscriminate Effects, opened for signature 13 October 1995, 1380 UNTS 370 (entered into force 30 July 1998) ('Protocol IV to the CCW').

[55] *North Sea Continental Shelf (Federal Republic of Germany v Denmark) (Merits)* [1969] ICJ Rep 3, 44.

[56] G. E. Marchant, B. R. Allenby and J. R. Herkert, (eds.), *The Growing Gap between Emerging Technologies and Legal-Ethical Oversight: The Pacing Problem*, vol. 7 (New York: Springer, 2011); G. E. Marchant, 'The Growing Gap between Emerging Technologies and the Law', in G. E. Marchant, B. R. Allenby and J. R. Herkert (eds.), *The Growing Gap between Emerging Technologies and Legal-Ethical Oversight* (New York: Springer, 2011), 19–33.

stability would result – even if it were even possible to keep up with recent rapid technological changes. As has already been noted, this challenge will most likely be better faced by technology than the lawyers themselves in coming years.[57] Furthermore, in contrast to laws governing commercially applied ENPs, developing the law governing the use of ENPs in armed conflict requires consensus among States for the adoption of treaty provision.

The rapid development of ENPs has not gone unnoticed by States or those working in the field of weapons regulation and disarmament. In 2002, in his role as leader of the Stockholm Institute Peace Research Institute (SIPRI) Chemical and Biological Warfare project, Zanders noted challenges to the Organisation for the Prohibition of Chemical Weapons (OPCW) mandate, singling out nanotechnologies in that 'the growing overlap between chemistry and biotechnology as well as the design and production processes based on nanotechnology may be of particular concern'.[58] In 2004, the Arms Control Association recognised these issues, and suggested the following:[59]

> Many of the international legal tools to prevent the development of these weapons are already in place, notably the [BWC] and the [CWC], which together ban military use of all of the weapons imagined here ... Consideration should also be given to a new convention that would prohibit the non-consensual manipulation of human physiology, to support and extend the provisions of the CWC, BWC and international humanitarian law.

A very early proposal for an agreement at the international level was articulated by Abbott et al. in 2006,[60] and further expanded upon in 2010.[61] Other calls have been made by experts in the field, noting that 'strong international action is needed to assess the threats from the new age of biological techniques'.[62] Despite these calls for action, there appear to be virtually no visible links between the military approaches and the lessons already learned in the

57 Richard Susskind, *The Future of Law: Facing the Challenges of Information Technology* (Oxford: Clarendon Press 1996).

58 J. P. Zanders, 'The Chemical Weapons Convention and Universality: A Question of Quality over Quantity?', *Disarmament Forum*, 4 (2002), 23.

59 M. Wheelis, 'Will the 'New Biology' Lead to New Weapons?', *Arms Control Today*, 34(6) (2004), 6.

60 K. W. Abbott, G. E. Marchant and D. J. Sylvester, 'A Framework Convention for Nanotechnology?' *Environmental Law Reporter*, 36 (2006), 10931–10942.

61 K. Abbott, Douglas S. and Garry M., 'Transnational Regulation: Reality or Romanticism?', in G. Hodge, D. Bowman and A. Maynard (eds.), *International Handbook on Regulating Nanotechnologies* (Cheltenham: Edward Elgar, 2010).

62 Malcolm Dando, 'Find the Time to Discuss New Bioweapons', *Nature*, 535 (2016), 9.

commercial sector relating to multi-party approaches for governing nanotechnologies.

In its more recent statement to the Eighth Review Conference of the BWC in 2016, the International Committee of the Red Cross (ICRC) suggested that the 'Review Conference establish an effective mechanism for assessing the implications of developments in science and technology for the BWC'.[63] This issue as to the applicability of the BWC to developments in technology is one of which States are acutely aware, but one where relationships with developments in the commercial sector require significant strengthening, to ensure consistent accountability and that advances in contemporary scientific knowledge are adequately taken into account.

While further consideration may be given to additional regulation, it is helpful to remember that either the BWC or the CWC will apply to any future use of any nanotechnologies in armed conflict (or in peacetime) where these developments result in toxicity or disease.[64] If nano-robots, or other nano-based delivery systems, were to be used for malevolent ends to apply biological substances, or to work as toxins or poisons, they would be prohibited under the Geneva Gas Protocol and the BWC. If the nanoscale science were to use the toxic properties of chemicals, then this use would fall foul of the customary international law prohibition on chemical weapons. Materials that might be capable of dual use would likely fall under the existing exceptions under the CWC if they are intended for purposes not prohibited by the Convention.[65] Nonetheless, statements by States in international fora that nanomaterials are included in existing prohibitions would help further to strengthen the existing governance framework.

In addition to the juridical question is the issue of ensuring compliance given the size and nature of ENPs. Chemicals under the CWC are divided into categories via 'Schedules', depending on their risk of use in violation of the CWC. Schedule 1 lists those chemicals which present a high risk of undermining the CWC if used. These include precursor chemicals used to produce

63 International Committee of the Red Cross, *Biological Weapons Review Conference: ICRC Statement* (8 November 2016) www.icrc.org/en/document/biological-weapons-review-conference-statement-icrc. The ICRC noted that States must keep up with 'fast-moving scientific and technological developments and their potential risks in order to prevent the development and use of biological weapons while ensuring that biological research for peaceful and beneficial purposes remains unhindered'.

64 The 1925 Geneva Protocol prohibits the use of 'asphyxiating, poisonous or other gases, and of all analogous liquids, materials or devices' and 'bacteriological methods of warfare'. The BWC and CWC, in effect, prohibit all use, transfer, stockpiling etc. of chemical and biological weapons.

65 Chemical Weapons Convention, Articles II (9), IV.

nerve agents, or mustard agents, and are identified by name and the CAS numbering system. Schedule 2 lists chemicals that are not usually produced in large quantities for non-military purposes. Schedule 3 specifically lists dual-use chemicals which may be a risk, but which also have legitimate purposes and are widely produced.[66] The three Schedules refer to the chemicals by CAS number and do not differentiate between those at the nanoscale and the same chemical manufactured at the non-nanoscale. Whether included in the Schedules or not, if used for purposes contrary to the Convention, even if compliant with the Schedules, the use of nanomaterials as a toxin or poison would be prohibited. However, for verification purposes, as was mentioned in relation to civilian applications, chemicals permitted at the regular scale might be toxic at the nanoscale, and verification regimes need to reflect this. Just as with the example of graphene given in connection with the commercial sector, it is important to understand that ENPs exhibit different toxicity profiles to their microscale counterparts, and may even vary depending on their shape and density at the nanoscale.

However, even if it were practicable to adopt a similar schedule system for nanomaterials, provisions seeking to control their import, export and use would be virtually unenforceable, as minute undetectable quantities with potentially lethal effect would escape existing verification measures. It is this aspect of enforcing the Geneva Gas Protocol, the BWC and the CWC that poses some of the biggest challenges, and where lessons learned from the definitions and regulatory efforts in the commercial sector may shine some light.

The scientific community is slowly developing a better understanding of the potential toxicity of materials at the nanoscale. As more information comes to light about the long-term effects of ENPs on the human body and on the environment,[67] more consideration will need to be given to how to manage this kind of matter, particularly given its scale and the minute quantities that would suffice to contravene the legal frameworks in question if used for hostile purposes, bearing in mind the difficulty of tracing hostile use. Links need to be made in a timely manner between published research in industrial,

66 Chemical Weapons Convention, Annex on Chemicals, www.opcw.org/chemical-weapons-convention/annexes/annex-on-chemicals.

67 OECD, *OECD Chemical Studies Show Way Forward for Nanomaterial Safety* (9 June 2015) www.oecd.org/chemicalsafety/nanosafety/news-nanomaterial-safety.htm; K. Rasmussen, et al., 'Review of Achievements of the OECD Working Party on Manufactured Nanomaterials' Testing and Assessment Programme: From Exploratory Testing to Test Guidelines', *Regulatory Toxicology and Pharmacology*, 74 (2016), 147–160.

commercial and academic sectors and national ministries of defence, the OPCW and BWC Secretariat.

Another relevant source of international law governing the use of ENPs is the Martens Clause, originally included in the 1899 Hague Convention II,[68] that further reiterates the relevance of the Law of Armed Conflict (LOAC) to future technologies:

> In cases not covered by this Protocol or by other international agreements, civilians and combatants remain under the protection and authority of the principles of international law derived from established custom, from the principles of humanity and from the dictates of public conscience.[69]

This principle was discussed in great detail by the International Court of Justice in its *Advisory Opinion on the Legality of Nuclear Weapons*, where it was held to be 'an effective means of addressing rapid evolution of military technology'.[70] As succinctly stated by Meron, '[w]here there already is some legal basis for adopting a more humanitarian position, the Martens clause enables decision makers to take the extra step forward'.[71] Although some consider the Martens Clause to be the ultimate legal contingency plan, we argue that the Martens Clause underpins the existing legal framework, and supports existing law, in addition to providing guidance where legal provision is lacking.[72] It is imaginable that genetic modification, also an alteration at the nanoscale, would be hard to reconcile with customary law principles, and thus with the sentiments expressed in this Clause, if used as a means or method of warfare.

68 Convention (II) with Respect to the Laws and Customs of War on Land and its Annex: Regulations concerning the Laws and Customs of War on Land, opened for signature 29 July 1899, 187 CTS 429 (entered into force 4 September 1900) ('Hague Convention (II)'), preamble. See also Convention (IV) with Respect to the Laws and Customs of War on Land and its Annex: Regulations concerning the Laws and Customs of War on Land, signed 18 October 1907, 205 CTS 277 (entered into force 26 January 1910) ('Hague Convention (IV)'), preamble.

69 Protocol Additional to the Geneva Conventions of 12 August 1949, and relating to the Protection of Victims of International Armed Conflicts (Protocol I), opened for signature 8 June 1977, 1125 UNTS 3 (entered into force 7 December 1979) ('API'), Article 1(2).

70 *Legality of the Threat or Use of Nuclear Weapons (Advisory Opinion)* [1996] ICJ Rep 226, [84]. In a slightly different context, Schmitt argues that the Martens clause only applies in the absence of treaty law in Michael Schmitt and Jeffrey Thurner, 'Out of the Loop – Autonomous Weapon Systems and the Law of Armed Conflict', *Harvard National Security Journal*, 4 (2013), 231, 275.

71 T. Meron, 'The Martens Clause, Principles of Humanity and Dictates of Public Conscience', *The American Journal of International Law*, 94(1) (2000), 78.

72 R. Sparrow, 'Ethics as a Source of Law: The Martens Clause and Autonomous Weapons' on *International Committee of the Red Cross Blog* (14 November 2017) http://blogs.icrc.org/law-and-policy/2017/11/14/ethics-source-law-martens-clause-autonomous-weapons/.

10.4 CONSIDERATION OF GOVERNANCE OF MILITARY USE OF BIOLOGICAL AND CHEMICAL NANOTECHNOLOGIES

As early as 2001 some authors forecast that specific challenges would be faced in attempting to regulate military applications of nanotechnology that would use both biological and chemical properties, with a recommendation to 'update' the BWC and CWC.[73] As early as 2003, Reynolds was addressing the challenging aspects of nano-robotics, noting that 'these nanodevices would not suffer from the constraints facing living organisms – they would not have to be made of protein or other substances readily extractable from the natural environment, nor would they have to be capable of reproducing themselves'.[74] Reynolds relies heavily on Eric Drexler's earlier work to extrapolate potential issues for regulation. Bowman and Hodge expressed similar concerns in that:

> nano-bots may in the future travel through the blood-stream seeking and killing off cancer cells, or may assist with the regeneration of healthy cells. At the opposite extreme, it may also be possible to use nano-bots for military purposes to detect motion in a field and transmit signals many miles away, or to achieve 'programmable' genocide.[75]

Review conference mechanisms help to ensure the continuing relevance of treaties, and provide additional guidance where the technology, such as nanoscale science, is moving quickly. The mere fact that the original drafters could not have foreseen the use of nanoscale science does not render the BWC or the CWC obsolete. More importantly, language contained in these treaties applies equally both to materials at the conventional scale and the nanoscale.

Further consideration of the appropriate legal regime may be required for ENPs that can asphyxiate without a chemical action (similar to asbestos fibres, many of which have at least one dimension at the nanoscale), or assemblies of atoms (nano-robots) that can behave like a pathogen.[76] Mathews suggests that the BWC and the CWC:

[73] G. Reynolds, 'Environmental Regulation of Nanotechnology: Some Preliminary Observations', *Environmental Law Reporter*, 31(6) (2001), 10681.

[74] G. Reynolds, 'Nanotechnology and Regulatory Policy: Three Phases', *Harvard Journal of Law and Technology*, 17(1) (2003), 179, 185 citing E. Drexler, *Engines of Creation* (Oxford: Oxford University Press, 1990), 56–63, as his reference for the technology.

[75] D. Bowman and G. Hodge, 'A Small Matter of Regulation: An International Review of Nanotechnology Regulation', *Columbia Science and Technology Law Review*, 8 (2007), 5.

[76] R. Mathews, 'Chemical and Biological Weapons', in R. Liivoja and T. McCormack (eds.), *Routledge Handbook of the Law of Armed Conflict* (London: Routledge, 2016), 229–230.

> prepare statements for endorsement at their respective Review Conferences which confirm that all such novel materials are captured by the prohibition of CBW, even if there are arguments that a particular material may not be covered by the definitions of BW and CW contained in BWC and CWC.[77]

Mathews further suggests that in the absence of such clarification, such substances would nevertheless be captured by the language of the 1899 and 1907 Hague Regulations which prohibit 'poison and poisoned weapons'. Mathews' approach is practical and in accordance with the existing customary international law prohibitions on poisons and chemical weapons. Arguing about whether a substance at the nanoscale acts on its chemical or physical properties (which often cannot be differentiated at the nanoscale) will seldom change the outcome of the conclusion of a legal review: weaponisation of nanoparticles targeting humans by any means flies in the face of existing customary international law standards regarding chemical or biological matter at regular scale,[78] and, depending on the characteristics of the weapon, potentially contravenes Protocol I of the CCW prohibiting non-detectable fragments,[79] Protocol III of the CCW regulating the use of incendiary weapons,[80] and Protocol IV of the CCW prohibiting blinding laser weapons.

10.5 CONSIDERATION OF ADEQUACY OF GOVERNANCE OF HOSTILE USE NANOTECHNOLOGIES

A substantial and growing body of literature exists that focuses specifically on the use of nanotechnologies by the armed forces. These types of analysis fall loosely into two camps. The first consists of those calling for a strengthening of the existing legislation.[81] The second consists of those calling for greater

77 Ibid.

78 If nanoparticles are used to target objects, they are subject to general legal principles, all of which remain in effect regardless of the size of the matter in question; the relevant principles and rules are discussed in more detail in Chapter 2.

79 Protocol on Non-Detectable Fragments (Protocol I) to the Convention on Prohibition or Restrictions on the Use of Certain Conventional Weapons Which May be Deemed to be Excessively Injurious or to Have Indiscriminate Effects, 10 October 1980, 1342 UNTS 171 (entered into force 2 December 1983).

80 Protocol on Prohibitions or Restrictions on the Use of Incendiary Weapons (Protocol III) to the Convention on Prohibition or Restrictions on the Use of Certain Conventional Weapons Which May be Deemed to be Excessively Injurious or to Have Indiscriminate Effects, opened for signature 10 October 1980, 1342 UNTS 171 (entered into force 2 December 1983).

81 J. Altman, *Military Nanotechnology: Potential Applications and Preventive Arms Control* (London: Routledge, 2006); J. Matsuura, *Nanotechnology Regulation and Policy Worldwide* (Boston and London: Artech House, 2006); M. Kosal, *Nanotechnology for Chemical and Biological Defense* (New York: Springer, 2009); K. Abbott, D. Sylvester and G. Marchant,

regulation, or new laws that specifically regulate the use of ENPs for military purposes.[82]

Enquiry and analysis has been undertaken by leading scientific commentators including Altmann, Gubrud and Kosal,[83] who focus on the potential applications and risks of the science for the armed forces.[84] Altmann and Gubrud suggest that advances in nanotechnologies:

> will probably require fundamental changes in the international system, particularly strengthening of law and political institutions, including international criminal law, and reorienting of the military mission from warfighting to organizing cooperative security.[85]

At the same time as changes based on nanotechnology emerge, other evolving technologies, in conjunction, will have a profound impact on society and how it functions as a whole. These other technologies include AI, computer networks, automated systems and drone capacities, to name just a few. It is important to consider the impact of ENPs within the broader context of developments in these other fields. More specifically, consideration should also be given to a new convention that would prohibit the non-consensual manipulation of human physiology, to support and extend the provisions of the CWC, BWC and international humanitarian law so that perceived gaps cannot be misused. Consider, for example, the way in which the 'sole object' language in the Hague Declaration (II) on the Use of Projectiles the Object of Which is the Diffusion of Asphyxiating or Deleterious Gases enabled the use

'Transnational Regulation: Reality or Romanticism?', in G. Hodge, et al. Maynard (eds.), *International Handbook on Regulating Nanotechnologies*; H. Nasu and T. Faunce, 'Nanotechnology and the International Law of Weaponry: Towards International Regulation of Nano-Weapons', *Journal of Law, Information and Science*, 22 (2009), 21–54, https://law.anu.edu.au/sites/all/files/users/u9705219/nanoweapons_int_law_0.pdf.

82 B. R. Allenby, 'Are new Technologies Undermining the Laws of War?', *Bulletin of the Atomic Scientists*, 70(1) (2014), 21; R. Pinson, 'Is Nanotechnology Prohibited by the Biological and Chemical Weapons Conventions?' *Berkeley Journal of International Law*, 22 (2004), 278; Reynolds, 'Environmental Regulation'.

83 M. Kosal, *Nanotechnology for Chemical and Biological Defense* (New York: Springer, 2009).

84 J. Altmann and M. Gubrud, 'Risks from Military Uses of Nanotechnology – The Need for Technology Assessment and Preventive Control' (Paper presented at Nanotechnology – Revolutionary Opportunities and Societal Implications, 3rd JOINTEC-NSF Workshop on Nanotechnology, Lecce, Italy, 31 January – 2 February 2002) 2–3; Jürgen Altmann, *Military Nanotechnology: Potential Applications and Preventive Arms Control* (Routledge, 2006).

85 Altman and Gubrud, 'Risks from Military Uses of Nanotechnology', 2–3, with the same concerns reiterated in J. Altmann, 'Preventing Hostile and Malevolent Use of Military Technology after 15 Years of the US National Nanotechnology Initiative', in M. Martellini and A. Malizia (eds.), *Cyber and Chemical, Biological, Radiological, Nuclear, Explosive Challenges* (New York: Springer, 2017), 49–72.

of combined effects munitions that employed such gases during World War I.[86] The purpose of any such extension in the scope of the BWC and CWC would be to seek to ensure that nanotechnology substances analogous to chemical or biological weapons but that fall outside the current definitions in both the BWC and CWC are nevertheless prohibited.[87]

Kosal explores the relationships among technology, strategy and governance, suggests an urgent need to contemplate measures to counter the use of biological and chemical ENPs given the likelihood of their potentially hostile use, and notes the difficulties of regulating, managing, tracking and policing their use. Referring to the CWC, Kosal points out the absence of domestic export control legislation and the lack of updates to the schedules of the CWC since 1997, arguing that further attention should be paid to these existing requirements to avoid misuse of nanomaterials.[88] Kosal also argues that the use of nanotechnologies may have profound implications for offensive and defensive strategic considerations, given the nature of ENPs and the difficulties confronting tracking and detecting them.[89] Kosal's approach is a practical one, ensuring that the existing multilateral framework is implemented in a timely and effective manner, rather than calling for new laws to address what is already well regulated under existing international treaties.

Wallach's approach is a similarly pragmatic one, in which he argues that 'functional equivalence' prohibits ENPs being used in any way that is not compliant with the existing legal framework. If nanomaterials are 'functionally

[86] Parties were to 'abstain from the use of projectiles *the sole object of which* is the diffusion of asphyxiating or deleterious gases.' Hague Declaration (II) on the Use of Projectiles the Object of Which is the Diffusion of Asphyxiating or Deleterious Gases, opened for signature 29 July 1899, 187 CTS 453 (entered into force 4 September 1900) (emphasis added).

[87] In 1915, Germany used canisters to release chlorine gas at the Battle of Ypres, relying heavily on winds to disperse their contents. Germany later argued that they did not use projectiles (which was technically true), and therefore were complying with the specific prohibition on projectiles. Later, when the Germans did switch to projectiles, Germany argued that the purpose of the projectiles was to injure personnel with their shrapnel and that the gas contained therein was incidental. Despite Germany's semantic justifications regarding their behaviour, they were held to have been in violation of the prohibition, although no individual was ever held responsible. The Commission on the Responsibility of the Authors of the War and on Enforcement of Penalties nonetheless included the 'use of deleterious and asphyxiating gases' 'as one of the war Crimes committed by Germany during World War I': 'Commission on the Responsibility of the Authors of the War and Enforcement of Penalties', *American Journal of International Law*, 14(1/2) (1920), 95, 115. H. Levie, 'Humanitarian Restrictions on Chemical and Biological Weapons', *University of Toledo Law Review*, 13 (1982), 1192 submits that this may have been incorrect, as the effect of the chlorine gas was totally reliant on 'fickle prevailing winds' and that the cylinders 'had not been made available for this purpose'.

[88] Kosal, *Nanotechnology*, 128.

[89] Ibid.

equivalent' to microscale materials prohibited under the Geneva Gas Protocol, BWC or CWC, then it can be argued that they remain prohibited regardless of their size if their function is equivalent to one that is prohibited.[90] This approach relies on the function of matter, rather than focusing on its size. Poisoning, or any hostile use of biological or chemical matter, regardless of size, is clearly prohibited under the existing law. The prohibition on poisoning does not specify the scale of the poison and includes all poisonous matter.

Altmann's approach regarding modification of the legal framework, like that of Kosal and Wallach, is equally compelling, noting that changing the wording of the CWC will be too arduous and suggesting clarification that:

> [t]oxic substances that are not of biological origin or are not produced by biological systems would not count as toxins, but would fall under the CWC, i.e. be prohibited if directed against humans or animals.[91]

Altmann's suggestion to clarify the definitions, rather than to redraft the text, makes sense because, at least in his view, the text already provides adequate oversight. It just needs clarification. Such clarifications are currently being considered by review bodies under both the BWC and the CWC.

The law governing the military use of ENPs that come within the definitions of biological weapons or of chemical weapons already exists. The repeated call for revised legislation by authors such as Reynolds to 'update' the BWC and CWC, without providing any suggestions as to how this could be done at best ignores the existing legal framework, and, at worst, undermines it.

Leading commentators such as Allenby,[92] Pinson[93] and Reynolds[94] have collectively suggested that the existing law is not adequate; that the law is outdated; that we need to renegotiate LOAC; or that we need to create new law to respond to the rapidly changing technology. We argue, however, that just as in the commercial context, existing law is not irrelevant. The relevant legal framework governs existing and emerging technologies but there may be a need for strengthening of understandings and of definitions. We also argue that semantics should not be used in a way that may undermine the principles of the existing law.

[90] D. Marcus, 'Famine Crimes in International Law', *American Journal of International Law*, 97(2) (2003), 245, 262–264. Marcus proposes that international law should extend to man-made famines as 'famines are often functionally equivalent to genocide': 248.

[91] Altmann, *Military Nanotechnology*, 171–172.

[92] Allenby, 'Are New Technologies Undermining the Laws of war?'.

[93] Pinson, 'Is Nanotechnology Prohibited'.

[94] Reynolds, 'Environmental Regulation'.

Finally, the legal requirement to review all new or modified weapons includes consideration of all limitations on how the new or modified weapon may be used. An effective review would require some common understandings about what risks (if any) the inclusion of ENPs poses,[95] including the effect on enemy combatants – in order to assess whether the superfluous injury/unnecessary suffering principle would be breached–and as to the effect on the environment. As ENPs are increasingly used across all sectors of society,[96] it can only be imagined that this increase is being mirrored, if not further accelerated, in non-commercial consumer products, such as military investment and applications. However, just as trade secrets may prohibit effective oversight in the commercial world, the cloak of secrecy shrouding development of new military capabilities makes knowledge and governance of inclusion of ENPs in military applications virtually impenetrable.

Nasu and Faunce suggest that 'current weapons development programmes are embedded in the existing weapons manufacturing infrastructures and integrated weapons systems based on conventional technologies', and that 'the weapons development programmes are by and large driven by the 'needs' of troops ... rather than by the 'possibilities' that nanotechnology may present'.[97] In other words, military interest often drives the development of the science. Sparrow refers to the 'depressingly high percentage of nanotechnology research ... funded by the military and ... directed towards military applications',[98] noting that 'insofar as any arms race is a

95 *API*, Article 36. For more detailed weapons review guidance, see International Committee of the Red Cross, 'A Guide to the Legal Review of New Weapons, Means and Methods of Warfare: Measures to Implement Article 36 of Additional Protocol I of 1977', *International Review of the Red Cross*, 88(864) (2006), 931; V. Boulanin, *Implementing Article 36 Weapon Reviews in the Light of Increasing Autonomy in Weapon Systems* (November 2015) SIPRI Insights on Peace and Security www.sipri.org/sites/default/files/files/insight/SIPRIInsight1501.pdf. For a more detailed discussion of the timing and complexity of Article 36 reviews pertaining to nanoenhanced weapons, means, or methods of warfare, see R. Liivoja, K. Leins and T. McCormack, 'Emerging Technologies of Warfare', in R. Liivoja and T. McCormack (eds.), *Routledge Handbook of the Law of Armed Conflict* (Routledge, 2016), 603, 620.

96 *Project on Emerging Nanotechnologies*, The PEW Charitable Trusts www.pewtrusts.org/en/archived-projects/project-on-emerging-nanotechnologies.

97 H. Nasu and Robert McLaughlin (eds.), *New Technologies and the Law of Armed Conflict* (The Hague: Asser Press, 2014) 145.

98 R. Sparrow, 'The Social Impacts of Nanotechnology: An Ethical and Policy Analysis', in A. Maynard and J. Stilgoe (eds.), *The Ethics of Nanotechnology, Geoengineering and Clean Energy* (Routledge, 2017), 179. Sparrow notes that referring to military research as having great 'spin-offs' 'neglects the fact that the purported benefits of this research could almost always have been secured more efficiently and quickly by civilian research, had the funding for such research been available'.

zero-sum game, military applications of nanotechnology have no social benefits'.[99] The Institute for Soldier Nanotechnologies at MIT is just one example, for which the 'mission is to help the Army dramatically improve the survivability of the Soldier by working at and extending the frontiers of nanotechnology through basic research and the transitioning of promising outcomes of that research in collaboration with our Army and industry partners'.[100]

Three major issues arise with these collaborations. One is not new, namely the issue of potential dual use.[101] The second is the issue of the requirement to review new weapons. The third, already touched upon, is the issue of secrecy surrounding developments that undermine or challenge the ability for new means or methods of warfare to be effectively reviewed.

Although nanotechnologies promise many interesting applications for the armed forces, for the purposes of applying the law of armed conflict, such 'nano-enhancements' do not necessarily raise new or novel legal issues. Better energy storage, more rapid computations, lower power consumption, or lighter and better materials with hydrophobic or hydrophilic properties do not present any immediate challenges to the existing legal framework. Some developments, however, may pose challenges, and it is these developments that will be explored in the next section.

99 Ibid.

100 John Joannopoulos and Francis Davis, *Institute for Soldier Nanotechnologies* (18 June 2015), Massachusetts Institute of Technology, http://web.mit.edu/annualreports/pres15/2015.18.06.pdf 1:

> This mission includes not only decreasing the weight that soldiers carry but also improving blast and ballistic protection, creating new methods of detecting and detoxifying chemical and biological threats, and providing physiological monitoring and medical treatment. The ultimate goal is to help the Army create integrated systems of nanotechnologies that combine high-tech protection and survivability capabilities with low weight, increased comfort, improved performance, and better compatibility with the end user... Approximately 30 faculty members from a dozen MIT academic departments, as well as more than 100 graduate students and postdoctoral associates, participate in ISN research.

The Institute for Soldier Nanotechnologies website breaks these areas of research into three, namely lightweight, multifunctional nanostructured materials and fibres, soldier medicine, and blast and ballistic threats.

101 J. B. Tucker (ed.), *Innovation, Dual Use, and Security: Managing the Risks of Emerging Biological and Chemical Technologies* (Cambridge, MA: Massachusetts Institute of Technology Press, 2012).

10.6 TIMING AND CONTENT OF THE ARTICLE 36 REVIEW

Before any technology is deployed in armed conflict, the legality of its use is required to be confirmed by legal review. Article 36 of Additional Protocol I requires that:

> [i]n the study, development, acquisition or adoption of a new weapon, means or method of warfare, a High Contracting Party is under an obligation to determine whether its employment would, in some or all circumstances, be prohibited by this Protocol or by any other rule of international law applicable to the High Contracting Party.[102]

The first important point to make is that not all research or advances in science and technology require review under Article 36 of Additional Protocol I. Determining when these advances constitute the 'study' or 'development' of a new weapon, means or method of warfare' is the first challenge. The LOAC only applies to a weapon, means or method of warfare, not advances in science and technology generally. As a result, it is only when the technologies discussed above are overtly under consideration as a means or method of warfare that they would be subject to this review process.[103]

Although not all research and advances in science have to be reviewed, the question of the timing of the review is becoming more and more complicated. Increasingly research in emerging technologies is the subject of military funding. States parties to API have an obligation to ensure that research that is directed at the development of a weapon or method of warfare is the subject of an Article 36 review. Even though the final decision regarding legality of a weapon will rest with the lawyers, it is also incumbent upon scientific experts to provide accurate information regarding the new technologies, and for those undertaking such weapon reviews to determine whether such developments comply with applicable international law. Where regulatory frameworks have been developed for ENPs in the commercial sector, it would be normal to assume that these same frameworks would also govern research and production under military auspices. Potentially applicable human rights frameworks applicable during armed conflict, and to military operations falling short of armed conflict, such as internal riot control, would also need to be considered in relation to technologies or systems intended for armed forces use in such

[102] API, Article 36. For more detailed weapons review guidance, see International Committee of the Red Cross, 'A Guide to the Legal Review'.

[103] For a more detailed discussion of the timing and complexity of Article 36 reviews pertaining to nano-enhanced weapons, means, or methods of warfare, see Liivoja, et al. 'Emerging Technologies of Warfare' , 620.

circumstances. One such example would be the use of optogenetics, which uses nanolasers to control emotional responses. It is imaginable that the use of optogenetics to control emotional responses would violate the human right to not perpetrate torture, as well as potentially violate the human right to health and mental health, and as such would be prohibited.[104]

In 2016, the Nobel Peace Prize for Chemistry was awarded for 'the design and synthesis of molecular machines'.[105] The development of 'nano-robots' is advancing quickly, and their uses multiply, from cleaning the ocean[106] to cleaning arteries,[107] but they raise urgent governance issues if used by armed forces. The ability to manipulate nanomaterials, coupled with an increased knowledge of the brain, may offer potential future applications of enhancement and degradation that will require careful consideration under the existing legal framework, and potentially require consideration of new governance mechanisms.[108]

In 2006, Doswald-Beck made note of the possibility of genetic modification being used as a method of warfare, and the attention that it had garnered from States. She noted that it was prohibited in law and that States had already been motivated in 2007 to take steps to prevent its use in armed conflict:[109]

> Mention must be made of a potential new method of warfare that is already prohibited in law but that could have horrific effects if developed, namely,

[104] S. Casey-Maslen, N. Corney and A. Dymond-Bass, 'The Review of Weapons under International Humanitarian Law and Human Rights Law', in S. Casey-Maslen (ed.), *Weapons under International Human Rights Law* (Cambridge: Cambridge University Press, 2014).

[105] *The Nobel Peace Prize in Chemistry 2016*, Nobelprize.org www.nobelprize.org/nobel_prizes/chemistry/laureates/2016/.

[106] L. Zyga, *Microbots can Clean Up Polluted Water* (11 April 2016) Phys Org https://phys.org/news/2016-04-microbots-polluted.html.

[107] H. Brueck, *This Tiny Robot Team Could Help Stop the No. 1 Killer in America* (12 January 2016) Fortune http://fortune.com/2016/01/12/artery-clearing-nanobot/.

[108] 'Fundamental questions that neuroscientists have previously approached with classical biochemical and electrophysiological techniques can now be addressed using optogenetics. The term optogenetics reflects the key programme of this emerging field, namely, combining optical and genetic techniques. With the already impressively successful application of light-driven actuator proteins such as microbial opsins to interact with intact neural circuits, optogenetics rose to a key technology over the past few years.' G. Dugué, W. Akemann and T. Knöpfel, 'A Comprehensive Concept of Optogenetics', in T. Knöpfel and E. Boyden (eds.), *Optogenetics: Tools for Controlling and Monitoring Neuronal Activity* (New York: Elsevier, 2012).

[109] Since 2006, some States, including the USA and the UK, have permitted modifications of germlines under some circumstances. L. Doswald-Beck, 'Humanitarian Law in Future Wars', in M. Schmitt and L. Green (eds.), *The Law of Armed Conflict: Into the Next Millennium* (Newport, RI: Naval War College, 2006) 139.

genetic weapons. The spectre of this as well as of new and obviously preliminary developments in bio-technology has already motivated States to begin negotiations for the development of verification methods for the Biological Weapons Convention.

Genetic modification, using a new technique called CRISPR-Cas9, was listed as the breakthrough of the year for 2015. CRISPR-Cas9 uses bacteria to modify genes. The use of a part of a bacterium for hostile purposes, i.e. not for 'prophylactic, protective or other peaceful purposes',[110] would clearly be prohibited under the BWC. In November 2017, new research indicated that genetic modification could be undertaken without using any bacteria, but rather simply by using roaming nanoparticles.[111] Such use would be prohibited to the armed forces under the CWC, but clarifying this at a Review Conference would arguably ensure that this common understanding is reflected and secure compliance.[112]

10.7 ABILITY TO INDEPENDENTLY PROVIDE REVIEW WITH NARROW FIELD OF EXPERTS

The second point concerns the designated authority responsible for the review. Traditionally, it has been legal advisors within the military wishing to adopt the new means or method of warfare, or military or civilian lawyers, or committees, who were tasked with the responsibility for undertaking any weapon review. As the science becomes more sophisticated, unilateral review processes will become less and less effective. In the past, these reviewers have been obliged to seek technical advice about the intended use and expected effects of a proposed weapons system. For example, if you imagine a kinetic weapon, with a particular kind of ammunition, the questions to be asked in the context of a review (as discussed in more detail in Chapter 2) would be reasonably straightforward. Now, however, the science is often multidisciplinary and increasingly complex. Some have suggested that those setting the requirements for projects need to have their initial specifications assessed to determine whether they can satisfy an Article 36 review, as the time of development

[110] Biological Weapons Convention, Article 1.

[111] H. Yin, et al., 'Structure-Guided Chemical Modification of Guide RNA Enables Potent Non-Viral in Vivo Genome Editing', *Nature Biotechnology* (2017), nature.com/articles/doi:10.1038/nbt.4005.

[112] Chemical Weapons Convention, Articles 1(b) and 1(c).

or procurement of the weapon, particularly if it involves advanced technologies (in which we would include nanoparticles) may be too late.[113]

Weapons reviewers are required to have an extensive legal understanding as well as an understanding of the 'engineering design, production and testing (or validation) methods and the ways in which the weapon might be employed on the battlefield'.[114] The practicalities of finding expertise in both the technical complexity of nanomaterials *and* an understanding of the nuances of international legal obligations will pose an increasingly significant challenge. The legal review of emerging technologies in general, and those involving nanomedicine in particular, raise similar issues.[115] Boothby notes that nanotechnologies are 'likely to mean that the weapons reviewer is likely to need more scientific information'.[116] We would suggest that beyond this, scientists also need to have a basic knowledge of the international legal frameworks to ensure that they are not developing science for purposes contrary to international law.

With the increasing complexity and rapid advances in technology, evaluations of new weapons systems will require intimate collaboration between the traditional reviewers, the lawyers, and the scientists and engineers, who must collectively be able to foreshadow potential applications of their use by the military that would be prohibited under the existing legal framework. In addition, lawyers conducting reviews will need to sufficiently familiarise themselves with the germane science and engineering not only to analyse compliance with the law, but also to be able to ask scientists the right questions to ensure that reviews comply with national legal obligations.

The increased use of nanomaterials in fabrics, materials, armour and munitions raises immediate questions about their long-term environmental impact, and the subsequent potential effects on public health. Any review

[113] Personal communication with Rob Sparrow, PREMT Conference, Future Wars and Public Conscience, 29–30 May 2017. Sparrow raised this point in the context of a conversation about the requirement for all technology to be reviewed by experts who were 'outside' of those proposing the technology, but who were familiar with it, to avoid conflicts of interest. It was discussed that with emerging technologies in general, finding experts with this level of objectivity and distance from their colleagues to perform such reviews is becoming increasingly difficult.

[114] A. Backstrom and I. Henderson, 'New Capabilities in Warfare: An Overview of Contemporary Technological Developments and the Associated Legal and Engineering Issues in Article 36 Weapons Reviews', *International Review of the Red Cross*, 94 (2012), 483.

[115] D. Bowman and J. Gatof, 'Reviewing the Regulatory Barriers for Nanomedicine: Global Questions and Challenges', *Nanomedicine*, 10(21) (2015), 3275.

[116] W. Boothby, 'How Will Weapons Reviews Address the Challenges Posed by New Technologies?', *Military Law and the Law of War Review*, 52(1) (2013), 58.

would need to consider long-term assessments of harm, taking into account both the characteristics of each nanomaterial and the intended circumstances of use. Maynard et al. regard two of the 'five grand challenges' for the platform technology to be to 'learn how harmful nano-materials are' and to 'evaluate the impact of engineered nanomaterials from cradle to grave'.[117] The million-dollar question is how to adequately protect people and the environment from the potential risks that ENPs may pose if used in armed conflict, with each individual ENP and its potential use requiring consideration.

10.8 CONCLUSIONS AND RECOMMENDATIONS

From this chapter's comparative analysis of the way that ENPs are regulated in the commercial and military sectors, three main points become clear. Firstly, although the use of ENPs is heralded as new and is said to challenge existing legal frameworks, this chapter has illustrated that existing governance frameworks, even if not explicitly identifying ENPs as being within their purview, still apply. From the discussion of both the commercial and military uses of ENPs, there is an existing framework governing their use. In some instances, clarification of definitions, or explicit inclusion of ENPs in existing frameworks, will need to occur. This is particularly relevant in the context of treaties and customary law governing the military use of weapons.

Secondly, the governance of ENPs in the commercial sector is bringing together multiple protagonists alongside the developments themselves. Policy makers, safety regulators, industry and NGOs have responded both traditionally and creatively to reduce potential risks to individuals and the environment. This approach of not just regulation, but also risk management frameworks, codes of conduct and reporting apparatus, crossing disciplines, sectors and geographical boundaries, has profoundly and permanently changed the way that governance is considered for ENPs. The authors suggest that this multi-disciplinary, multi-pronged approach to governance is the way forward to govern ENPs in the military sector as well. Greater transparency is required to ensure that multi-disciplinary, multi-pronged governance approaches are possible. Public scrutiny is required for ongoing research, and legal reviews of military use of ENPs should occur each time that the weaponisation of an ENP, whether a new ENP or an existing substance, is studied.

[117] Maynard, et al., 'Safe Handling', 267. Maynard and Bowman regard two of the 'five grand challenges' for nanotechnology to be to 'learn how harmful nano-materials are' and to 'evaluate the impact of engineered nanomaterials from cradle to grave'.

Thirdly, this type of approach will be needed not only for ENPs, but for other technologies as they rapidly advance. Developments surrounding the interplay between multiple technologies, such as AI, manipulating data and the Internet of Things, have made the need to consider the efficacy of existing legal frameworks for effectively governing the individual technologies even more urgent. As the technologies develop rapidly and become more complex, the legal frameworks that govern them require clarification, and, at times, may require additional governance and/or interpretative tools, given their seemingly novel nature.[118] While this chapter discusses the significance, and the complexity, of regulating new and emerging technology in our lives, more significantly it highlights the importance of experts across applications communicating to find creative approaches to provide adequate governance frameworks. While the conversation regarding these developments has extended to automated systems and cyber and their use in armed conflict, the consideration of the governance of nanotechnologies has lagged behind that of its counterpart technologies in the military context. Governance frameworks of ENPs in the commercial sector provide ample learning opportunities for governance frameworks in the military context. The time for the governance of military ENPs to be strengthened is the present. Even more broadly, governance frameworks for emerging technologies should not be considered as stand-alone, and lessons learned from previous governance frameworks–in this case ENPs–may provide an excellent precedent for governance approaches in relation to future technologies.

[118] A. Shahani, *Microsoft President Urges Nuclear-Like Limits on Cyberweapons* (16 May 2017) National Public Radio www.npr.org/sections/alltechconsidered/2017/05/16/528555400/microsofts-president-reflects-on-cyberattack-helping-pirates-and-the-nsa.

11

Naval Technologies

WOLFF HEINTSCHEL VON HEINEGG

11.1 INTRODUCTION

As regards manned naval surface and subsurface platforms there are no technological developments that give rise to doubts as to their compliance with international law in general or the law of armed conflict in particular. While the basic design and tasks have remained largely unchanged, there are, of course, numerous developments that have contributed, or are about to contribute, to an increased efficiency of naval platforms in combat operations. These developments include stealth,[1] multi-function weapons[2] and enhanced radars.[3] Other relevant improvements include enhanced countermeasure resistance, improvements in the ability of vessels to defend against various threats, advanced defence systems in particular against drones and missiles,[4] reduced noise performance and improved security of data transmission.[5] Further developments in naval platform technologies have enabled considerable reductions in vessels' crew numbers, which are also of economic relevance.[6] In the same vein, the development of naval weapon systems also aims

[1] For the newest generation of destroyers (DDG 1000) see US Navy, Fact File, available at www.navy.mil/navydata/fact_display.asp?cid=4200&tid=900&ct=4.

[2] See Report to Congress on US Navy Destroyer Programs, October 5, 2017, available at https://news.usni.org/2017/10/05/report-congress-u-s-navy-destroyer-programs-2. See also R. O'Rourke, 'Navy Frigate (FFX[X]) Program: Background and Issue for Congress', 8 December 2017, available at https://fas.org/sgp/crs/weapons/R44972.pdf; www.naval-technology.com/projects/standard-missile-6/.

[3] *Ibid.*

[4] 'Report to Congress on Navy Laser, Railgun and Hypervelocity Projectiles', 26 February, 2018, available at https://news.usni.org/2017/12/05/report-congress-navy-laser-railgun-hypervelocity-projectiles-2.

[5] K. Wong, 'Australia Seeks Guidance on Quantum Technologies', 11 January 2018, available at www.janes.com/article/77006/australia-seeks-guidance-on-quantum-technologies.

[6] See US Navy, Fact File.

at cost reduction (e.g., low-cost munitions[7]), improved efficiency (e.g., anti-ship missiles and naval strike missiles[8]) and enhanced resilience against threats (e.g., multi-shot mine neutralisation[9], laser weapons, railguns, hypervelocity projectiles[10], and laser warning systems[11]).

From the perspective of international law, in particular of the law of armed conflict, those developments do not require a reconsideration of the existing rules or the adoption of new rules because they merely increase the efficiency of capabilities whose compliance with international law has already been settled. The defensive systems of surface platforms, sophisticated and advanced as they may be, are designed to protect them against various threats and, thus, to achieve the protection of the vessels and their crews. There are, of course comparatively new threats naval platforms are exposed to, such as cyber operations in the form of GPS spoofing,[12] which have – for the time being – been adequately addressed in the Tallinn Manual 2.0,[13] discussed in Chapter 5 of the present volume and which will not be dealt with here. Offensive naval weapons – guns, missiles, torpedoes and naval mines – are lawful means of warfare.[14] The fact that, and the manner in which, those means of naval warfare have become more sophisticated rather enhances their legality under the existing law. Accordingly, the lawfulness of their employment merely depends on whether the way in which they are used is in compliance with the rules and principles of targeting law. The technological developments that have enhanced their precision and discrimination capabilities are such that there can be no serious doubt about the capability of these weapons to comply with the principle of distinction.[15]

7 M. Eckstein, 'Navy Releases Details of New FFG(X) Guided Missile Frigate Program in Request to Industry', 10 July 2017, available at https://news.usni.org/2017/07/10/navy-releases-details-of-new-ffgx-guided-missile-frigate-program-in-request-to-industry.

8 See www.naval-technology.com/projects/naval-strike-missile-nsm/.

9 G. Fein, 'Surface Navy 2018: Saab's NuMNS Extends MCM Operations', 11 January 2018, available at www.janes.com/article/77005/surface-navy-2018-saab-s-mumns-extends-mcm-operations.

10 'Report to Congress on Navy Laser'.

11 See https://saab.com/naval/electronic-warfare/naval-laser-warning-systems/nlws/.

12 See, *inter alia*, D. Hambling, 'Ships Fooled in GPS Spoofing Attack Suggest Russian Cyberweapon', available at www.newscientist.com/article/2143499-ships-fooled-in-gps-spoofing-attack-suggest-russian-cyberweapon/.

13 M. N. Schmitt (gen. ed.), *Tallinn Manual 2.0 on the International Law Applicable to Cyber Operations*, Rules 45 *et seq*. (Cambridge: Cambridge University Press, 2017).

14 L. Doswald-Beck (ed.), *San Remo Manual on International Law Applicable to Armed Conflicts at Sea*, paras 78–92 (Cambridge: Cambridge University Press, 1995).

15 See W. Heintschel von Heinegg, 'The International Law of Mine Warfare at Sea', 23 *Israel Yearbook of Human Rights*, 53–76 (1993).

However, two developments deserve closer attention: unmanned maritime systems and underwater infrastructure (including military devices and systems deployed at sea or on the seabed). The former are increasingly used for a variety of civilian and military purposes without international law providing adequate rules as to their status, their navigational rights and obligations, or their legality as means of warfare. The latter are partly regulated by international law insofar as there are rules relating to submarine cables and pipelines. States have a right to lay such cables and pipelines and States are obliged to enact domestic legislation dealing with the civil and criminal liability of their nationals and of vessels flying the State's flags for causing damage to submarine cables. In particular, due in large measure to their important role in carrying much Internet traffic, submarine cables have increasingly come into focus as potential targets for States and most likely also for terrorists. Moreover, their protection has not been adequately addressed by current law. The law of armed conflict merely regulates the destruction of submarine telegraph cables connecting occupied with neutral territory and is silent where other cables and pipelines are concerned. Other underwater infrastructure and devices have remained unregulated under international law although they are of increasing importance for civilian and military purposes.

11.2 UNMANNED MARITIME SYSTEMS

During the last decade and a half international attention has tended to focus on unmanned aerial vehicles. The remarkable development and increased use of unmanned maritime systems (UMS)[16] have remained largely unnoticed for some time. Recently, this has changed. Nevertheless, with the exception of a few publications,[17] the legal issues concerning UMS have remained largely unaddressed.

[16] In addition to the references below see Office of the Under Secretary of Defense for Acquisition, Technology, and Logistics, *Study on Next-Generation Unmanned Undersea Systems* (Washington, DC, October 2016), available at www.acq.osd.mil/dsb/reports/2010s/Next-Generation_Unmanned_Undersea_Systems.pdf; K. Wong, 'Sea Sentinels: Chinese Unmanned Maritime Systems Gain Traction', available at www.janes.com/images/assets/219/77219/Sea_sentinels_Chinese_unmanned_maritime_systems_gain_traction.pdf.

[17] A. Norris, *Legal Issues Relating to Unmanned Maritime Systems*, available at www.iqpc.com/media/1002182/50661.pdf; M. N. Schmitt and David S. Goddard, 'International Law and the Military Use of Unmanned Maritime Systems', 98 International Review of the Red Cross 1–26 (August 2016).

According to the US DoD, 'UMS comprise unmanned maritime vehicles (UMVs), which include both unmanned surface vehicles (USVs) and unmanned undersea vehicles (UUVs), all necessary support components, and the fully integrated sensors and payloads necessary to accomplish the required missions'.[18] Although those definitions seem to suggest that UMVs/USVs/UUVs are but components of UMS it would not be correct to hold that they do not qualify as 'systems' in their own right because they are composed of various subsystems.[19] Since a distinction between 'systems' and 'vehicles' does not prove helpful it seems to be correct to consider the terms 'UMS' and 'UMV' as largely synonymous.

UMS comprise surface, semi-submersible and undersea vehicles of various sizes.[20] They are either remotely operated/controlled or – at least partly – 'autonomous'. Many of the systems in use today are remotely operated or controlled but they 'capitalize on automation in extreme circumstances, such as a lost link condition, to automatically perform a pre-programmed set of instructions.'[21] It may be added that, at present, there is no unmanned maritime system which may be considered as fully autonomous in the strict sense, although, e.g., gliders may travel considerable distances without being remotely controlled/operated.

UMS can perform a wide variety of missions or tasks. It is important to bear in mind that UMS, in particular UUVs, are today used for the performance of the following important civilian/non-military tasks: (1) Offshore oil and gas missions; (2) undersea cable deployment and inspection; (3) commercial salvage; (4) aquaculture; (5) science missions, such as oceanography and marine archaeology;[22] (6) protection of the marine environment;[23] and (7) maritime search and rescue.[24] According to the DoD Roadmap, current military missions performed by UMS (USVs and UUVs) include 'mine warfare, mine neutralization, reconnaissance, surveillance, hydrographic surveying, environmental analysis, special operations, and oceanographic

[18] US Department of Defense, *Unmanned Systems Integrated Roadmap FY 2013–2038*, p. 8.

[19] For example, the major UUV subsystems are: the pressure hull, the hydrodynamic hull, ballasting, power and energy, electrical-power distribution, propulsion, navigation and positioning, obstacle avoidance, masts, manoeuvre control, communications, locator and emergency equipment, and payloads. See National Defense Research Institute, *A Survey of Missions for Unmanned Undersea Vehicles*, p. 46 *et seq.* (RAND Corp. 2009).

[20] For an overview see DoD 'Roadmap', p. 8.

[21] *Ibid.*, p. 15.

[22] RAND, p. 41 *et seq.*

[23] A robot called 'COTSbot' has been developed, which is capable of selecting targets (i.e. starfish that pose a threat to coral reefs) and using lethal force (by injecting bile) without any human involvement. See *New Scientist* (16 January 2016), p. 35 *et seq.*

[24] See www.intechopen.com/books/search-and-rescue-robotics-from-theory-to-practice/unmanned-maritime-systems-for-search-and-rescue.

research'.[25] Similarly, the UUV Master Plan identifies nine specific mission categories and prioritises them as follows:[26] (1) Intelligence, surveillance, and reconnaissance (ISR); (2) mine countermeasures (MCM); (3) anti-submarine warfare (ASW); (4) inspection/identification; (5) oceanography; (6) communication/navigation network node (CN3); (7) payload delivery; (8) information operations (IO); and (9) time-critical strike (TCS).[27]

In view of the foregoing and in order to distinguish them from advanced naval weapon systems, UMS may be defined as: self-propelled or remotely-navigated craft that are normally recoverable and designed to perform functions at sea by operating on the surface, semi-submerged or undersea. UMS either are remotely operated or remotely controlled, or they perform some or all of their functions independently from a human controller or operator.

11.2.1 *Legal Status of UMS*

Although UMS navigate in sea areas they cannot, without difficulty, be considered 'ships'. The international law of the sea lacks a uniform definition of the term 'ship'. UNCLOS[28] uses the terms 'vessel' and 'ship' interchangeably, without providing a definition of either term. In other treaties a 'ship' is defined as either (1) 'any sea-going vessel of any type whatsoever, including floating craft, whether self-propelled or towed by another vessel, making a sea voyage';[29] (2) 'a vessel of any type whatsoever operating in the marine environment [...] includ[ing] hydrofoil boats, air-cushion vehicles, submersibles, floating craft and fixed or floating platforms';[30] (3) 'every description of water craft, including non-displacement craft and seaplanes used or capable of being used as a means of transportation on water';[31] or (4) 'any self-propelled sea-going vessel used in international seaborne trade for the transport of goods, passengers, or both'.[32] While those treaties do not necessarily prohibit treating UMS as vessels or ships, in particular UNCLOS seems to be limited to

[25] DoD Roadmap, p. 109.

[26] US Department of the Navy, *The Navy Unmanned Undersea Vehicle (UUV) Master Plan* (November 2004), p. 16.

[27] For a detailed description of those missions/tasks see RAND Report, p. 13 *et seq*.

[28] United Nations Convention on the Law of the Sea of 10 December 1982, 1833 UNTS 3, 397.

[29] 1962 Amendments to the 1952 Convention for Prevention of Pollution of the Sea by Oil, 11 April 1962, Annex, Article 1(1), 600 UNTS 332.

[30] International Convention for Prevention of Pollution from Ships of 1973, as amended by the 1978 Protocol of 17 February 1978, 1340 UNTS 61, 184.

[31] Convention on the International Regulations for Preventing Collisions at Sea (COLREGS), of 20 October 1972, 1050 UNTS 15.

[32] United Nations Convention on Conditions for Registration of Ships (not in force), Article 2(4).

manned platforms. In view of these difficulties, States seem to be hesitant to characterise UMS as ships or vessels and rather refer to them as 'craft'.[33] It is quite probable that a considerable number of States is not prepared to either recognise as, or to assimilate, UMS to ships/vessels. Nevertheless, States are not prevented from contributing to a progressive development of international law by, for instance, listing their military UMS in their registries of warships. Additionally, like-minded governments could agree on a joint statement according to which they consider military UMS to be warships. That might render a formal modification or amendment of the existing international law unnecessary. Thus, if States in general were to make such a determination, it would then be possible to characterise UMS as warships, if they are operated by the armed forces of a State for exclusively non-commercial governmental purposes and if they are either remotely controlled or have been pre-programmed by those who are under regular military discipline.

The fact that there is, at present, no agreement on whether UMS qualify as ships or warships may not be misunderstood as barring the application of the flag State principle or as denying sovereign immunity to military UMS. All UMS, whether civilian or governmental, do have a nationality. They either belong to nationals or corporations of a given State or they are operated by State agencies, including the armed forces. Accordingly, and if they are marked by the national emblem of a given State, they are subject to the exclusive jurisdiction of that State. If UMS are operated by the armed forces or any other government agency of a State, they may not necessarily qualify as warships or State ships, but in view of the fact that they either constitute State property or serve exclusively non-commercial governmental functions, they do enjoy sovereign immunity and may only be interfered with by other States in very exceptional circumstances (e.g., in an international armed conflict). Accordingly, it is correct to hold that 'USVs and UUVs engaged exclusively in government, non-commercial service are sovereign immune craft.'[34] It is important to note that an independent legal status of sovereign immunity applies to UMS operating independently from another platform. 'USV/UUV status is not dependent on the status of its launch platform.'[35] If a UMS is tethered to a controlling platform, it may be difficult to attach to the former an independent legal status. However, in view of the distinction between Remotely Operated Vehicles (ROVs) and Autonomous Unmanned Vehicles

[33] US Navy/US Marine Corps/US Coast Guard, *The Commander's Handbook on the Law of Naval Operations* (NWP 1–14M), paras 2.3.4–2.3.6 (Edition August 2017).

[34] NWP 1–14M, n. 33, para 2.3.6.

[35] *Ibid.*

(AUVs) becoming increasingly obsolete, the legal status of UMS should not necessarily be based on, or linked to, that of the controlling platform.

11.2.2 *UMS Operations in Different Sea Areas*

It is not entirely clear whether and to what extent coastal States are prepared to recognise independent navigational rights of foreign UMS in the sea areas in which they enjoy either territorial sovereignty (i.e. territorial sea, archipelagic waters) or certain functionally limited sovereign rights (i.e. Exclusive Economic Zone (EEZ) and continental shelf). It is, however, a fact that UMS do navigate in the sea areas of other States. According to NWP 1–14M, USVs and UUVs retain independent navigation rights[36] and they, thus, are considered as enjoying the same navigation rights as manned surface vessels and submarines. Other States have not (yet) made official statements to that effect although they in fact make use of UMS for governmental, scientific and commercial purposes. Hence, it is safe to conclude that UMS enjoy the right of freedom of navigation in the high seas and in the EEZ as well as the rights of innocent passage, transit passage and archipelagic sea lanes passage. Nevertheless, there remains a certain degree of uncertainty and States are advised to take the steps necessary for the recognition of independent navigational rights of UMS, either implicitly by listing them in their registries of ships/warships or explicitly by unilateral or joint statements.

In the context of navigational rights of UMS, there are two issues that require special attention. The first relates to UMS navigating in the territorial sea of another State while the second concerns their employment for the purposes of information gathering at sea.

As regards UMS operations in the territorial sea of another State, there should be no considerable disagreement about their obligation to comply with the conditions of innocent passage[37] and it should be similarly accepted that underwater UMS are required to navigate on the surface and to show their flag.[38] There is, however, no agreement on the consequences of military UMS not complying with those conditions. Some hold that the coastal State's options would be limited to a request for compliance and, if that request is disregarded, to a demand that the UMS leave the territorial sea immediately, as provided for in Article 30 UNCLOS. According to that position, non-innocent passage is not considered a violation of the coastal State's territorial

[36] Ibid., paras 2.5.2.5. (innocent passage) and 2.5.3.2 (transit passage).
[37] Article 19, UNCLOS.
[38] Article 20, UNCLOS.

sovereignty.[39] That position seems to be untenable. According to Article 2(1) UNCLOS, the 'sovereignty of a coastal State extends [. . .] to the territorial sea'. The right of innocent passage other States enjoy therein is but a limitation of that sovereignty and the coastal State is obliged to tolerate its exercise only insofar as passage is in fact innocent. Accordingly, 'a coastal State may take affirmative action in [. . .] its territorial sea to prevent passage that is not innocent, including where necessary, the use of force'.[40]

Information gathering at sea has always met with suspicion by certain coastal States. Such activity would not be in compliance with the right of innocent passage, transit passage or archipelagic sea lanes passage if it exceeds what is necessary for ensuring safety of navigation or for the normal mode of operation. According to Article 56(2) lit. (b)(ii) UNCLOS, coastal States have, in their Exclusive Economic Zone (EEZ), jurisdiction with regard to marine scientific research. Hence, information gathering, in the Exclusive Economic Zone (EEZ) of another State, may not amount to marine scientific research. UNCLOS does not define the term 'marine scientific research' (MSR), but it may be concluded from Article 243 UNCLOS that it refers to 'those activities undertaken in ocean space to expand scientific knowledge of the marine environment and its processes'.[41] A systematic interpretation of the different UNCLOS provisions leads to the conclusion that MSR must be distinguished from 'survey activities'.[42] Accordingly, MSR does not include hydrographic surveys,[43] including military surveys, or operational oceanography.[44] Military surveys in foreign EEZs have been conducted by numerous States, including Russia, Japan, Australia, South Africa, China, the United Kingdom and

39 For instance, James Kraska holds that 'vessels not in innocent passage may be foregoing their privileged status and right to be present in the territorial sea, but that does not necessarily mean they are acting in violation of the law of the sea or international law more broadly. These vessels, such as submerged submarines, are merely legally unprotected or unprivileged. If they are caught they may be required to leave the territorial sea.'

40 NWP 1–14M, para 2.5.2.1.

41 G. Walker (ed.), *Definitions for the Law of the Sea*, p. 241 (Leiden/Boston: Martinus Nijhoff, 2012).

42 See Articles 19 (2)(j), 21 (1)(g) and 40 UNCLOS. See also J. Ashley Roach and Robert W. Smith, *Excessive Maritime Claims*, 3rd edn ((Leiden/Boston: Martinus Nijhoff, 2012).

43 For the importance of hydrographic surveys and nautical charting, including electronic charting, see UNGA resolution A/RES/66/231 of 24 December 2011.

44 For an in-depth analysis of the different forms of marine data collection and for further references see Roach and Smith, n. 42, p. 413 *et seq*. See also R. (Pete) Pedrozo, 'Military Activities in the Exclusive Economic Zone: East Asia Focus', 90 International Law Studies (2014), 514–543, 525 *et seq*.

other NATO States.[45] Such surveys cannot be considered MSR and they are not covered by the exclusive sovereign rights enjoyed by coastal States pursuant to Article 56(1) lit. (b)(ii) UNCLOS. Rather, any form of marine data collection that is not covered by the term 'MSR' is a right granted under 'other pertinent rules of international law',[46] which may neither be impeded nor interfered with by the coastal State.[47]

11.2.3 *Compliance with the 'Rules of the Road'*

The COLREGS[48] or 'Rules of the Road' *inter alia* provide for steering and sailing rules but, in view of the Convention's definition of the term 'vessel',[49] it is far from clear whether the COLREGS apply to UMS. There are indications that some States appear 'to have implicitly recognized that at least some UMSs satisfy the broad COLREGS definition of 'vessel'',[50] but that would not justify the conclusion that all (civilian or military) surface UMS must navigate in compliance with the 'Rules of the Road'. Moreover, it may not be left out of consideration that State vessels and, thus, military and other State UMS are not as such bound by the COLREGS. As in the case of manned warships, it would, therefore, significantly add to clarity if States, either individually or within the IMO framework, made a statement identifying the exact scope of applicability of the COLREGS to UMS.

Apart from those general observations, it must be borne in mind that 'UMS command and control will be transmitted via radio frequency links between the control station and the unmanned system. For safe operations of an unmanned maritime system, highly reliable radio communications between the UMS and the maritime control station are required to support sense and avoid functions. In the end, unrestricted and autonomous unmanned maritime systems operations will rely on critical communications'.[51] Accordingly, governments as well as civilian entities operating UMS ought to undertake the

45 See the references by J. Kraska, *Maritime Power and the Law of the Sea* (Oxford: Oxford University Press, 2011), 270 *et seq.*

46 Article 58(2) UNCLOS.

47 See N. Klein, *Maritime Security and the Law of the Sea*, (Oxford: Oxford University Press, 2011), 217 *et seq.*

48 N. 31.

49 According to Rule 3 (a) COLREGS, the 'word "vessel" includes every description of water craft, including non-displacement craft, [Wing in Ground effects] WIG craft and seaplanes, used or capable of being used as means of transportation.'

50 Norris, *Legal Issues Relating to Unmanned Maritime Systems*, 49.

51 T. Kidd and S. Ward, 'A New Century for Unmanned Maritime Systems, October-December 2012, available at www.doncio.navy.mil/chips/ArticleDetails.aspx?ID=4195.

necessary efforts to integrate them, as far as operational requirements permit, into the global maritime communication network and/or to otherwise prevent their undue interference with innocent shipping. In view of the increased use of UMS for scientific, commercial or governmental purposes, the greatest possible efforts should be made in order to guarantee the safety of surface vessels.

11.2.4 *UMS and the Law of Naval Warfare*

Under the law of naval warfare, enemy military UMS qualify as lawful military objectives because, by nature, they make an effective contribution to the enemy's military action. The fact that their status as warships is not yet settled is irrelevant for their qualification as lawful targets. Civilian UMS, whether enemy or neutral, may become liable to attack if they, by location, use or purpose, make an effective contribution to the enemy's military action. Whereas enemy civilian UMS are always liable to capture as prize, neutral civilian UMS may be captured only if they are engaged in un-neutral service. While these findings are based on well-established rules of the law of naval warfare,[52] it is not entirely clear (1) whether military UMS may be employed for the exercise of belligerent rights; and (2) whether they comply with weapons law and targeting law.

11.2.4.1 UMS as Warships

At present, State practice as reflected in the increased development and procurement of military UMS[53] seems to suggest that military UMS are considered lawful means of warfare that may be employed for the exercise of the full range of belligerent rights. However, the uncertainty as to whether UMS can be assimilated to, or qualify as, ships could have wide-ranging consequences with regard to the legality of their employment in times of international armed conflict.[54] Since the 1856 Paris Declaration,[55] the exercise

[52] See *San Remo Manual*, n. 14, paras 40, 59–60, 65–66, 67–68, 135, 141, 146.

[53] See, inter alia, the references in footnote 14 *et seq.*

[54] The limitation of the exercise of belligerent rights to warships, as defined in Article 29 UNCLOS, does not apply in non-international armed conflict. See W. Heintschel von Heinegg, 'Methods and Means of Naval Warfare in Non-International Armed Conflicts', in K. Watkin and A. J. Norris, eds., *Non-International Armed Conflict in the Twenty-First Century*, 211–236, at 219 International Law Studies Vol. 88. Newport 2012).

[55] *Declaration Respecting Maritime Law*, Paris, 1 American Journal of International Law Suppl. 89–90 (1907).

of belligerent rights, i.e. attack, visit, search and capture, has been limited to warships, as defined in Article 29 UNCLOS and the corresponding customary law.[56] If military UMS are not classified as warships because they are unmanned, their use for the exercise of belligerent rights would probably be contrary to the law of naval warfare. This is yet another reason why States should clarify their status under both domestic and international law. However, pending a clarification of their legal status under the law of the sea and the law of naval warfare, UMS should be rendered identifiable as being operated by the regular armed forces of the respective State. Moreover, those controlling or pre-programming them should be under regular armed forces discipline in order to ensure compliance with the law of armed conflict. This is of particular relevance if UMS are used for the exercise of belligerent rights in high seas areas, i.e. outside neutral waters. Otherwise, neutral States and neutral vessels could consider the exercise of belligerent rights as either privateering or piracy, both of which are prohibited under long-standing rules of international law.[57]

11.2.4.2 UMS as Means of Warfare

Whereas the majority of UMS are used for intelligence, surveillance or reconnaissance operations (ISR) or for oceanography, some are designed for attack purposes, such as those employed for ASW, MCM or mine-laying. If and to the extent UMS are employed for the purposes of attack as defined in Article 49(1) API, they qualify as means of warfare[58] and their employment is subject to weapons law and targeting law.

According to Article 35(2) API, it is prohibited to employ means or methods of warfare of a nature to cause superfluous injury or unnecessary suffering and according to Article 35(3) API, it is prohibited to employ methods and means of warfare which are intended or may be expected to cause widespread, long-term and severe damage to the natural environment. In addition, Article 51(4) lit. b) and c) of API prohibits the employment of a means of combat which

[56] See, *inter alia*, *San Remo Manual*, n. 14, para 13(g). See further R. W. Tucker, *The Law of War and Neutrality at Sea*, p. (Washington, DC, 1955), 38 *et seq.*; L. Oppenheim, *International Law*, Vol. II, 7th edn *et seq.* (7th ed. by Hersch Lauterpacht, London 1952); C. John Colombos, *The International Law of the Sea*, p. 507 *et seq.* (6th ed., London 1967).

[57] Privateering has been prohibited by the 1856 Paris Declaration. Pirates have since times immemorial been considered 'enemies of mankind'. The crime of piracy is defined in Article 101 UNCLOS.

[58] As defined in the HPCR Manual on International Law Applicable to Air and Missile Warfare, Rule 1 (t) (Bern, 15 May 2009); in the following referred to as 'AMW Manual'.

cannot be directed at a specific military objective or whose effects cannot be limited as required by API. Apart from the rule on the natural environment, those prohibitions are customary in nature.[59] They are, however, not overly relevant for the employment of current military UMS as means of warfare. As regards the capability of being directed at a specific military objective, it must be borne in mind that UMS are equipped with weapons that use the same or similar technologies as, for instance, modern naval mines or torpedoes. They carry weapons that home onto targets that have been identified by magnetic, electromagnetic, acoustic or other signatures allowing for a sufficiently reliable identification of the targets as lawful military objectives. Regularly, the collateral effects on civilians and civilian objects of current naval weapons systems will not be excessive in relation to the military advantage anticipated. In this context it is important to take into consideration the environment in which they are employed, in particular the characteristics of saline waters, and the fact that they will in most cases be used against targets which are not surrounded by civilian or specially protected objects. Moreover, such considerations are not particularly relevant to the lawfulness of the means of warfare as such under weapons law but, rather, are relevant to their employment under the law of targeting. The same holds true with regard to the protection of the marine environment. No conventional naval weapons system in existence is to be expected to inflict on the marine environment the kind and degree of damage prohibited by Article 35(3) API. It is therefore safe to conclude that UMS are lawful means of warfare, whether they are remotely operated/controlled or whether they operate with a certain degree of autonomy. Of course, a UMS that operates with some degree of autonomy must be capable of being used in a manner that complies with targeting law.

If UMS are employed for attack, the same rules apply as in the case of manned platforms qualifying as means of warfare. The target must be a lawful military objective, expected collateral damage must not be excessive in relation to the military advantage anticipated and, finally, the required precautions in attack must be taken. With regard to the last of these, it must be borne in mind that, at sea, different standards apply than in land warfare. According to Article 57(4) API, military operations at sea are subject to 'all reasonable precautions'. Irrespective of the question whether the difference

[59] See Rules 5, 88 and 89 AMW Manual, n. 58. See also W. H. Boothby, *Weapons and the Law of Armed Conflict*, 2nd edn (Oxford: Oxford University Press, 2016), 21 *et seq.*, 46 *et seq.*, 60 *et seq.*, 74 *et seq.*

between the standards of feasibility and reasonableness is to be considered a 'tenuous nuance'[60] (a view not shared by the present author), it follows from the wording of Article 57(4) API that the standard of reasonableness only applies to naval attacks that may have an effect on the civilian population or civilian objects on land without being directed against targets on land. Sea-to-sea, sea-to-air and air-to sea operations whose effects may not extend to land continue to be governed by the customary obligation to take precautions in attack. Such precautions arguably may include the feasibility standard,[61] although the wording of Article 57(4) API seems to suggest that the standard is more liberal. Accordingly, and without prejudice to the duty of constant care,[62] the employment of UMS for purposes of attack must be in accordance with the following requirements:

- The system and/or operator must be able to verify the legality of the target.[63]
- The system and/or operator must be capable of determining the probability of collateral damage.[64]
  - If collateral damage may be expected to be excessive in relation to the military advantage anticipated, the attack may not be executed.[65]
  - If the collateral damage, though not excessive, is still to be expected, there is an obligation to minimize or avoid collateral damage by all feasible precautions.[66]
  - Unless circumstances do not permit, an advance effective warning must be given.[67]

The obligation to avoid or minimise collateral damage by the choice of a different weapon ('weaponeering') will regularly have no effect in view of the standard of 'feasibility' (or of a more liberal standard), because UMS do not, and probably will not, carry more than one type of weapon. The obligation to give an advance warning could come into effect, if UMS are employed for mine-laying purposes, in particular within the EEZ of a neutral State.[68]

[60] ICRC, *Commentary on the Additional Protocols of 8 June 1977 to the Geneva Conventions of 12 August 1949*, para 2230 (Geneva 1987).
[61] *San Remo Manual*, n. 14, para 46; *AMW Manual*, n. 58, Rules 30–39.
[62] Article 57(1) API.
[63] Article 57(2) lit. (a)(i) API; *San Remo Manual*, n. 14, para 46(b).
[64] *San Remo Manual*, n. 14, para 46(a).
[65] Article 57(2) lit. (a)(iii) API; *San Remo Manual*, n. 14, para 46(d).
[66] Article 57(2) lit. (a)(ii) API; *San Remo Manual*, n. 14, para 46(c).
[67] Article 57(2) lit. (c) API.
[68] *San Remo Manual*, n. 14, para 35.

11.2.5 *UMS and the Law of Neutrality*

For reasons of completeness, it may be added that in times of an international armed conflict the employment of UMS is subject to the law of maritime neutrality (unless the use of force has been authorised by a Chapter VII decision of the UN Security Council).[69] It is important to emphasise that the law of neutrality applies to all military UMS whether they are used for attack purposes or not. Accordingly, UMS may not engage in hostile actions in neutral waters or use neutral waters as a sanctuary or base of operations.[70] In the EEZ of a neutral State, they must be employed with due regard for the rights and duties of the coastal State.[71] The due regard rule also applies to the (hostile) use of UMS on the high seas.[72]

Although it is an unsettled issue, according to the position taken here, belligerent UMS enjoy the rights of transit passage and archipelagic sea lanes passage.[73] Accordingly, neutral States may not suspend, hamper, or otherwise impede those rights.[74] Belligerent UMS enjoy the right of innocent passage, unless the neutral coastal State has, on a non-discriminatory basis, conditioned, restricted or prohibited passage through its territorial sea.[75] The right of non-suspendable innocent passage may not be suspended.[76]

The parties to an international armed conflict must respect the sovereign immunity of neutral UMS, which are employed for non-commercial governmental purposes. Neutral civilian UMS may only be interfered with, if they qualify as lawful military objectives. It needs to be stressed that the capture of neutral civilian UMS under prize law will, if at all, apply in very exceptional situations only. Usually, UMS are not used for the transport of cargo that could constitute contraband.

These findings are, again, closely linked to the – unsettled – legal status of UMS and their navigational rights and obligations. Hence, the law of neutrality is yet another reason why those issues ought to be clarified as soon as possible.

69 *San Remo Manual*, n. 14, para 7; *AMW Manual*, n. 58, Rule 165.
70 *San Remo Manual*, n. 14, paras 14–16.
71 *San Remo Manual*, n. 14, para 34.
72 *San Remo Manual*, n. 14, para 36.
73 *San Remo Manual*, n. 14, para 28.
74 *San Remo Manual*, n. 14, para 29.
75 *San Remo Manual*, n. 14, para 19.
76 *San Remo Manual*, n. 14, para 33.

11.2.6 *UMS: Preliminary Conclusions*

The legality of the use of UMS, whether in times of peace or war, primarily depends on their legal status which, in turn, continues to be unsettled. Although it seems that States consider the use of UMS to be in accordance with both the law of the sea and the law of naval warfare, such implicit recognition is not sufficient and is counter-productive. The seizure of a US UMS by the Chinese Navy in 2016[77] must, of course, be understood in light of Chinese interests in, and illegal claims to, the South China Sea but the uncertainties as to the legal status of UMS and their navigational rights has most likely induced the Chinese Navy to disrespect the sovereign immunity of the US UMS. Accordingly, governments are urged to contribute to legal clarity which is important for safeguarding navigational and other rights of UMS, including their use as means of warfare in times of international armed conflict. Additionally, governments should take the necessary steps with a view to ensuring that the operation of UMS, including those operated for non-governmental purposes and whether civilian or military, does not unduly interfere with the navigational rights and safety of innocent shipping.

11.3 UNDERSEA INFRASTRUCTURE

The term 'undersea infrastructure' as understood here applies to submarine cables and pipelines as well as to infrastructure devices and systems deployed beneath the surface of the sea or on the seabed. While the laying or deployment of such undersea infrastructure does not seem to cause any considerable legal problems, their protection against undue interference has been, and continues to be, a neglected issue of international law.

11.3.1 *Relevance of Undersea Infrastructure*

Although the laying of undersea pipelines is a challenging and comparatively expensive undertaking such pipelines are often the best means for the transport of oil or gas over long distances. For instance, the pipeline connecting Nyhamna in Norway and Easington in England has a length of 1,166 km and it delivers 26 billion cubic meters of natural gas to the UK each year. Oil and gas

[77] 'Chinese Warship Seizes US Underwater Drone in International Waters', available at www.theguardian.com/world/2016/dec/16/china-seizes-us-underwater-drone-south-china-sea.

pipelines have been laid in all continents and are an important factor for energy security.[78]

Submarine cables include high voltage power cables, telegraph and telephone cables and, in particular, 'fibre-optic cables that carry over 95% of the world's international voice, data, and video traffic, including almost 100% of transoceanic Internet ocean traffic and, because of their bandwidth, the lower cost and longer lifespan [they] have surpassed satellites as the principal means of delivering international telecommunications traffic.'[79] According to other sources, '99 percent of the world's long-distance communications travel through fibre links. The remaining 1 percent are [...] satellite-based.'[80] It may be added that even communication networks that start with satellite systems (such as ships at sea) often are dependent on seabed cables to connect satellite ground stations with end users. Although used for commercial and governmental purposes, most of these cables and cable networks are owned and operated by consortia of private carriers. Many contemporary security and military operations, in particular the use of unmanned aerial systems, depend upon global network reliability. The necessary bandwidth can only be provided by submarine cables that provide fast, large and inexpensive connection.

Underwater systems and devices are used for a wide variety of civilian and scientific purposes, such as maintenance and operation of offshore installations, submarine pipelines and cables, preservation of the marine environment and protection against natural disasters (tsunamis). They also serve a wide variety of military or security purposes. Sensors and other devices, such as SOSUS,[81] have for a long time been used for the detection of submarines and for the protection of certain parts of the coastline against potential enemies, criminals or terrorists. Today, such systems have become increasingly complex and sophisticated. For instance, the PLUSNet system[82] 'consists of an

[78] For the location of the various pipelines, see the map provided by www.theodora.com/pipelines/world_oil_gas_and_products_pipelines.html.

[79] L. Carter, D. Burnett, S. Drew, G. Marle, L. Hagadorn, D. Bartlett-McNeill and N. Irvine, *Submarine Cables and the Oceans: Connecting the World*, at 8 UNEP-WCMC Biodiversity Series No. 31, ICPC / UNEP-WCMC (2009), available at www.iscpc.org/documents/?id=132; L. R. Wrathall, 'The Vulnerability of Subsea Infrastructure to Underwater Attack: Legal Shortcomings and the Way Forward', 12 *San Diego Int'l Law Journal* (2010–2011), 223–261, at 228. See also M. P. Green and D. R. Burnett, 'Security of International Submarine Cable Infrastructure – Time to Rethink?', in M. H. Nordquist, et al. eds., *Legal Challenges in Maritime Security*, (Dordrecht/Boston: Martinus Nijhoff, 2008), 557–583, at p. 559.

[80] M. Sechrist, *Cyberspace in Deep Water: Protecting Undersea Communication Cables By Creating an International Public–Private Partnership* (Harvard Kennedy School, 23 March 2010), available at http://belfercenter.hks.harvard.edu/files/PAE_final_draft_-_043010.pdf.

[81] SOSUS is the abbreviation for Sound Surveillance System.

[82] PLUSNet is the abbreviation for Persistent Littoral Undersea Surveillance Network.

undersea network of "sea gliders" and long-endurance Unmanned Undersea Vehicles (UUVs). The UUVs carry highly capable sensors and perform as autonomous vessels with long underwater dwell times. The sea gliders are smaller autonomous vessels that collect the UUV data, and return to the surface to transmit that data to a shore-based collection and processing station. The PLUS system is designed for easy deployment from any ship with a winch and crane and sufficient storage capacity'.[83] A similar concept is Seaweb, an acoustic communications network which uses through-water acoustic modems to connect up a large number of sensor nodes to a surface gateway node, allowing real-time digital communication between data gathering sensors and land-based command centres.[84] The development of those and other underwater surveillance systems is characterised by an increased use of mobile rather than fixed components. In the near future, the independent (autonomous) operation of UMS will be considerably enhanced by underwater charging stations that are based on resonant wireless power transfer.[85]

11.3.2 *Peacetime Legal Framework*

Despite the considerable vulnerability of undersea infrastructure vis-à-vis various threats,[86] including by States,[87] the existing international legal framework is surprisingly incomplete.

[83] See https://subseaworldnews.com/2013/10/28/persistent-littoral-undersea-surveillance-plus-system-completes-initial-in-water-testing/.

[84] For an in-depth description see Joseph Rice, 'Seaweb Acoustic Communication and Navigation Network', available at https://calhoun.nps.edu/bitstream/handle/10945/40396/rice_seaweb.pdf?sequence=1&isAllowed=y.

[85] See K. Owens, 'Wireless Underwater Charging Tech Will Bring a New Level of Autonomy for UUVs', September 1, 2017, available at https://defensesystems.com/articles/2017/09/01/navy-drone-uuv.aspx.

[86] See, inter alia, W. Heintschel von Heinegg, 'Protecting Critical Submarine Cyber Infrastructure: Legal Status and Protection of Submarine Communications Cables under International Law', in K. Ziolkowski (ed.), *Peacetime Regime for State Activities in Cyberspace* (Tallinn: NATO, CCD COE, 2013), 291–318.

[87] For instance, naval units are capable of intercepting data traffic routed through submarine fibre-optic cables. See 'New Nuclear Sub Is Said to Have Special Eavesdropping Ability', *New York Times*, 20 February 2005, available at www.nytimes.com/2005/02/20/politics/20submarine.html?_r=0. Recently, NATO was concerned about the operation of Russian warships in the vicinity of important submarine communications cables in the North Atlantic. See Michael Birnbaum, 'Russian Submarines are Prowling Around Vital Undersea Cables. It's Making NATO Nervous', available at www.washingtonpost.com/world/europe/russian-submarines-are-prowling-around-vital-undersea-cables-its-making-nato-nervous/2017/12/22/d4c1f3da-e5d0-11e7-927a-e72eac1e73b6_story.html?utm_term=.c8392188f79e.

11.3.2.1 Submarine Cables and Pipelines

The International law of the sea focuses on the right to lay undersea pipelines and cables and it regulates the relations between coastal States and the States laying them in a rather general manner, by distinguishing between the different sea areas.[88] Articles 113–114 deal with the breaking or injury of a submarine cable or pipeline, and Article 115 provides for indemnity for loss incurred in avoiding injury to a submarine cable or pipeline.

However, UNCLOS is silent on the legal status of submarine pipelines and cables and on the question as to whether and to what extent such cables and pipelines are subject to the jurisdiction of the States that own them or whose nationals have laid and operate them. Article 113 UNCLOS does not make it clear which other States also have jurisdiction over the breaking or injury to submarine cables and pipelines beyond the territorial sea. It is, however, questionable whether there is a presumption that only a non-flag State can exercise judicial jurisdiction against foreigners over the breaking of cables or pipelines outside its own maritime zones. Article 113 UNCLOS merely deals with the obligation of the flag State or the State of nationality of the perpetrator to penalise the breaking or injury of submarine cables or pipelines. This does not imply the exclusion of an exercise of jurisdiction in other than criminal matters. States have accepted the obligation to enact domestic criminal legislation because they agree that submarine cables and pipelines must be protected. Implementing the obligation to legislate may not be considered a waiver of exercising jurisdiction in other matters or of taking the measures necessary to protect submarine cables and pipelines against malicious interference. Hence, States having laid submarine cables and pipelines, or whose nationals have laid and/or operate them, retain the right to exercise their jurisdiction in accordance with well-established principles of international law, i.e. under the passive nationality and protective principle. The fact that the jurisdiction of coastal States is limited by the rights enjoyed within the EEZ does not mean that coastal States that have laid, or are connected to a submarine cable or pipeline, are no longer entitled to take protective measures. Finally, it would make little sense if, on the one hand, the right to lay submarine cables and pipelines is again and again emphasised and preserved but if, on the other hand, the States making use of that well-established right were legally barred from taking the necessary measures to protect them against threats. Hence, Article 113 UNLOS merely excludes the

[88] See the following provisions in UNCLOS: Article 51(2) – archipelagic waters; Article 58(1) – EEZ; Article 79 – continental shelf; Arts. 87(1) lit. (c), 112 – high seas.

exercise of criminal jurisdiction by States other than the flag State or State of nationality. Those States that have a valid interest in protecting 'their' submarine cables and pipelines may not only exercise jurisdiction in other than criminal matters, but may also take all measures necessary for their protection.

11.3.2.2 Undersea Systems and Devices

There are no provisions in UNCLOS or in other treaties explicitly addressing the installation of undersea systems and devices. Of course, within their territorial sea or, where applicable, archipelagic waters, costal and archipelagic States are entitled to install and operate such systems, whether civilian or military in nature. Coastal and archipelagic States may also install and operate undersea systems and devices, which are necessary for the exercise of their sovereign EEZ or continental shelf rights and duties, including marine scientific research. As in the case of submarine cables and pipelines, coastal States are entitled to maintain, repair and protect such undersea systems and devices.

The law is less clear with regard to the right of other than coastal States to install and operate such systems in the EEZ of other States. Insofar as they are used for marine scientific research they may not be installed and operated in the EEZ of another State. UNCLOS does not define the term 'marine scientific research', but it may be concluded from Article 243 UNCLOS that it refers to 'those activities undertaken in ocean space to expand scientific knowledge of the marine environment and its processes.'[89] A systematic interpretation of the different UNCLOS provisions leads to the conclusion that MSR must be distinguished from 'survey activities'.[90] Accordingly, MSR does not include hydrographic surveys,[91] including military surveys, or operational oceanography. Military surveys in foreign EEZs have been conducted by numerous States, including Russia, Japan, Australia, South Africa, China and NATO States. They cannot be considered MSR and they are not covered by the jurisdiction enjoyed by the coastal State in accordance with Article 56 (1)(b)(ii) UNCLOS. Rather, any form of marine data collection that is not covered by the term 'MSR' is a right granted under 'other pertinent rules of international law', which may neither be impeded nor interfered with by the coastal State. Hence, it is safe to conclude that survey activities and

[89] G. Walker (ed.), *Definitions for the Law of the Sea*, 241.

[90] See Articles 19 (2)(j), 21 (1)(g) and 40 UNCLOS.

[91] For the importance of hydrographic surveys and nautical charting, including electronic charting, see UNGA resolution A/RES/66/231 of 24 December 2011.

operational oceanography not qualifying as marine scientific research are lawful in a foreign EEZ according to customary international law and thus according to Article 58(2) UNCLOS, which explicitly refers to 'other pertinent rules of international law'. The States that installed and are operating such systems or devices, or whose nationals have installed and operate them, are entitled to maintain, repair and protect them. The above finding also applies to the continental shelf.

In the high seas the right to install and operate such systems and devices is recognised by customary international law, i.e. by 'other rules of international law' as provided for in Article 87(1) UNCLOS.

It needs to be stressed that the foregoing findings are not limited to civilian or scientific systems and devices. The peaceful uses clause in Article 88 UNCLOS, which according to Article 58(2) UNCLOS also applies in the EEZ, does not prohibit military uses of the seas that do not qualify as a use or threat of force.

11.3.3 *Status under LOAC*

11.3.3.1 Submarine Cables and Pipelines

The law of armed conflict is silent on submarine pipelines. Submarine cables are only dealt with insofar as they are connecting occupied territory with a neutral territory. For those submarine cables Article 54 of the 1907 Hague Regulations provides that they 'shall not be seized or destroyed except in the case of absolute necessity' and that 'they must likewise be restored and compensation fixed when peace is made.' It is, therefore, safe to conclude that submarine cables are not protected against seizure or destruction if they are connecting enemy territory, which is not occupied, with neutral territory. Of course, cables that only connect neutral territories would be protected by the law of neutrality.

The San Remo Manual provides: 'Belligerents shall take care to avoid damage to cables and pipelines laid on the sea-bed which do not exclusively serve the belligerents.'[92] According to the associated Explanations, 'cables and pipelines exclusively serving one or more of the belligerents might be legitimate military objectives.'[93]

Those rules and findings seem to correctly reflect the *lex lata* insofar as submarine pipelines and submarine high voltage cables are concerned. If they

[92] *San Remo Manual*, n. 14, para 37.

[93] *San Remo Manual* – Explanations (Cambridge: Cambridge University Press, 1995), p. 111.

qualify as lawful military objectives, they may be seized or destroyed, provided the principle of proportionality and the obligation to take feasible precautions are observed.

It is, however, doubtful whether those rules and findings also apply to submarine communications cables. Other than telegraphic cables, modern submarine communications cables are the backbone of global data traffic. While they may physically connect the territories of two States, it will only in rare circumstances be possible to determine that they are 'exclusively serving one or more belligerents' or one or more neutral States. Today's submarine communications cables are interconnected. Hence, data packages will travel over routes that are unpredictable. Accordingly, it is important to distinguish between submarine communications cables and other submarine cables.

11.3.3.2 Undersea Systems and Devices

The law of armed conflict is equally silent on the legal status of undersea systems and devices. Of course, if they directly contribute to the enemy's military action by nature, location, purpose or use, they are liable to destruction or capture, if that offers a definite military advantage. Hence, many of the military systems referred to above will qualify as lawful military objectives. The same holds true for civilian enemy and neutral systems and devices that directly contribute to the enemy's military action.

Whereas neutral systems and devices will be protected by their neutral status it is not entirely clear whether and to what extent enemy civilian undersea systems and devices are protected by the law of armed conflict. Of course, as civilian objects they are protected against direct attack and they must be taken into consideration in the proportionality analysis and in the application of precautions in attack. This, however, does not necessarily mean that they may not be captured. At first glance, there seems to be no cogent reason to treat such systems and devices differently than enemy merchant vessels or enemy cargoes, which may be captured outside neutral waters.[94] It may, however, not be left out of consideration that some undersea systems and devices serve important scientific or humanitarian functions. Therefore, the same principles as those underlying the 1907 Hague Convention (XI)[95] should be applied. Accordingly, undersea systems and devices used for non-military

[94] *San Remo Manual*, n. 14, para 135.

[95] Convention (XI) Relative to Certain Restrictions with Regard to the Exercise of the Right of Capture in Naval War, The Hague, 18 October 1907, 2 AJIL Suppl. 167–174 (1908).

scientific missions should be excluded from the right of capture.[96] Equally systems exclusively employed for the prevention of pollution incidents,[97] such as those monitoring and repairing submarines pipelines, and for the advance warning of natural disasters (tsunamis) should not be liable to capture. Of course, that protection is subject to their innocent employment.

11.3.4 *Undersea Infrastructure: Preliminary Conclusions*

The international legal status of undersea infrastructure may be considered a neglected issue. In view of the economic and military importance of submarine cables, pipelines, systems and devices this is an unsatisfactory situation. Accordingly, the existing international legal regime is in desperate need of either an amendment or clarification. As a minimum, the peacetime rules ought to be interpreted as implicitly recognising the right of States to protect 'their' undersea infrastructure against undue interference, whether by State or by non-State actors. The law of armed of armed conflict is even less developed. This may be acceptable as far as undersea systems and devices are concerned, because they can be easily classified as either protected civilian objects or lawful military objectives in accordance with the established principles and rules of targeting law. It may be likewise acceptable with regard to undersea pipelines, provided they only connect two States that are parties to an international armed conflict. However, the provision of Article 54 of the 1907 Hague Regulations on submarine telegraph cables can no longer be considered as adequately reflecting the overall importance of modern submarine communication cables.

11.4 CONCLUDING REMARKS

Most of the issues concerning modern naval technologies relate to the widely undetermined legal status of UMS and undersea infrastructure. It may be that some governments prefer legal ambiguity because they believe that it serves their interests by providing them with a considerable margin of discretion and freedom of action. According to the position taken here, such an approach might be acceptable in situations in which States are confronted with comparatively weak States or non-State actors. It is, however, far from settled that confrontations between major powers will not occur in the near future. In order to avoid any confusion or potential of escalation, at least like-minded governments should take the steps necessary to clarify their position vis-à-vis UMS and undersea infrastructure.

96 See also *San Remo Manual*, n. 14, para 136(e).

97 *San Remo Manual*, n. 14, para 136(g).

12

Outer Space

MELISSA DE ZWART

Military use has been an element of space activities since the beginning of the space age.[1]

12.1 INTRODUCTION

The capacity for human beings to put an object in space, which became a reality for the United States and USSR in the 1950s, and further to place human beings in space and on celestial bodies, which was achieved in the 1960s, brought with it hitherto uncontemplated threats to the security and safety of the Earth as well as incredible joy and wonder about human perseverance, braveness and ingenuity.[2] Never before had it been possible for any nation to occupy such a strategic 'high ground'.[3] Never before had it been possible to place a weapon in orbit around our planet or to maintain surveillance of foreign powers with such accuracy and continuity. Moreover, at that time, of course, nobody contemplated that space would within a relatively short time become vital to communication, navigation and virtually every other service on Earth. We are now so dependent upon a range of space infrastructure that the defence forces of many nations

[1] K.-U. Schrogl and J. Neumann, '[Outer Space Treaty] Article IV', in S. Hobe, B. Schmidt-Tedd and K.-U. Schrogl (eds.) *Cologne Commentary on Space Law Volume 1; Outer Space Treaty*, Cologne, Carl Heymanns, 2009, 71 (hereinafter 'Cologne Commentaries').

[2] It should also be noted that the Union of Soviet Socialist Republics (USSR) sent a dog, Laika, to space in 1957. She was used to test whether a living being could survive the conditions of launch. Laika survived for only a few hours in space and is believed to have deceased before her fourth rotation of the Earth due to stress and overheating, D. Whitehouse 'First Dog in Space Died Within Hours' BBC News, 28 October 2002, http://news.bbc.co.uk/2/hi/science/nature/2367681.stm.

[3] M. N. Schmitt 'International Law and Military Operations in Space', in A. von Bogdandy and R. Wolfrum (eds.) *Max Planck Yearbook of United Nations Law*, volume 10, 2006, 89 at 94.

regularly hold exercises such as 'A Day Without Space' to ensure that they are prepared for the possible destruction of or serious interference with such services as Global Navigation Satellite System (GNSS), Internet and satellite communications, navigation and timing.[4]

This chapter will consider the international framework that was developed during the Cold War to ensure that neither of the spacefaring nations at that time was able to leverage their space capabilities to place a weapon of mass destruction in orbit around the Earth and further to set the rules for what was perceived as a largely military and scientific environment that would only be used and exploited by nation States. In fact, to use a much over-cited statement, space has now become 'congested, competitive and contested'[5] and this is due largely to the massive growth in the civilian applications of space-based technology. The chapter will then explore the current explosion in New Space applications of space and consider the public–private partnerships that have evolved both to enable States to continue to engage in expensive space activities and now to realise the potential for commercial space uses, including communication, Internet, tourist and mining opportunities. It will address the problems of dual use technologies and the application of international law to outer space. Some conclusions will then be drawn regarding what steps need to be taken to assure the security of outer space for the benefit of the whole of humankind.

The disparity between the massive growth in global reliance on space technology and the outdated and fragile rules that attempt to regulate this environment particularly exemplifies the difficult relationship between technological growth and regulatory lag that this book addresses. Further, the complicated relationship between the military and civil sector in both the exploration and exploitation of space highlights the lack of clarity in this area regarding military and non-military uses of space.

Particular complexities that arise with respect to the military uses of space have been identified by Tronchetti as fourfold: (1) the fact that military applications and purposes, such as surveillance, are particularly sensitive and States are therefore reluctant to reveal their existence for fear of relinquishing

[4] J. M. Cosumano Jr, 'A Day Without Space: Ensuring It Doesn't Happen', Summer 2002, *Army Space Journal*, 2–3. Of course the impact now upon the banking, Internet, transport and communication sectors would be just as disruptive to civilian services as to defence.

[5] The term appears to have been first used by Ambassador Gregory L. Schulte at the 2011 National Space Symposium, 'Schulte: Space is Congested, Contested, Competitive' *Space Watch*, June 2011 (Volume 10, Issue 6) www.spacefoundation.org/news/space-watch-june-2011-volume-10-issue-6 but has been used many times since, for example, see further, R. G. Harrison 'Unpacking the Three C's: Congested, Competitive, and Contested Space' (2013) 11.3 *Astropolitics*. 123.

a technological and strategic advantage; (2) the absence of a single unitary body of laws governing military uses of outer space; (3) the lack of clarity resulting from the need to apply a range of inconsistent and overlapping rules and the absence of an 'understanding of key terms and concepts'; and (4) the fact that most space technology is capable of dual military and civilian use.[6] These factors reflect both the unique nature of the space environment and of human activities there, inherently difficult, hostile to human exploration, expensive and dependent upon sophisticated technological advancements, and the fact that its use affects the whole of humanity. Further, the intertwining of civilian and military uses gives rise to specific complexities in terms of the application of international law to the use of space in times of peace and of hostility.

This chapter will consider emerging space technologies and, where possible, will identify whether such uses may be categorised as military or non-military. As will be explained, many of the current uses of space upon which we routinely rely were developed initially for a military purpose. Similarly, the US Air Force, in particular, continues to perform tasks that may be more suitably undertaken by the commercial or civilian sector.[7] As elsewhere in this work, this chapter will ask what those dealing with this technology in times of hostility can learn from the application of that technology for non-combat purposes and vice versa.

12.2 THE SPACE RACE AND THE OUTER SPACE TREATIES

The space race between the United States and the USSR in the 1950s and 1960s was a time of great national pride and wonder at the technological achievements of humankind and the bravery of those pioneering astronauts and cosmonauts. It was also, however, a time of great tension and fear when the global threat that would potentially be posed by the weaponisation of outer

6 F. Tronchetti, 'Legal Aspects of the Military Uses of Outer Space', in F. von der Dunk and F. Tronchetti (eds.) *Handbook of Space Law*, (Cheltenham: Edward Elgar, 2015), 331–332.

7 In particular, reference here is to the role that the US Air Force, together with NASA, plays in monitoring space debris. The Air Force currently tracks around 23,000 pieces of space debris and issues proximity warnings with space assets as it identifies emerging risks. This technology is being expanded, in conjunction with Lockheed Martin, to create the 'Space Fence' which will be capable of tracking much smaller pieces of debris and thus increase coverage to over 200,000 pieces. This service is provided to all space operators and, with the exponential growth in commercial uses of space, may in time be seen as a service more logically provided by a commercial operator: R. Mola 'How Things Work: Space Fence: The New Early-Warning System to Protect Spacecraft from orbiting junk', *Air & Space Magazine*, February 2016, www.airspacemag.com/space/how-things-work-space-fence-180957776/.

space was first appreciated. As Vlasic notes, the launch by the USSR of Sputnik in 1957 evidenced not a 'triumph for world science' but rather a demonstration of a 'revolutionary military instrument with potentially awesome strategic consequences'.[8] That time of tension was largely resolved by the realisation by the two spacefaring nations that they had a common interest in the avoidance of the placement of weapons of mass destruction in outer space and, indeed, any use of space for war-fighting.[9] This is not to conclude, however, that the use of space to support military activities was rejected outright; the focus was, rather, on any uses that would irrevocably harm the space domain and wreak uncontrolled and unanticipated damage on Earth and on the vulnerable space environment. Further, it was recognised that satellites and personnel of all spacefaring nations were equally vulnerable to the discharge of nuclear weapons in space, and thus to ensure their own national space programs, a fragile equilibrium was reached.

On 14 November 1957, the UN General Assembly (UNGA) adopted the first resolution directed at the use of outer space. That resolution used the phrase 'exclusively for peaceful purposes', language that was to be repeated in many subsequent resolutions, declarations or treaties.[10] This was followed with UNGA Resolution 1348 (XIII) on 13 December 1958, which concerned 'Questions of the Peaceful Use of Outer Space' and established an ad hoc Committee on the Peaceful Uses of Outer Space (UNCOPUOS). The UN General Assembly established UNCOPUOS as a permanent committee of the UNGA on 12 December 1959.[11] UNCOPUOS operates largely by consensus and by its mandate specifically deals only with peaceful uses of outer space. Matters relating to military uses of outer space are left to the UN Conference on Disarmament.

The first specific reference to outer space in a multilateral treaty was in the Partial Test Ban Treaty of 1963, which provided that each Party undertook 'to prohibit, to prevent, and not to carry out any nuclear weapon test explosion... in the atmosphere ... including outer space.'[12] Both the USA and USSR announced compliance with this provision by refraining from placing any

[8] I. A. Vlasic, 'Space Law and the Military Applications of Space Technology', in N. Jasentuliyana (ed.) *Perspectives on International Law*, (Kluwer Law International, London, 1995), 385.

[9] A. G. Quinn 'The New Age of Space Law: The Outer Space Treaty and the Weaponization of Space', (2008) 17 *Minnesota Journal of International Law* 475, 479–480.

[10] UNGA Resolution 1148 (XII) 14 November 1957. See further, Vlasic, 'Space Law', 390.

[11] UNGA Resolution 1472 (XIV) 12 December 1959.

[12] Treaty banning nuclear weapon tests in the atmosphere, in outer space and under water (done 5 August 1963, entered into force 10 October 1963) 480 UNTS 43 (Partial Test Ban Treaty) Article I.

weapons of mass destruction (WMD) or nuclear weapons in outer space. This was followed by UNGA Resolution 1884 (XVIII) of 17 October 1963 which further called upon States:

(a) To refrain from placing in orbit around the earth any objects carrying nuclear weapons or any other kinds of weapons of mass destruction, installing such weapons on celestial bodies, or stationing such weapons in outer space in any other manner;
(b) To refrain from causing, encouraging or in any way participating in the conduct of the foregoing activities.[13]

As the authors of the Cologne Commentaries observe, much of this language is drawn from President Eisenhower's suggestion to the UN General Assembly on 22 September 1960:

1. We agree that celestial bodies are not subject to national appropriation by any claims of sovereignty.
2. We agree that the nations of the world shall not engage in warlike activities on these bodies.
3. We agree, subject to appropriate verification, that no nation will put into orbit or station in outer space weapons of mass destruction. All launchings of space craft should be verified in advance by the United Nations.
4. We press forward with a programme of international co-operation for constructive peaceful uses of outer space under the United Nations.[14]

Whilst the reference to 'peaceful uses' was later adopted in Article IV of the Outer Space Treaty, the requirement of 'appropriate verification' was not. This concept of openness and transparency with respect to space activities is articulated by Eisenhower and others as being central to the very nature of the development of space technology. However, as will be discussed further below, technological advantage remains a key element of the use of space, both in the civilian and military domains, and requirements of information sharing and verification run counter to the need to maintain technological superiority.

[13] UNGA Res 1884 (XVIII) 17 October 1963. See further discussion in Schrogl and Neumann, 'Cologne Commentaries', n. 1, 74.

[14] Schrogl and Neumann, 'Cologne Commentaries', 75, cited from B. Cheng, 'Military Use of Outer Space: Article IV of the 1967 Space Treaty Revisited', in C. J .Cheng and D. H. Kim (eds.) *The Utilization of the World's Air Space and Free Outer Space in the 21st Century* (The Hague: Kluwer, 2000) 315. Vlasic also notes that Eisenhower had used the language 'only for peaceful purposes' with respect to use of outer space by the United States and USSR in January 1958, US Senate Committee on Aeronautical & Space Sciences, Documents on International Aspects of the Exploration and Use of Outer Space, 1954–1962, 88th Congress, 1st Session, 53 (1963) at 52, cited Vlasic, 'Space Law', 390 at n. 24.

The Treaty on Principles Governing the Activities of States in the Exploration and Use of Outer Space, including the Moon and Other Celestial Bodies (the 'Outer Space Treaty')[15] remains the major achievement of UNCOPUOS. Of the five major Space Treaties,[16] the Outer Space Treaty has the largest number of ratifying parties, standing currently at 105 ratifying parties and 25 signatories.[17] This high adoption rate supports the Treaty's status in international law, with a number of the principles embodied in the Outer Space Treaty regarded as codifications of customary international law.[18] It provides the overarching constitution for the use of outer space and adopts many of the principles articulated in the earlier Resolutions of the UN General Assembly. Given the Cold War context of its formulation and the fact that space activities were at that time contemplated as being only within the capacity of states, the reach and scope of the Outer Space Treaty provides little practical guidance with respect to modern day military, civilian and commercial uses of space. Its provisions evidence a guarded effort to maintain balance and transparency to ensure continued access to space by the nations that were at that time spacefaring, i.e. USA and USSR. Later treaties reflect an interest by non-spacefaring and developing countries to be able to access the

[15] Treaty on Principles Governing the Activities of States in the Exploration and Use of Outer Space, including the Moon and Other Celestial Bodies, January 27 1967, 18 U.S.T. 2410, 610 U.N.T.S 205 (entered into force 10 October 1967) (the 'Outer Space Treaty').

[16] The five United Nations space treaties are: Outer Space Treaty (Treaty on Principles Governing the Activities of States in the Exploration and Use of Outer Space, including the Moon and Other Celestial Bodies, opened for signature 27 January 1967, 610 UNTS 205, entered into force 10 October 1967); Rescue and Return Agreement (Agreement on the Rescue of Astronauts, the Return of Astronauts and the Return of Objects Launched into Outer Space, opened for signature 22 April 1968, 672 UNTS 119 entered into force 3 December 1968); Liability Convention (Convention on the International Liability for Damage Caused by Space Objects, opened for signature 29 March 1972, 961 UNTS 187, entered into force 1 September 1972); the 'Registration Convention' (Convention on Registration of Objects Launched into Outer Space, opened for signature 14 January 1975 entered into force 15 September 1976); Moon Agreement (Agreement Governing the Activities of States on the Moon and Other Celestial Bodies, opened for signature 18 December 1979, 1363 UNTS 3 entered into force 11 July 1984) (collectively the 'UN Space Treaties').

[17] Committee on the Peaceful Uses of Outer Space: Legal Subcommittee, Status of International Agreements relating to activities in outer space as at 1 January 2017, 56th sess, Agenda Item 5, UN Doc A/AC.105/C.2/2017/CRP.7 (23 March 2017), 12.

[18] H. Quzhi, 'The Outer Space Treaty in Perspective' (1997) 40 *Proceedings on the Law of Outer Space* 51, 53, but see M. N. Schmitt 'International Law and Military Operations in Space', in A. von Bogdandy and R. Wolfrum (eds.) *Max Planck Yearbook of United Nations Law*, vol. 10, 2006, who states that the accepted body of customary international law applicable to outer space extends only to 'the free use of space by all states, a prohibition on claims of sovereignty over space, free exploration of space' and, possibly, 'the obligation to rescue astronauts in distress', at 99.

growing benefits of access to the space environment. However, much uncertainty now arises in the application of the OST and the other space treaties to the modern uses of space.

Art I of the Outer Space Treaty provides that:

> The exploration and use of outer space, including the Moon and other celestial bodies, shall be carried out for the benefit and in the interests of all countries, irrespective of their degree of economic or scientific development, and shall be the province of all mankind.

The wording of this article, as with most of the Outer Space Treaty, is highly aspirational. It draws upon UNGA Resolution 1962 (XVIII) which included in its preamble the intention 'that the exploration and use of outer space shall be carried on for the benefit of mankind and for the benefit of States irrespective of their degree of economic or scientific development.'[19] UNGA Resolution 1962 (XVIII) further stated that:

> ...the principles of freedom, exploration and use of outer space shall be carried on for the benefit and the interest of all mankind whereby outer space and the celestial bodies shall be free for exploration and use by all states on the basis of equality and in accordance with international law.[20]

The reference to the benefit of all countries and province or interest of all mankind in fact runs counter to the highly secretive and competitive manner in which space technology generally, and the space race of the 1950s and 1960s in particular, had been conducted. Despite the admirable principles of all nations sharing in the benefits of space discovery and exploration, States were engaging covertly in development of space based applications and technologies, including weaponry.

Article III of the OST further picked up on the wording and intentions of UNGA Resolution 1962 (XVIII), restating the application of the principles of international law:

> States Parties to the Treaty shall carry on activities in the exploration and use of outer space, including the Moon and other celestial bodies, in accordance with international law, including the Charter of the United Nations, in the interest of maintaining international peace and security and promoting international cooperation and understanding.

[19] UNGA Resolution 1962 (XVIII) 'Declaration of Legal Principles Governing the Activities of States in the Exploration and Use of Outer Space' 13 December 1963. See further, S. Hobe '[Outer Space Treaty] Article I', Cologne Commentaries, n. 1, 25 at 29.

[20] UNGA Resolution 1962 (XVIII), ibid.

Again, it is unclear to what extent the interests of international peace and security and international cooperation must truly be promoted under the Treaty. Given the highly secretive and highly competitive nature of the space race both before and after the Treaty, it seems that the level of cooperation envisaged was minimal to say the least. This principle of sharing has most notably been adopted in the development and use of the International Space Station, albeit subject to a number of complex agreements regarding access, funding, intellectual property and confidentiality of information.[21] However, it would be more accurate to say that it is a principle observed mostly in the breach.

As noted by the Cologne Commentaries, Article IV 'is commonly regarded as the focal point in the Outer Space Treaty dealing with the military uses of Outer Space'.[22] It provides:

> States Parties to the Treaty undertake not to place in orbit around the Earth any objects carrying nuclear weapons or any other kinds of weapons of mass destruction, install such weapons on celestial bodies, or station such weapons in outer space in any other manner.
>
> The Moon and other celestial bodies shall be used by all States Parties to the Treaty exclusively for peaceful purposes. The establishment of military bases, installations and fortifications, the testing of any type of weapons and the conduct of military manoeuvres on celestial bodies shall be forbidden. The use of military personnel for scientific research or for any other peaceful purposes shall not be prohibited. The use of any equipment or facility necessary for peaceful exploration of the Moon and other celestial bodies shall also not be prohibited.

Article IV therefore contains two key provisions:

1. the prohibition upon placement of weapons of mass destruction in space; and
2. the requirement that the Moon and other celestial bodies be used exclusively for peaceful purposes.

[21] 'As of January 2018, 230 individuals from 18 countries have visited the International Space Station' E. Howell, 'International Space Station: Facts, History & Tracking' *Space.com* (online) 7 February 2018 www.space.com/16748-international-space-station.html. The main partners are US, Russia, Europe, Canada and Japan, but other nations are able to participate in the scientific work of the station through a range of projects. The Agreements include the Intergovernmental Agreement, various Memoranda of Understanding and a large number of subsidiary agreements, see further European Space Agency, *International Space Station Legal Framework*, www.esa.int/Our_Activities/Human_Spaceflight/International_Space_Station/International_Space_Station_legal_framework.

[22] Schrogl and Neumann, 'Cologne Commentaries', n. 1, 71.

Notably the provision has not been interpreted as preventing the transit of WMD *through* outer space. As recent events have highlighted, non-permanent transit through an orbital trajectory is not regarded as a violation of the treaty.[23] Nor does it prevent the placement or use of (non-WMD) weapons in outer space or the placement of military space stations in space itself as opposed to on a celestial body.[24] Further, it leaves open the conduct of military space based exercises and (non-WMD) weapons testing. Again, therefore it seems that these quite aspirational statements of the Outer Space Treaty are interpreted restrictively in practice.

Of course, a key issue for the consideration of military uses of outer space is the meaning of the term 'peaceful purposes', contained both in Article IV and in the Preamble to the OST which states: '*Desiring* to contribute to broad international cooperation in the scientific as well as the legal aspects of the exploration and use of outer space for peaceful purposes'. As noted above, the term 'peaceful purposes' was originally used by US President Eisenhower in 1960. However, despite the heritage of these words, their meaning remains hotly contested. The issue, of course, is whether such language is intended to be non-military or non-hostile. As noted at the outset of this chapter, outer space has been used for military activities by many nations since the very first launches into space. These uses include surveillance, communications and navigation, and the transit of weapons through space, so an interpretation of the term that prohibits all military uses of outer space would sit very uncomfortably with the practice of many States following the adoption of the treaty.

Article VI provides that States bear state responsibility for activities undertaken by government and non-government (commercial) users:

> States Parties to the Treaty shall bear international responsibility for national activities in outer space, including the moon and other celestial bodies, whether such activities are carried on by governmental agencies or by non-governmental entities, and for assuring that national activities are carried out in conformity with the provisions set forth in the present Treaty. The activities of non-governmental entities in outer space, including the moon and other celestial bodies, shall require authorization and continuing supervision by the appropriate State Party to the Treaty. When activities are carried on in outer space, including the moon and other celestial bodies, by an international organization, responsibility for compliance with this Treaty shall be borne both by the international organization and by the States Parties to the Treaty participating in such organization.

23 See T. Paine 'Bombs in Orbit? Verification and Violation under the Outer Space Treaty', *The Space Review*, 19 March 2018, www.thespacereview.com/article/3454/1.

24 Schmitt, 'International Law and Military Operations', 104.

This provision becomes particularly important in the context of space activities being undertaken by non-government, commercial entities. States remain responsible for activities undertaken by commercial entities and are obliged by the Treaty to provide continuing supervision for such activities. In addition, the Liability Convention provides that State liability extends to all launches which are made from that State's territory or facility.[25]

Under Article IX of the Outer Space Treaty:

> In the exploration and use of outer space, including the moon and other celestial bodies, States Parties to the Treaty shall be guided by the principle of co-operation and mutual assistance and shall conduct all their activities in outer space, including the moon and other celestial bodies, with due regard to the corresponding interests of all other States Parties to the Treaty. States Parties to the Treaty shall pursue studies of outer space, including the moon and other celestial bodies, and conduct exploration of them so as to avoid their harmful contamination and also adverse changes in the environment of the Earth resulting from the introduction of extraterrestrial matter and, where necessary, shall adopt appropriate measures for this purpose. If a State Party to the Treaty has reason to believe that an activity or experiment planned by it or its nationals in outer space, including the Moon and other celestial bodies, would cause potentially harmful interference with activities of other States Parties in the peaceful exploration and use of outer space, including the Moon and other celestial bodies, it shall undertake appropriate international consultations before proceeding with any such activity or experiment. A State Party to the Treaty which has reason to believe that an activity or experiment planned by another State Party in outer space, including the Moon and other celestial bodies, would cause potentially harmful interference with activities in the peaceful exploration and use of outer space, including the Moon and other celestial bodies, may request consultation concerning the activity or experiment.

Again, the aspirational language of this provision provides little by way of clear guidance regarding how space actors should behave. Terms such as 'harmful interference' and 'due regard' lack clear scope and meaning in terms of when

[25] Convention on International Liability for Damage Caused by Space Objects, opened for signature 29 March 1972, 961 UNTS 188 (entered into force 1 September 1972). See also Outer Space Treaty, Article VIII regarding jurisdiction over spacecraft and personnel on any spacecraft launched from the relevant state's territory: 'A State Party to the Treaty on whose registry an object launched into outer space is carried shall retain jurisdiction and control over such object, and over any personnel thereof, while in outer space or on a celestial body. Ownership of objects launched into outer space, including objects landed or constructed on a celestial body, and of their component parts, is not affected by their presence in outer space or on a celestial body or by their return to the Earth. Such objects or component parts found beyond the limits of the State Party to the Treaty on whose registry they are carried shall be returned to that State Party, which shall, upon request, furnish identifying data prior to their return.'

such thresholds have been reached and as to what level of response is required in order to ensure that the requirement of due regard is met.

Notably there is no definition of 'outer space'. As the Tallinn Manual notes: 'The boundary of outer space has been discussed by governments, the UN, and the scientific and legal communities for decades, without resolution. A 1967 conclusion that 'no clear scientific or technical criteria could be found which would permit a precise and lasting definition of outer space and which would be acceptable to all States' remains accurate today.'[26] The absence of a clear vertical delineation of space leaves open the question of how and by whom jurisdiction may be exercised over the space domain or, perhaps more precisely and accepting that no State may exercise jurisdiction over the space domain as such, leaves open the question of precisely where the boundary lies between airspace, where notions of territorial sovereignty apply, and outer space, where they do not.

All of these gaps and uncertainties leave considerable scope for the development of commercial space industries. Further, they leave wide open the question of the legality of use of outer space for the military purposes not explicitly addressed in the Outer Space Treaty. As many space uses are now likely also to be dual use, the scope becomes wider still, but so too does the debate regarding the legitimacy (and potential vulnerability) of such uses.[27] Whilst the other space treaties provide further details with respect to international law obligations regarding rescue and return of astronauts and space objects, liability for damage caused by space objects, registration of space objects and (less universally accepted) rules relating to use of the Moon, many key terms remain uncertain. To highlight but one example, it is not clear if commercial space passengers will be treated as 'astronauts' for the purposes of the Rescue and Return Agreement.[28] Despite this lack of certainty, commercial developments of space continue apace.

12.3 THE EVOLUTION OF SPACE WEAPONRY AND TECHNOLOGICAL APPLICATIONS

Weapon technology and space technology have been inextricably linked since the dawn of the space age. The capacity to launch satellites into space was

[26] M. N. Schmitt (ed), *Tallinn Manual 2.0 on the International Law Applicable to Cyber Operations* (Cambridge: Cambridge University Press), 271, citing Report of the Scientific and Technical Subcommittee on the Work of its Fifth Session, para 36, UN Doc. A/AC.105/39 (6 September 1967).

[27] See, for example, the Australian Optus C-1 Satellite which provides both military and civilian communication services, Australian Government Department of Defence, 'Department of Defence extends Optus Satellite contract', 1 June 2017 www.minister.defence.gov.au/minister/marise-payne/media-releases/department-defence-extnds-optus-satellite-contract.

[28] S. Hobe, 'Legal Aspects of Space Tourism', (2007–2008) 86 *Nebraska Law Review*, 439.

made possible by the rocket technology developed by German scientists in World War II. Post-war rocket research using German technology was conducted by the UK, largely at the Woomera Rocket Range,[29] commissioned for this purpose in outback Australia, by the United States and by the Soviet Union. A Soviet intercontinental ballistic missile (ICBM) launched Sputnik I, the world's first satellite, into orbit on 4 October 1957. Thus 'the evolution of satellite technology followed the evolution of missile technology'.[30] Of course, it was recognised that ICBMs could also be used to carry a nuclear weapon and on 27 August 1957 the first nuclear test detonation in space was undertaken by the Soviets. The US followed suit with nuclear explosions over the South Pacific at an altitude of 200–300 miles.[31] A number of other weapons tests followed into the early 1960s.

This same ICBM technology is still in use and was even contemplated by SpaceX founder Elon Musk for his planned Mars project.[32] It was only Musk's frustration with the purchase of such a rocket that caused him to pursue other, far more novel, technologies which have resulted in the development of the SpaceX reusable launch vehicle.[33] SpaceX and other NewSpace operators have completely disrupted pre-existing business and technology models in the space industry, bringing novel materials and approaches to launch vehicles, fuel and design.

Early military uses of satellites were focused on reconnaissance and observation as well as early warning systems to track, in the case of the United States, any launch of Soviet ICBMs. Following on from reconnaissance, the next use of space for military purposes was in connection with communication and navigation. These uses then expanded to weather monitoring, intelligence gathering and surveillance. The earliest space surveillance system involved the Corona satellites (operating under the guise of a space exploration programme, entitled Discoverer) which carried a camera, between 5 and 9 feet long, that used special 70mm film made by Kodak which enabled wide, high resolution shots. Each satellite had to carry several miles of film, which after being exposed was returned in sections back to Earth in specially

[29] M. de Zwart and D. Stephens, 'Non-Military Space Testing in the Woomera Prohibited Area: Opportunities for the Australian Space Industry?', (2017) 45 *Federal Law Review*, 39.

[30] R. Ramey, 'Armed Conflict on the Final Frontier: The Law of War in Space', (2000) 48 *Air Force Review*, 1, 13.

[31] Ramey, ibid., 12.

[32] A. Vance, *Elon Musk, Tesla, SpaceX and the Quest for a Fantastic Future*, (Harper Collins, 2015) 104–106; T. Junod 'Elon Musk: Triumph of His Will' 15 November 2012, *Esquire* (online) www.esquire.com/news-politics/a16681/elon-musk-interview-1212/ .

[33] L. Grush, 'SpaceX Makes Aerospace History with Successful Launch and Landing of a Used Rocket', 30 March 2017, *theverge.com* (online), www.theverge.com/2017/3/30/15117096/spacex-launch-reusable-rocket-success-falcon-9-landing.

designed heat resistant buckets, which were dropped from the satellite and then collected by a plane at 15,000 feet.[34] This project, which entailed both novel technology, including the film and camera developed by a commercial provider, and significant skill and dexterity in the capture of the buckets, paved the way for close collaboration between defence and civilian contractors.

The development of the US Global Navigation Satellite System (GNSS) Global Positioning System (GPS) began in the 1970s and was originally planned for military use.[35] However, civilian use rapidly outstripped military use, creating unique issues regarding security, reliance and predictability. The Russian GNSS Glonass was also developed for military use and has been adapted for civilian use. Other GNS systems have been developed by China, India and the European Union.[36] Despite this, the GPS remains the most heavily utilised system by non-military users both inside and outside the US. This is a significant example of a military technology responsible for the creation of enormous commercial innovation and exploitation, being uses that were not foreseen by the developers of the original system. It is now virtually unthinkable by most citizens of the developed world that they could need to survive without GPS-enabled maps, weather forecasting, time recording, Internet facilities and navigation. The system has spawned a vast range of applications and consumer and business services.

The 1991 Gulf War (Operation Desert Storm) has been described as 'the first space war'[37] with the use of satellite imaging power, signals intelligence and GPS providing superior knowledge to the coalition forces. Notably, in addition to these US Department of Defense resources, imagery was acquired from commercial remote sensing satellites.[38] Again, it is now commonplace for the armed forces to acquire high resolution imagery from commercial providers such as Planet.[39]

12.4 THE GROWTH OF THE COMMERCIAL SPACE INDUSTRY

For many years after the Cold War, the United States held significant technological advantages in the space domain. Carter et al. note that the United

[34] D. Oberhaus, 'The First Spy Satellites Had to Drop Gigantic Buckets of Film Back to Earth', 22 April 2017, *Motherboard* (online), https://motherboard.vice.com/en_us/article/jpzpkg/the-first-spy-satellites-had-to-drop-gigantic-buckets-of-film-back-to-earth.

[35] P. Larsen, 'Issues Relating to Civilian and Military Dual Uses of GNSS', (2001) 17 *Space Policy*, 111.

[36] L. Chengzhi, 'The Chinese GNSS System Development and Policy Analysis', (2013) 29 *Space Policy*, 9.

[37] US Air Force Chief of Staff, Merrill McPeak, quoted in Vlasic, 'Space Law' ', 388.

[38] Ibid., 388.

[39] www.planet.com.

States developed an 'offset strategy' whereby US technological superiority would offset the power of sheer numbers of their military and political rivals. This strategy has comprised two elements: 'The first was to field superior technology through aggressive pursuit of military R&D, and developing a high-technology defense industrial base. The second was to deny opponents that technology through a system of export controls and protection of technological secrets.'[40] This strategy of the US Department of Defense has been to remain 'at least one technology generation ahead of any foreign or commercial space power'[41]

Leadership in space technology and innovation is now vested in new space ventures such as Virgin Galactic, Blue Origin, Planet, Rocket Lab and SpaceX. Many of these New Space entrepreneurs have entered the space technology industry from a background in Internet and information technology. Their disruptive approach to use of new fuels, new methods of design including 3D printing, and an emphasis on low-cost reusability has resulted in an explosion of space start-ups. Similarly, the development of cube sat and nano sat technology offers new flexible and low-cost entry points to space situational awareness and surveillance.[42]

Militaries worldwide are increasingly turning to commercial providers for a range of space related services and technology.[43] In addition, national governments are actively encouraging the development of domestic space industries. Thus impediments are gradually being reduced or removed. National security concerns, however, remain important in this context. Further, space industries were hampered for many years by the strict application of ITAR and government contracting rules. The United States, for example, has recently passed a range of legislation aimed at supporting the growth of private commercial space businesses, including those directed at exploration, tourism and mining. This includes the US Commercial Space Launch Competitiveness Act, passed by United States Congress in 2015, which has the explicit purpose to 'facilitate a pro-growth environment for

[40] A. B. Carter, M. Lettre and S. Smith, 'Keeping the Technological Edge', 129. See also H. Kleinberg, 'On War in Space', (2007) 5 *Astropolitics*, 19.

[41] US Department of Defense, *Quadrennial Defence Review*, 6 February 2006, 55–56, cited in Schmitt, 'International Law and Military Operations', 92.

[42] A 'cube sat' is a small satellite measured in the standard size of 1 Unit (or 1U) being 10 cm × 10 cm × 10 cm. This size can be scaled up in modules consisting of multiples of this standard unit size. A nanosat weighs between 1 and 10 kg, see NASA 'What are SmallSats and CubeSats?' www.nasa.gov/content/what-are-smallsats-and-cubesats.

[43] E. Waldrop 'Integration of Military and Civilian Space Assets: Legal and National Security Implications' (2004) 55 *Air Force Review* 157. See also Schmitt, 'International Law and Military Operations', 98.

the developing commercial space industry by encouraging private sector investment and creating more stable and predictable regulatory conditions'.[44] Other countries actively pursuing the development of a commercial space industry through legislative reform include Luxembourg,[45] New Zealand,[46] the UK[47] and the United Arab Emirates,[48] to name but a few.

12.5 CURRENT ISSUES WITH THE REGULATION OF SPACE

Space has a number of key differences from other operating domains. First, as addressed above, there is likely to be a great distance between the operator and

[44] US Commercial Space Launch Competitiveness Act, 114th Congress, Public Law 114–190. See further, M. Schaefer 'The Contours of Permissionless Innovation in the Outer Space Domain' (March 28, 2017). *University of Pennsylvania Journal of International Law*, Vol. 39, No. 1, 2017, available at https://ssrn.com/abstract=2942526.

[45] See 'Luxembourg aims to Contribute to the Peaceful Exploration and Sustainable Utilization of Space Resources for the Benefit of Humankind': www.spaceresources.public.lu/en.html#, which states that: 'Luxembourg's vision is built on support for advanced research activities and technological capabilities, drawing on the country's existing expertise in the space sector and its ongoing strategy of economic diversification into future-oriented high-tech industries.'

[46] New Zealand has recently enacted the *Outer Space and High-altitude Activities Act* 2017 in order to facilitate the development of a launch site in New Zealand for one company, Rocket Lab. Rocket Lab is a United States based company, with a New Zealand subsidiary, focused on delivering low-cost launch services for smaller primary payloads. Rocket Lab completed their launch facility, Launch Complex 1, on the Mahia Peninsula on 27 September 2016, with the first test launch occurring on 25 May 2017. Notably the US origins of the Rocket Lab technology have created unique issues in terms of export to and use of that technology in New Zealand, even by its own subsidiary. To accommodate these issues the *Outer Space and High-altitude Activities Act* includes specific terms that can be related to the bilateral *Technology Safeguards Agreement* between New Zealand and the United States executed in 2016. This Agreement provides an ability to export rocket technology which would otherwise be prevented by the American Missile Technology Control Regime. These political and national security-related issues remain a significant impediment to global technology sharing and the development of multi-national space corporations.

[47] Space Industry Act UK (2018). See further, UK Gov. UK Press Release 'Rocketing to success - Space Industry Act to unlock billions for the UK economy', 16 March 2018, www.gov.uk/government/news/rocketing-to-success-space-industry-act-to-unlock-billions-for-the-uk-economy, which claims that the recently enacted Space Industry Act, 2018 'is the most modern piece of space industry regulation anywhere in the world', which 'will open up the universe to businesses, allowing them to better compete in the commercial space race using UK spaceports, and taking advantage of future developments, including hypersonic flight and high-speed point-to-point transport.' Further, it will facilitate the UK 'government's plans to add billions of pounds to the UK economy and create hundreds of high-skilled jobs by unlocking opportunities in the new space age'.

[48] J. Foust, 'United Arab Emirates to Establish Human Spaceflight Program', *Space News* (online) http://spacenews.com/united-arab-emirates-to-establish-human-spaceflight-program, 2 October 2017.

the satellite. With the exception of the very few missions which involve human personnel, such as those conducted on the International Space Station, most space operations, such as those connected with surveillance or communications, involve a satellite in space and a human operator on the ground. Space technology generally provides support for operations conducted on the ground by facilitating communications, targeting-related activities and surveillance.

Second, another element of the space domain is that 'space operations are always global by their very nature'.[49] Objects in orbit around the Earth are moving at incredibly fast velocities and have very limited manoeuvrability. Satellites move rapidly around the Earth in a path determined by their specific orbit, such that there is usually no capacity to adjust the specific territory which a satellite covers and indeed, large areas will be inaccessible, either permanently or until the satellite's orbit next brings it within range of the relevant location. Only satellites located in Geosynchronous Orbit (GEO) maintain what appears from Earth to be a stable presence above a territory. Attacks on space assets may therefore generally be more predictably achieved by damage to the ground stations associated with such satellites either by kinetic or cyber means. Similarly, at the moment there is no Death Star weapon capable of maneuvering to target a specific place on Earth. From a military perspective space might therefore be the 'ultimate high ground'[50] but it is a very fragile and vulnerable high ground. The difficulty of concealing and protecting space-based assets is very high and there is no territory as such in space which may be defended. Rather, as noted above, space operations are by their very nature dependent upon technology and the retention of such 'high ground' therefore critically depends upon the maintenance of technological superiority.

However, the current superiority of US space capabilities represents both an advantage and a vulnerability for the United States. As with any technological superiority, however, scientific advance must be constantly pursued and the technology must be updated in order to maintain the advantage. Further, the reliance of NASA and the armed forces on developments in the commercial sector has complicated the maintenance of that advantage. This reliance on the commercial sector brings significant challenges and the need for adaptation.[51] The majority of new technologies are now being produced in the commercial sector, with many of them being developed primarily with a view

[49] H. Kleinberg, 'On War in Space' (2007) 5 *Astropolitics*, 1 at 12.
[50] Kleinberg, ibid., 18; Schmitt, 'International Law and Military Operations', 94.
[51] Carter, et al., 'Keeping the Technological Edge', 130.

to commercial, rather than defence, purposes. The number of commercial space technology providers is rapidly increasing, and while NASA and the Department of Defense remain the major customers of such businesses, the commercial applications of space-related technology are escalating daily. It is therefore unlikely in the medium to long term that the US government will remain the most important customer of such businesses.[52]

12.6 RULES APPLICABLE TO MILITARY CONFLICT IN OUTER SPACE

As the authors of the Tallinn Manual 2.0 have noted in their consideration of the relationship between the closely related space and cyber domains:

> The Experts observed that from the beginning of the space age, outer space has been used for military purposes such as reconnaissance and surveillance. Moreover, a great deal of space technology has had a 'dual nature' in the sense that civilian space capabilities often stem from military space developments and many civilian space applications, such as commercial satellite imagery, are used for military purposes. Therefore, the Experts rejected the premise that any purported limitation on the use of outer space to peaceful purposes should be interpreted as 'non-military'.[53]

With respect to the military uses, it is currently acknowledged that outer space differs from other domains of conflict in that, at least currently, there will be distance between the space weapons and the human combatants.[54] Space is currently employed in support of military uses which have their impact upon land, sea and in the air, as opposed to being a domain of war-fighting itself, although it is expected that this will change in the near future.

What would constitute permitted military operations in space remains to be clarified. As noted above, there is considerable debate regarding the attributes and definition of a military use of outer space for the purposes of the outer space treaties, particularly as to the uses that on this basis are not permitted. In this regard, Tronchetti and other commentators have long drawn a distinction between 'militarisation' versus 'weaponisation' of outer space.[55]

[52] Ibid., 136.

[53] M. N. Schmitt (ed.), (2017) *Tallinn Manual 2.0 on the International Law Applicable to Cyber Operations* (Cambridge: Cambridge University Press), 275 (footnote omitted).

[54] R. A. Ramey, 'Armed Conflict on the Final Frontier: The Law of War in Space', (2000) *The Air Force Law Review*, 2.

[55] Tronchetti, 'Legal Aspects of the Military Uses of Outer Space', 333–334. See also, A. G. Quinn 'The New Age of Space Law: The Outer Space Treaty and the Weaponisation of Space', (2008) 17 *Minnesota Journal of International Law* 475.

Given that activities conducted in space have historically been, and currently still are, conducted by armed forces of various nations, the distinction should be drawn between 'militarisation' and 'weaponisation'[56] of outer space, with an acknowledgement that international law permits non-aggressive military uses of space.[57]

However, as noted above, it is highly contested whether the five outer space treaties reflect this reality. How then can commercial and military uses of space evolve in the context of these treaties? As Meyer notes, despite the fact that the Prevention of an Arms Race in Outer Space (PAROS) has been a standing item on the agenda of the Conference on Disarmament (CD) since 1982, 'no significant action has been undertaken multilaterally on space security for decades.'[58]

In the absence of any likelihood of updating the outer space treaties, various governments have looked to multilateral treaties and alternatives such as 'soft law' instruments, including the European Code of Conduct for Activities in Outer Space, formally proposed in December 2008 and the Draft Treaty on the Prevention of the Placement of Weapons in Outer Space, the Threat or Use of Force Against Outer Space Objects, formally tabled by Russia and China in 2008 in the Committee on Disarmament and revised in 2014. Neither of these proposals, which seek to place limits on technological advancements of space technology capable of being weaponised or otherwise applied to potential conflicts, has been adopted.

In addition, the UN has supported Transparency and Confidence Building Measures since 2005. The UN Group of Governmental Experts (GGE) formed in 2011, issued its Final Report on "Transparency and Confidence-building Measures in Outer Space Activities' which was presented to the UN General Assembly in December 2013.[59] The General Assembly encouraged Member States to implement the measures detailed in the report in their relevant domestic mechanisms on a voluntary basis.

Mention should also be made of The Convention on Prohibition of Military or Any Other Hostile Use of Environmental Modification

[56] Ramey, 'Armed Conflict', 6. See further Vlasic, 'Space Law', 386 (at fn 6): militarisation 'is used to indicate the use of outer space by a significant number of military spacecraft' whereas weaponisation 'refers to the placing in outer space for any length of time any device designed to attack man-made targets in outer space and/or in the terrestrial environment.'

[57] Ramey, 'Armed Conflict', 7.

[58] P. Meyer, 'Dark Forces Awaken: The Prospects for Cooperative Space Security', (2016) *The Nonproliferation Review*, 23 3–4, 495, 496.

[59] UN General Assembly 'Group of Governmental Experts on Transparency and Confidence-building Measures in Outer Space Activities' A/68/189, Sixty-eighth session, July 29, 2013.

Techniques, (ENMOD)[60] which prohibits the 'military or any other hostile use of environmental modification techniques having widespread, long-lasting or severe effects as the means of destruction, damage or injury to any other State Party'.[61] Environmental modification techniques are defined as 'any technique for changing – through the deliberate manipulation of natural processes – the dynamics, composition or structure of the earth, including its biota, lithosphere, hydrosphere, or of outer space.' It is likely that environmental issues will become more pressing, with greater use being made of the space environment. In this vein, it may also become more vital for voluntary protocols regarding space debris to become more legally binding.[62]

Particular concern has been expressed about the impact of such soft-law measures on space research and development programs.[63] As noted above, US space policy has relied upon maintaining a technological advantage. With the increasing cooperation between military and commercial space activities, any soft law instrument which imposes limitations on development of space technology which may have a military application may also impede further development of dual use technologies. There is therefore a concern that compliance with a non-binding Code of Conduct may have more significant impact of technological development than the binding international space treaties.[64]

12.7 TECHNOLOGY AND SPACE POWER THEORY

As has been discussed above, until the last decade the use and development of space technology has been undertaken largely by the United States and the former USSR/Russian Federation. Now, many countries including China,

[60] Convention on Prohibition of Military or Any Other Hostile Use of Environmental Modification Techniques, May 18, 1977, 31 U.S.T. 333 (entered into force 5 October 1978) ('ENMOD').

[61] ENMOD, Article III(1).

[62] In 2007, the General Assembly of the United Nations endorsed the Space Debris Mitigation Guidelines of the Committee on the Peaceful Uses of Outer Space. These Guidelines provide a framework of what States should be considering in the design, manufacturing, and operational phases of their space activities in order to minimize debris and the risk of creating more debris. However, in order to be effective, these Guidelines require implementation in domestic space law.

[63] J. Keuter 'Rules of the Road in Space: Does a Code of Conduct Improve US Security?' quoted in J. Beard 'Soft Law's Failure on the Horizon: The International Code of Conduct for Outer Space Activities' (2017) 38.2 *Pennsylvania Journal of International Law* 335 at 376; J. Foust 'Debating a Code of Conduct for Space' *The Space Review* (online) 7 March 2011, www.thespacereview.com/article/1794/1.

[64] Beard, ibid., 377–379.

India and France have advanced space programs. As space becomes increasingly commercialised and the United States in particular becomes more dependent upon private commercial providers for its technological advances, it also becomes more vulnerable. There are far more operators in space and they do not all share the same vision of space security, operate at the same technological level or comply with space laws and norms. Kleinberg therefore concludes that just as technological superiority is vital to space power, that same technological superiority could deter any conflict in space.[65]

As noted above, it is this issue of technological asymmetry which causes particular complexity in the proposals for further treaties or 'softlaw' instruments to deter an arms race in outer space.

Such a unique domain calls for its own regulatory theory and body of space laws. For example, 'spacepower theory' attempts to 'describe, explain, and predict how individuals, groups, and states can best derive utility, balance investments, and reduce risks in their interactions with the cosmos.'[66] Hays notes that despite the fact that humanity has now been in space for over 50 years we have not developed a sophisticated and predictable spacepower theory model which could be used to guide us with respect to the dilemmas which are rapidly advancing as space becomes more commercialised and potentially more militarised and weaponised.[67]

As new laws relating to outer space are increasingly being enacted at the national, domestic law, level, this seems to be an opportune time to review the applicability and relevance of the outer space treaties. In the absence of any universal will to revise such treaties or formulate new ones, it may be necessary and sensible to provide a gloss or updated understanding of the laws we already have.

12.8 CURRENT INITIATIVES

There are currently several important initiatives to develop regulatory guidelines applicable to the military uses of outer space.

One such project is the Woomera Manual on the International Law of Military Space Operations (the Woomera Manual). It brings together experts from all over the world to identify and articulate the law applicable to military activities associated with the space domain. The Manual will consider the

[65] Kleinberg, 'On War and Space', 20.

[66] P. L. Hays 'Spacepower Theory', in K.-U. Schrogl, et al. (eds.) *Handbook of Space Security*, (New York: Springer, 2015), 58.

[67] Hays, ibid., 57.

particular laws that apply to the space domain, including the definition of 'outer space', and what may constitute 'peaceful purposes', as well as the relevant laws relating to use of force and international humanitarian law. It is intended that this Manual will define the limits of what is legal in the context of conducting military and security activities in outer space and will be in the tradition of the San Remo Manual of International Law Applicable to Armed Conflict at Sea, the Harvard Manual on International Law Applicable to Air and Missile Warfare and the Tallinn Manual on International Law Applicable to Cyber Warfare. The intended publication will be useful to government legal advisers, military lawyers and operators in the field, as well as to other, broader audiences.

The project involves leading academics, military lawyers and operators, non-governmental organisations and government advisers from a diverse range of backgrounds, countries and cultures. It will be completed by 2020. The aim of the project is to articulate the current law in a readily understandable and accessible fashion to ensure the ongoing peace in and sustainability and security of outer space into the future for the benefit of all users.

12.9 CONCLUSIONS

Space is becoming an increasingly vital aspect of our day-to-day lives. Like the Internet, it has evolved in unanticipated and life-altering ways from its military beginnings. Military and commercial applications now develop hand in hand with one another and governments are increasingly reliant on the cheaper and more innovative solutions offered by New Space entrepreneurs who are cutting the time and costs involved in getting to space and making use of space resources. The flexibility of low-cost commercial solutions, such as cube sats providing space situational awareness, ensures that good use is made of ever-reducing government budgets. These commercial providers have different motivations and principles from those embraced by governments. Thus the space domain faces new challenges from businesses that are looking to make a profit from space operations and that are accountable to shareholders and corporate regulators rather than national governments and from new spacefaring nations that are bringing new security challenges to the domain.

In order to ensure continued access to and beneficial uses of space, it is important that space security be maintained. The UN space treaties appear to be an outdated remnant of an optimistic, albeit successful, attempt to avert the risks that would have been associated with the detonation of WMD in outer space. These treaties, with their deliberately vague language, now represent an obstacle to the further development of space. New rules need to be adopted that will enable explorers, tourists, settlers and miners to go beyond the Moon to Mars.

Projects such as the Woomera Manual will undertake valuable work to translate and apply existing law to the ever-expanding military, civilian and commercial uses of space. The rules that such projects recognise and develop will provide a clearer understanding of the application of international law, particularly the law as to the use of force and international humanitarian law, to the space environment. These rules will take account of the fact that many nations are now part of the global space industry, whether as launch providers, service providers or by virtue of their engagement in research and development. As use and users of space continue to increase, the potential for conflict will inevitably become more prevalent. As commercial operators establish mines, tourist ventures and colonies in space, we will need to grapple with how to operate, defend and protect such sites.[68] It is possible that without clear legal rules private militia will be created to protect the significant investment involved in creating such capabilities. Protection of intellectual property and trade secrets will remain as important as protection of the infrastructure and personnel associated with the operation of such facilities.

While space is the next frontier for human exploitation and exploration, access to space is already vital for many terrestrial services. It is therefore essential that clear rules are identified now to ensure the safety and success of that future.

[68] M. de Zwart, 'GOOGLE in Space? How Will Space Governance Accommodate Non-State Actors?', (May 27, 2016), RUMLAE Research Paper No. 16–21, available at https://ssrn.com/abstract=2809409.

13

Synthetic Brain Technologies: Beyond Artificial Intelligence

DAVID P. FIDLER

'Beneath the skull lies the next frontier.'
The Economist, 4 January 2018

13.1 INTRODUCTION

The brain has long been a source of fascination and inspiration for those wanting to make machines perform like sentient life. In 1943, Warren McCulloch and Walter Pitts proposed translating the behaviour of neurons in the brain and nervous system into computational logic,[1] which suggested 'the human brain could be thought of as a computing device.'[2] In 1948, Alan Turing's work on 'intelligent machinery' envisioned a machine with a 'child-brain' capable of learning.[3] Turing likened 'the cortex of an infant' to an 'unorganised machine' that, through training, organises itself to learn and develop intelligence. He sketched ideas for building an 'unorganised machine' to simulate what happens in the human brain. These ideas were precursors of what became known as 'neural networks' – networked layers of computer processors designed to mimic how neurons function – developed to support artificial intelligence technologies.[4]

1 W. S. McCulloch and W. Pitts, 'A Logical Calculus of the Ideas Immanent in Nervous Activity', *Bulletin of Mathematical Biophysics* (1943); 5: 115–133.

2 L. Hardesty, 'Explained: Neural Networks', *MIT News*, 14 April 2017, available at http://news.mit.edu/2017/explained-neural-networks-deep-learning-0414.

3 A. Turing, 'Intelligent Machinery: A Report', National Physical Laboratory (1948), available at www.alanturing.net/intelligent_machinery/.

4 Hardesty, 'Explained: Neural Networks': 'Modeled loosely on the human brain, a neural net consists of thousands or even millions of simple processing nodes that are densely interconnected'.

Progress in artificial intelligence and advances in neuroscience have generated increasing interest in melding mind and machine in unprecedented ways. In particular, research on creating artificial brains is underway,[5] as are efforts to produce physical and virtual interfaces between the human brain and mechanical devices.[6] Such work on synthetic brain technologies overlaps with the evolution of artificial intelligence capabilities, but ambitions in synthetic brain research reach beyond making machines faster at being smarter. As computing devices and strategies more closely approximate what makes the brain integral to the human condition, practical and profound questions emerge.

Synthetic brain technologies could, as their advocates often claim, be transformative and provide people with assistance on a host of problems, from improving human health to managing the consequences of climate change.[7] However, transformative technologies are often disruptive. Synthetic brain technologies have the potential to outperform and displace people in tasks and jobs long considered beyond the ability of machines, including machines that have artificial intelligence capabilities. Achieving human-level machine intelligence potentially takes disruption into more mind-binding territory. Can brain-machine interfaces meld human personality with machines? Can consciousness emerge from machine experiences made possible by an artificial brain? Do synthetic brain technologies promise a new kind of intelligence that we do not politically, legally and ethically know how to manage?

This chapter explores legal questions that synthetic brain technologies might raise in peace and war. It begins with a description of some of the technologies associated with developing artificial brains and brain-machine interfaces. The objective is to communicate basic features of the technologies and research-and-development efforts in this realm. A comprehensive

[5] 'Special Report: Can We Copy the Brain?', *IEEE Spectrum* (2017), available at https://spectrum.ieee.org/static/special-report-can-we-copy-the-brain.

[6] S. Levy, 'Brain-Machine Interface Isn't Sci-Fi Anymore', *Wired*, 13 September 2017, available at www.wired.com/story/brain-machine-interface-isnt-sci-fi-anymore/?mbid=BottomRelatedStories (noting that '2017 has been a coming-out year for the Brain-Machine Interface [BMI], a technology that attempts to channel the mysterious contents of the two-and-a-half-pound glop inside our skulls to the machines that are increasingly central to our existence').

[7] G. Zorpette, 'The Benefits of Building an Artificial Brain', *IEEE Spectrum*, 31 May 2017, available at https://spectrum.ieee.org/computing/hardware/can-we-copy-the-brain (arguing that 'machines that have something like human cognition ... will be our intellectual partners in solving some of the great medical, technical, and scientific problems confronting humanity').

examination of this technological field is beyond this chapter's scope, and any such examination soon would be obsolete given the scale, pace and diversity of ongoing research and development activities.

The chapter's next two sections examine the possible legal implications of synthetic brain technologies. The analysis in these sections is necessarily speculative. Given the current state of research and development on these technologies, whether the promise many people see in them will mature into applications relevant to law in peace and war remains uncertain. Further, how societies adjust law to manage other technological advances, such as the continued development of artificial intelligence, might well affect the regulatory context in which synthetic brain technologies are exploited. However, speculation can still provide some insights and inform the kinds of questions that need to be asked as synthetic brain technologies develop.

An overarching question for peacetime and armed conflict settings is whether existing law and legal systems – legacy rules – can accommodate the emergence of synthetic brain technologies or whether societies will have to develop new legal strategies for them. For example, the law of armed conflict evolved to apply no matter what technologies belligerents deployed on battlefields. In peacetime, principles of human rights law create obligations on governments to respect, protect and fulfil rights whatever the technological context. New health technologies often require highly regulated clinical trials to test their safety and efficacy in people, and technologies that involve physically interfacing the brain and computers will not escape such regulatory regimes.[8]

Moreover, the close proximity to human cognition that distinguishes synthetic brain technologies from other types of machine intelligence should mean such technologies have little difficulty learning the requirements of the law, whether the law is old or new. The harder, and perhaps more interesting question, is whether synthetic brain technologies might produce non-human challenges to the legitimacy of existing laws that assume human agency. To reframe Turing's famous question,[9] can machines think justly in peace and war?

[8] A. Palmer, 'Thought Experiments: Brains and Machines', *The Economist*, 4 January 2018, available at www.economist.com/technology-quarterly/2018-01-06/thought-experiments (noting that, in connection with brain–computer interfaces, '[t]here are no shortcuts to clinical trials and regulatory approval').

[9] A. Turing, 'Computing Machinery and Intelligence', *Mind* (1950); 49: 433–460 ('I propose to consider the question, 'Can machines think?').

13.2 REPLICATING AND INTERFACING WITH THE HUMAN BRAIN

The human brain has phenomenal capabilities that no current technology, including the most impressive 'deep learning' happening in artificial intelligence, yet comes close to matching.[10] Despite its compact size, it contains approximately 100 billion neurons, with each neuron having around 10,000 synapses, producing on the order of 100 trillion synaptic connections between neurons.[11] Unlike electricity-hungry server farms that support powerful computers and computing networks, the brain requires very little energy – about what it takes to power a low-wattage light bulb – to perform its astounding feats.[12] This dynamic biological, chemical and electrical network processes sensory information, sustains memory and facilitates learning in a manner far more sophisticated than the most impressive supercomputer or artificial intelligence algorithms. In addition, the brain is essential to what we understand as human individuality, personality, dignity and consciousness.

The proliferation of interest in developing artificial brains and brain-computer interfaces reflects how attractive the brain's capabilities are for purposes beyond the functions the brain developed during human evolution.[13] The interest also represents the perception that progress in neuroscience, computer science and data science now makes mapping, exploring and tapping into the brain's wonders more scientifically and technologically possible.[14] Although much of how the brain operates remains a mystery,

[10] Generally, 'deep learning' refers to improved machine learning that happens as a computer analyses data through many layers of networked computer processors – neural networks – designed to mimic how neurons work in the brain. Advances in mathematical algorithms, computing power and speed, and access to massive amounts of data have allowed neural networks to produce increasingly impressive achievements in machine learning.

[11] Blue Brain Project, *Digital Reconstruction*, available at https://bluebrain.epfl.ch/cms/lang/en/pid/59963.

[12] J. Hasler, 'We Could Build an Artificial Brain Right Now', *IEEE Spectrum*, 1 June 2017, available at https://spectrum.ieee.org/computing/hardware/we-could-build-an-artificial-brain-right-now (noting the human brain runs on about 20 watts of power).

[13] J. Richardson, 'Inside the Race to Hack the Human Brain', *Wired*, 16 November 2017, available at www.wired.com/story/inside-the-race-to-build-a-brain-machine-interface/?mbid=BottomRelatedStories (noting that 'financial analysts are forecasting a $27 billion market for neural devices within six years, and countries around the world are committing billions to the escalating race to decode the brain').

[14] See, e.g., R. Gonzalez, 'Neuroscientists Just Launched an Atlas of the Developing Human Brian', *Wired*, 7 December 2017, available at www.wired.com/story/neuroscientists-just-launched-an-atlas-of-the-developing-human-brain/?mbid=BottomRelatedStories. Researchers are also applying other areas of science to the challenge of understanding and replicating how the brain functions. For example, nanotechnology has been used to construct a synthetic synapse out of carbon nanotubes. See C. Dillow, 'A Synthetic Brain Synapse is Constructed

neuroscience research has improved our understanding of the brain.[15] Like other branches of science, neuroscience has exploited advances in computing, digital technologies, data analytics and other scientific disciplines.[16] Likewise, these advances help researchers see new possibilities for the knowledge neuroscience generates, such as informing how computer scientists can build on developments in artificial intelligence by replicating brain functions entirely through and on computers.

13.2.1 *Replicating the Human Brain*

The synergies between neuroscience, computer science and data science can be seen in the ambitions of the Human Brain Project, a high-profile research project supported by the European Union. Among its objectives are replicating the brain's architecture and its activities on computers, testing brain simulations with robots and developing neuromorphic, or brain-inspired, computing strategies.[17] Similarly, the Swiss-based Blue Brain Project 'aims to build comprehensive digital reconstructions (computer models) of the brain' and use the reconstructions to run simulations of neuron and synapse activity.[18] Military and intelligence agencies are also funding and participating in brain-related research.[19]

from Carbon Nanotubes', *Popular Science*, 22 April 2011, www.popsci.com/science/article/2011-04/synthetic-synapse-constructed-carbon-nanotubes. On nanotechnologies generally, see Chapter 10.

[15] See, e.g., Allen Institute for Brain Science, available at www.alleninstitute.org/what-we-do/brain-science/ (working to discover the brain's mysteries and share valuable resources with the global community of neuroscientists); and University of Southern California Neuroimaging and Informatics Institute, Human Connectome Project, available at www.humanconnectomeproject.org (working to provide 'an unparalleled compilation of neural data, an interface to graphically navigate this data and the opportunity to achieve never before realized conclusions about the living human brain').

[16] The neuroscience application of computing, digital technologies and data analytics has produced the field of 'neuroinformatics', defined as 'a highly interdisciplinary field employing methods and approaches from computer science, information systems and integrative biology to identify, analyze, digest and compute neuroscience data.' S. Polavaram and G. Ascoli, 'Neuroinformatics', *Scholarpedia*, 2015, doi: 10(11):1312, available at www.scholarpedia.org/article/Neuroinformatics. Scientists can also use nanolasers to manipulate neurons and study their functions. See K. Leins, 'Shining a Regulatory Spotlight on New Lasers: Regulation of the Use of Nanolaser Technologies in Armed Conflict', *Jurimetrics* (2016); 56 261–278, at 267 (noting how nanolasers enable 'scientists to manipulate or move matter at the single neuron level, isolating particular responses and behaviors generated within the brain').

[17] Human Brain Project, *Overview*, available at www.humanbrainproject.eu/en/about/overview/.

[18] Blue Brain Project.

[19] In the United States, see Intelligence Advanced Research Projects Activity (IARPA), Machine Intelligence from Cortical Networks (MICrONS), available at www.iarpa.gov/index.php/

As noted above, the development of artificial intelligence has looked to neuroscience in building large, multi-layered neural networks that enable computers to engage in deep learning. Artificial intelligence technologies are achieving impressive results. As one example, in December 2017, NASA announced that researchers developed and used a neural network to identify previously unknown planets orbiting stars by processing massive amounts of data collected by the Kepler Space Telescope. A Google engineer built a neural network and trained it to recognise planets using a database of planets previously identified by astronomers studying Kepler Space Telescope data.[20] NASA has a vast database generated from the Kepler Space Telescope's hunt for planets, which is much too big for astronomers to search themselves. Data from nearly 700 star-systems was run through the neural network, which identified two planets orbiting two different stars. The researchers plan to apply the neural network to the Kepler Space Telescope's full data set on 150,000 stars in the search for more planets.[21]

However, artificial intelligence technologies, including neural networks, 'are still a world away from being intelligent, from being able to understand and act in the world the way that we do.'[22] These technologies do not replicate what happens in the brain because 'they use an overly simplified neuron model, one that omits essential features of real neurons, and they are connected in ways that do not reflect the reality of our brain's complex architecture'.[23] The researchers who used a neural network to find planets cautioned against reading too much into their achievement because '[n]eural networks may be called artificially intelligent, but the algorithm is only as good as it was taught to be, and it cannot generalize what it's learned in the same way that humans can.'[24]

research-programs/microns; and Defense Advanced Research Projects Agency (DARPA), Systems of Neuromorphic Adapted Plastic Scalable Electronics (SyNAPSE), available at www.darpa.mil/program/systems-of-neuromorphic-adaptive-plastic-scalable-electronics.

20 On Google's work on advancing artificial intelligence, see Google, *Google Brain Team*, available at https://research.google.com/teams/brain/.

21 NASA, *Artificial Intelligence, NASA Data Used to Discover Eighth Planet Circling Distant Star*, 14 December 2017, available at www.nasa.gov/press-release/artificial-intelligence-nasa-data-used-to-discover-eighth-planet-circling-distant-star.

22 J. Hawkins, 'What Intelligent Machines Need to Learn from the Neocortex', *IEEE Spectrum*, 2 June 2017, available at https://spectrum.ieee.org/computing/software/what-intelligent-machines-need-to-learn-from-the-neocortex.

23 Hawkins, 'What Intelligent Machines Need to Learn'.

24 M. Young, 'Neural Network Finds 8-Planet System', *Sky & Telescope*, 15 December 2017, available at www.skyandtelescope.com/astronomy-news/neural-network-find-rivals-solar-system/.

Creating an artificial or synthetic brain seeks to go beyond what artificial intelligence technologies achieve and how they produce their results. Achieving this objective requires 'reverse engineering' the human brain *in silico* – reconstructing or replicating the brain's structure and functions through digital technologies.[25] Developing this capability is both tantalising in its potential applications and disturbing in its possible implications.

An artificial brain functioning at the level of human cognition would be an astonishingly powerful and versatile tool that could be applied to a vast array of tasks people perform. An artificial brain's utility would surpass the capabilities of machinery and devices exploiting automation, autonomous technologies or artificial intelligence. Indeed, an artificial brain could be more powerful than a human brain because it would have the attributes of human cognition, access to the phenomenal data-processing capabilities of computers, and the ability to operate continuously without the biological requirements that constrain human performance. Presumably, as an *in silico* technology, an artificial brain could be copied and upgraded on a large scale, making its transformative possibilities widely available across economies, governments and societies.

These almost endless possibilities excite researchers pushing different scientific fields towards replication of the human brain.[26] Harnessing artificial brains would provide societies with a new capability to help humans address difficult problems from the individual to the planetary level. Building and operating artificial brains could help identify potential causes of degenerative brain diseases, such as Alzheimer's and other types of dementia and could facilitate the search for better elder care, pharmaceutical treatments, cures and preventive measures.[27] Artificial brains could be exploited to process the massive quantities of digital data produced in an increasing number of ways, including by the expanding human dependence on digital devices and by the growing use of terrestrial and space-based imaging and

[25] J. J. DiCarlo, 'To Advance Artificial Intelligence, Reverse-Engineer the Brain', *Wired*, 3 February 2018, available at www.wired.com/story/to-advance-artificial-intelligence-reverse-engineer-the-brain/.

[26] See, e.g., US National Institutes of Health, Brain Research through Innovative Neurotechnologies (BRAIN) Initiative, available at www.braininitiative.nih.gov (public–private partnership that aims 'at revolutionizing our understanding of the human brain. By accelerating the development and application of innovative technologies, researchers will be able to produce a revolutionary new dynamic picture of the brain that, for the first time, shows how individual cells and complex neural circuits interact in both time and space').

[27] Artificial intelligence technologies are also being developed to diagnose diseases. See, e.g., A. Ram, 'Deep Mind Develops AI to Diagnose Eye Diseases', *Financial Times*, 4 February 2018, available at www.ft.com/content/84fcc16c-0787-11e8-9650-9c0ad2d7c5b5.

sensor technologies.[28] Such data-processing prowess could be applied, for example, to finding patterns and insights in data to inform policy making in every political realm, including disease prevention, cybersecurity, precision agriculture and environmental protection. Plugging artificial brains into the workforce could improve economic productivity by adding cognitive firepower to enterprises, freeing human employees from performing time-consuming data processing and other tasks, and creating new types of jobs.

The optimism associated with artificial brain technology has, however, a familiar and disconcerting ring to it. In the past, transformative technologies have always come with adverse consequences, both foreseeable and unpredictable. For example, when it emerged into widespread use in the mid-1990s, the Internet was widely viewed in optimistic ways, especially as a technology that would liberate human expression and communication, empower and expand democratic forms of political life and hasten the demise of authoritarian governments. Today's cyberspace comes nowhere close to approximating this vision of the Internet's implications for freedom and democracy in human affairs.[29] Similarly, the potential for malicious use of artificial intelligence is already a significant concern.[30] Thus, replication of the human brain could also generate serious problems that societies might struggle to handle.

For example, an artificial brain would be a dual-use technology – a technology used for both civilian and military purposes. In the United States, defence and intelligence agencies are funding research and development on synthetic brain technologies with the purpose of finding ways to exploit such technologies in the competition for geopolitical advantage and during armed conflict.[31] Just as with the Internet, there is nothing intrinsically peaceful or warlike about artificial brain technology. How governments use such a dual-use technology will not, in all likelihood, be determined by artificial brains.

Politically, artificial brains could produce many kinds of disruptive effects. As experienced with other technological leaps, disputes about equitable access to artificial brain technologies would arise and generate friction within

[28] On the increasing importance of data derived from space-based monitoring, see National Academies of Sciences, Engineering, and Medicine, *Thriving on Our Changing Planet: A Decadal Strategy for Earth Observation from Space* (2018), available at www.nap.edu/catalog/24938/thriving-on-our-changing-planet-a-decadal-strategy-for-earth.

[29] See, e.g., D. P. Fidler, 'Requiem for the Internet Freedom Strategy', *Net Politics*, 17 October 2017, available at www.cfr.org/blog/requiem-internet-freedom-strategy.

[30] See, e.g., M. Brundage, et al., *The Malicious Use of Artificial Intelligence: Forecasting, Prevention, and Mitigation* (February 2018), available at https://img1.wsimg.com/blobby/go/3d82daa4-97fe-4096-9c6b-376b92c619de/downloads/1c6q2kc4v_50335.pdf.

[31] See sources cited in footnotes 19 and 37.

societies and among high-, middle- and low-income countries. The political fallout and social damage associated with the spread of robotic automation in manufacturing could be repeated, but on an even larger scale, by pervasive use of artificial brain technologies in economic activity.[32] Previous technological revolutions have stressed or shattered institutions, customs and belief systems that contributed to political and social stability. What widespread application of artificial brains might do to individual relationships, community bonds, religious institutions and the interplay between the governed and the government is anybody's guess, which is hardly a comforting thought.

Functioning artificial brains would also generate ethical and philosophical questions and controversies. If a device replicates human cognition, is it something more than a magnificent artificial intelligence application? Is the device's ability to learn like a person evidence that it is more than a machine but not human? Will artificial brains develop idiosyncratic traits we could liken to personalities? Reactions we could associate with emotions? Output we could see as creativity? Preferences and beliefs we could analogise to ideology, dignity, honour or conscience? Behaviour we could say exhibits consciousness?[33]

13.2.2 *Interfacing with the Human Brain*

Brain–computer interfaces constitute a different strategy for harnessing the human brain in new ways for novel purposes. The objective is not to copy the brain *in silico*. Rather, the goal is to connect the human brain directly with computers and/or devices in order to facilitate a wide array of possible tasks. The most familiar and advanced example of a brain–computer interface involves physically implanting electrodes or electronic devices into the brain in order to allow a paralysed person to control computer-enabled prosthetic limbs, move damaged arms or legs, or operate machines, such as a personal computer. As *The Economist* noted in early 2018:

> Since 2004, 13 paralysed people have been implanted with a system called BrainGate … An array of small electrodes … is implanted into the motor cortex, a strip of the brain that governs movement. These electrodes detect

[32] Similarly, the impact on economic activity, and especially employment, of the accelerating progress in, and expanding adoption of, artificial intelligence technologies is a growing concern.

[33] See, e.g., C. Koch and G. Tononi, 'Can We Quantify Machine Consciousness?', *IEEE Spectrum*, 25 May 2017, available at https://spectrum.ieee.org/computing/hardware/can-we-quantify-machine-consciousness.

> the neurons that fire when someone intends to move his hands and arms. These signals are sent through wires that poke out of the person's skull to a decoder, where they are translated into a variety of outputs, from moving a cursor to controlling a limb.[34]

Electronic brain implants have also been used to treat brain disorders, including epilepsy and Parkinson's disease,[35] and to improve memory.[36] Ambitions for brain–computer interfaces extend, however, beyond this approach and this purpose. At the farthest edge of these ambitions is the creation of non-invasive brain–computer interfaces that facilitate telepathic communication between people and machines and, with computers as the technological medium, between people. As with artificial brain technology, brain–computer interface technologies are attracting increasing interest from governments (including military forces),[37] researchers and the private sector. Perhaps the biggest development in this area is the interest and money companies and entrepreneurs in Silicon Valley and beyond are devoting to this research and development challenge (see Table 13.1).

Although not an attempt to reverse engineer the human brain, achieving better brain–computer interfaces would create its own expansive universe of applications. The cutting-edge research pushing the field towards non-invasive interfaces opens possibilities for linking human brains with virtually any digital device, and, through networked devices, for brain-to-brain connections – including connections with artificial brains. Human brains could become, like cars and home appliances, part of the burgeoning 'Internet of things.' This connectivity is part of what inspires advocates of developing brain–computer interfaces to believe such interfaces could help humans become more intelligent,[38] an outcome some believe is necessary as artificial intelligence continues to become more powerful.[39]

[34] Palmer, 'Thought Experiments'.

[35] Palmer, 'Thought Experiments'.

[36] B. Carey, 'The First Step Toward a Personal Memory Maker?', *New York Times*, 12 February 2018, (reporting 'that scientists had developed a brain implant that boosts memory').

[37] US Defense Advanced Research Projects Agency (DARPA), Towards a High-Resolution, Implantable Neural Interface, available at www.darpa.mil/news-events/2017-07-10.

[38] Z. Niemtus, 'The Disrupters – Humans 2.0: Meet the Entrepreneur Who Wants to Put a Chip in Your Brain', *The Guardian*, 14 December 2017, available at www.theguardian.com/small-business-network/2017/dec/14/humans-20-meet-the-entrepreneur-who-wants-to-put-a-chip-in-your-brain (quoting the founder of Kernel as wanting 'to expand the bounds of human intelligence').

[39] Palmer, 'Thought Experiments', noting that the founder of Neuralink, Elon Musk, thinks brain–computer interfaces are necessary if humans 'are not to be left in the dust by AI'; and C. Koch, 'To Keep Up with AI, We'll Need High-Tech Brains', *Wall Street Journal*, 27 October 2017.

TABLE 13.1 *Companies Investing in Brain–Computer Interfaces*

Company	Description
Facebook	Conducts research on a 'silent speech' brain–computer interface that would allow people to type text directly from the brain.
Kernel	Develops invasive and non-invasive technologies to understand and treat neurological diseases and to create technologies for human cognitive advancement.
Neurable	Develops non-invasive brain–computer interfaces that enable people to control computers through cerebral activities.
Neuralink	Develops ultra-high bandwidth, non-invasive brain-machine interfaces to connect computers and people.
Neuropace	Makes an implantable electronic device that detects brain activity associated with epileptic seizures and delivers electronic treatment to prevent seizures.
Openwater	Uses opto-electronics to create a wearable device that generates high-resolution imaging of the brain and body for detecting diseases (e.g., cancer) and brain disorders and for, eventually, communicating through thoughts.
Paradomics	Develops high-bandwidth neural interface using hundreds of thousands of microwires to enable biomedical device companies to address neural prosthetic devices.

Sources: Company websites.

In theory, and like artificial brains, such brain–computer interface capabilities could be applied in any field of human endeavour. Such technologies could revolutionise how people interact with digital devices and how people communicate with each other. In both cases, human dependence on and vulnerability to, digital technologies would likely increase, raising questions about privacy, ownership of data flowing between brains and machines, the security of information transmitted across brain–computer interfaces, the impact on human relationships and the effects on conceptions of human personality, autonomy and dignity.

In all likelihood, the disruptive political, economic and social potential of brain–computer interfaces will depend foremost on whether research and development advance non-invasive interfaces. The actual and perceived risks of physically embedding electrodes, wires or chips in the brain will limit adoption of invasive brain–computer interfaces. Those with the need or interest in having invasive brain–computer interfaces will remain a small proportion of the population that suffers from specific conditions, such as paralysis or specific brain-related disorders. As companies with invasive techniques, such as Neuropace's embedded device for epilepsy prevention, have discovered, the medical and legal requirements for invasive interfaces impose economic and regulatory costs that constrain wider adoption of the technologies.

13.3 SYNTHETIC BRAIN TECHNOLOGIES AND THE LAW FOR PEACETIME

13.3.1 *Preliminary Observations on Synthetic Brain Technologies and Law*

In many contexts, technological developments emerge rapidly and are adopted widely, leaving law and policy struggling to keep pace with the technologies themselves and their impact on societies. Synthetic brain technologies have not matured sufficiently for this pattern to have appeared. The optimism associated with artificial brain research and brain–computer interfaces typically comes with acknowledgment that these technologies remain decades from approximating the ambitions of researchers, developers, entrepreneurs and visionaries. With synthetic brain technologies the policy and legal urgency seen with other dual-use technologies, such as autonomous weapons,[40] is not present.[41]

This timeframe makes it difficult to assess the scientific, technological, policy and legal landscape that might be in place when synthetic brain technologies might start to fulfil the potential many claim they have. Literature on synthetic brain technologies typically identifies the many gaps that remain in scientific understanding of how the brain works. Producing a machine with human-level cognition will require closing, if not eliminating, these gaps, and no one knows when science will reach the point at which it can replicate such cognition. The challenges go beyond continuing to decipher how neurons and synapses work. They include figuring out how billions of neurons and trillions of synaptic connections function at lightning speeds with minimal energy consumption in producing the profound cognitive capabilities humans have.

[40] On autonomous and automatic weapons, see Chapter 6.

[41] Advances in neuroscience are having other, more immediate impacts on law, captured by the emerging areas of 'neurolaw' and 'neurolegislation.' See, e.g., F. X. Shen, 'Law and Neuroscience 2.0', *Arizona State Law Journal* (2016); 48: 1043–1086; F. X. Shen, 'The Overlooked History of Neurolaw', *Fordham Law Review* (2016); 85: 667–695; F. X. Shen, 'Neurolegislation: How U.S. Legislators are Using Brain Science', *Harvard Journal of Law & Technology* (2016); 29: 495–526; and A. J. Kobler, 'Will There Be a Neurolaw Revolution?', *Indiana Law Journal* (2014); 89: 807–845.

As research on artificial brains and brain–computer interfaces proceeds, the technological, policy and legal landscapes might shift in response to other developments. Continued progress in artificial intelligence technologies might lessen interest in, and demand for, pushing the boundaries of artificial brain research.[42] Government and military interest in better brain–computer interfaces to improve prosthetic technologies might decline as the counter-insurgency conflicts in Iraq and Afghanistan that produced limb injuries on a large scale fade farther into the past.[43] Governments and societies might begin to recoil at the impact of artificial intelligence and start regulating it in ways that constrain the development and application of synthetic brain technologies. In short, conclusions drawn from applying the law of peace and war as it exists today to the current state of synthetic brain technologies may well have a very short shelf-life.

Thinking about synthetic brain technologies and the law also requires appreciating how such technologies differ from other dual-use technologies that have raised concerns under peacetime and wartime laws. For example, fears about autonomous vehicles include legal and ethical trepidation about removing humans from decision-making processes critical to the safety of passengers and transportation systems.[44] Brain–computer interfaces would not create this concern because a human would remain in the decision-making loop of an interface used to guide a vehicle. Using an artificial brain capable of human-level cognition to make decisions that might, for example, affect the enjoyment of human rights in peacetime might not generate serious anxieties. Presumably, such a brain can, like a human's, be trained to take into account the requirements of applicable law. In other words, the very things that distinguish artificial brains and brain–computer interfaces from other technologies, especially artificial intelligence, potentially increase the prospects that use of these technologies can comply with the law.

Synthetic brain technologies will be vulnerable, however, to cyber intrusion, interference and manipulation because they are cyber-based technologies.[45] Like other cyber-enabled technologies, artificial brains and brain-computer interfaces would become part of, and subject to the cybersecurity problems associated with, the Internet of things. The utility of synthetic brain

[42] On potential legal impacts of artificial intelligence, see generally H. Surden, 'Machine Learning and the Law', *Washington Law Review* (2014); 89: 87–115.

[43] See, e.g., D. Wallace, 'Trends in Traumatic Limb Amputation in Allied Forces in Iraq and Afghanistan', *Journal of Military and Veterans Health* (2012); 20:2, available at http://jmvh.org/article/trends-in-traumatic-limb-amputation-in-allied-forces-in-iraq-and-afghanistan/.

[44] For discussion of autonomous vehicles, see Chapter 6.

[45] For discussion about cybersecurity challenges, see Chapter 5.

technologies, and how people perceive them, could be undermined if cybersecurity vulnerabilities are not addressed.

Unauthorised access to, and interference with, computers is illegal in most national systems of law, and, for some countries, cybercrime treaties supplement national laws.[46] However, as seen in connection with societal dependence on cyber technologies, legal prohibitions of unauthorised access and interference with computers and other digital devices have not deterred State and non-State actors from engaging in this behaviour on a massive scale. Neither have laws intended to protect an individual's privacy or a corporation's intellectual property provided much protection in cyberspace. Further, cyber interference with brain–computer interfaces that involves corrupting or deleting information the interfaces produce creates other questions for cybercrime law, including whether 'the intentional wiping of someone's stored memories should be a recognized crime' and whether a cyber attack on a brain–computer interface should be considered the same as 'a physical assault on a victim.'[47]

In addition, artificial brain technologies might have a very large cyber 'attack surface' because they would need high-speed access to diverse computer systems and networks in order to receive, analyse, learn from and transmit information in replicating human-level cognition. 'Hardening' synthetic brain technologies against cyber threats might constrict their ability to function like human brains by filtering information streams, restricting access to certain data sources, or embedding security software that slows down learning. Generally, at both national and international levels, law has not, so far, proved an effective instrument in efforts to improve public and private-sector cybersecurity. Nothing about synthetic brain technologies suggests they will somehow make law a more powerful way to strengthen cybersecurity.

Not only could artificial brains and brain–computer interfaces be hacked but they could also be manipulated by the dissemination of propaganda and 'fake news.' Widespread concerns about terrorist groups, foreign governments, and domestic politicians exploiting social media platforms for divisive and/or malicious ends reflect many things, including the susceptibility of many humans to misleading and false information. Synthetic brain technologies may also prove susceptible to such manipulation. Countries with legal systems

46 See, e.g., Council of Europe Convention on Cybercrime, 23 November 2001, European Treaty Series No. 185.

47 S. S. Wu and M. Goodman, 'Neural Devices Will Change Humankind: What Legal Issues Will Follow?', *American Bar Association SciTech Lawyer* (Winter 2012), available at www.americanbar.org/groups/science_technology/publications/scitech_lawyer/2012/winter/neural_devices_will_change_humankind_what_legal_issues_will_follow.html.

that protect against government censorship are struggling with the problem of social media manipulation and, to date, have tried to rely on non-binding public–private partnerships rather than legal measures to address terrorist propaganda, foreign information operations and the proliferation of 'fake news.'[48] Whether synthetic brain technologies might tip the balance more in the direction of legal mandates is impossible at the moment to predict.

13.3.2 *Artificial Brains and the Law for Peacetime*

An artificial brain functioning with human-level cognition could be used for almost any intellectual task humans now undertake – a scope that would surpass the potential impact of autonomous technologies.[49] Thus, the scope of peacetime law potentially implicated by artificial brain technologies is comprehensive. Use of artificial brains by governments or the private sector would implicate a host of legal regimes, including those associated with utilisation of digital technologies, such as privacy protections, data processing regulations and cybersecurity requirements. Again, in theory, there is no reason to believe an artificial brain would be incapable of following legal requirements applicable to the tasks it is performing. For example, an artificial brain tasked to evaluate job candidates should be able to do so in conformity with laws that prohibit gender, racial or age discrimination. As with humans performing this task, the legal requirements of non-discrimination law in employment contexts would be necessary inputs for the decision-making process of an artificial brain. The purposes behind non-discrimination rules, and other legal regimes, would not tolerate exempting artificial brains from complying with the law.

Of course, many areas of law require judgement calls – rules that call for balancing different factors and interests in making decisions. But an artificial brain could learn from how such judgement calls have been made in the past, just as a lawyer, judge or government official does in accessing an array of electronic sources providing data on decisions of courts, administrative agencies and legislatures. An artificial brain could also seek advice from specialists, such as other artificial brains and human lawyers specifically trained in the issue at hand, and ask questions if the inputs it has do not produce sufficient clarity. It could send 'close calls' to human decision-makers, much

[48] However, patience with the effectiveness of non-legal approaches is running thin. See, e.g., C. Stupp, 'France Eyes EU Law to Crack Down on Terrorists' Use of Social Media', *Euractiv*, 16 February 2018, available at www.euractiv.com/section/digital/news/france-eyes-eu-law-to-crack-down-on-terrorists-use-of-social-media/.

[49] For discussion of the peacetime implications for law of autonomous technologies, see Chapter 6.

in the same way hard decisions often move up decision-making hierarchies in organisations.

Having a machine capable of performing with human-level cognition perhaps raises more questions about authority and responsibility in the law than it does about the machine's ability to comply with legal requirements. For example, should a government body authorise an artificial brain to practise law or undertake other activities that require legal authorisation, such as practising medicine or supervising prisoners and prison conditions? Should an artificial brain be able to enter legally binding contracts as an agent for a principal? Should a company be able to designate an artificial brain as an officer (e.g., company secretary) under corporate law? Should vehicles that transport people by land, sea and air be piloted by licensed artificial brain technologies?

Admittedly, these questions sound odd, but the strangeness arises because law (as opposed to science fiction) developed without having to grapple with machines possessing human-level intelligence. Progress in artificial intelligence has already sparked debates about the future legal status of thinking machines. For example, in February 2017, the European Parliament called on the European Commission 'to explore, analyse and consider the implications of creating a specific legal status for robots in the long run, so that at least the most sophisticated autonomous robots could be established as having the status of electronic persons responsible for making good any damage they may cause, and possibly applying electronic personality to cases where robots make autonomous decisions or otherwise interact with third parties independently.'[50] This call created controversy and prompted more than 150 European experts from various fields to oppose the creation of legal personality for robots.[51] The ethical and legal reasons for such opposition came from models of legal status that require the existence of a human being.[52]

Artificial brain technologies raise the possibility that governments and private-sector actors could authorise under law, or delegate legal authority to, such technologies to perform tasks historically only ever directly performed or controlled by people.[53] The current emphasis in policy and law on

[50] European Parliament, Resolution of 16 February 2017, para. 59(f).

[51] See J. Delcker, 'Europe Divided Over Robot "Personhood"', *Politico*, 11 April 2018, available at www.politico.eu/article/europe-divided-over-robot-ai-artificial-intelligence-personhood/.

[52] *See* Open Letter to the European Commission on Artificial Intelligence and Robotics, April 2018, available at https://g8fip1kplyr33r3krz5b97d1-wpengine.netdna-ssl.com/wp-content/uploads/2018/04/RoboticsOpenLetter.pdf.

[53] For analysis of artificial intelligence and administrative decision making, see C. Coglianese and D. Lehr, 'Regulating by Robot: Administrative Decision making in the Machine-Learning Age', *Georgetown Law Journal* (2017); 105: 1147–1223.

maintaining human control over autonomous weapons, aircraft, cars or vessels (e.g., 'person on the loop' systems) reflects the unease artificial intelligence already creates on this issue.[54] Artificial brain technologies would represent a leap in machine-based capabilities beyond what experts anticipate from artificial intelligence, and these capabilities would generate authority questions the law simply has not confronted before.

Problems about legal authority could arise with artificial brains even if no authority is delegated to such brains. For example, a law firm could operate an artificial brain in a way that replicates tasks performed by a human lawyer who has a licence to practise law. How do we handle a machine that is, in fact, practising law without a licence to practise law?[55] The increasing use of artificial intelligence in legal services has stimulated question about lawyers' use of machine learning, with the answer generally being that there must be a 'lawyer on the loop' – licensed lawyers must understand the technology they use and exercise independent legal judgment over everything in the legal advice they provide to clients, including contributions from artificial intelligence technologies.[56] While an understandable and manageable rule in early phases of integrating machine learning into peacetime economic and social activities, the development of human-level intelligence in artificial brains would more seriously challenge an approach that holds that only people can provide legal advice to people.

Behaviour sanctioned through law comes with responsibility under the law. Artificial brains could create questions in this realm of law as well. Many areas of law, including criminal and tort law, apply a 'reasonable person' test in determining whether a person has violated the law. Would a corporation's or person's reliance on inputs from an artificial brain, with capabilities beyond other types of machine-learning approaches, satisfy the reasonableness test? Should this test be applied to independent decisions taken by artificial brains, or is some other standard (e.g., a 'reasonable artificial brain' test) required that is disconnected from how we perceive the reasonableness of human behaviour? Could a corporation or person be held liable for the consequences caused by the independent decisions made and controlled by an artificial

[54] *See* Chapter 6, specifically Section 6.10 (examining civil aviation, surgical procedures and driverless cars).

[55] On digital technologies and the unauthorised practice of law generally, see C. M. Garlic, 'Guidance by Clicks: The Unauthorized Practice of Law in a Technology Driven World', *Trinity Law Review* (2014); 19: 267–307.

[56] See, e.g., Lois D. Mermelstein, 'Ethics Update: Lawyers Must Keep Up with Technology', American Bar Association (2013), available at www.americanbar.org/publications/blt/2013/03/keeping_current.html.

brain? When should it be considered unreasonable not to involve an artificial brain in making decisions and taking action?

Artificial brains would have less disruptive impact on responsibilities the law assigns to governments. For example, international human rights law imposes obligations on governments to protect, respect and fulfil certain human rights.[57] These obligations remain with the government no matter how it chooses to comply with them, and regardless of the technologies deployed in governmental activities.

Government delegation of authority to artificial brains as part of complying with human rights law will not relieve the government of its legal responsibility for violations of individual rights that occurred through such delegated authority. A government is responsible under international law for an internationally wrongful act committed by an entity, which could include an artificial brain, empowered by that government to exercise governmental authority, even if the entity acting under such authority exceeds its authority or contravenes instructions.[58] A government is also responsible for an internationally wrongful act committed by any person or groups acting on the instructions of, or under the direction or control of, that government and this rule could apply to 'proxy artificial brain' scenarios.[59]

International law on State responsibility only applies if a State has committed an internationally wrongful act.[60] Acts of espionage by States are not considered illegal under international law, even espionage that collects sensitive intelligence on corporations or individual persons.[61] Thus, under present international law, efforts to hack an artificial brain and conduct espionage against it would not be an internationally wrongful act.[62] Further, the advent of artificial brains would not likely change this legal reality, especially when efforts to limit the legal scope of espionage in the digital age under

57 Whether corporations and other non-State actors have obligations under international human rights law remains controversial.

58 International Law Commission, Responsibility of States for Internationally Wrongful Acts (2001), Articles 5 and 7.

59 International Law Commission, Article 8.

60 International Law Commission, Article 1.

61 A. Deeks, 'Confronting and Adapting: Intelligence Agencies and International Law', *Virginia Law Review* (2016); 102(3): 599–685.

62 Since the autumn of 2015, major countries have agreed in bilateral and plurilateral contexts not to engage in economic cyber espionage, creating a non-binding cyber norm against this activity. Whether this norm crystallises into a rule of customary international law, or is integrated into treaty law, remains to be seen.

human rights law, including the right to privacy, have had little, if any, impact on State practice.[63]

The development of the technologies used in artificial brain research might have impact on criminal law before artificial brains become a reality. Advances in neuroscience, computer science, data analytics and genomics are converging to produce the field of neuroforensics – the use of brain-related information in legal proceedings and contexts, including criminal law. Progress in 'neurotechnologies and associated databases' is producing 'attempts to introduce neuroscience and behavioural genetic evidence into legal proceedings', and:

> neuroscience evidence ... might conceivably be used by law enforcement, the courts, regulatory agencies and others as factors in predicting dangerousness; assessing competence to stand trial; assessing volitional control over actions; revealing mitigating factors relevant to sentencing; predicting recidivism; distinguishing pain from malingering; verifying intent; and manipulating memories.[64]

The ability of artificial brains to think, learn and make independent decisions might pose its own problems for criminal law. An artificial brain could make decisions that result in acts prohibited by criminal law, but, having made them through its own cognitive processes, it would be difficult to impose criminal liability on a person (e.g., a developer or supervisor) who lacks the requisite mental state (*mens rea*) for the crimes in question. Trying to apply the *mens rea* requirement to the artificial brain would serve little purpose without the criminal law having punishments for crimes committed by thinking machines.

Maturation of artificial brain technologies would replay controversies experienced in other technological contexts over equitable access to the fruits of scientific research and development. International law recognises the human right to enjoy the benefits of scientific progress and its applications.[65] According to the UN Special Rapporteur in the Field of Cultural Rights, the right to enjoy such benefits includes the 'core principle' that 'innovations

63 International Group of Experts, *Tallinn Manual 2.0 on the International Law Applicable to Cyber Operations*, 2nd edn (M. Schmitt, gen. ed.) (2016) [hereinafter *Tallinn Manual 2.0*], at 169.

64 US National Academies of Sciences, Engineering, and Medicine, *Neuroforensics: Exploring the Legal Implications of Emerging Neurotechnologies – A Workshop*, available at www.nationalacademies.org/hmd/Activities/Research/NeuroForum/2018-MAR-6.aspx.

65 International Covenant on Economic, Social, and Cultural Rights, 16 December 1966, 993 UNTS 3, Article 15(1)(b).

essential for a life with dignity should be accessible to everyone, in particular marginalised populations.'[66] However, the scope of this right and what qualifies as compliance with it have long been sources of disagreement, which arises with each transformative technology that appears. The UN Special Rapporteur captured how old and new this problem is by asserting that more human rights attention is needed for 'scientific advances likely to have a significant impact on human rights, such as electricity, information and communication technologies, nanotechnology and synthetic biology[.]'[67]

Including artificial brain technologies in international discussions on the right to enjoy the benefits from scientific progress and its applications will not reduce the controversies that surround the scope or content of this right. In fact, artificial brain technologies will, in all likelihood, be expensive and complex capabilities that only a few countries will be able to exploit effectively for many years. Calls for more equitable access to artificial brain technologies will probably generate the same disagreements that calls for more equitable access to other new technologies have provoked.

What might be possible with artificial brain technologies has prompted philosophical and ethical questions that have implications for law. For example, one fascinating possibility associated with artificial brains is whether, in replicating human cognition, such brains can also replicate, or exhibit, human traits, such as consciousness. Although a profound question, the law does not address how awareness of self and the world arises and functions through brain activities.[68] In the legal realm, a more relevant question is whether an artificial brain can develop its own thoughts, opinions, beliefs and conscience.[69] International human rights law protects individual privacy and the freedoms of thought, opinion, belief and conscience because these freedoms are considered integral to human dignity.[70]

[66] Report of the Special Rapporteur in the Field of Cultural Rights, *The Right to Enjoy the Benefits of Scientific Progress and Its Applications*, UN General Assembly Doc. A/HRC/20/26, 14 May 2012, at 9.

[67] Report of the Special Rapporteur in the Field of Cultural Rights, at 9.

[68] Generally, even the legal determination of when a person is 'brain dead' focuses on measurable indicators of brain activity and not on whether a person retains consciousness.

[69] On whether artificial brains would require legal protections, see generally R. Dowell, 'Note and Comment: Fundamental Protections for Non-Biological Intelligences or How We Learn to Stop Worrying and Love Our Robot Brethren', *Minnesota Journal of Law, Science & Technology* (2018); 19: 305–336. On the impact of the right of freedom of thought from various cognitive enhancement technologies, see M. J. Blitz, 'Freedom of Thought for the Extended Mind: Cognitive Enhancement and the Constitution', *Wisconsin Law Review* (2010): 1049–1117.

[70] International Covenant on Civil and Political Rights, 16 December 1966, 999 UNTS 171, Articles 17, 18 and 19.

Should the law recognise the right to privacy in the workings of artificial brains and impose requirements on governmental and non-governmental use of information such brains create, store, process and transmit?[71] Should the law protect thought, opinion, belief and conscience when artificial brains think, for example, in terms of justice? The objective would not, of course, be the protection of human dignity. However, if artificial brains can develop, through machine-based learning, their own sense of right and wrong and beliefs based on that sense, that ability could affect how an artificial brain evaluates rules of law and legal activities it encounters in functioning. Calls for justice often involve criticism of existing law as harmful, unfair or illegitimate. What should happen if an artificial brain conscientiously objected to the mechanical application, the cynical manipulation or malicious abuse of existing law? Would there be social value in protecting in law such objections from a thinking machine?

For example, many countries protect government workers who disclose illegal or other wrongdoing under so-called whistleblower' laws.[72] We could imagine an artificial brain technology operating under government auspices identifying and disclosing illegal or improper activities. However, as typically written, whistleblower laws apply to government employees not government-owned machines.[73] Without protection, disclosure of illegal acts or other misconduct by an artificial brain technology functioning within government could be ignored and possibly deleted from electronic records, which could allow identified abuses of governmental power to continue.

13.3.3 *Brain–Computer Interfaces and the Law for Peacetime*

With brain–computer interfaces, a human brain is at the centre of activity, which means a human remains involved in what the interface does. The interface functions to enhance human performance rather than to replicate human cognition outside the human body, as an artificial brain would do.[74]

[71] Even the very protective rules in the European Union's General Data Protection Regulation only apply 'to the protection of *natural persons* with regard to the processing of personal data' (emphasis added). General Data Protection Regulation, Regulation (EU) 2016/679, 27 April 2016, *Official Journal of the European Union*, L 119/1, Article 1.

[72] See generally G. Thüsing and G. Forst (eds.) *Whistleblowing: A Comparative Perspective* (Dordrecht: Springer, 2016).

[73] See, e.g., US Whistleblower Protection Act of 1989, as amended by the Whistleblower Protection Enhancement Act of 2012, available at www.gpo.gov/fdsys/pkg/BILLS-112s743enr/pdf/BILLS-112s743enr.pdf.

[74] On other human enhancement technologies, see Chapters 7 and 8.

Legally, this feature of brain–computer interfaces means we are working with the traditional subjects and objects of the law – human beings – rather than disembodied, thinking machines. Thus, generally speaking, the potential legal implications of brain–computer interfaces in peacetime are less radical than artificial brain technologies.[75]

As already noted, invasive brain–computer interface technologies do not escape existing legal regimes that apply to implantation of medical devices.[76] Surgeons implanting neural devices into human brains are still subject to liability under tort law. Nor do interfaces intended to provide health benefits avoid laws designed to ensure the safety and efficacy of medical devices. The implanted device for epilepsy prevention developed by Neuropace had to obtain legal approval from the US Food and Drug Administration and remains subject to various federal regulations.[77] Manufacturers of brain-computer interface technologies could be sued under tort, product liability and other consumer protection laws. The more invasive the brain–computer interface technology, the more regulated it will be under existing law.

The aperture for legal impact does not necessarily widen significantly with non-invasive brain–computer interfaces. Any non-invasive means of connecting the brain to a computer will have to be safe for humans to use for any purpose or it will not survive legal challenges, including tort suits by people allegedly injured by non-invasive technologies. Non-invasive interfaces can be designed for health purposes, such as understanding and treating neurological diseases. Many jurisdictions would treat such a non-invasive interface as a medical device subject to regulatory approval and oversight. For example, a non-invasive brain–computer interface technology intended to diagnose disease or cure, mitigate, treat or prevent human disease would be a 'medical device' under US law and regulated by the US Food and Drug Administration.[78]

75 On the legal implications of brain implants and other neural devices, see generally Wu and Goodman, 'Neural Devices Will Change Humankind'.

76 See generally G. Matthieu and B. Williams-Jones, 'Examining the National Regulatory Environment of Medical Devices: Major Issues in the Risk-Benefit Assessment of High-Risk Devices', *McGill Journal of Law and Health* (2015–2016); 9: 17–67.

77 Letter dated 14 November 2013 from the US Food and Drug Administration to Neuropace, Inc. approving the commercial distribution of the RNS System device, available at www.accessdata.fda.gov/cdrh_docs/pdf10/P100026A.pdf. See generally E. Chan, 'The Food and Drug Administration and the Future of the Brain-Computer Interface: Adapting FDA Device Law to the Challenges of Human-Machine Enhancement', *John Marshall Journal of Computer and Information Law* (2007); 25: 117–164.

78 A non-invasive brain–computer interface technology intended for use in disease diagnosis or the cure, mitigation, treatment or prevention of human disease would be a 'medical device'

Non-invasive brain–computer interfaces could be developed for non-medical purposes, including enhancement of human performance in commercial settings where employees face intensifying competition from artificial intelligence technologies. However, depending on the jurisdiction, such interfaces might still be regulated like medical devices. For example, under US law, any device intended to affect any function of the human body through non-chemical means is a medical device regulated by the US Food and Drug Administration.[79] Under this legal regime, a non-invasive brain-computer interface designed to improve human performance or intelligence in any setting would be affecting a function of the human body – the brain – without using chemical substances.

Unlike artificial brain technologies that function outside the human body and independently of a human brain, brain–computer interfaces operate with humans at the centre of the interaction. For this reason, laws that apply to human activities in different fields of endeavour will also regulate human behaviour enhanced by brain–computer interfaces. Laws do not cease to apply when humans get smarter, faster or stronger, however those improvements are achieved.

Brain–computer interfaces might generate difficult issues under criminal law. Information produced by such interfaces might flow into neuroscience research and databases attractive to the emerging field of neuroforensics noted earlier.[80] This possibility raises questions about how law enforcement agencies should treat information flowing between the brain and a computer interface under rules regulating government search and seizure. In terms of the Fourth Amendment of the US Constitution, would an individual have a reasonable expectation of privacy in brain–computer communications happening through the interface, especially if the communications are stored on a computer controlled or operated by a third party?[81] Could a law enforcement agency obtain a warrant or court order to 'wiretap'[82] or 'lawfully

under US law and would be regulated by the US Food and Drug Administration. *See* US Food and Drug Administration, *Is the Product a Medical Device?*, available at www.fda.gov/MedicalDevices/DeviceRegulationandGuidance/Overview/ClassifyYourDevice/ucm051512.htm.

79 US Food and Drug Administration, *Is the Product a Medical Device?*.

80 *See* footnote 64 and associated text.

81 Under US constitutional jurisprudence, a person has no reasonable expectation of privacy under the Fourth Amendment for information the person voluntarily shares with a third party. *See United States* v. *Miller*, 425 US 435 (1976), and *Smith* v. *Maryland*, 442 US 735 (1979). However, under the European Convention on Human Rights, information shared with third parties remains protected by the right to privacy. European Convention for the Protection of Human Rights and Fundamental Freedoms, 4 November 1950, Article 8.

82 US Wiretap Act, 18 U.S.C., Sections 2510–2522.

hack'[83] a brain–computer interface for law enforcement or counter-intelligence purposes? Should information created and stored through a brain–computer interface be protected from law enforcement access and use by the Fifth Amendment right against self-incrimination?

Like artificial brain technologies, brain–computer interfaces might generate concerns about equitable access to their use and benefits. It is not hard to imagine that non-invasive brain–computer interface technologies might long be available only to affluent individuals, well-resourced companies and high-income countries. The rich could exploit this technological advantage in ways that widen the gap between the wealthy and the rest in society. As noted above, the right to enjoy the benefits from scientific progress and its applications in international law has not been an effective legal basis for addressing technological inequities within countries or between them. Nor is it clear that national legal systems that attempt to reduce inequities involving access to technologies would make more equitable access to brain–computer interfaces a priority.

Brain–computer interfaces might not create the kind of philosophical, ethical and legal questions artificial brain technologies could raise concerning consciousness and freedoms of thought, opinion, belief and conscience.[84] A brain–computer interface could become so important that an individual considers it integral to his or her dignity. What this means for law is not entirely clear and would depend on the circumstances of any given individual and his or her brain–computer interface.

A person could physically or mentally depend on a brain–computer interface to function in society. Consider, for example, such an interface created by implanted medical devices that treat or regulate brain diseases or that allow a person to control prosthetic limbs. In such cases of functional dependence, protecting the person's access to, and use of, the interface could be important in the application of law, especially criminal law.[85] However, non-invasive interfaces used to enhance performance of healthy people in professional and

[83] S. M. Bellovin, M. Blaze, S. Clark and S. Landau, 'Lawful Hacking: Using Existing Vulnerabilities for Wiretapping on the Internet', *Northwestern Journal of Technology and Intellectual Property* (2014); 12(1): 3–64.

[84] For an argument that brain–computer interfaces form part of neuroscience support for a theory of human freedom, see N. A. Farahany, 'A Neurological Foundation for Freedom', *Stanford Technology Law Review* (7 December 2011): 11–53.

[85] See, e.g., American Bar Association, *Standards on Treatment of Prisoners* (2010), Standard 23–6.10: Impairment-Related Aids, available at www.americanbar.org/publications/criminal_justice_section_archive/crimjust_standards_treatmentprisoners.html#23-6.10 ('Prisoners whose health or institutional adjustment would otherwise be adversely affected should be provided with medical prosthetic devices or other impairment-related aids, such as eyeglasses, hearing

personal settings might not be a persuasive basis for arguing such interfaces become critical to human dignity in the application of law.

13.4 SYNTHETIC BRAIN TECHNOLOGIES AND THE LAW ON THE USE OF FORCE AND ARMED CONFLICT

13.4.1 *Preliminary Observations*

Generally, international law on the use of force (*jus ad bellum*) and armed conflict (*jus in bello*) applies regardless of the technologies militaries and combatants deploy.[86] Both sets of rules largely focus on the material consequences experienced by people and property in determining the legality of actions. For example, in applying *jus ad bellum* to cyber operations, the Tallinn Manual 2.0 argued that '[a]cts that injure or kill persons or physically damage or destroy objects are uses of force.'[87] Similarly, the Tallinn Manual 2.0 stated that the 'crux of the notion' of 'attack' in *jus in bello* 'lies in the effects that are caused', with any cyber operation 'reasonably expected to cause injury or death to persons or damage or destruction to objects' being an attack under the law of armed conflict.[88] This approach means that *jus ad bellum* and *jus in bello* would apply to any use of synthetic brain technologies that rises to the level of a use of force or an attack during armed conflict.

Whether the law of armed conflict applies in cyberspace has generated controversy in intergovernmental efforts seeking to clarify international law's application in cyberspace.[89] This controversy is relevant to synthetic brain technologies given their reliance on cyber capabilities. However, the

aids or wheelchairs, except when there has been an individualized finding that such an aid would be inconsistent with security or safety').

86 International law also contains rules that apply to specific types of weapon technologies, such as the treaties prohibiting the use of biological and chemical weapons in armed conflict.

87 *Tallinn Manual* 2.0, at 333.

88 *Tallinn Manual* 2.0, at 415.

89 After adopting consensus reports in 2013 and 2015, the UN Governmental Group of Experts on Information and Communication Technologies in the Field of International Security failed in 2017 to adopt a consensus report. This failure occurred in part because certain countries were not willing to state clearly that *jus ad bellum* and *jus in bello* apply in cyberspace. *See* A. M. Sukumar, 'The UN GGE Failed. Is International Law in Cyberspace Doomed as Well?', *Lawfare*, 4 July 2017, available at https://lawfareblog.com/un-gge-failed-international-law-cyberspace-doomed-well (noting the refusal of certain countries to affirm ' the application of the UN charter principles on the use of force and international humanitarian law' because that outcome 'would result in the 'militarisation' of cyberspace'). On efforts of the UN Group of Governmental Experts, see Chapter 5, specifically Section 5.5.

controversy mainly arises from geopolitical concerns a few countries, most notably China, have with the current advantages the United States might enjoy in armed conflict because of its advanced military cyber capabilities. Most States, including Western countries with serious military forces, acknowledge that *jus in bello* applies to means and methods of cyber warfare. In addition, the military use of offensive cyber capabilities in the armed conflict against the Islamic State by Australia, the United Kingdom and the United States means that State practice supports the application of the law of armed conflict to cyber weapons.[90] This outcome reinforces the long-standing position that *jus in bello* applies to any technology deployed during armed conflict.

In and of themselves, synthetic brain technologies are not, and cannot be, weapons, which makes them different from other dual-use technologies. For example, biological agents, chemical substances and radioactive material have peaceful uses, but each of these technologies can be used directly as a weapon to cause injury, death, damage and destruction. Launching an artificial brain at an enemy target would not, by itself, create these consequences. To do so, the artificial brain would have to be integrated with some kinetic, electromagnetic, cyber or other payload. It makes no sense to launch a brain–computer interface as a weapon because the interface could only support, control or enhance through human control some other weapon technology.

In addition, given the many ways synthetic brain technologies could be utilised, military use of an artificial brain or brain–computer interface would also not, by itself, produce consequences regulated by *jus ad bellum* and *jus in bello*. Civilian and military leaders could, for example, consult an artificial brain on whether using force in self-defence would be justified in specific circumstances or on the best ways to fulfil obligations created by military occupation of enemy territory. Military forces could undertake non-combat activities, such as moving supplies on military bases or delivering foodstuffs to civilian populations, by using brain–computer interfaces to control robots.

Civilian and military leaders could exploit synthetic brain technologies in making decisions whether to use force or launch attacks during armed conflict. Input from an artificial brain into such decisions would not create problems under *jus ad bellum* and *jus in bello* as long as humans maintain control over the input and make the actual decisions to use force or launch attacks. Controversies would arise if civilian and military leaders permitted an artificial brain to make such decisions independently and the legal implications of this scenario for *jus ad bellum* and *jus in bello* are discussed below.

90 A. Grigsby, 'Year in Review: Militaries Got More Cyber in 2016', *Net Politics*, 26 December 2016, available at www.cfr.org/blog/year-review-militaries-got-more-cyber-2016.

Integrating synthetic brain technologies into weapon systems could also be possible. For example, an artificial brain could be wired into a kinetic weapon, creating a thinking weapon with capabilities beyond the existing generation of smart or autonomous weapons. Brain–computer interfaces could potentially allow soldiers on the battlefield directly to control powerful robotic weapon platforms. The law of armed conflict would apply to the integration of synthetic brain technologies into weapons and to the use of such weapons in armed conflict.

13.4.2 *Synthetic Brain Technologies and the Law on the Use of Force*

International law prohibits States from using force in their international relations, unless a State uses force in self-defence or under a Security Council decision to authorise the use of force.[91] Other justifications for using force, such as humanitarian intervention, are, and have long been, controversial. Legally, as noted above, whether a State has violated the prohibition on the use of force is determined by analysing the consequences of a State's actions, rather than identifying the technologies in play in the situation in question. If a State uses certain kinds of technology against another State, such as powerful kinetic weapons, then such facts would inform the consequence-based analysis. Artificial brains and brain–computer interfaces could be exploited in deploying different weapons, including kinetic weapons. Even so, the use-of-force analysis would look at the intended, foreseeable or actual consequences to people and property in assessing whether a State committed an illegal use of force regardless of which technologies were involved.

Similarly, a Security Council authorisation to use force is grounded in the Council's determination that a threat to international peace and security exists – a determination that emerges from an assessment of the material consequences produced by the actions of States and/or non-State actors. Likewise, whether a State is permitted to use force in self-defence under the UN Charter requires assessing whether the consequences experienced by that State are serious enough to qualify as an 'armed attack.' In both the Security Council and self-defence cases, the assessment focuses on the adverse consequences to persons and objects rather than the different technologies that might have enabled such consequences to occur.

An adversary's possession of certain weapon technologies, such as weapons of mass destruction, has been used to argue that the right to use force in

[91] UN Charter, Articles 2(4), 42 and 51.

anticipatory self-defence justifies preemptive military action to prevent military threats from becoming imminent.[92] Such arguments have not produced consensus among States. Unlike the technologies used in weapons of mass destruction, an adversary's possession of synthetic brain technologies poses no direct threat to life or property that would support the preemptive self-defence argument. For weapons that integrate synthetic brain technologies, the preemptive self-defence position would still have to focus on the supported weapon systems because the synthetic brain components would not increase the lethality or destructiveness of the weapons themselves.

Under a Security Council authorisation to use force, UN member States can use force as long as the means and methods deployed comply with international law, including the law of armed conflict. No rule or instrument of international law prohibits States from using synthetic brain technologies in weapon systems, so a State could use weapons that utilise such technologies in a use of force pursuant to Security Council authorisation. How the law of armed conflict applies to the use of synthetic brain technologies in armed conflict is the focus of the next section.

A final *jus ad bellum* issue involves the scenario where a State permitted artificial brain technologies to make the decision to use force in a context without a Security Council authorisation. Why or when a government would delegate such momentous decision-making authority to an artificial brain is not clear. Artificial brain technologies could perhaps be delegated authority to launch weapons in self-defence in response to an armed attack, but the political, military or legal advantages gained by such a delegation are not clear. Any such delegation would not relieve a State of its obligation only to use force when justified under, and in compliance with, international law. Under the international law on State responsibility, any State delegating a use-of-force decision to artificial brain technologies would remain legally responsible for any such decision and its consequences.[93]

13.4.3 *Synthetic Brain Technologies and the Law of Armed Conflict*

International law regulates what weapons States can deploy in armed conflict and how States use legal weapons in armed conflict.[94] These rules would only

[92] See, e.g., White House, *The National Security Strategy of the United States of America* (September 2002), available at www.state.gov/documents/organization/63562.pdf, at 15.

[93] International Law Commission, Responsibility of States for Internationally Wrongful Acts (2001), Article 4.

[94] *See* Chapter 2 (on the law on weapons) and Chapter 3 (on the law regulating the use of weapons).

apply to synthetic brain technologies if governments integrated such technologies into weapon systems. As noted earlier, neither artificial brains nor brain-computer interfaces are, as such, weapons or weapon systems. Military forces could operate synthetic brain technologies during war for many purposes, such as intelligence gathering and data analytics, that do not violate or even trigger the application of the law of armed conflict.

The integration of synthetic brain technologies into weapon systems does not, by itself, raise legal red flags. Under international law, States would have to conduct a legal review of any weapon system that uses synthetic brain technologies to ensure the system's compliance with international law, including the law of armed conflict.[95] The synthetic brain components in a weapon system are unlikely to cause problems in a legal review.

First, none of the treaties that ban certain weapons apply to synthetic brain technologies.[96] Second, building artificial brain technologies or brain-computer interfaces into weapon systems does not make the nature of such systems likely to cause superfluous injury or unnecessary suffering, produce indiscriminate effects, or result in widespread and severe environmental damage. Indeed, having a weapon system wired with human-level intelligence or with connections to a human brain could potentially give the system additional capabilities for compliance with the law of armed conflict. Such capabilities do not guarantee such compliance or ensure mistakes will never be made. The fog of war would confound synthetic brain technologies as much as it does human intelligence, but that reality would not sway a legal review to reject a weapon system utilising synthetic brain technologies.

The use of any weapon system with synthetic brain capabilities will have to comply with the law of armed conflict's rules on launching attacks on enemy forces, including the principles of military necessity, distinction and proportionality.[97] In terms of military necessity, the determination that attacking a target produces a military advantage is unlikely to be driven by the availability of a weapon system with synthetic brain components. The principle of distinction's prohibition on intentionally attacking civilian targets applies regardless of weapon technology, and there is nothing about synthetic brain

[95] On the legal review of a new weapon system, see Chapter 2 , specifically Section 2.2 on the requirement for a legal review, Sections 2.3–2.4 on the principles applied in a legal review and Section 2.6 for a summary of the criteria applied in a legal review.

[96] On the prohibitions concerning specific types of weapons, see Chapter 2, specifically Section 2.5. See also the discussion of proposed bans on autonomous weapon systems operating without sufficient human involvement in Chapter 6, specifically Section 6.9.

[97] *See* Chapter 3 for a detailed treatment of the law of armed conflict's rules on conducting hostilities.

technologies that makes compliance with this prohibition problematic.[98] Similarly, synthetic brain technologies integrated into weapon systems would not create radical implications for determining under the proportionality principle whether an attack causes excessive incidental damage to civilian life and property in relation to the military advantage the attack is expected to produce.

One important caveat to these observations arises in connection with delegating targeting decisions to artificial brain technologies integrated into, or in control of, weapon systems. The advent of lethal autonomous weapon systems has generated controversies about the legal and ethical propriety of allowing machines to make targeting decisions and execute attacks without human decision-makers involved.[99] These controversies are relevant to permitting artificial brain technologies in weapon systems to select targets and launch attacks without human involvement. This issue does not arise with brain–computer interfaces because these interfaces involve direct human participation.

In connection with autonomous technologies that might permit a weapon system to make targeting and attack decisions without human involvement, whether international law requires a determination that such a weapon system can make the assessments and decisions mandated by the law of armed conflict has generated disagreement.[100] Some experts hold that any weapon system that makes independent decisions on targeting and launching attacks must be able to take such actions in compliance with the law of armed conflict. The US government has rejected this position, arguing that international law does not require weapon systems to be able to make legal determinations, even when the system makes targeting and attack decisions without human involvement.[101]

Existing autonomous weapon systems do not, and cannot, make independent targeting and attack decisions, which makes this disagreement somewhat conceptual at the moment. However, artificial brain technologies operating with human-level intelligence could potentially have the capability to apply

98 The same would be the case with rules on attacking other protected categories of people, objects or activities, including medical personnel, installations containing dangerous forces and cultural property.

99 See, e.g., Campaign to Stop Killer Robots, available at www.stopkillerrobots.org (arguing that '[g]iving machines the power to decide who lives and dies on the battlefield is an unacceptable application of technology' and that '[h]uman control of any combat robot is essential to ensuring both humanitarian protection and effective legal control').

100 See Chapter 6, specifically Section 6.5.

101 See Chapter 6, specifically Section 6.5, footnote 25.

the law of armed conflict in making target and attack decisions without human involvement. The presence of this capability would not resolve the disagreement described above, but it would make the legal question more pressing.

In addition, the capability to make legal decisions could also result in an artificial brain technology in a weapon system refusing to obey an order to launch what it determines is an illegal attack. The law of armed conflict imposes a duty on every combatant to disobey manifestly unlawful orders.[102] This obligation seems relevant for artificial brain technologies functioning with human-level intelligence that operate in weapon systems. Thus, an artificial brain independently controlling a weapon system must also have the capacity to refuse manifestly unlawful orders in order to comply with the law of armed conflict.

The use of artificial brain technologies in weapon systems creates other potential issues under the law of armed conflict. Decisions such technologies are programmed or permitted to make without human involvement could result in adverse consequences, including the death of civilians, that raise questions of compliance with the law of armed conflict and legal responsibility for what happened. Just like humans, artificial brain technologies operating in weapon systems could make mistakes, such as causing a kinetic weapon to hit a civilian area rather than a military target. Such mistakes do not necessarily involve a violation of the law of armed conflict, but they create obligations on the military forces involved to investigate the incident and take steps to correct what produced the mistake.

What if artificial brain technologies in a weapon system went 'rogue' and intentionally directed a weapon to attack a civilian population? Such an attack constitutes a war crime in international law, and military personnel who engage in war crimes face individual criminal liability for their actions.[103] In addition, commanders who knew, or should have known, about the commission of war crimes and failed to take appropriate action can be held criminally responsible.[104] With an artificial brain, there is no individual soldier to prosecute for committing the war crime. Holding commanders criminally liable in such situations might prove very difficult.[105] The State that deployed the

[102] International Committee of the Red Cross, *Rule 154 – Obedience to Superior Orders*, Customary International Humanitarian Law Database, available at https://ihl-databases.icrc.org/customary-ihl/eng/docs/v1_rul_rule154.

[103] See Rome Statute of the International Criminal Court, 17 July 1998, 2187 UNTS 3, Article 8.

[104] Rome Statute, Article 28.

[105] Questions of command responsibility under the law of armed conflict could also arise if commanders used brain–computer interface technologies. See J. M. Prescott, 'Building the

weapon would be responsible under the international law of State responsibility, but this responsibility is not criminal in nature. Thus, this scenario identifies the possibility that criminal sanctions for war crimes could have no effective application concerning weapon systems operating with artificial brain technologies.

The law of armed conflict includes other rules under which the use of synthetic brain technologies would not likely cause major problems. For example, the law of armed conflict contains rules on the treatment of wounded and sick combatants,[106] prisoners of war,[107] and civilian populations under military occupation.[108] States could use synthetic brain technologies in various ways to help them comply with the obligations imposed by these rules. Brain–computer interfaces could perhaps help military police personnel operate ground and aerial vehicles more effectively over more territory, thus contributing to fulfilment of an occupying force's duty to maintain law and order. Artificial brains could have a wide scope of application in these situations, including supporting efforts to care for wounded combatants, treat prisoners of war humanely, and fulfil the myriad responsibilities an occupying military force faces. As in other contexts, the most difficult questions would arise from delegating authority for artificial brains to make independent decisions in contexts regulated by the law of armed conflict, including whether artificial brain technologies could demonstrate the ability to comply with the law.

13.5 CONCLUSION

Unlike other technologies analysed in this volume, synthetic brain technologies are very much over-the-distant-horizon possibilities. The research-and-development distance to the horizon is significant and daunting, especially for artificial brain technologies. The timeframe for reaching the horizon is, even

Ethical Cyber Commander and the Law of Armed Conflict', *Rutgers Computer and Technology Law Journal* (2014); 40: 42–77, at 52 (pointing out the problems commander use of brain–computer interfaces could create under the law of armed conflict, including command responsibility).

106 Geneva Convention (I) on the Amelioration of the Condition of the Wounded and Sick in Armed Forces in the Field, 12 August 1949, 75 UNTS 31; and Geneva Convention (II) for the Amelioration of the Condition of the Wounded, Sick and Shipwrecked Members of the Armed Forces at Sea, 12 August 1949, 75 UNTS 85.

107 Geneva Convention (III) Relative to the Treatment of Prisoners of War, 12 August 1949, 75 UNTS 135.

108 Geneva Convention (IV) Relative to the Protection of Civilian Persons in Time of War, 12 August 1949, 75 UNTS 287.

in optimistic scenarios, measured in decades. In the meantime, technological, policy and legal developments, particularly concerning artificial intelligence, might change what the horizon for synthetic brain technologies can be and might alter the incentives for advancing towards that shifting objective. This context understandably makes analysis of synthetic brain technologies under the existing laws applicable in peace and war a curious exercise.

The most potent ideas associated with synthetic brain technologies are the possibilities to replicate human-level intelligence *in silico*, and to enhance human intelligence and performance through direct, non-invasive brain-computer communications. Realising these ideas would be a technological marvel for the ages, but, for law, the implications range from the pedestrian to the transformative. Law already regulates the application of human intelligence in all fields of human endeavour. The advent of artificial brains or brain–computer interfaces will not change the means and ends of a great deal of existing law. Societies are unlikely to allow artificial brains or people operating brain–computer interfaces to act in ways ordinary humans cannot.

The transformative potential primarily arises from how artificial brains force jurisprudence to contemplate expanding legal authority/rights and responsibility/duties beyond humans. In history, humans have been the only agents of law. This reality has stimulated theoretical debates about human agency in law, which have involved, among other things, different perspectives on human nature, natural law and the relationship of morality and law.[109] The possibility of artificial brains that function with human-level intelligence creates the opportunity to re-think legal agency in a very different sense. The notion is, indeed, fantastical. It means we would have to take seriously the need, in law, to protect cognitive activities generated entirely through *in silico* methods. It means we would have to contemplate the need, in law, to expect those cognitive activities to meet obligations without human intervention.

We have time to ponder these unprecedented questions for the rule of law. About the only other thing that is certain as we gaze towards that distant horizon is that these questions will be too important for only lawyers to try to answer.

109 See, e.g., H. L. A. Hart, *The Concept of Law* (Oxford: Oxford University Press, 1961); and L. Fuller, *The Morality of Law* (New Haven, CT: Yale University Press, 1969).

14

Biometrics

WILLIAM H. BOOTHBY

14.1 INTRODUCTION

The *Concise Oxford English Dictionary* defines biometrics as 'the application of statistical analysis to biological data',[1] which, in many ways, lays bare the challenge of writing' this chapter. The topic, like many of the others addressed in this book, is very broad. The issues raised by the use of biometrics are varied and numerous, spanning different fields, including human rights, data protection and the law of armed conflict to name but three. The danger, as when addressing any such broad area, is of scratching the surface while failing to address important aspects. The additional point to acknowledge is that biometrics, in one form or another, has been in use for many more decades than many of the other technologies addressed in this volume (excepting manipulation of nanomaterials),[2] but further development of these technologies,[3] their potential for employment in novel applications, and their capacity to influence the performance or facilitate the use of other relevant technologies suggests that to fail to include biometrics would be remiss. Nevertheless, an attempt will be made to keep this chapter relatively brief, partly at least because to do otherwise might distort the balance of the book or blunt its intended focus as a whole on recently developed, cutting edge, and at or over the horizon technologies.

[1] *Concise Oxford English Dictionary*, 11th edn (2006), 136.

[2] Consider the use of photographs by police forces to record the facial features of criminals and the use of fingerprints to detect the perpetrators of crimes.

[3] Consider, for example, the Biometrics Exploitation Science and Technology Program discussed on the US Director of National Intelligence website at www.iarpa.gov/index.php/research-programs/best and research into the development of improved facial recognition under the Janus Program, discussed at www.iarpa.gov/index.php/research-programs/janus.

Confirming somewhat that this is by no means an 'at or over the horizon technology', the European Network and Information Security Agency (ENISA) published a report on behavioural biometrics some years ago.[4] The ENISA report refers to behavioural biometrics that are capable of being used in an information security context to identify individuals by using their unique features or the unique activities that they perform. The ENISA report explains, in terms that are satisfactory for present purposes, what biometrics is really all about. Examples of biometric features that can be used for such purposes include but are not limited to: blinking patterns; aspects of electromagnetic signals generated by the heart; aspects of electromagnetic signals generated by the brain; the way in which an individual walks; the way in which an individual types:[5] aspects of a person's strategy in certain games; linguistic features; voice characteristics;[6] the face;[7] fingerprints;[8] palm prints;[9] iris recognition;[10] hand geometry; signature; DNA; hand veins; car driving style and credit/debit card usage patterns.

The aspects that certain traits associated with the human body have in common is that they tend to distinguish the particular person and that they have a permanent quality. However, not all traits that are used for biometric purposes are necessarily either unique to an individual or permanent features of that individual or his or her behaviour. Different traits will therefore have differing reliability as identifiers of a person so using multiple traits will tend to

[4] G. Hogben (ed.), 'ENISA Briefing: Behavioural Biometrics', 18 February 2010, available at www.enisa.europa.eu/publications/behavioural-biometrics.

[5] Consider, for example, the Plurilock behavioural biometrics platform which reportedly identifies people by the way they use their desktops and their mouse, builds a biometric profile unique to the user, verifies that the user is authorised to access the system and automatically challenges intruders; Plurilock website available at www.plurilock.com.

[6] Distinguish voice recognition from speech recognition. In the latter case, it is the content of what is said that is analysed.

[7] For a comprehensive discussion of facial recognition technologies and of the applications of it ranging from fraud detection and shoplifting prevention to checking whether persons have taken their medication, see K. Sennaar, 'Facial Recognition Applications: Security, Retail and Beyond', *Techemergence*, 16 November 2017, available at www.techemergence.com/facial-recognition-applications/ and for an example of facial recognition technology, see MXSERVER presented at www.tygart.com.

[8] Consider, for example, Verifinger, 'a fingerprint identification technology designed for biometric systems developers and integrators'. It operates on one-to-one and one-to-many bases; source Neurotechnology website www.neurotechnology.com/prices-verifinger.html.

[9] The PalmSecure system uses palm vein authentication technology and does not require that the hand be in contact with the device; www.fujitsu.com/jp/group/frontech/en/solutions/business-technology/security/palmsecure/sensor/.

[10] Note, for example, EyeLock's nano NXT access control system using iris recognition technology, described at www.eyelock.com/index.php/products/nano-nxt.

improve the reliability of a biometric identification.[11] In addition to these metrics, already heralded over a decade ago, there are new ways of collecting data, as well as the ability for mobile devices constantly to track, and monitor, their users' biometric data. Such data is often somewhat mindlessly signed over in exchange for the ability to access an app or service, or the ability to genetically map an individual, technologies which now open up the potential to alter genes remotely and without much intervention. The development of the use of nano-bots, the ability to manipulate genetic data (including genetic material that may be passed to subsequent generations) and the ability to manipulate the physical brain all offer immense opportunities and challenges to ensuring ethical use of biometric data. Further, genetic mapping may enable entire families or family generations to be associated, tracked and categorised for military purposes.

Some biometric testing methods involve the use of complex equipment while others require no equipment.[12] Some are intrusive, while other methods can be undertaken remotely and unobtrusively. Error rates vary and, as the discussion in the present chapter will reveal, biometric techniques can be applied for a variety of purposes, including verifying entitlement to access systems or locations, determining whether a person poses a threat and establishing the identity of an individual to name but three.

The data that these techniques collect may indicate sensitive characteristics about individuals; the ENISA report cites, as one example, that biometric information as to a person's gait may be an indicator for depression. There has been research into the analysis of brain signals for the more precise identification of individuals as compared to the use of other biometric methods[13] and one can only speculate as to what other aspects of a person's physical or mental health such a technique might disclose. All biometric data is arguably sensitive as it reflects information specific to an individual. As our ability to process this information increases (through AI), the information as it exists

[11] For an account of the evolution and applications of biometrics, see A. K. Jain and A. Kumar, 'Biometric Recognition: An Overview', in E. Mordini and D. Tzovaras (eds.) *Second Generation Biometrics: The Ethical, Legal and Social Context*, The International Library of Ethics, Law and Technology, vol. 11.(Dordrecht: Springer, 2012, 49–78).

[12] As examples of the former, consider iris recognition, fingerprint and palm print technologies. As an example of the latter, consider observation of a person's gait. For an explanation of machine recognition technology as a 'cascade of neural firings', see A. Greenfield, *Radical Technologies* (London: Verso, 2017), 215.

[13] Under the research, the electroencephalography signals generated by mental tasks are acquired to extract the distinctive brain signature of an individual; see, e.g., C. R. Hema, A. Elakkiya and M. P. Paulraj, 'Biometric Identification Using Electroencephalography', 106 *International Journal of Computer Applications* 15 (November 2014), 17.

becomes more valuable and useful to those who are interested in it. The sensitivity of this information should have significant implications for the obtaining and other processing of particular types of biometric data, and the governance of these processes. The ease with which access may be obtained using, e.g., AI technologies suggests that suitable technical measures must be taken and careful controls must be applied and rigorously enforced if the data protection and human rights obligations set forth in this chapter are to be complied with.

The collection of data and subsequent identification of an individual without that person's consent poses legal issues, which become even more problematic if personal data of a sensitive nature is involved. If unobtrusive methods are being used to obtain the biometric data it is unlikely, but by no means impossible, that consent will have been obtained in advance of the processing. Absent such consent, the circumstances in which the processing will be lawful will be limited,[14] and will be considered later in the chapter. It is also clear that personal data lawfully obtained in this way may only be used for the lawful purposes for which it was obtained, so security arrangements and access control arrangements need to be in place to seek to ensure that such data is not otherwise processed. The more numerous the agencies and/or individuals that have access to the data, the greater the degree of control that must be exercised to seek to ensure that processing is limited to that which is permitted by applicable law. Spoofing and other system attack methods[15] and erroneous conduct of the screening processes will adversely affect the reliability of the results that are obtained. The focus of this chapter is, however, not so much on their reliability as on some of the legal issues that some biometric techniques pose. It suffices for present purposes to recall, as noted earlier, that spoofing and the danger that an erroneous identification is made may in part be addressed by the contemporaneous use of multiple biometric techniques.

[14] Note, however, that metadata is capable of being used largely as governments see fit, and that such metadata may, depending on the particular context, be highly indicative of the personal data to which it relates. Maintaining the security of such metadata will therefore be a priority, e.g., for government departments and agencies that have such databases.

[15] The Intelligence Advanced Research Projects Activity in the United States has announced a multi-year research effort to develop and evaluate biometric presentation attack detection technologies to ensure the integrity of biometric security systems: 'IARPA launches "Odin" Program to harden biometric technology against attacks', Office of Director of National Intelligence, News Release 22–17, dated 19 October 2017, available at www.dni.gov/index.php/newsroom/press-releases/item/1808-iarpa-launches-odin-program-to-harden-biometric-technology-against-attacks.

14.2 THE MILITARY EMPLOYMENT OF BIOMETRICS

In the military context, biometrics can be and are employed to check that persons entering military facilities have proper authorisation to do so. Such control of entry activities in which biometrics either assist human operatives or in which the whole process relies on biometrics seems likely to continue to be a vitally important use of the technology. Bernice Baker supports this view, pointing out that the military use of these technologies is not limited to identifying persons who represent threats to the mission but also involves facilitating the recognition of friendly personnel.[16]

The digital environment and the changing nature of armed conflict are likely to increase military requirements for biometric solutions to identification problems.[17] Certainly, recent operations demonstrate a need for improved, even individualised, targeting intelligence to enable the identification and engagement of specified persons. The inherent nature of counter-insurgency operations renders the targeting of individuals a key component in the campaign, and the reliable identification of such individuals is therefore of vital importance. Specifically, 'identity verification emerged as one of the major technical challenges of the campaigns in Iraq and Afghanistan' particularly in the context of population-centric counter-insurgency operations in locations where the population often does not possess or carry reliable identity documents and in which adverse fighters do not distinguish themselves from the civilian population.[18] Operating such systems involves the generation of substantial databases recording the biometric details of potentially large numbers of individuals. It is therefore clear that biometric information is likely to be used in the military context, inter alia, to differentiate between friendly and adverse forces or personnel and to support targeting decisions.

[16] B. Baker, 'Booming Biometrics: How Human Data Is Transforming the Battlefield', *Army Technology*, 18 March 2014. Iris recognition seems to be among the particularly useful technologies and by using multi-modal methods information can be built around an identity by using multiple sources.

[17] Consider, for example, MForce developed by Human Recognition Systems Ltd, which is reportedly a 'multi-modal biometric platform that can be deployed through fixed, portable and mobile devices to deliver the benefits of positive identification to organisations working in remote, expansive or hostile environments'; MForce website available at www.hrsid.com/product-mforce.

[18] G. Voelz, 'Catalysts of Military Innovation: A Case Study of Defense Biometrics', 23 *Defense ARJ* 2, April 2016, 178, 185. As to the use of the Biometric Automated Toolset and of Handheld Interagency Identity Detection Equipment to record multi-modal (fingerprint, iris and face) biometric data in respect of detainees during armed conflict and counter-insurgency operations, see ibid., 186–187.

'Targeting decisions' as used here is not intended to presuppose the use of lethal force but may also consist of decisions as to the undertaking of any kind of military action, including non-forceful action, directed at individuals.[19]

Biometric technologies are therefore likely to be of vital significance in enabling the use of other novel technologies in the military sphere. Specifically, it is likely to be biometrics that will be employed by highly automated and autonomous weapon systems of the sort discussed in Chapter 6 to locate, identify and verify individuals as lawful targets for hostile or other action. The reliability with which the biometric technology distinguishes such lawful human targets will be a vitally important aspect when determining whether particular highly automated or autonomous weapon systems can lawfully be used in specific circumstances and, of course, in determining whether the technology meets the military requirement.

Technologies of this nature that enable targets to be located and recognised using biometric data are currently being developed.[20] Automatic Target Recognition, for example, will enable targets to be acquired by weapon systems using visual characteristics and patterns and these are the developments that, as pointed out in the previous paragraph, have obvious potential relevance to the use of remotely controlled, automated or autonomous weapon systems. Specifically, the employment of such technology would enable a weapon system to locate and engage a target, potentially a human target, autonomously basing target recognition either on comparison with a database loaded onto the weapon or on comparison with a centralised database.[21]

These developments should, perhaps, be considered in the context of a change in the nature of armed conflict towards counter-insurgency operations in which, in order to prevail, it is vital to identify individuals who, taken together, comprise the adversary, and to take those individuals out of the fight. It is thus important to be able to distinguish those individuals from the

[19] The targeting law that regulates the conduct of attacks and of military operations causing harm is to be found in API, Articles 48–67, in the rules as to special protection of particular kinds of person and object and in customary law. Military operations in which harm is not caused do not generally fall within these rules.

[20] Consider, for example S. Z. Li, et al., 'Biometrics at a Distance', in M. Tistarelli, S. Z. Li and R. Chellappa (eds.), *Handbook of Remote Biometrics: for Surveillance and Security* (New York: Springer, 2009), 1–22, and see W.-N. An, Z. Chen, et al., 'Remote Target Recognition and Tracking Based on Biometric Characteristics', 2017 Conference on Electrical Engineering and Automation Control, available at http://dpi-proceedings.com/index.php/dtetr/article/viewFile/10741/10291.

[21] Consider A. R. Dombe, 'Biometric Target Recognition', 19 March 2017, available at www.israeldefense.co.il/node/28881 where CIA use of facial recognition, body shape data, gait and walk are among the methods contemplated to enable identification of human targets.

peaceful members of the domestic population among whom they frequently deliberately conceal themselves. The technology does not, however, address all the complexities of such warfare, and may have the effect of over-simplistically classifying individuals as being permanently either civilians or fighters, when in reality, their role may change in the course of the conflict.

It is the reality of that kind of warfare that has driven the quest for ever more reliable biometric methods of target identification. In this context, US military doctrine developed notions of Biometrically Enabled Intelligence and of Identity Operations, combining biometrics with forensics and other capabilities.[22] Evolving US military thinking sees a role for biometrics in connection with a number of military activities including raids, checkpoint operations, border control, maritime interdiction, force protection and what is described as human terrain mapping.[23] The US Defense Biometric and Forensic Office has issued a strategy document in which it sees battlefield support roles for biometrics as including counter-improvised explosive device activities, targeting and force protection.[24]

Instructions governing the collection, processing and use of biometric information should seek to ensure that its collection, storage, use and other processing are in accordance with the applicable legal rules, that appropriate precautions are taken to protect biometric records from hacking operations and that misuse of the data is prevented to the maximum extent possible. It is worthy of note that the United States armed forces have issued publicly available, joint service tactics, techniques and procedures detailing the procedures to be employed in connection with the use of biometrics.[25] In a report to Congressional Committees, however, the US Government Accountability Office concludes that while the US DoD has made significant progress in addressing its long-term requirements for deployable biometric capabilities, e.g., by issuing new doctrine, it still lacks a strategic plan and there are concerns in relation to the use of acquisition protocols and as to the lack of a back-up capability for the biometric database.[26]

[22] Voelz, 'Catalysts of Military Innovation', 189. Consider in this regard the notion of Identity Intelligence, or 'I2', connecting biometrics with other sources of data to connect individuals with other persons, places, activities or materials.

[23] Voelz, 'Catalysts of Military Innovation', 190.

[24] Defense Biometric and Forensic Office Research, Development, Test and Evaluation Strategy, 6 January 2016, 3.

[25] Multi-Service Tactics, Techniques, and Procedures for Tactical Employment of Biometrics in Support of Operations, May 2016.

[26] US GAO, Report to Congressional Committees, DoD Biometrics and Forensics: Progress made in establishing long-term deployable capabilities, but further actions are needed, August 2017.

So, as we have seen, biometric techniques have considerable relevance to military activities outside the scope of targeting. Whether by recording the biometric details of detainees, by collecting biometric information during surveillance and other information gathering operations or otherwise, biometric data collection will clearly be an important part of the wider information gathering and evaluation task in support of the military campaign as a whole. If information is key, in the context of modern kinds of armed conflict, biometrics is an essential element of that information campaign.

14.3 LEGAL ISSUES ASSOCIATED WITH MILITARY USE OF BIOMETRICS

The use by the armed forces of biometrics techniques to control entry to sensitive military facilities will likely involve the maintenance of a database recording the biometrics of those who are permitted to have access. On the assumption that the persons and their biometric details included in that database are recorded there with the consent of the persons concerned, that the biometric details are not processed for any purpose not reflected in that consent, that required safeguards are applied to the processing of any personal, including biometric, data obtained from visitors, that the accuracy of the database is maintained, that data is removed from the database when no longer required for the purposes for which it was put there and that the requirements of data protection legislation are otherwise fully complied with, there would seem to be no legal objection to such use of biometric techniques and, thus, of the corresponding data.

Where human rights law is concerned, the mere act of using biometric methods in accordance with data protection legislation to seek to verify the identity of an individual seeking entry to a facility should not per se raise human rights concerns. Of course, particular actions taken in connection with the obtaining of biometric data, such as unlawful deprivation of liberty or failure to exercise required respect for privacy or correspondence, may involve breaches of human rights.[27]

[27] The tension that arises is between the personal and permanent nature of the features that the biometric processes are analysing and the right of the affected individual to privacy. These concerns are particularly significant when the biometric processes are being employed for surveillance purposes. The right to privacy may also be relevant if the information is used in a manner for which consent has not been given; see A. Willoughby, 'Biometric Surveillance and the Right to Privacy', 36 *IEEE Technology and Society Magazine* 3, September 2017, available at http://ieeexplore.ieee.org/document/8038135/.

During periods of armed conflict the collection of information to inform targeting decisions is not likely to prove controversial under the law of armed conflict. Indeed, that body of law requires that those who plan or decide upon attacks 'do everything feasible to verify that the objectives to be attacked are neither civilians nor civilian objects and are not subject to special protection'.[28] Doing everything feasible in the present context would definitely include obtaining all reasonably available biometric data in order to verify that the intended human objective is in fact a lawful target. Doing everything feasible would also likely include taking reasonably available steps to seek to verify that the sources of the biometric identification are reliable, that any associated equipment is working properly and, if this can be done reasonably easily, checking that the biometric test results that are being used to inform targeting decisions have not been corrupted by, e.g., malfunction, interference or spoofing. It seems sensible to suggest that the more that classification of a person as a target depends on biometric data, the greater the care that is required in seeking to ensure the accuracy of the data and the reasonableness of the decisions that are being based upon it. If this is correct, it suggests that where biometric tests are the basis for autonomous or automated targeting decision-making, human oversight of some kind should be available to check that the biometric data collection and evaluation processes are operating as they are designed to do.[29]

New biometrics-enhanced weapons or methods of warfare would require legal review under the arrangements discussed in Chapter 2. This involves determining whether the new weapon or method is of a nature to cause superfluous injury or unnecessary suffering, whether it is indiscriminate by nature, whether it is intended or expected to have prohibited impact on the natural environment or uses the environment as a weapon and whether there is any ad hoc rule of the law of armed conflict that prohibits or limits the circumstances of lawful use of the weapon. In addition, the interpretation and use of the data would almost certainly be computer-driven. As such, it would not suffice to rely on a black-box analysis.[30] Nevertheless and as we have seen,

[28] API, Article 57(2)(a)(i).

[29] Note that if decisions as to the targeting of individuals are to be based entirely on biometric techniques, such techniques would have to be able to make all of the differentiations required by the law of targeting. The technology would therefore have to be able to distinguish between combatants/fighters and civilians, between able-bodied combatants/fighters and those who are *hors de combat*, and between combatants and non-combatants. It would also, e.g., need to be able to detect when the status of an individual has changed such that the individual is no longer a lawful target.

[30] Consider L. Bennett Moses, 'Sui Generis Rules', in G. E. Marchant, B. R. Allenby and J. R. Heckert (eds.), *The Growing Gap between Emerging Technologies and Legal-Ethical Oversight: The Pacing Problem*, Vol. 7, International Library of Ethics, Law and Technology (New York:

compliance with the rules of targeting law is, if anything, likely to be facilitated by the use of biometric information subject to the considerations discussed above.

Complex issues may arise if biometric information is transferred by one State to another with a view to its use by the recipient State to inform targeting decisions. Those issues may be minimal if both the transferring and recipient States are consistently seeking to comply with the principles and rules of the law of armed conflict and if they interpret those rules similarly. If, by contrast and for example, the recipient State were routinely shown to be failing to comply with targeting law, the transferring State may find itself accused of aiding the commission of breaches of the law of armed conflict and, on such a basis, may attract to itself a measure of legal responsibility if, for example, the information that is supplied is, to the knowledge of the transferring State, going to be used to facilitate attacks that are likely to be unlawful. There are in this respect questions as to liability that are not yet fully resolved and that will tend to depend on the particular circumstances.[31]

Legal issues discussed below in relation to the commercial/domestic security use of biometrics will also apply to equivalent military uses. These would, for example, include situations in which the armed forces are operating in circumstances not amounting to armed conflict and in support of the civil authorities and any other circumstances in which military operations are subject to domestic and applicable human rights law. Those legal issues will be considered in the context of commercial/domestic security use of biometrics and the points made in that discussion do not need to be repeated here.

14.4 USE OF BIOMETRICS IN THE COMMERCIAL/DOMESTIC SECURITY CONTEXT

As we saw earlier in the chapter, there are numerous kinds of biometric data that might be employed in connection with a selection of tasks, so it seems sensible to shape the legal discussion in the next section of the chapter by seeking to identify some particular categories and examples of peacetime biometric activity that are, it is thought, representative. The first category might, for example, take the forms of: iris recognition to permit a person to

Springer 2011) 77 and L. Bennett Moses and J. Chan, 'Using Big Data for Legal and Law Enforcement Decisions: Testing the New Tools' (2014) 37(2) *University of New South Wales Law Journal* 643.

[31] Consider, for example, the obligation under common Article 1 to the 1949 Geneva Conventions to 'respect and to ensure respect' for the Conventions and the similar obligation in respect of API set forth in Article 1(1).

pass through the immigration control area at an airport; the use of fingerprint and keystroke biometrics to verify whether an individual is entitled to have access to a specific computer system; or the use of face, iris and/or fingerprint recognition technologies to determine, e.g., whether a person should be permitted within a secure area. We could sensibly summarise these processes as identification in support of access control, or category 1. The second category could, for example, involve monitoring the pulse, body movements, body temperature and blinking patterns of individuals to determine, for example, whether there are bodily indications of an intent to perform a criminal or violent act during a flight. We could summarise this as biometrics of intent, or category 2. The third category of biometric activity would involve, e.g., the use of face recognition technology to seek to identify individuals who have committed criminal acts, which we could summarise as the biometrics of detection, or category 3.

Examples of category 1 biometric testing have been employed in numerous countries for some years and do not per se raise challenging legal issues. The stored biometric data relating to the individuals who are nationals or habitual residents of the country applying the immigration controls, or of the persons whose routine access to the relevant computer system or restricted area is permitted will have been obtained with the knowledge and consent of those individuals. Provided that the use to which that data is put is limited to the purposes for which it was acquired and that other data protection requirements are adhered to, there is unlikely to be a legal problem associated with the use of their data, with the technology and with the maintenance of the associated database. One issue worthy of mention concerns the use of the biometric data that will be obtained from persons seeking access who are not reflected in the database. Their consent might be obtained at the point of collection of the data, and both data protection rules and practical considerations will require that their data is only retained for as long as is required in connection with the control of entry function.

Category 2 biometric monitoring does raise potential issues. Work on the development of biometric techniques in the essentially peacetime, domestic environment to seek to differentiate between persons who represent a threat, and those who do not, has been proceeding for a number of years. If these are the objectives of this 'biometrics of intent' technology, then that technology has its shortcomings. There is reported difficulty in distinguishing between different emotional states so while the technology may be technically able to differentiate between individuals who should be subject to further more specific scrutiny and those for whom that is not indicated, it would seem to be premature to try to use the technology as the basis of a positive assertion that an individual poses a threat.[32] This should not be interpreted as suggesting that use of biometrics for such purposes will never be feasible in the future. That will depend on whether future refinements of the biometric tests render the use of such biometric tests a sufficiently reliable indicator of specific intent. For the time being this is reportedly not the case but the technology does appear to be capable of indicating which persons ought to be subjected to further investigation. It is then on the basis of those further investigations that more reliable decisions can be taken e.g as to whether to refuse permission to board a passenger aircraft.

Future Attribute Screening Technology, or FAST, employs sensors and cameras located, e.g., at checkpoints to monitor natural signals such as breathing, heart rate, body temperature, fidgeting and eye movement. Taken together such features are considered to indicate whether an individual may intend to do harm. Reports suggest that such technology could be employed, e.g., at stadia, public buildings and transport hubs such as airports. The technology could also, it seems, be used to monitor individuals, e.g., when they are being asked significant questions. Indeed, it is apparent that there is the potential to use metadata generally to predict movement, behaviour or intent and that when such data is processed from a number of sources, a more extensive, potentially more intrusive but arguably more useful 'biometric picture' is likely to become available.

The sponsor of the FAST programme, the US Department of Homeland Security (DHS), has published a factsheet[33] where it is noted that security checkpoint officers currently use visual inspection techniques and devices, such as X-ray machines, to detect individuals who have weapons or other harmful objects. To identify individuals with what the factsheet describes as 'malintent', the officials must use what are seen as more simplistic techniques such as verifying credentials, random selection for questioning and visual observation, often operating under time pressure with insufficient information. The Science and Technology Directorate of the DHS intends to use

[32] See comments attributed to Dr Oshin Vartanian, reported in Researchers Developing Biometric Intent Scanners, Homeland Security News Wire, 29 March 2011. As to the dangers of seeking to infer intent from words used, e.g., on social media and for criticism based on civil liberties concerns associated with predictive algorithms, consider Greenfield, *Radical Technologies*, 226–233.

[33] US Department of Homeland Security, Future Attribute Screening Technology Factsheet, 2 March 2016, available at www.dhs.gov/publication/future-attribute-screening-technology. Early testing reportedly achieved 81 per cent accuracy using contact sensors, the nature of which is not specified, while 71 per cent to 74 per cent accuracy has been achieved using remote, non-contact sensors.

FAST to screen individuals for malintent. Where the employment of such biometric testing in the airport security context is concerned, the idea seems to be that persons whose biometrics raise concerns will then be the subject of more thorough consideration to determine whether they do or do not represent a threat.

Importantly, the DHS factsheet describes FAST as 'gender, culture and age-neutral and it does not use profiling techniques or collect personally identifiable information.' Recognising the potential relevance of privacy and civil liberty rights, the Factsheet asserts 'the system is intentionally designed to only examine physiological and behavioural cues so that no privacy or civil rights are violated', a potentially excessively bold statement. Somewhat more realistically, the Factsheet emphasises that 'FAST does not connect physiological data to an individual, nor does it permanently store collected data once the analysis is complete.' This of course implies that such data is temporarily stored, raising questions as to the duration of that temporary storage, the security of the storage arrangements, what other processing takes place while the data is so stored and whether relevant consents are sought and obtained.

Turning now to category 3 biometrics to which we referred earlier, the use of face recognition technology to try to identify the perpetrators of criminal acts is not without controversy. Typically, security cameras may have recorded footage involving images of persons of interest in connection with a criminal investigation. Putting names to the images on such footage of certain individuals who are assessed as being of interest may materially advance the investigation and will therefore be a priority activity. London's Metropolitan Police have used particularly gifted human beings, known as 'super-recognisers', to recognise relevant persons based on their ability to recognise almost everyone they have ever seen and their linked ability to put names to the faces that they see on the screen. This method has reportedly achieved impressive success.[34]

The question then arises as to how computer-based facial recognition technology performs against the human alternative. Raouf Hamzaoui, Professor of Technology at De Montfort University, is quoted as expressing the view that 'with low quality pictures, typical of CCTV, where there is darkness, facial coverings, blurring and so on, the software struggles and the human does better'.[35]

The use by the Metropolitan Police in the UK of facial recognition technology during the Notting Hill Carnival in 2017 attracted comment by

[34] L. Potts, 'The Police "Super-Recognisers" Putting Names to Faces', BBC News, 19 October 2015, available at www.bbc.co.uk/news/uk-england-34544199.

[35] Ibid.

Paul Wiles, Biometrics Commissioner for the UK. His considered comments deserve careful attention. He notes that there is public benefit if such methods can be shown to help the police to catch offenders or prevent crime. Tests for facial matching in large crowds have so far had poor results, so what is taking place at the moment can best be described as trials of the technology.[36] The Commissioner considers that police forces should work together to agree on a single facial recognition system that is proved to work 'and government needs to create a legislative framework for its use, with independent oversight to provide public assurance'. The Commissioner then identifies what is the essential challenge for regulation. He points out: 'The use of facial images, especially in public places, is very intrusive of individual freedom, especially because images can be captured without the subject being aware. The public benefit of the use of such an intrusive technology must outweigh the interference in individual privacy. Such a difficult balance between public benefit and individual privacy should not be decided by the police but is best decided by Parliament through informed debate and legislation. As is currently the case for DNA and fingerprints the legislation should include independent oversight to reassure the public that their privacy is being properly protected.'[37] It is evident from these remarks that, certainly so far as the Biometrics Commissioner of the UK is concerned, new legislation is needed to strike the required balance between the privacy interests of individuals and the public or national interest in seeing to it that crime is prevented and criminals are apprehended and convicted for their wrongdoings. There is an obvious similarity of challenge here between these concerns and those that were identified in Chapter 5 in connection with the Internet of Things.

Two points should be made here. First, facial and voice recognition are not new techniques; they were used effectively in the German Democratic Republic in the 1980s. What is new is the ability of artificial intelligence to use multiple sources of data to corroborate the identification and location of individuals with greater reliability than would be possible using a single source of individual biometric data. Second, the use of biometric data of this kind must of necessity be considered in the context of information from other sources that may be available, such as mobile phone technologies and other data.

36 The Commissioner sensibly argues that such experiments should be properly designed and evaluated, involving external experts and that the results should be published.

37 P. Wiles, 'Metropolitan Police's Use of Facial Recognition Technology at the Notting Hill Carnival, 2017', Press Release, 23 August 2017.

14.5 SOME LEGAL CONSIDERATIONS ASSOCIATED WITH PEACETIME EMPLOYMENT OF BIOMETRICS

Mention has already been made of data protection obligations and these will be discussed in this section. We should, however, start by assessing the position under human rights law. Under the International Covenant on Civil and Political Rights[38] privacy, family and home rights are protected in the following terms: '1. No one shall be subjected to arbitrary or unlawful interference with his privacy, family, home or correspondence, nor to unlawful attacks on his honour and reputation. 2. Everyone has the right to the protection of the law against such interference or attacks.'

Where regional human rights treaties are concerned, the American Convention on Human Rights provides: 'No one may be the object of arbitrary or abusive interference with his private life, his family, his home, or his correspondence, or of unlawful attacks on his honor or reputation.'[39] The European Convention on Human Rights stipulates that '[e]veryone has the right to respect for his private and family life, his home and his correspondence.'[40] Article 8(2) goes on to provide: 'There shall be no interference by a public authority with the exercise of this right except such as is in accordance with the law and is necessary in a democratic society in the interests of national security, public safety, or the economic well-being of the country, for the prevention of disorder or crime, for the protection of health or morals, or for the protection of the rights and freedoms of others.'

There is a clear link between these Article 8 rights and the data protection rules that will be discussed below. The link is that a breach of data protection legislation involving the inappropriate processing of personal data is liable, subject to the particular circumstances, also to be regarded as a breach of Article 8.[41] Nevertheless, the European Court of Human Rights does recognise that to be able to carry out law enforcement tasks the police must be in a position to establish the identity of citizens.[42]

Access control biometric testing of the sort referred to in category 1 in the previous section and which is undertaken in compliance with the data

[38] International Covenant on Civil and Political Rights, New York, 16 December 1966.

[39] American Convention on Human Rights, San Jose, 22 November 1969, Article 11(2). Article 11 (3) recognises that everyone has the right to protection of the law against such interference or attacks.

[40] European Convention for the Protection of Human Rights and Fundamental Freedoms, Rome, 4 November 1950, Article 8(1).

[41] Consider, e.g., X v. *Germany*, No 8334/78, 24 DR 103 (1981).

[42] *Vasileva* v. *Denmark*, 40 EHRR 681, para. 39.

protection rules discussed below is unlikely to be regarded as breaching the rights to privacy or to a private life discussed here. It can easily be imagined, however, that processing of the kind of data required for category 1 purposes beyond the requirements of the function for which the data was provided could, again depending on the exact nature of the additional processing, breach privacy/private life rights.

Where category 2 biometrics of intent testing is concerned, one can imagine that the biometric data that will be collected will often be very sensitive. If such data is anything other than strictly anonymous, i.e. if there is a way in which the data can be linked with the identity of the person who provided it, particular care will be required to ensure that its security is maintained and that it is only processed to the extent strictly required by the purpose in obtaining it. The nature of the specific data and of the way in which it is stored will influence the kinds of protection and precautionary measures that will be required. It is not possible to be prescriptive as to specific required measures. It must be emphasised, however, that a high degree of care will be needed both in the design of the systems in which the data is held, and the linkage arrangements between those systems and other systems and persons that jointly have access to that information as well as in the processing of the biometric data that is held in those various systems.

Category 3 biometrics of detection data may not have been obtained with the consent of the person concerned as consent is defined in applicable data protection legislation. Care will be required to ensure that legal rules as to the collection and processing of such data are adhered to and a failure in that regard might be interpreted as a breach of the human right to respect for privacy/private life.

The discussion now turns to consider specifically data protection issues and will continue to be by reference to the categories of biometric activity identified in the previous section. Taking category 1 first, therefore, the principal concern will likely be to ensure that data protection requirements are complied with in relation to persons whose data is stored on a long-term basis within the system and those whose biometrics are collected in association with their attempt to achieve access. In the immigration context, the former are likely to consist of the members of the domestic population who have passports issued by the State authorities and habitual residents whose biometric details appear on an established database; the latter will be other persons seeking entry into the country. Likewise, where security arrangements for access to a computer system are concerned, the former will be those who are permitted access and whose biometric details are therefore stored within the relevant security systems, while the latter will be the biometrics of any other person seeking access. Equally, a biometric system to limit access to a secure area will necessarily require the retention of biometric data of those permitted such access and will acquire some biometric information about others seeking such access.

Under European Union law, the General Data Protection Regulation[43] (GDPR) has applied to all members of the European Union, including the UK, since 25 May 2018. It is an extensive and complex provision and it is only therefore possible to highlight one or two of its provisions in order to seek to show some of the legal issues that are likely to arise in relation to the processing of biometric data about individuals. The Regulation identifies 'personal data' as 'any information relating to an identified or identifiable natural person ('data subject')'. An 'identifiable natural person' is then stated to be 'one who can be identified, directly or indirectly, in particular by reference to an identifier such as a name, an identification number, location data, an online identifier, or to one or more factors specific to the physical, physiological, genetic, mental, economic, cultural, or social identity of that natural person.'[44] The identifiers referred to in this definition clearly will encompass most, perhaps virtually all, forms of biometric data. However, it is perhaps noteworthy that the individual must be identifiable by reference to the listed aspects, presumably alone or in combination with one another or with other data.

'Processing' is defined for the purposes of the Regulation as 'any operation or set of operations which is performed on personal data or on sets of personal data, whether or not by automated means, such as collection, recording, organisation, structuring, storage, adaptation or alteration, retrieval, consultation, use, disclosure by transmission, dissemination or otherwise making available, alignment or combination, restriction, erasure or destruction'.[45] This has the effect that virtually any kind of handling of personal data constitutes processing. Consent, for the purposes of the Regulation, must be 'freely given, specific, informed and unambiguous [...] agreement to the processing of personal data relating to' the relevant person.[46] Biometric data is then specifically defined as 'personal data resulting from specific technical processing relating to the physical, physiological or behavioural characteristics

[43] European Union, General Data Protection Regulation, Regulation 2016/679, was adopted on 27 April 2016 and became effective on 25 May 2018 (GDPR). At the time of writing, the Data Protection Bill is pending before the UK Parliament; it will implement aspects of the GDPR as adjusted to suit national requirements.

[44] GDPR, Article 4(1).

[45] GDPR, Article 4(2).

[46] GDPR, Article 4(11).

of a natural person, which allow or confirm the unique identification of that natural person, such as facial images or dactyloscopic data'.[47]

Six principles must be complied with in the processing of personal data. In heavily summarised form, they are:

1. that the processing must be fair and transparent;
2. that the personal data must be collected for specified, explicit and legitimate purposes and must not be further processed in an incompatible manner;
3. that the personal data must be adequate, relevant and limited to its purpose;
4. that the personal data must be accurate and kept up to date;
5. that the personal data can be stored in identifiable form for no longer than necessary for the purposes for which it was processed; and
6. that appropriate security of the personal data must be ensured.

These principles will also therefore have to be complied with in relation to the processing of all biometric data that constitutes personal data.

For the processing of personal data, one of six criteria must be satisfied. These, again in summarised form, are: that the data subject consents to the processing; that the processing is necessary for the performance of a contract to which the data subject is a party; that processing is necessary to comply with a legal obligation binding the controller; that processing is necessary to protect the vital interests of the data subject or someone else; that processing is necessary for the performance of a task carried out in the public interest or in the exercise of authority vested in the controller; or where there are third party interests, unless these are overridden by the interests, rights or freedoms of the data subject.

Article 9 of the GDPR is of vital importance. It provides, so far as strictly relevant, that 'the processing of genetic data, biometric data for the purpose of uniquely identifying a natural person, data concerning health [...] shall be prohibited'. This prohibition is, however, subject to a complex list of exemptions. So, for example and in broad terms, there is an exemption where the data subject has explicitly consented to the processing for specified purposes. Other exemptions apply, e.g., where the data has manifestly been made public by the data subject or where there are certain health- and employment-related considerations involved. More specifically, there is an exemption where 'processing is necessary for reasons of substantial public interest, on the basis

47 GDPR, Article 4(14).

of Union or Member State law which shall be proportionate to the aim pursued, respect the essence of the right to data protection and provide for suitable and specific measures to safeguard the fundamental rights and the interests of the data subject'.[48]

So, under the GDPR the processing of biometric data to identify a person is only going to be lawful if the relevant person gives free and informed consent to the processing in question or if, irrespective of such consent, substantial public interest renders such processing necessary provided that the processing is proportionate to its purpose, that essential data protection rights are respected and that the essential rights of the data subject are safeguarded. While it remains to be seen how these provisions are going to be interpreted and applied in practice, the following would therefore seem to be the critical questions to ask under the GDPR about biometric identification processes of the sort that were referred to in the previous section under category 3:

Has the person who is the subject of the biometric data processing given his or her explicit, free and informed consent to the processing of his or her biometric data?

If not, is there a substantial public interest that renders the undertaking of that biometric data processing necessary?

If so, is the data processing in question proportionate to its purpose; have the data protection and other rights of the data subject been sufficiently safeguarded?[49]

The problem with all of this of course is the potential for hacking operations to be employed to gain access to such data, whether by private individuals acting on their own initiative or by States' agencies or by persons whose actions are otherwise attributable to States. It is Article 32 of the GDPR that addresses security issues. Paragraph 1 provides as follows:

> 'Taking into account the state of the art, the costs of implementation and the nature, scope, context and purposes of processing as well as the risk of varying likelihood and severity for the rights and freedoms of natural persons, the

[48] GDPR, Article 9(2). Note that the processing of photographs should not systematically be considered to be processing of special categories of personal data as they are covered by the definition of biometric data only when processed through a specific technical means allowing the unique identification or authentication of a natural person. Such personal data should not be processed, unless processing is allowed in specific cases set out in the Regulation; Recital 51 to the Regulation.

[49] These matters would all, in the event of legal challenge, fall to be considered by the tribunal exercising jurisdiction in the case and should therefore be addressed by those responsible for deciding whether particular processing of biometric data should take place.

controller and the processor shall implement appropriate technical and organisational measures to ensure a level of security appropriate to the risk, including inter alia as appropriate:

a) the pseudonymisation and encryption of personal data;
b) the ability to ensure the ongoing confidentiality, integrity, availability and resilience of processing systems and services;
c) the ability to restore the availability and access to personal data in a timely manner in the event of a physical or technical incident;
d) a process for regularly testing, assessing and evaluating the effectiveness of technical and organisational measures for ensuring the security of the processing.'

It is worth appreciating some relevant points here. It will be the prevailing level of knowledge and of technical understanding that will be relevant when the adequacy of the security measures that have been taken is considered. The purpose for which the data was being processed (remember that processing includes its retention), the risks associated with unauthorised access to it and the likelihood of such hacking operations are among the factors to be considered. The precautions must be taken to ensure an appropriate level of security and the technical and organisational measures that are taken must be appropriate for that purpose. The responsibility rests with the controller of the data and with the person doing the processing and the list of possible measures is illustrative not exhaustive.

Compliance with a code of conduct may help to show that reasonable steps have been taken. When assessing the appropriate level of security, account must be taken of particular risks associated with the processing, e.g., risks of unlawful loss, destruction alteration or disclosure.

There are, however, important exclusions from the application of GDPR. Article 2(2) stipulates that the Regulation does not apply to the processing of personal data 'in the course of an activity which falls outside the scope of Union law'. Recital 16 to the Regulation repeats this exclusion and cites as an example of excluded matters 'activities concerning national security'.[50] Under Recital 19, processing of personal data by competent authorities 'for the purposes of the prevention, investigation, detection or prosecution of criminal offences, including the safeguarding against and the prevention of threats to public security…' lies outside the scope of GDPR and is regulated by

[50] Similarly excluded by the same Recital are activities concerning the Common Defence and Security Policy of the European Union.

Directive 2016/680.[51] Under that Directive, Member States are required to protect rights and freedoms including the right to the protection of personal data. Importantly, the Directive applies to the processing of personal data wholly or partially by automated means and specifically defines biometric data.[52]

The Directive applies similar principles to those discussed earlier in connection with GDPR and requires Member States, e.g., to make provision as to time limits for storage of personal data and as to the limitation of processing to that which is necessary. Article 10 requires, inter alia, that biometric data processed uniquely to identify a natural person be only permitted when strictly necessary, must be subject to safeguards for the rights and freedoms of the data subject but would in principle be permitted if authorised by Member State law.[53] Article 11 of the Directive is of particular relevance to the present discussion. It stipulates:

> 'Member States shall provide for a decision based solely on automated processing, including profiling, which produces an adverse legal effect concerning the data subject or significantly affects him or her, to be prohibited unless authorised by Union or Member State law to which the controller is subject and which provides appropriate safeguards for the rights and freedoms of the data subject, at least the right to obtain human intervention on the part of the controller.'[54]

So it is the domestic law of the Member States that must achieve the safeguards stipulated here in the Directive. That domestic legislation will no doubt clarify what for these purposes constitutes 'adverse legal effect' and significant effect, and will specify the safeguards that are deemed to be appropriate. It remains to be seen how these exclusions will be interpreted and, specifically, what breadth of activities will be regarded as concerning national security.

[51] Directive of the European Parliament and of the Council dated 27 April 2016, 2016/680 (the Directive).

[52] See Articles 1–3 of the Directive. Under Article 3(13) biometric data is defined as 'personal data resulting from specific technical processing relating to the physical, physiological or behavioural characteristics of a natural person, which allow or confirm the unique identification of that natural person, such as facial images or dactyloscopic data'.

[53] Note that UK, Ireland and Denmark have an opt-out from the Directive. If exercised, this means that neither the GDPR nor the Directive will apply for those States to the activities noted earlier as excluded from the scope of the GDPR.

[54] The Directive, Article 11(1). Article 11(2) provides that: 'Decisions referred to in paragraph 1 of this Article shall not be based on special categories of personal data [which would seemingly include biometric data] referred to in Article 10, unless suitable measures to safeguard the data subject's rights and freedoms and legitimate interests are in place.'

It must be emphasised that only very limited aspects of the GDPR and of the Directive have been addressed in this section as to do otherwise would go beyond the intended scope and focus of the chapter. It is, however, clear that those seeking to employ biometric testing technologies must, before doing so, have careful regard to the GDPR and to the Directive not just in relation to the use of biometric techniques designed to identify persons but also in relation to the broader, data protection obligations associated with the use of personal data derived from biometrics.

In the United States and where category 2 biometric activity, as described in the previous section, is concerned, in the airport security context specific concerns have been expressed as to the private and personal nature of the data that is being observed and there are doubts as to the efficacy of the technology.[55] Under United States law, it is the Fourth Amendment to the United States Constitution that provides that '[t]he right of the people to be secure in their persons, houses, papers, and effects against unreasonable searches and seizures, shall not be violated…' Christopher Rogers has conducted a careful analysis of the FAST scheme in light of the Fourth Amendment and associated case law, noting that the governmental interest in preventing terrorism must be weighed against an individual's right to privacy. Relevant privacy interests would comprise 'personal medical data, to which the Supreme Court has given heightened protection, and mass data collection' which some Courts, it seems, have found to breach the Fourth Amendment.[56]

Rogers identifies circumstances in which FAST scans would not breach the Fourth Amendment as being: '(1) the person subject to the scan is on notice that the scan it about to take place, (2) the scan is limited in scope, and (3) the scan is narrowly tailored to advance the governmental interest.' If such safeguards were to be applied, e.g., to FAST screening at airports, the searches involved in the FAST screening may be regarded as reasonable and thus as not breaching the Amendment. However, deployment of FAST elsewhere to detect crime before it is committed would likely require a warrant and there is the concern that, by being configured to detect 'malintent', FAST screening would go beyond screening specifically in relation to terrorism and thus would not be 'narrowly tailored to the regulatory scheme that it is promoting.'[57]

55 For a brief description of the technology and an indication of some concerns, including complaints that the technology violates privacy rights, see P. Benson, 'Will Airports Screen for Body Signals: Researchers Hope So', CNN, 7 October 2009.

56 C. A .Rogers, 'A Slow March towards Thought Crime: How the Department of Homeland Security's Fast Program Violates the Fourth Amendment', 64 *American University Law Review* 2 (2014), 337, 384.

57 Ibid.

Another form of biometric testing could involve depicting significant images, related for example to terrorism, while monitoring changes in breathing, cardiac rhythm and/or body temperature.[58] The open question is whether detaining an individual for the purposes of conducting such testing breaches the human rights of the individual concerned. The exceptions to the right to liberty and security under the European Convention do not include a right to detain on security grounds. Similarly, there is no mention of a right to detain in order to apply the biometric tests with a view to determining whether an individual intends to commit a criminal act.[59] Equally, having applied such tests, if they were to strengthen the suspicions as to the intent of the relevant person, Article 5 of the Convention would not permit the detention of that individual to enable those suspicions to be further investigated. Perhaps the answer here is that the tests are undertaken with the consent of the individual, and those individuals who decline to consent will be refused permission to board their flight, being action that would not per se involve a breach of Convention rights.

Interestingly, the International Covenant provides, so far as relevant, that '[e]veryone lawfully within the territory of a State shall, within that territory, have the right to liberty of movement . . .' and that '[e]veryone shall be free to leave any country, including his own.'[60] Although the relevant checks might take place 'air-side' in an airport, this will still almost certainly be regarded as 'within the territory' of the particular State. While a person is indeed free to leave a State, that freedom does not necessarily imply a freedom to leave by means of a particular flight, or indeed to leave by any flight. The rights in Article 12 are not subject to any restrictions 'except those which are provided by law, are necessary to protect national security, public order (ordre public), public health or morals or the rights and freedoms of others, and are consistent with the other rights recognised in the present Covenant.'[61] So any restriction of the freedom of movement right must be reflected in legal provision, e.g., that authorises the conduct of the biometric testing, and such biometric testing must be necessary on one at least of the grounds mentioned in Article 12(3) and must be consistent with other human rights.

58 D. Talpalariu, 'Airport Speech and Brain Scans Identify the Terrorist in You', *Softpedia*, 3 December 2008.

59 See European Convention, Article 5(1).

60 International Covenant, Article 12(1) and (2).

61 International Covenant, Article 12(3). Article 12(4) provides that no one shall be arbitrarily deprived of the right to enter his own country.

14.6 COMPARISON OF THE MILITARY AND COMMERCIAL APPLICATIONS

When discussing the peacetime/commercial application of biometric technologies, we identified three categories of usage, namely control of access (category 1), biometrics of intent (category 2) and biometrics of detection (category 3). In the peacetime environment we treated immigration controls at an airport, control of access to computer systems and restriction of access to secure areas as examples of category 1. The security concerns giving rise to such testing are also apparent in the military environment where there is a constant need to limit access to sensitive areas to those with the appropriate security clearances, where there is a vital interest in trying to make sure that only authorised persons have access to specified computer systems and where there is a need to ensure that those having any kind of access to military equipment of diverse kinds are the persons who are permitted, indeed often required, to be there.

Biometrics of intent are likely to become relevant during an armed conflict, and during insurgency and other military security situations if, for example, civilians or persons who have the appearance of civilians, are believed to be taking a direct part in hostilities, or to be about to do so. Taking a direct part in hostilities is a significant act for a civilian as it deprives that civilian of his or her protection from being made the object of attack.[62] If, in the circumstances in which they would be used in connection with hostilities during armed conflict, biometrics of intent can differentiate between civilians that are and intend to remain peaceful and civilians that intend to undertake hostile acts, this may be a precautionary measure of the sort contemplated in Article 57 of API, as to which see the discussion in the next section.

A vitally important practical aspect will, however, be the reliability of the technology when used in the challenging circumstances of hostilities. Concerns have, as we have seen, been expressed as to the reliability of using machines to undertake such facial recognition biometric analysis in the context of public disorder. Those concerns seem likely to be even more relevant in circumstances of armed conflict, when individuals are being observed often remotely, frequently using sensors located at a considerable distance, perhaps in circumstances where the person being observed may be partially obscured by clothing or personal military equipment and in a context

62 API, Article 51(3).

in which other factors such as smoke, dust, dirt or camouflage may be further challenging the accurate use of the biometric technology.[63]

Nevertheless, the military parallel for the biometrics of detection is likely to consist of any use of biometric techniques to determine that a particular person has become a lawful target on whatever basis. Where the peacetime application may, for example, involve the police in using CCTV footage to identify the ringleaders in a violent, riotous situation, in warfare such technology might as we have seen be employed to identify the persons who have taken part in hostilities, who have a combat role in an organised armed group and who have shown themselves to be lawful targets.

It is also important to recognise that the armed forces may be deployed otherwise than in connection with an armed conflict, e.g., to assist the domestic law enforcement authorities to deal with internal security situations, disasters and other events of that kind. In such situations, they are likely to be using biometric techniques in similar circumstances to the civilian police and other personnel that they are assisting, as in border control operations, at security checkpoints, to control entry to public buildings and sensitive facilities and generally to maintain public security.

The point that emerges, therefore, is that there are considerable similarities, or at least parallels, in the kinds of situation in which the armed forces and civilian authorities will tend to employ biometric methods. In the next section we discuss the similarities and differences in the applicable law.

14.7 COMPARISON OF THE APPLICABLE LAW

If the armed forces are employed in internal security operations, for example assisting the domestic law enforcement agencies, their use of biometrics will be subject to the same data protection and human rights-based rules as bind the civil authorities in peacetime. Likewise, and in the US context, employment of the armed forces in support of the civil authorities will attract the application of the Fourth Amendment in circumstances in which it would have applied to corresponding activities by law enforcement personnel.

It is generally recognised that human rights law continues to apply during periods of armed conflict.[64] However, in determining whether there has been

[63] Note, however, current research such as the Janus Program: see footnote 3.

[64] ICJ Advisory Opinion on the Legality of the Threat or Use of Nuclear Weapons, 110 *Int'l Law Rep* 163 (1996), para 25; ICJ Advisory Opinion on the Legal Consequences of the Construction of a Wall in the Occupied Palestinian Territory, 9 July 2004, *ICJ Reports* (2004) 136, para 106; ICJ, *Case Concerning Armed Activities in the Territory of the Congo (DRC v. Uganda)* 2005 ICJ 116, 19 December 2005, paras 178–180, 216–217.

a human rights law breach in connection with the conduct of hostilities, it is the *lex specialis* of the law of armed conflict that is applied. So, if biometric testing is being undertaken as part of the hostilities, e.g., to differentiate between lawful targets of attack and persons entitled to be protected, in the view of the present author it is the law of armed conflict that will provide the criteria to determine the lawfulness of such biometric testing, not the language of human rights treaties. Contrast a situation in which, at a time when an armed conflict is taking place, biometric testing is taking place within a State that is a party to the armed conflict but for purposes that are not linked with the conflict. Such processing will be subject to the human rights and data protection rules that have been discussed above.

Where the use of biometric testing by armed forces in connection with the conduct of hostilities is concerned, API, Article 57(2)(a)(i) is the provision that requires everything feasible, that is practicable or practically possible, to be done to verify that the objective to be attacked is a lawful target. When it ratified API, the UK made the now widely recognised point that '[m]ilitary commanders and others responsible for planning, deciding upon or executing attacks necessarily have to reach decisions on the basis of their interpretation of the information from all sources which is reasonably available to them at the relevant time'.[65] Doing everything feasible and the notion of reasonably available information, taken together, indicate that if biometric information can reasonably be obtained and would be relevant to the targeting decision, all practicable steps should be taken to obtain and consider that information. Herein, perhaps, lies the important distinction between the legal approach in the law of armed conflict to such a use of biometric information, and the rather more restrictive arrangements under the GDPR and, to an extent, under the Directive and the Fourth Amendment.

14.8 WHAT CAN WE LEARN FROM THE LAW APPLICABLE TO THE TWO CONTEXTS?

The discussion in this chapter shows that the applicability of the respective legal regimes depends not so much on whether it is armed forces or internal security personnel that are involved as on the nature of the situation that is being addressed. Even then it has been recognised that in situations of armed conflict, circumstances can arise that will be regulated by human rights law, by data protection regulations and by provisions of applicable domestic law.

[65] Statement (c) made on ratification of API by the UK on 28 January 1998.

So, it is the circumstances that will determine the applicable law, not whether it happens to be the armed forces or civil police personnel who are taking action. That said, there is a fundamental difference in approach between the law of armed conflict on the one hand and GDPR, the Directive and the Fourth Amendment on the other. The former, in very general terms, seems to view biometrics as enhancing the decision-making process whereas the latter, in like terms, applies restrictions and conditions to the handling of biometric data.

Where armed forces and civil law enforcement agencies are employing biometrics for similar security purposes, the same legal rules will apply except where Recitals 16–19 and Article 2(2) and (3) of GDPR exclude the application of the Regulation. That provision, however, only has the effect, as we have seen, of applying the Directive for most EU States. For UK, Ireland and Denmark it may have the effect that the domestic law of those States will govern the relevant activities.

The other significant point is that where biometric testing is being employed as an element in the conduct of hostilities, and its use in targeting decision making was cited as an example, the law of armed conflict rules will, as *lex specialis*, be the basis for determining whether particular action constitutes a human rights breach.

All of this seems to suggest that the regulation of biometrics technologies in military and civilian contexts is subject only to the same *lex specialis* distinction that applies in general to the law relating to conflict. This in turn leads one to believe that there is no specific lesson as such that peacetime regulation can learn from the restrictions and rules applicable in armed conflict or vice versa. The two bodies of law in their application to biometrics seem, indeed, to be relatively harmonised.

14.9 BIOMETRICS IN THE CONTEXT OF OTHER TECHNOLOGIES

The purpose of these closing paragraphs of the chapter is to note a few of the other technologies considered in this book that appear to have particular relevance to biometrics and/or vice versa.

There is an obvious link between computer networks, AI and biometrics. So, for example, it would likely be by cyber means that an attempt would be made adversely to affect biometric testing by corrupting databases on which biometric procedures are based, by interfering with the processing of biometric data or by otherwise impeding its collection, analysis and evaluation. As we have noted, research programmes are currently under way seeking to render the relevant processes robust. That said, such systems, unless using quantum computing, will always have vulnerabilities.

Biometric identification methods are likely to be critical to the accurate identification by autonomous weapons systems of lawful human targets and the reliable performance of such biometric identification systems will therefore be essential if such autonomous weapon systems are to be used lawfully against such target sets. This implies heavy reliance on black-box technologies, and assumes their reliable performance, which may or may not be realistic. Similar considerations apply in relation to some of the highly automated systems discussed in Chapter 6. There is an equally important linkage working in the opposite direction. So artificial intelligence of the sort discussed in Chapter 6 is clearly going to be an essential element in the further development and refinement of biometrics analysis tools.[66] Having an ability to peer inside these black-boxes of decision-making will remain critical to ensuring that their employment is consistent with applicable law and ethics.

The application of brain–computer interfaces to the analysis of biometric data or to associated decision-making seems unlikely to raise legal issues simply because of the involvement of a human brain in such a process. It will however be important to ensure the continuing integrity of the brain-computer interface. Employing 'beyond the horizon' artificial brain technologies to undertake biometric data collection and/or analysis tasks or to decide courses of action based on biometric data inputs would, however, raise profound issues of the kind discussed in Chapter 13.

In the peacetime environment there is an obvious linkage between biometrics and the maintenance of security at many levels and in many environments. Computer assistance with the identification of people, with determining which people pose a threat and with seeking to prevent unauthorised intrusion into places and systems on which the security and wellbeing of the population at large depend will continue to be an essential requirement. Some of the concerns associated with the Internet of Things might in part be addressed by biometrics. Is it possible that biometric technology, which in some applications arouses so many concerns in relation to the maintenance of privacy, may become essential to the maintenance of privacy in a future characterised by big data and the Internet of Things?

It remains to be seen what new human enhancement technologies are developed and whether applications of these might include the employment

[66] Consider, for example, Onfido's use of machine learning technology to validate identity documentation and compare it with facial biometrics; https://onfido.com/about viewed on 17 January 2018.

of sensors that enable the future individual combatant, e.g., to determine the intent of a potential adversary. Sensors are already being fitted on an experimental basis to small drones and then used to perform facial recognition tasks. The day is fast approaching when minute cameras fixed to the uniforms of, say, police officers or armed forces personnel, can be used to produce images that can then be compared against databases of known criminals or, in the case of the armed forces, enemy combatants, commanders or fighters. The integration of this kind of technology with drones, whether remotely piloted, automated or autonomous or by means of human enhancements such as embedded chips is becoming an increasingly realistic prospect.[67] So, while biometric technology as such is not a particularly new notion, its novelty lies in its particular relevance to certain other cutting edge technologies and in the similarity of the legal challenges it poses with some at least of the challenges applying to other technologies discussed in this book.

[67] A. Conte, 'Drones with Facial Recognition Technology Will End Anonymity, Everywhere', *Business Insider*, 27 May 2013, available at www.businessinsider.com/facial-recognition-technology-and-drones-2013-5?IR=T.

PART III

Conclusions

15

So, What Do We Make of All This?

WILLIAM H. BOOTHBY

15.1 OPENING REMARKS

It is difficult to know exactly how to bring a book of this nature to a satisfactory conclusion. Each reader will inevitably have his or her own view as to which parts of the book ought to be emphasised as being of overall importance and which points can satisfactorily remain within the confines of the chapter where they were identified. It is impossible to foresee exactly how these technologies will interact, or at what rates they will evolve. These final pages of the book, therefore, are not intended to represent the definitive statement of what the book has achieved, because that is a matter for the audience, not the performers. Equally, these concluding pages do not in any sense diminish the importance of the very many points that the contributing authors may have made and that for whatever reason are not repeated here. Rather, this chapter sets forth a few of the editor's personal reflections on some aspects of the substantive chapters and offers to the reader some personal views which, of course, the reader is invited to reject in favour of his or her own reactions, ideas and opinions.

Chapters 1–4 aim to provide the background statements of law that represent collectively important aspects of the baseline against which particular technologies and the legal and ethical challenges they pose must be judged. They reflect the law as it is at the time of writing, and it remains an open question whether, and to what extent, the law will change to counter, address or otherwise respond to the challenges that the new technologies discussed in the book will pose. Where appropriate, the chapters have suggested legal approaches to address identified legal and/or ethical problems. Those suggestions amount to *lex ferenda* but the main focus of the book throughout has been to explain existing law and its application, or likely application, to techniques and capabilities a number of which remain but twinkles in development engineers' and scientists' eyes.

While one of the main purposes of the chapter is to bring some of the main lessons to be derived from each of the substantive chapters together, inevitably we will start in the next section by identifying what the main lessons of each chapter consist of. In the section thereafter we will consider matters of convergence or composition, both in relation to international law and as to the technological developments that the book is considering. We will then address some points of general application and, by contrast, some important differences in the regulation of specific technologies in the military/armed conflict and in the peaceful contexts and will bring the book to an end with some, it is hoped helpful, concluding remarks.

15.2 THE SUBSTANTIVE CHAPTERS

In this section we consider what the substantive chapters of the book show about the regulation of military and peaceful applications of the various technologies and activities and, in doing so, try to bring the important strands of this discussion together. Chapters 1–4, it will be appreciated, have been designed to set the legal scene, by providing the context in which the discussion of the specific technologies must be set.

Chapter 5 addresses cyber operations and capabilities against a background of considerable international controversy. The Tallinn Manuals, largely the product of like-, or at least similarly-minded expert work, are most valuable, important and authoritative statements of how the principles and rules of international law with which we are familiar are capable of being applied in the cyber context. The UN GGE reports of 2013 and 2015 and the failure of the 2016–2017 UN GGE to achieve consensus around a report show the limited and considerably less granular provisions that a more broadly-based and geographically inclusive group of experts have been prepared to acknowledge as having application in relation to cyber activities. The international differences of view in this field are also reflected in the failure of the Code of Conduct, proposed by Russia, China and certain other States, to achieve more broadly based support. There is little or no prospect of detailed treaty provision achieving general international consensus and the somewhat secretive nature of much of the national security-related activity in cyberspace renders the achievement of customary rules a rather distant prospect.

In relation to peacetime State activities in cyberspace, Chapter 5 recognised the need for a balance between notions of freedom of speech and expression, the desire for privacy and the intrusiveness required to support national security. It will be for each State to determine where this balance lies and to set its laws accordingly. We saw that encryption technologies, and the perceived national need for 'back doors', arouse particular controversy in this regard. Another considerable source of challenge is the technology that, collectively, represents the Internet of Things. The ongoing need to patch vulnerabilities that appear a long while after sale may be expected to conflict with the manufacturers' likely unwillingness to continue to support, and thus to patch, developing vulnerabilities associated with out-dated yet still 'smart' appliances and facilities. Whether this will result in enforced limitations of the active lives of such appliances or whether some other way of addressing the problem, such as by means of codes of practice, will be found remains to be seen. At any event, if at the beginning of the paragraph we considered freedom of expression, privacy and national security to be in potential mutual conflict, we can now add commercial interests and data protection legislation to this increasingly complex mix.

Maybe the common, rather more general theme that ran through Chapter 5 was this need to achieve a balance between apparently conflicting interests. In the peacetime context they seem to be the interests identified in the first sentence of the preceding paragraph. In the broader context of determining which principles and rules apply to activities in general in cyberspace, they are the conflicting interests of States. Maybe if consensus around treaty provision is indeed unachievable and if customary rules are unlikely to crystallise in the short to medium term, voluntary best practice and guidelines agreed perhaps on a regional basis may be the most sensible way of moving forward. As to the future, it will be interesting to see whether what we described as the compelling logic of the Tallinn Manuals attracts more general international acceptance and it will be intriguing to observe where the seemingly unavoidable conflicts between privacy and security interests will eventually find their equilibrium.

Chapter 6 grapples with the complexities associated with man in the loop, man on the loop and man out of the loop, i.e. entirely autonomous, technologies and seeks to clarify the meanings to be given to the associated notions of automation, of autonomy and of swarms. Defining the terms is always going to be critical to developing an accurate understanding of the legal issues involved. Taking human beings out of attack decision making poses challenges under the law of armed conflict, partly at least because that body of law requires attackers to take precautions that cannot be fully implemented by autonomous systems using currently available technology. The problem lies in the legal need for evaluative decisions to be taken and, as new weapon technologies may only be fielded if they can be used in a manner that complies with the requirements of existing law, we concluded that the circumstances in which autonomous offensive attack weapons that employ currently available technology can lawfully be used are limited.

Chapter 6 reinforces the view that it is not appropriate to consider technological developments in isolation from one another. The implications of cyber hacking and interference techniques should be borne in mind by those planning to develop and employ the remote attack technologies considered in the chapter, and it is clear that complex responsibility issues are likely to arise in this regard in the future. The adequacy of current law to deal with emerging automated and autonomous weapon technologies seems to be a matter of current controversy, and at the time of writing it remains to be seen what legal arrangements, if any, will emerge from the discussions currently taking place under the auspices of the Conventional Weapons Convention.

The chapter then considered regulation associated with the employment of automation and of autonomy in three non-military applications, namely civil aviation, surgery and driverless cars. Safety emerged as a prime consideration underpinning the approach to such regulation. Where driverless cars are concerned, it is interesting that these safety concerns appear to involve a somewhat cautious approach to the introduction of the new technology when, on most reckonings, the use of driverless cars seems likely to bring about a substantial reduction in road collisions and thus in deaths and injuries caused on the roads. One is left with the impression that it is the possibility of deaths or injuries being caused in an accident that would not have occurred had a human being been in control that is prompting this degree of caution.[1] If collective safety were to be the over-riding criterion, arguably the technology would be introduced more rapidly and more extensively.[2] At first glance it might seem that there is another important factor at work here, namely

[1] Consider the 2016 road accident in the United States, in which the autopilot sensors on a model S Tesla motor vehicle that was operating autonomously failed to detect a large white tractor trailer against a bright sky. The occupant of the vehicle, Mr Joshua Brown, was killed in the accident: D. Yadron and D. Tynan, 'Tesla Driver Dies in First Fatal Crash While Using Autopilot Mode', *The Guardian*, 1 July 2016, available at www.theguardian.com/technology/2016/jun/30/tesla-autopilot-death-self-driving-car-elon-musk, and consider the fatal accident that occurred on 18 March 2018 in Tempe, Arizona, involving a pedestrian and a self-driving Uber vehicle: D. Wakabayashi, 'Self-Driving Uber Car Kills Pedestrian in Arizona, Where Robots Roam', *New York Times*, 19 March 2018, available at www.nytimes.com/2018/03/19/technology/uber-driverless-fatality.html.

[2] Consider, however, the criticism that it was a lack of safeguards to prevent irresponsible use of a Tesla vehicle fitted with autonomous driving technology that contributed to a crash in which the operator of the vehicle died in 2016: N. E. Boudette and B. Vlasic, 'Tesla Self-Driving System Faulted by Safety Agency in Crash', *New York Times*, 12 September 2017, available at www.nytimes.com/2017/09/12/business/self-driving-cars.html.

consumer confidence, but how could that work in favour of an existing technology that involves so many road deaths and injuries? Perhaps the explanation is simply the all too familiar human suspicion of 'something new', such that the transition from the traditional to the novel must be undertaken in carefully chosen steps or stages if popular confidence in the new arrangements is to be achieved, a confidence that is seen to be critical to the long-term success of the new technology.

Arguably the closest parallel between the concerns associated with the military application of autonomy and those associated with its peaceful use arises in relation to automated surgery. We observed that it is the inability of machines to make decisions that do not boil down to mathematical calculations that renders the human surgeon irreplaceable. It is that feature that also renders the human being an essential participant in attack decision-making. Perhaps we can conclude that only when machines can make reliable decisions that go beyond the mathematical, what in Chapter 6 we have described as evaluative decisions, can machines be placed in ultimate control of surgical procedures, of offensive attack decision-making and, no doubt, of other activities that require such complex decisions to be taken. It remains to be seen when that level of technical sophistication will become available. However, caution will be required and the longer-term implications must be considered. After all, human skill fade seems to be a likely consequence if autonomous processes deprive human operators of sufficient experience to establish or maintain their skills. This may be expected to apply in surgery, driving and weapon operation as well as in other fields of activity. The danger then is that human expertise may not be available to detect when autonomous processes are ceasing to perform to the required standard or in the appropriate way.

Chapter 7 demonstrates that human enhancement technologies can take very many forms. In the uniquely different environment that is armed conflict combat, human enhancement emerges as yet another way in which the parties to the conflict seek to achieve and maintain an advantage over one another. The employment of such technologies in connection with armed conflict hostilities must be assessed by reference to the *lex specialis* of the law of armed conflict, and the chapter identified potential issues there. However, and depending on the particular technology involved, it would seem more likely that legal concerns over military human enhancement technologies will not so much be linked to that body of law as to applicable domestic and human rights law. Where intrusive enhancement techniques are concerned, the right of the individual to maintain the integrity of his or her own body will be a vital element. When enhancement techniques start to affect the functioning of the human brain, a whole new set of intractable challenges emerge, as the chapter clearly shows. One cannot help thinking that international engagement is needed with a view to developing a consensus as to what is, and respectively is not, acceptable in terms of human enhancement in the armed forces. Perhaps there are lessons to be learned here from some elements of the civilian anti-doping arrangements, e.g., the proscription of specified substances or techniques.

The chapter draws attention to the ethical challenges that a number of human enhancement technologies pose in their military application and questions whether these ethical concerns will translate into an international reluctance to deploy them. Ethical issues are shown also to arise in the peacetime environment, where, for example, notions of fairness between competitors motivate rules proscribing the taking of specified substances that would enhance performance. Human enhancement in the industrial and commercial sectors is also noted, and here also ethical issues are shown to be involved. As the chapter comments, it is an open question whether these ethical challenges will be addressed in regulations and guidelines.

The sometimes blurred distinction between performance enhancement and the treatment of disability and illness is another potentially challenging aspect, while the enhancement of some military personnel is shown to have potential benefits, such as reduced personnel requirements for certain kinds of operation, as well as notable drawbacks, such as adverse potential impacts on unit cohesion, individual problems associated with the ending of certain enhancement treatments and so on. If in the military context notions of freely given, informed consent sometimes seem difficult to apply, in the peacetime environment one can foresee similar difficulties. Did the athlete fully understand the long term health effects of a substance the trainer was supplying; was the perhaps young athlete realistically able to withhold consent; were the side-effects of the relevant substance adequately researched before their use in the employment sphere etc? In these respects there are clear parallels to be drawn between the legal concerns associated with, respectively, military and civilian applications of human enhancement. That there are ethical concerns is another point of similarity although their precise nature will tend to differ.

The point remains, however, that there are fundamental differences between the civilian and armed forces contexts where human enhancement is concerned. To grossly over-simplify, while 'all is fair in war', in peaceful applications, the maintenance of fairness in, e.g., sporting or commercial competition is the concept that underpins, indeed drives, regulation.

However, perhaps the most important aspects to emerge from the chapter are the sheer range of technologies that can be characterised as human enhancement and the extent of the scientific effort that is being devoted to their development. This spectrum of technologies, many of which are dual use in character, seems likely to witness rapid advances in the course of the next few years.

Chapter 8 poses explicitly a question that arises also elsewhere in this book, for example in the context of the artificial brain discussion in Chapter 13, namely 'how far can we change before we cease to be ourselves'? That cutting edge technological developments bring into question our status as human beings emphasises the importance and complexity of the issues being discussed. The legal implications are indeed ground breaking. Which rights apply, when do certain human rights cease to apply, and are additional rights required to cater for the peculiar requirements and vulnerabilities of enhanced individuals. While currently available enhancement technology does not seem to change the fundamental humanity of enhanced individuals, intriguing issues emerge as to the relationship between that which is considered to be human and that which is considered to be an object. This arises most readily in the context of the development of advanced prostheses with osseointegration and brain–computer interfaces, but that may not be the only context in which such issues fall to be addressed.

Bearing in mind the potential for and kinds of cyber threat discussed in Chapter 5, one might speculate that such threats might include hacking into and corrupting or interfering with the interface. Perhaps the practicability of such hacking would depend, inter alia, on the way in which the interface is set up, on the protective measures taken and so on. Whether a specific act of interference would qualify as an attack for the purposes of API, Article 49(1) would depend on whether it amounts to an act of violence, which in turn will depend on whether identifiable, non-minimal harm results. Whether such an act of hacking in the peacetime context would amount to a criminal assault will also likely depend on whether the effect on the relevant individual can be described as the application of violence. Interestingly, here there is an element of coherence between the law of armed conflict and criminal law approaches. As the chapter notes, if the prosthesis alone is affected, perhaps criminal damage would be the appropriate charge.

Accepting that the essential human status of enhanced individuals will normally, and for the foreseeable future, remain unaffected, the use of enhancements in the armed conflict context, absent the most unusual, frenzied circumstances, seems likely also to be designed to enhance, or to have the effect of enhancing, the ability to comply with the distinction principle and its associated rules. Characterising the use in armed conflicts of enhancements as methods of warfare would oblige States party to API legally to review such use in advance, which would be a positive development. The treatment following capture of enhanced individuals poses problems the practical solution of which, in some cases at least, would seem to require disclosure of the nature of the enhancement by the State to which the enhanced individual belongs to the Detaining Power. There is likely to be significant resistance to making such disclosures, so while this would seem to be a topic on which the law might profitably be developed, it is doubtful whether any such legal progress will in fact be achieved.

Overall, it is likely that the legal issues that emerge in the field of human enhancement will be specific to the particular kind of enhancement and to the actual context or circumstances of its employment. The strategic objective of emerging law will be to safeguard the human rights and human integrity of all those involved. The challenge, as the technology becomes ever more sophisticated, will lie in determining where the human element in 'human rights' finishes, and something rather different begins.

Chapter 9 explores the somewhat complex relationship between human degradation technologies and the notion of nonlethal weapons, both of which are ideas grounded in military doctrine. A core challenge that permeates the chapter is to determine how exactly to categorise different kinds of human degradation by reference to the established IHL lexicon of military operations and attack. Does an activity that is intended to degrade but that unintentionally causes physical or psychiatric injury or illness amount to an attack? The prevailing view is that it does, but there remain dissenting voices. Equally interestingly, some human degradation techniques sit at the ill-defined and not thoroughly understood divide between military operations that are regulated by Articles 48, 51(1) and 57(1) of API, and activities that do not amount to military operations and that therefore are not addressed by the law of targeting.

If, as the contemporary discourse would have us believe, military operations are characterised by the infliction of harm to persons or objects, how should pain, immobility, confusion or subjection to excessively loud music, or perhaps a combination of one or more of these features, be classified? Specifically, does the infliction of pain that lasts only as long as the targeted individual is exposed to the relevant technology amount to the infliction of harm? If it is, then arguably the mechanism used to produce that outcome is a weapon and would require review on that basis under Article 36 of API and, in the case of States not party to API, under the implied rule requiring States to undertake such reviews. If the technology is not considered to inflict harm, it follows that its use constitutes neither an attack nor a military operation with the consequence that the equipment used is not a weapon. This would mean that a State that is not a party to API is only required by the implied rule legally to review the technology before using it if the technology is to be used in combination with other technologies to cause harm or to undertake attacks. A State that is party to API would only be required to conduct a weapon review if the use of the particular technology, alone or in combination with other technologies, will cause harm, will constitute an attack or can properly be regarded as a method of warfare.

The use of chemical substances for human degradation purposes raises complex issues which the chapter neatly illustrates by reference to the Moscow Theatre siege. In the armed conflict sphere, the distinction between riot control agents as defined in the Chemical Weapons Convention and incapacitating chemical agents is of vital importance and it remains to be seen whether the ICRC insistence that the 'law enforcement exemption' is limited to riot control agents (RCAs) is supported by States. To extend the exemption to a wider selection of toxic chemicals has the potential to erode the overall protection afforded by the Convention, so perhaps States will indeed support the ICRC position; but that would leave States with the challenge of how to deal with the kind of situation that we saw in Moscow. In truth there are no easy answers here.

This is, of course, where the human rights and peacetime law, including applicable domestic law, kick in. Recognising that the use of some degradation technologies can have fatal consequences, the right to life and, in particular the very limited circumstances in which a deprivation of life would not breach that right, are all-important considerations. Equally important, for example, is the prohibition of inhuman or degrading treatment, a right that the law of armed conflict also accords to those not taking an active part in the hostilities. It is clear that if a State does not take the measures that human rights law requires with a view to protecting human life, that State is likely to be found to have breached the right to life. Equally if the State that has taken life fails adequately to investigate the circumstances after the event, a similar breach is also likely to be found. Where inhuman and degrading treatment is concerned, however, things do not seem to be quite so clear-cut. The chapter draws attention to flaws in the ECtHR's analytical approach to such issues and points out that the focus should instead be on both design purpose and on the intended effect in the circumstances of actual use.

More generally the chapter discloses inconsistencies in the threshold of application of weapons law, targeting law and human rights law to these technologies, and identifies intended use, design purpose or harmful effects as the features that are relevant to each. The chapter argues cogently that the actual physical and psychological impact on the degraded person in the context of the prohibition on cruel, inhuman and degrading treatment may present a more promising focus for future regulation of human degradation technology.

Chapter 10 draws an interesting contrast between the plethora of initiatives at the policy, safety and industry levels, including codes of conduct and reporting mechanisms associated with the commercial/industrial implementation of nanotechnology, and the secretiveness associated with its application to defence, where arguably greater transparency is required in relation to the same protective and safety aspects. The apparent absence of transparency in research as between the commercial/industrial sphere and the defence sphere makes it that much more difficult to gain an understanding of the implications of nanotechnology substances for health and the environment, a concern that should transcend spheres. Moreover, the 'multi-disciplinary, multi-pronged' approach to regulation witnessed in the commercial/industrial sector would seem to be a useful way ahead for the defence sector as well as a way of avoiding otherwise potentially catastrophic risks of environmental and human poisoning.

Recognising the comparatively lengthy history of nanotechnology, whether under that or a different label, the chapter notes the applicability to it in the commercial/industrial sector of traditional regulatory regimes, a feature that has parallels with the application in the defence context of existing law of armed conflict principles and rules, as noted later in the chapter. Interestingly, certain chemical substances pose apparently corresponding challenges to the regulatory regimes associated respectively with the commercial/industrial and defence spheres. Defining nanotechnology and related concepts poses the sort of challenge also seen in other domains, such as autonomy, but the importance of such definitions in support of a regulatory regime is clearly recognised. It is perhaps unsurprising in this regard that the commercial/industrial sphere may be achieving greater maturity than the defence sphere, although even in the former the need for ad hoc regulation is not universally accepted.

In the defence sector, the chapter notes the wide range of potential applications of nanotechnology and emphasises the difficulty of fitting certain of the emerging substances within the structure of the existing regulatory regimes, in particular by reference to the Biological and Chemical Weapons Conventions. The need for greater coordination between the two sectors, e.g., in relation to the applicability of the Biological Weapons Convention and in order to ensure that developments in scientific knowledge are adequately reflected in evolving regulation, is not only pressing but has obvious resonance with other technologies; indeed, it is at the core of the philosophy that drove the writing of this book. Scientific understanding must, ultimately, inform how regulatory regimes develop, because it is scientific knowledge that will disclose the threats and challenges that regulation will seek to address. That said, the relevance of much existing law in both sectors is also apparent and it is, after all, always good to recognise what you already have before considering what else you might need.

As the chapter notes, nanotechnology must not be viewed in isolation but as potentially contributing to developments in other technologies. So, it must be recognised that scientific progress relating to those other technologies, and regulation governing them may both have implications for nanotechnology and vice versa. Specifically, nanotechnology may have implications for the adequacy of legal or policy provision in respect of those other technologies, which again feeds into notions of convergence or composition that will be addressed in the next section of the present chapter. This cross-cutting effect is increased by the 'dual use' character of the technology, i.e. that it has peacetime and military/armed conflict applications, although that is a feature that nanotechnology has in common with the other technologies discussed in this book.

The worrying thought is that as technology in the 'nano' and other areas becomes more complex, the ability to determine whether particular applications are legally compliant potentially becomes increasingly problematic. New methods of governance must be sought that can respond in a timely fashion to the advance of knowledge regarding toxicity.

Chapter 11 starts by pointing out that of the many developments in technology associated with the maritime environment, the considerable majority are satisfactorily regulated under existing legal rules. That framework of rules was discussed in Part I of the present volume and does not require further articulation here. There are, however, two evolving maritime technologies that are of great importance and that do raise important, indeed pressing, legal concerns. Where the first of these is concerned, there are obvious links between the unmanned maritime systems discussed in Chapter 11 and some of the automated and autonomous technologies considered in Chapter 6. It is therefore sensible to suggest that technical understanding and legal interpretations that emerge from analysing such technologies in respect of one operational environment can be potentially applied in another. This 'lateral thinking' or cross-cutting approach again reinforces the appreciation that technological advance and the issues it raises transcend operational domains. That the definitions of 'warship' and of 'vessel' call into question the legal status of unmanned naval craft as well as the belligerent, navigational and other rights and obligations associated with their use has its parallel in the air domain. As we have seen, if nano air vehicles are so small that applying the markings required of military aircraft becomes impractical, they will similarly be ineligible for the exercise of belligerent rights. This leads one to wonder whether existing definitions of this kind operate as a welcome signpost to operational adjustments that must be made so that new technologies will fit the existing legal regime, or as a potential legal impediment to desirable progress. The answer will frequently depend, inter alia, on the particular context of the rule, on the protections that the rule is designed to afford, and on whether those protections remain relevant to the new technology.

It is clear that there are significant gaps in the legal regulation of unmanned maritime system operations and, as we have seen with other technologies, it remains to be seen whether States will be prepared to fill those gaps in the law. Chapter 11 has charted the course for States to follow. What does, however, clearly emerge is that established rules of the law of targeting and of weapons law apply to these emerging unmanned technologies but that the former body of law does so with some, perhaps limited adjustment in the maritime domain. This seems to be another example of the operational context necessarily driving limited adjustment of applicable legal rules.

Gaps in the law also surface when addressing undersea infrastructure. While, as with outer space, the idea of such infrastructure is not new, legal provision remains inadequate and incomplete, e.g., as to the status at law of undersea pipelines and cables. As with outer space, so in the undersea infrastructure context, we have facilities that are of vital global importance and yet for which international law provision is very much less than satisfactory. Arguably, in the cyber context that vulnerability forms a part of the vulnerability of cyber activities in general to which attention was drawn in Chapter 5. The abject failure of law adequately to protect undersea infrastructure arguably also represents a global weakness. It will be interesting to see whether States will, belatedly, be prepared to take the required action.

Chapter 12 points out the considerable discrepancy between the huge global reliance that is now placed on outer space capabilities and the outdated and limited nature of the applicable rules; this is yet another example of the considerable lag that exists in so many fields between technological advance and regulatory provision. The global threats posed by the potential weaponisation of space, and more specifically by the potential deployment and use of weapons of mass destruction in outer space, constitute the vital context here. The common interest of the superpowers then involved, namely the US and the USSR, drove the evolution of a fragile and distinctly limited legal framework. It is interesting that regulations and arrangements associated with peaceful and hostile activities in outer space are seen to have emerged somewhat in parallel. Another theme that resonates with points made elsewhere in this book is the conflict between transparency/openness and the wish to exploit technological advantage. So it is not perhaps surprising that aspirational language in early treaties is interpreted restrictively in later practice and seems to provide only limited guidance as to acceptable contemporary conduct in outer space. This resulting regulatory vagueness is the key legal backdrop to the rapid commercial exploitation of outer space noted in the chapter.

The close inter-relationship between military and peacetime applications and the emergence of commercial uses as being of increasing importance, both have obvious parallels for example in the cyber and biometric fields. While tourism, mining and exploration are evident ultimate goals of commercial developments in space, the vulnerability of space assets, e.g., to cyber interference seems likely to necessitate some kind of new regulation, and yet there seems to be little international stomach for the negotiation of new treaties. This might suggest the recognition of a paradox here, namely that regulations to protect valuable commercial investments in outer space activities have the potential, including in the military sphere, to impede that very commercial development and progress. It remains to be seen whether the public policy considerations that would justify regulatory limits on certain kinds of military activity in outer space will also be regarded as justifying limitation of corresponding but commercially undertaken activity.

Certainly, advances in outer space technology and increasing dependence on commercial providers breed national vulnerabilities owing to national dependence on the space environment, a feature that we have already seen in the cyber context. So, as with biometrics and perhaps nanotechnology, while the existence of the outer space environment is not new, it is changes in the nature of applications of it and in the way in which those applications are employed that pose legal challenges for the future, challenges which international manual projects and ultimately the actions of States need to address. One can only hope that the valuable Woomera Project will provide at least some of the much-needed clarity here.

Chapter 13 examines the hugely complex topics of artificial brain technology and brain computer interfaces, identifies the yawning gap between the capabilities of human brains and of currently available artificial intelligence and looks to enticing 'over the very distant horizon' possibilities. While there is an inevitable element of speculation associated with such a consideration of the relatively distant future, it is clear that such technologies are destined to pose a broad selection of political, societal, institutional, personal, governmental, ethical, legal, human rights and other challenges. Some of these will mirror concerns associated with other technologies discussed elsewhere in this volume. Many of the challenges will, however, be novel or will be more intractable versions of difficulties of a sort with which we are already fairly familiar. Most such challenges are likely to be far-reaching, with serious, indeed troubling, implications.

Artificial brain technologies and brain computer interfaces seem likely to be applicable to a very broad range of activities, so it follows that there will be implications for the law that regulates all the fields associated with such activities. The issues discussed in Chapter 13 will therefore be of considerable consequence by reference to both peacetime and warfare contexts.

The chapter draws interesting parallels with Chapter 6. Artificial brain technology raises issues some of which are similar to those linked with autonomy. Brain computer interfaces, by contrast, seemingly avoid those issues because of the continuing involvement of the human being but trigger a whole load of other concerns, such as whether the technology will need to be able to refuse to comply with manifestly unlawful instructions from a human being. So while there are some similarities in the issues raised across the technological domains being considered in the book, artificial brain and brain–computer interface technologies seem to intensify existing challenges while posing a raft of new areas of legal and ethical difficulty.

The chapter illustrates the linkages that exist between the legal issues posed by these brain technologies and: the legal challenges associated with human enhancement; the vulnerabilities and insecurities arising from the use of cyber technologies; the legal and ethical concerns associated with some applications of autonomous technologies; and the data protection issues brought about by the emergence of the Internet of Things.

Working out how the brain works and then synthesising such processes may well take decades during which relevant law may change to address challenges arising from other but linked technologies. There is, however, a possibility that brain computer interfaces may address some of the problems arising from artificial intelligence. This aspect points again to one of the central themes of this book, namely that emerging technologies should not be considered in isolation from one another, and that those promoting legal initiatives should take the evolving technological landscape as a whole into account when considering the need for, and contours of, new legal and policy arrangements.

Wisely, the chapter raises the very many questions that these technologies generate. These are the issues that must be addressed and it seems likely that time will be available to do so before the technologies will mature to the point of potential implementation. It is, however, at present unclear whether the international community has both the capacity and the will to address these issues. The track record in relation to cyber and autonomy is, as we saw respectively in Chapters 5 and 6, less than encouraging. That said, the fundamental importance of the questions raised suggests that legal and/or policy arrangements need to be formulated as the technologies mature. While the notions discussed in the chapter are indeed 'fantastical', they require careful national and international attention on a multi-disciplinary basis.

Chapter 14 tackles a selection of technologies many of which are not new but whose methods of employment are being adapted to modern requirements. The point is made that the law to be applied will be determined by the operational context, not by whether it is armed forces, law enforcement personnel or some other agency that is employing the biometric technology. It is therefore interesting to observe that both armed forces personnel and law enforcement, border security and similar agencies will be bound by applicable human rights, data protection and constitutional laws when undertaking what are fundamentally similar tasks such as control of access, biometrics of intent and biometrics of detection operations in a peacetime setting. It is only when the law of armed conflict applies that the regimes differ as between the conduct of hostilities and peacetime activities, but this is really only a reflection of the application of the law of armed conflict to hostilities in general. There are clear linkages between biometrics and cyber operations, and biometrics will be essential to the development of autonomous anti-personnel weapon systems while artificial intelligence will be a vital element in the development of future biometric analysis tools. Embedding biometric tools into future enhanced soldiers may seem futuristic, but that seems to be the direction in which technology may be expected to lead us.

The maintenance of security is a core requirement of modern society. The workplace, the leisure setting, transport hubs, public buildings, online and financial institutions provide just some of the environments in which security in the future seems inevitably to require the operation of increasingly reliable biometric technologies. If, as the author believes to be the case, security is essential to the proper operating of a modern, democratic and essentially free society, the future challenge will be that identified by the Biometrics Commissioner for the UK, namely achieving legislation that strikes the right balance between privacy interests of individuals and the national interest in ensuring that crime is prevented and criminals are apprehended. In a sense this will in the future boil down to a trade off between the intrusiveness that biometrics will inevitably involve and the freedoms that the biometric activity is designed to protect.

Paradoxical prospects seem to abound when considering biometric technology, but it is well to recall that these challenges between the interests of security and the preservation of the freedoms that the security is designed to protect are nothing new, they are just surfacing in yet another context.

So, having identified a few takeaways from the substantive chapters of the book, we will now look a little more generally at legal and technological relationships. How should one properly describe the relationship between the bodies of law to which we have referred in the preceding chapters and wider notions of international law? In a similar vein, how should we describe the relationship between the various advances in technologies discussed here and indeed among other evolutions in technology not addressed in this book?

15.3 CONVERGENCE, OR IS IT COMPOSITION?

As is apparent from this book and, indeed, upon the most casual observation of technological development in the world at large, scientific advances are achieved in many different fields of endeavour contemporaneously. To consider any one such field, and the changes that are taking place in it, in isolation is just as unrealistic as to consider such changes against a backdrop of the snapshot state of affairs in other technological areas as seen on any particular date or time. The more realistic approach notes that all technologies advance together, not necessarily at the same rate across all fields, and that it is this dynamic of developments in different technologies that provides the realistic context against which change in one particular activity must be considered. Moreover, emerging technologies are often employed in combination with one another to generate an integrated system that provides, e.g., significant commercial advantage, even dominance,[3] so, again, it is logical to consider together the various ingredients of such integrated systems.

Some technologies may have direct implications for another parallel technology, whereas in other fields the implications may be less direct. How we view a particular technology may be influenced by developments in another, loosely associated activity and so in this section we strive not to interpret each technology in isolation but, instead, to consider the composite effect of these interactions and mutual influences and developments. In short, we should recognise that developments in numerous branches of technology are linked

3 Consider e.g,. the systems employed by Amazon, including plans to use autonomous trucking, mobile warehousing and drone-based delivery systems in combination with network analysis and demand-anticipation algorithms; A. Greenfield, *Radical Technologies* (London: Verso, 2017), 278–280: 'The truly transformative circumstances will arise not from any one technology standing alone, but from multiple technical capabilities woven together in combination: ibid., 273.

with one another, so it is neither appropriate nor realistic to view them as a series of barely related, somewhat independent stovepipes.[4]

An obvious example is the interaction between cyber capabilities and remote, automated and autonomous weapon systems. Computer capabilities are critically important for the operation of such systems, and yet (or perhaps accordingly) cyber interference activities in one form or another have the potential to threaten the proper performance of such systems or to render their operation unreliable, a threat which will usually be countered by the employment of other, computer-based facilities. So it is immediately apparent that remote, automated and autonomous technologies and cyber capabilities are inextricably linked with one another and that maintaining the security of underwater infrastructure in the form of submarine communication cables is of vital importance in this regard. In Chapter 12 we also saw that cyber interference is among the threats to outer space technologies on which so many activities on Earth depend. As with cyber, current international law provision in relation to outer space, unmanned maritime systems and underwater infrastructure is limited and, frequently, open to conflicting interpretations.

Similarly, it emerged from Chapter 13 that there is a clear association between concepts of artificial intelligence and the development of brain computer interfaces and of artificial brain technologies. Artificial intelligence will be essential to the achievement of both of these brain technologies and it is by identifying common threads of this nature that it is possible to identify where specific legal issues are likely to have common application to different but more or less closely related technologies.

While there is no legal obligation to making a weapon system robust against external intrusion/interference there is a military interest in maintaining the integrity of weapon systems. There are also risks that, if interference precipitates erroneous attacks, confidence in such weapon systems and broader popular confidence in activities of the armed forces may suffer. Indeed, to some casual observers, erroneous attacks may have the appearance of

4 Consider in this regard the comments of Professor Genevieve Bell at the Australian National University College of Engineering and Computer Science following her appointment as the 3A Institute Director, in a talk entitled 'Managing the Machines: Building a New Applied Science for the 21st Century' (2018), available at https://cecs.anu.edu.au/3a-institute. Noting that different technology requires different sets of knowledge, Professor Bell identifies core issues such as: what does it mean for us to acknowledge that a thing is, like us, autonomous; if technology has the capacity to do things, how much agency do we allow it; and are we safe in the vicinity of these technologies – specifically how do notions of safety work in relation to algorithms and what values are embedded in algorithms?

deliberate, unlawful attacks. So, the vulnerabilities arising from the relationship between cyber, remote, automated or autonomous attack technologies and submarine communication cables suggest the need for States to develop the security of the weapon systems and of the infrastructure on which their reliable operation depends and to ensure that relevant international and domestic law provision is adequate.

We have seen that effective governance of artificial intelligence will be required if autonomous weapon systems are to be capable of lawful use, partly at least because of the need for the evaluative decisions prescribed by the law of targeting to be taken when such weapon systems are being employed. Governance of an evolving and self-learning algorithm provides its own unique challenges. If a highly automated or autonomous weapon system is intended for anti-personnel targeting purposes, it will need to be able to distinguish between persons who are and, respectively, persons who are not lawful targets, and we have seen the critical importance that biometric technologies will have in enabling such automated or autonomous determinations to be undertaken by the machine. We also saw how biometrics seem likely to be the essential ingredient in certain types of human enhancement, particularly those enhancements that are designed to enable the enhanced soldier to determine the intent of an adversary. It follows from the preceding paragraphs of the current section that to view autonomy, cyber, outer space technologies, particular maritime technologies, biometrics and certain kinds of human enhancement entirely independently from one another fails to recognise the degree to which they depend upon one another and/or have the capacity adversely to affect one another. In a sense, however, each technology develops along its own track, so perhaps what we are really talking about here is the need that when we consider the implications of each technology and of the action that must be taken in order to seek to achieve its optimal use, a degree of lateral thinking is required of the sort that is discussed in the present section.

The vital role of computer systems in supporting outer space military operations renders it likely that during periods of armed conflict, those computer links will be the vulnerability that an adverse party is most likely to seek to exploit, and similar considerations will apply in the maritime environment. How particular weapon systems are integrated onto a specific platform, and how the computer systems that support them are set up for operational use will be factors that influence how the cyber threat can be addressed. Autonomous systems will be employed in both environments, for example to protect valuable platforms or to undertake any of the tasks noted in Chapter 11, so again to view maritime, outer space, cyber and autonomy technologies as being independent of one another is to fail to appreciate the inter-related way in which technology will be employed in military operations. The cyber threat may, for example, render increasing levels of autonomy more desirable in the maritime and outer space domains if the consequence would be to reduce somewhat the scope for cyber interference.

In current discussions, meaningful human control is seen by some experts as part of a potential answer to the ethical and other criticisms of autonomous weapon technology. Viewing the issue more laterally, however, might involve considering whether an interpretation of meaningful human control that involves certain constraints on the autonomous aspect of the technology might also have the effect of increasing the scope for adverse party interference in, or of increasing the risk of other malfunctions of, sophisticated weapon systems thus raising the spectre of misdirected attacks resulting in excessive civilian casualties.

Taking our lateral thinking approach one stage further, the very idea of 'meaningful human control' raises a panoply of issues when considered in relation to brain computer interfaces and artificial brain technologies. At what level of sophistication is an artificial brain considered to have the required characteristics and behaviours such as to render it capable of satisfying the 'meaningful human control' requirement, or does the word 'artificial' mean that such technology will always be regarded as incapable of addressing the concerns that would underpin a 'meaningful human control' requirement? This is a point that has application not just in the context of the lethal autonomous weapon system discussions at the CCW in Geneva, discussed in Chapter 6, but also to ideas of driverless cars, autonomous civil aviation, autonomous surgery and to the range of possible uses of some emerging brain technologies. Indeed, is it imaginable that a limit will be identified to the distance that it is realistic to expect an astronaut to travel or to the stresses that it is realistic for an astronaut to cope with in outer space when undertaking military, civilian and/or commercial activities there and will such limits incentivise the employment of artificial brain technologies and other kinds of artificial intelligence on board? If any of such possibilities were to become a realistic prospect, one can imagine a number of the challenges and issues noted in Chapter 13 also becoming live issues in relation to such outer space operations.

In all likelihood, 'meaningful human control' would be satisfied by the kinds of brain–computer interfaces contemplated in Chapter 13, but would not be satisfied by the employment of an artificial brain. It will, perhaps, be for applicable regulations and policies to specify the answers to these issues, and it is imaginable that such answers may vary depending on the nature and context of the relevant decisions, on the environment in which the technology is being employed and on the particular characteristics of the technology concerned.

Some technological changes pose questions as to the continuing relevance of established legal rules in what may at first glance appear to be somewhat different circumstances or fields of activity. The increasing reliance on outer space and the much broader selection of activities now undertaken there suggest that the limited and out-dated rules developed during the Cold War urgently need revision. Likewise, as nanotechnology starts to enable the development of nano vehicles and specifically of nano air vehicles, this will have implications for the technologies considered in Chapter 6. As military aircraft are increasingly miniaturised, compliance with the traditionally accepted definition of 'military aircraft' becomes somewhat problematic. Does it make sense to persist in an external marking requirement when the markings themselves would not be visible? Does the adaptation of the command and crewing requirements of the definition, as put forward in the AMW Manual, reflect the practice of States and does the AMW Manual provision operate sensibly as a condition precedent for the exercise of belligerent rights? Perhaps more fundamentally, can computerised support technology that permits the proper conduct of all targeting law decision-making be built into a nano-sized air vehicle and can the legal challenges associated with operating such vehicles in swarms of whatever configuration be properly met? So in technological terms, certain applications of nanotechnology should not be viewed in isolation but should be seen as integral to, or at least as profoundly affected by, the technologies addressed in Chapters 5 and 6.

In each of Chapters 5–14 we have considered the differences and similarities between the peacetime and armed conflict applications of the respective technologies, and how the associated rules, regulations and guidelines relate to one another. That is also an interaction process because, although the rules, regulations and guidance that apply to the military or to armed conflict use of these technologies are frequently not the same as those that apply to their peacetime applications, there are important linkages between the various legal provisions as they apply in these diverse contexts. Moreover, we should consider how these international rules, regulations and guidance fit into the more broadly based framework of international law and, indeed, how that framework can most accurately be described.

Judge Christopher Greenwood remarked in a dissenting judgment in the *Diallo* case, "[i]nternational law is not a series of fragmented specialist and self-contained bodies of law, each of which functions in isolation from the others; it is a single, unified system of law and each international court can, and should, draw on the jurisprudence of other international courts and tribunals…"[5] These strong arguments are increasingly being put forward against the characterisation of international law as fragmented and Judge Greenwood's view is often described in terms of what has come to be known as 'convergence'.[6] If international law is a single, unified system and if international courts should draw on each other's jurisprudence, so also, the argument might go, should the international law rules that respectively regulate the use of technologies in armed conflict and in the peacetime setting be seen as parts of a broader whole.

In the present author's view, the observer of international law is not confronted with a stark choice between two alternative interpretations, the 'series of fragmented, self-contained, isolated bodies' approach and the 'single, unified system' appreciation. There is a respectable view that international law exhibits elements of both of these positions. Undoubtedly, the international law rules relating to civil aviation are neither closely linked with nor fundamentally inconsistent with the law as to the use of force. Equally, where the law of armed conflict and the law of human rights are concerned, there are aspects such as their shared notion of non-refoulement that are coherent, but there are contexts in which their approaches differ, such as aspects of the use of lethal force and of detention.

Moreover, as Anthea Roberts has commented: 'Instead of constituting a single, uniform field, international law is an amalgamation of multiple, partially overlapping fields.' In her seminal treatise, Dr Roberts points out that '[i]nternational law is best understood as a transnational legal field made up of multiple, partially overlapping fields'.[7] Indeed, when considering the degree to which international law is really converging, the factors addressed by Dr Roberts in her book must be carefully considered, including but by no means limited to the centrality of language, differing emphasis where notions of sovereignty are concerned, the role of academic institutions and of academics and their writings and, perhaps above all, the profound, indeed overwhelming, influence of Anglo-Saxon common law influences over contemporary international law interpretations. It is arguable that the geo-strategic developments that Dr Roberts identifies in Chapter 6 of her work may result in the evolution

5 ICJ, Ahmadou Sadio Diallo (*Republic of Guinea* v. *Democratic Republic of Congo*) 2012 ICJ 324, Judgment of 19 June 2012, Declaration of Judge Greenwood, 394.

6 See, for example, M. Andenas, 'Reassertion and Transformation: From Fragmentation to Convergence in International Law', 46 *Georgetown Journal of International Law* (2015) 685–734.

7 A. Roberts, *Is International Law International?* (Oxford: Oxford University Press, 2017), 6 and 25.

of competing appreciations of international law centred on what the present author would see as the emerging powers that will join the United States in being of global significance, particularly China, Russia and potentially India and certain others. These evolving trends also, therefore, call into question notions of convergence. Is the proper interpretation that ideas of convergence were rooted in the globalisation that was a feature of thought and events in the 1990s and 2000s? If so, perhaps the coming decades will see a tendency not so much to fragmentation as to a coalescing of legal thought around collections of States that are not so much defined by reference to geographical regions but, rather, that are characterised by groupings of the like-minded at the centre of each of which will be a global power such as the United States, Russia and China. Perhaps the experience in relation to cyber security, as outlined in Chapter 5, is something of a foretaste of such a tendency, with the interpretations of two of the 'like-minded groups' competing with the approach adopted by the Western group. This thought amounts, of course, to speculation as to future events, and the future is, as we all know, inherently uncertain. An alternative interpretation is that convergence is indeed what we are witnessing, for example in relation to cyber activities. From somewhat opposing views as between Western States and China, the differences in their respective interpretations as to sovereignty, as to the militarisation of cyberspace, as to the need for ad hoc law in relation to cyberspace and as to Internet governance seem to be reducing somewhat.[8] Whether this will continue to a degree such as to amount to convergence and whether Russia will choose to adopt a similar course is not at present clear, but it must be acknowledged that any notion of convergence or composition for the purposes of the present discussion is multi-faceted, and includes the respective elements of international law, the positions that States adopt as to international law and its development and the evolution of the novel technologies we have been discussing.

Where lawyers, as opposed to law, are concerned, as Roberts observes, '[i]f international lawyers operate in silos, either domestically or internationally, they risk failing to connect with, and understand the perspectives of, those coming from diverse backgrounds and holding diverse perspectives'.[9] Arguably, Roberts' book demonstrates inter alia that international lawyers do

8 Consider Z. Huang and K. Mačák, 'Towards the International Rule of Law in Cyberspace: Contrasting Chinese and Western Approaches', 16 *Chinese Journal of International Law* (2017) 271, where the apparent convergence of, or reduction of the differences between, the positions of, respectively, China and the United States and other States that are like-minded is assessed.

9 A Roberts, *Is International Law International*, 323.

often, but not always, operate in those silos and this is a factor that also contributes to the nature of international law. Were there to be evidence of these silos starting to crumble and give way to an arrangement that is more receptive to divergent views and approaches, this might indeed be evidence of convergence among lawyers. The present author rather doubts that there is yet much evidence of such a development.

So perhaps we can conclude that international law is, in truth, a composite body of law, made up of numerous elements, not all of which contain identical provisions at the points where they nudge up against one another but which collectively strive to regulate activity at the international level with a broadly coherent general strategic purpose, namely the orderly conduct of international affairs. Interpretations of the various principles and rules may differ both at the geostrategic level and among individual practitioners and academics. So it is doubtful whether convergence, based on the verb 'to converge' or 'come together from different directions so as eventually to meet'[10] properly encapsulates this.

Moreover, we have seen that the various kinds of technology that we have been discussing in this book should also be seen as elements of a composite body of technology that is the collective product of scientific advance in its various fields, not as individual advances in specific fields of endeavour that have little or nothing to do with one another. Reflecting a little further on the position in relation to international law, the author wonders whether the word 'composition', in the sense of 'the nature of the ingredients or constituents of a whole or mixture'[11] more accurately describes both the relationship between the various constituent elements of international law and the most sensible way of considering advances in technology. In the end it is for the reader to reach his or her own conclusion as to whether a single descriptor can properly be applied to both contexts and, if so, as to what the most appropriate descriptor really is. Those who would wish to emphasise what they see as the incompatibility of certain of the international law elements with one another might prefer some other term. It will be a matter of personal interpretation and taste whether the reader regards international law as consisting of a 'whole' or as comprising a 'mixture'. Either way, international law does have its various constituent parts and the reader may feel, on balance, that 'composition' somewhat more accurately describes the relationship between them, avoiding as it does any bold assertion as to their level of coherence while equally avoiding any overt suggestion or inference of mutual incompatibility.

[10] *Concise Oxford English Dictionary*, 11th edn (Oxford: Oxford University Press, 2006), 312.

[11] Ibid., 294.

If applying ideas of 'composition' to international law seems at first glance to make sense, the other potential 'composition' circumstance that we should consider is the relationship between peacetime applications of the different technologies that we have addressed. Again, there will be numerous linkages, and in this paragraph only a couple of fairly obvious kinds of interaction are considered. So, the first example might be the artificial intelligence that would be required to enable the development of driverless cars and of autonomous civil aircraft movements. Such artificial intelligence would rely heavily on computer technology and is likely to be susceptible to cyber interference and malfunction. The reliable performance of driverless cars will require that sufficient robustness against such cyber challenges be built into the support systems and achieving that robustness is likely to involve the employment of computer-based systems.

An alternative example concerns biometric technology which might be combined with artificial intelligence to facilitate the efficient performance of security screening processes. Either the biometric or the artificial intelligence aspect of the composite system could prove to be a point of failure, and acceptable reliability of both such aspects is therefore necessary. The point here is that it is only by considering the 'composite' effects of developments in both or all relevant technologies that one can properly evaluate the possibilities associated with each of them.

Where the law applicable to the peacetime employment of novel technologies is concerned, it is interesting to note the similarity in the data protection and human rights law issues posed respectively by the Internet of Things and certain applications of biometric technologies. This to a degree reflects certain aspects that the two technologies have in common, specifically that each involves access to the personal data of individuals and that in each case the obtaining of the consent of the data subject may not always be easy to achieve or demonstrate. Just as the Wharton School of Business was established after the Industrial Revolution, perhaps it is time to establish a new school of thought and of science to respond to and govern this 'Fourth Revolution'[12] in a way that encompasses the potential applications of emerging science in both peacetime and warfare settings.

[12] This 'Fourth Industrial Revolution' is characterised by 'a fusion of technologies that is blurring the lines between the physical, digital and biological spheres'", is evolving ' at an exponential rather than linear pace' and 'the breadth and depth of these changes herald the transformation of entire systems of production, management and governance': K. Schwab, 'The Fourth Industrial Revolution: What it Means, How to Respond', World Economic Forum, Geneva, 14 January 2016, available at www.weforum.org/agenda/2016/01/the-fourth-industrial-revolution-what-it-means-and-how-to-respond/.

We therefore seem to be identifying that there are at least two strands of composition at work here. First, there is the association between the technologies that we have considered. Second, there is the recognition that the bodies of international law that we have been considering, the law of armed conflict, data protection law, human rights law, the law applicable to certain kinds of civil aviation, the law that regulates activities in outer space, the law as to the use of drugs in sport, the law applicable to unmanned maritime systems and to underwater infrastructure and other law applicable to specific technologies in particular contexts are all facets of a composite body of international law. Those facets may not always be entirely consistent with one another, but that does not alter their membership, as it were, of the composite whole. We have seen that there are similarities in some of the challenges that arise respectively in armed conflict and in peacetime. Some such challenges are addressed in a similar way in the respective contexts, often because the same legal rules apply to the particular activity in armed conflict and in peace. Where different rules apply it tends to be because the situation at hand demands different rules. That is, however, only logical. After all, in any body of law differing situations will involve the application of rules that differ according to the requirements of the respective situation.

So, for example, the law of armed conflict prioritises the protection of victims whereas in a peacetime setting other priorities, such as privacy protection, are at play. In some senses it makes no difference whether one regards the laws applying in the different contexts as distinguishable parts of a fragmented pattern of international law or whether one sees international law as a singular whole with rules that differ in accordance with the differing circumstances for which they are designed to cater. What matters from a pragmatic perspective is that the various rules are relevant to the circumstances for which they were made. It is, moreover, logical that the employment of novel technologies in both warfare and peace must comply with the existing applicable body of law and with such further legal provisions as may be developed to deal with the technology in question.

It has in the past been tempting to view international law as fragmented with elements of it apparently competing with one another. Reconciling the challenges associated with the contemporaneous application of human rights law and the law of armed conflict is a context in which it would be easy to regard the two as discrete, and occasionally conflicting, bodies of law. That those bodies of law apply concurrently in armed conflict situations and that, in relation to the conduct of hostilities, the law of armed conflict provides the yardstick for determining whether a human rights law breach has occurred, is now widely accepted and tends to support the composition approach. That the law of armed conflict achieves answers that sometimes conflict with those that reflect human rights law does not necessarily imply that they are different bodies of law – rather it suggests that some de-conflicting adjustments are required in the relationship between two elements of the single corpus of international law.

15.4 SOME CONCLUDING REMARKS

When writing a book of this kind, it is only ever going to be possible to take a snapshot of the state of technology and the state of the law as at the date of publication. In the present volume, we have attempted to peer over the horizon and tried to see what technological innovations seem likely to materialise over the coming years, possibly decades. Where the law is concerned, peering over that horizon is even less easy. The law has difficulty coping with new and emerging technologies, partly at least because the precise contours of that technology are by definition unknown, and partly because of doomed or unsatisfactory early attempts to regulate technologies that were then novel, such as aerial bombardment, explosive bullets and asphyxiating gas weapons. The last of these examples, however, illustrates the danger of failing to regulate, or at least of regulating adequately. It was the adoption in The Hague in 1899 of a treaty text that did not regulate asphyxiating gas with sufficient thoroughness that rather demonstrates the major challenge that confronts international law and those charged with the development of the law, a challenge that can be characterised in the old expression: 'you are damned if you do and you are damned if you don't'. Differentiating between emerging technologies that represent the sort of challenge that was posed by asphyxiating gas as the nineteenth century drew to a close and the kind of technology that offers opportunities of improvement that ought not to be snuffed out at birth is always going to be a tricky call, particularly when the opportunities and risks are insufficiently known.

And yet that is the very call that policy makers, treaty negotiators, legislators and others in positions of authority will have to make. They will be informed by such scientific knowledge as may be available to them. There will be powerful, sometimes emotive arguments deployed by lobbyists of one kind or another. Those with the loudest voices will not necessarily have the most cogent, the most far-sighted or indeed ultimately the most balanced arguments. Those developing the law will be influenced by diverse factors, including public opinion, financial and economic development interests, ethical and humanitarian concerns, worries about the potential for future legal challenge and anxieties about the capacity of the new to render existing ways of doing things obsolete, with all that that implies for regional employment and financial well-being of domestic populations. Complex policy balances must be struck, e.g., between technical progress and human rights, between national security and personal privacy, between the interests of the individual and those of the State and, e.g., in the outer space context, between the need properly to regulate and the desirability of maintaining that level of operating freedom that incentivises technological advance.

Balancing of conflicting interests is nothing new for policy makers, treaty negotiators and legislators. However, where some new technologies are concerned some potentially more perverse trade-offs are in play. As we have seen, the introduction of driverless cars will on any reckoning bring an immediate reduction in the toll of deaths and serious personal injuries arising from road use. Therefore, the idea that introducing driverless cars must be delayed at all on safety grounds seems at first sight to be somewhat perverse. The apparent purpose of the delay is thoroughly to assess the risk of collisions that would be attributable to the new technology but that would not have occurred in its absence. One is left with the impression that the deaths occurring during the period of such delay that would not have happened had driverless cars already been in operation are seen as an acceptable price to pay, perhaps in part to promote public confidence in the new arrangements, a confidence without which such new arrangements might never materialise.

Ethical concerns will be a powerful driver for future regulation of emerging technologies in particular fields. Human enhancement in athletics and other fields of sport is already the subject of careful regulation and increasingly universal enforcement. In that sense the ethical imperative is to try to achieve and then maintain perceptions of fairness on the field of play. There will be ethical worries over the employment of biometric technologies to obtain intrusive and inherently private data about people. Where the ethical line of acceptability will lie in keeping back doors open in relation to vulnerabilities in the Internet of Things and with regard to 'end to end' encryption of such media as WhatsApp and Snapchat remains to be seen. The point is that society, probably on a global scale, has yet to reach a conclusion as to what kinds and extent of intrusion by the State in the interests of collective security is acceptable. Perhaps the problem is that there is no fixed line here. In the aftermath and context of nasty terrorist incidents, those closely affected by them, or thereby placed in fear, are more likely to favour intrusion than citizens who are remote from the front line of such violence.

If the technologies mentioned in the previous paragraph raise ethical concerns, and they do, these will pale into relative insignificance compared with the ethical challenges posed by the development of brain computer interfaces and of artificial brain technologies. Chapter 13 mapped out for us where many of these challenges lie. It remains to be seen if and how the international community will respond to them. While time is available for such a response, there is nevertheless a need for collective thought about these issues, and as technological development continues, the requirement for thought will increasingly be accompanied by a need for appropriate action.

We have considered notions of data protection in the context of biometrics and challenges associated with the Internet of Things and have noted the considerable significance attached by the regulatory framework to notions of consent to data processing.[13] While the idea of consent was certainly in the past meaningful, what, one might wonder, does consent really consist of in an era when most of us, when signing up for new or updated Internet capabilities, readily, indeed routinely, click the 'I agree' button without having even skimmed through that to which agreement is being given and thus without the slightest notion of the permissions being granted to those to whom consent is being sent.[14] What is consent when the consenter has not ploughed through the vast pages of very small print, and when even if he or she were to do so, not all of the possible processing would be specifically spelt out there in readily understandable form? Increasingly, uninformed or at least under-informed consent seems to be equated with actual consent. Actual individual control of the uses to which one's personal data is put would probably imply a requirement for informed consent,[15] but one rather feels that such a requirement, thoroughly implemented, would likely impede the efficient functioning of the relevant systems and in a contest between such technical efficiency and privacy safeguards, it is the former that seems likely in the future increasingly to win.

[13] As to the link between biometrics and the Internet of Things, see Greenfield, *Radical Technologies*, 35–36.

[14] Note in this regard the reported suggestion that there should be a common blockchain holding all personal information in duly encrypted form, accessible to all legitimate public or private actors who retrieve the information to which they are permitted access: Greenfield, *Radical Technologies*, 160–161.

[15] Consider, however, the obligation under GDPR to give information as to data processing to the data subject in concise, transparent, accessible and easily intelligible form and the obligation of the data controller to facilitate the exercise of the data subject's rights; GDPR, Articles 12 and 13. Note in particular, however that '[w]here the controller intends to further process the personal data for a purpose other than that for which the personal data were collected, the controller shall provide the data subject prior to that further processing with information on that other purpose and with any relevant further information' as referred to in Article 13(2); GDPR, Article 13(3). It is an open question whether this represents effective protection of the individual in an environment in which processing of data takes place at digital speed and given that the safeguards seem to require the affected individual to take the initiative.

Staying with the personal data theme for a moment, one wonders whether the very concept of what comprises an individual's personal space will loosen as a result of numerous contemporary developments, including social media, the Internet of Things and biometrics to name but three. These developments may well indicate that intrusions into our private space will continue, maybe even intensify, but one cannot help feeling that such a process will only go so far and that sooner or later there will be something of a reversal. While mankind certainly consists of 'social animals', they like their individuality, and if the socialisation basis seems to threaten that individuality, one cannot help feeling that the individuality will be reasserted. As for human enhancement, will human beings be prepared to sacrifice their bodily and/or mental integrity for what may, in the case of some enhancements, be rather short-term gains? The answer in some cases will probably be yes. However, a full understanding of the implications of such enhancements seems more likely to cause human beings to make rational choices as to which enhancements to accept and which ones to reject. The requirement in this respect for informed consent has obvious parallels with the discussion of personal data in the previous paragraph.

Some of the technologies we have considered in these chapters go to the very root of what we are as human beings. Some ideas of human enhancement involve contemplation of what might be described as 'super-humans', with characteristics and capabilities that go beyond what the human frame, intellect and endurance can normally be expected to achieve. Science seems to be on the cusp of synthesising at least some aspects of brain functioning, and of directly linking the human brain with mechanical instruments and, indeed, with other human brains in ways that raise complex and profound ethical and moral questions, not least in relation to ideas of personal identity. At what point does a machine take on sufficient human characteristics that it should be subject to some at least of the regulatory provisions to which human beings are subject, and which regulations would be affected in this way? Even more fundamentally, we must strive to address the 'Frankensteinesque' spectre that instead of man's hand guiding his tools to achieve his or her goals, the roles are reversed and the tools start to take control of man. There are those who consider that autonomy takes the tools out of man's hand. Others point out that it is man who builds the autonomous system, it is man who loads the task-critical data onto the system, it is man who gives the instruction that the mission shall take place and it is man who actually initiates the autonomous activity, whether in the peacetime or warfare context. Perhaps the concerns start to mount when learning enables the machine to act outside the constraints imposed by man. Undoubtedly, the concerns would grow rapidly if a human being's control function were to be performed by an artificial brain, particularly if that artificial brain were to have the capacity to develop itself and to formulate its own responses in ways that go beyond those considered when the system was created. Perhaps brain–computer interfaces will provide part at least of the answer here, by enabling direct human engagement with the operation of computer systems, although as the required pace of decision making increases, the probable inability of the human brain to match that increased pace requirement will be an important factor to bear in mind.

The boundary between activities that are within and outside human-imposed constraints may also be difficult to define because the constraints themselves may of necessity be less than precise. Ultimately, anxiety is likely to increase as machines transition from being the servant to becoming the master, and it will be for States, individually and collectively, to determine where on that road the border lies between the acceptable and the unacceptable.

15.5 AND LASTLY

Public acceptance of emerging technologies is of course going to be crucial to their success. As Andrew Maynard rightly points out, '[f]or a new technology to be accepted and adopted within society, communities that are potentially impacted by it need to understand the benefits and risks to them, and how they can play a role in increasing the former while reducing the latter'.[16] The public debate will be informed by scientific journals, by informed documentary presentations in the media and by news releases and will be misinformed by scare stories, sensational reporting and false assertions.

In discussing options for the restrictive regulation of new technologies, it is important to think as logically, rationally and laterally as possible. It would, for example, be a mistake to conclude, if the emergence of ad hoc international law provision whether in the form of a treaty or of customary law rules seems impractical, that the only alternative available option will involve the use, say, of the criminal law. Alek Emery perceptively observes:

> 'By manipulating the economic considerations within the market through civil liability based regulation, there is the opportunity for increased responsibility within the marketplace without the additional costs of direct agency oversight. . . . Moreover, while such regulations are domestic in nature, they

[16] A. Maynard, 'Frankenstein Foods, Nanotech and the Trouble with Communicating Technology', World Economic Forum, 10 October 2016, available at www.weforum.org/agenda/2016/10/the-trouble-with-technology-innovation.

> can influence economic considerations in the global marketplace, and hopefully reach actors who would otherwise fall beyond the reach of traditional regulatory methods.[17]

So, imposition of tortious liability may, depending on the circumstances and on the nature of the concerns raised by the particular technology, be an option worth considering.

The core point is that the technologies we have discussed in this book are diverse, the legal challenges they seem destined to pose are many and varied and the ways in which those challenges can most advantageously be addressed will necessarily differ from technology to technology and from context to context. The task for policy makers and others involved in the vitally important task of addressing these challenges is to find the most effective and globally acceptable approach to addressing each challenge; approaches which balance carefully the numerous competing interests that we have discussed in the foregoing pages are the approaches that are most likely to match this requirement.

States, scientists, officials, experts, indeed all those with the required understanding and influence have the onerous responsibility of finding these balances. Achieving that purpose will involve hard work, multilateral engagement, good will, as much transparency as possible and, quite probably, the willingness to compromise. Simply implementing emerging technologies without giving proper consideration to both the risks and the opportunities involved is going to be unacceptable. Unfortunately, and as this book has shown, international law does not have a universally impressive track record when dealing with novel technologies. So it is now for the international community to 'step up to the plate' and to take the lead in maximising net global benefit from the products of scientific advance. It is to be hoped that this volume will provide those whose task this will become with some useful insights in that regard.

[17] A. C. Emery, 'Zero-Day Responsibility: The Benefits of a Safe Harbor for Cyber Security Research', 57 *Jurimetrics* 4, Summer 2017, 483, 503.

Select Bibliography

K. W. Abbott, G. E. Marchant and D. J. Sylvester, 'Transnational Regulation: Reality or Romanticism?', in G. Hodge, D. Bowman and A. Maynard (eds.), *International Handbook on Regulating Nanotechnologies* (Cheltenham: Edward Elgar, 2010).

K. W. Abbott, G. E. Marchant and D. J. Sylvester, 'A Framework Convention for Nanotechnology?' *Environmental Law Reporter*, 36 (2006), 10931–10942.

K. Abuya, 'Kenya's Computer and Cybercrimes Bill 2016 Gets Approval', techweez, 7 April 2017, available at www.techweez.com/2017/04/07/computer-cybercrimes-bill-2017-kenya-approved/.

Academy of Medical Sciences, the British Academy, the Royal Academy of Engineering and the Royal Society, 'Human enhancement and the future of work', Report from a joint workshop hosted by the, November 2012, available at www.acmedsci.ac.uk/viewFile/publicationDownloads/135228646747.pdf.

S. Ackerman, 'Army's Fingerprint and Iris Scan Databases Head for the Cloud', *Wired*, 30 May 2013, at www.wired.com/2013/05/army-biometrics-cloud/.

E. Ackerman, 'FAA Unveils Drone Rules: Autonomy is In, Drone Delivery is Out', *IEEE Spectrum*, 16 February 2015, available at http://spectrum.ieee.org/automaton/robotics/drones/faa-proposed-commercial-drone-rules.

Airbus Corporation, 'CityAirbus demonstrator passes major propulsion testing milestone', available at www.airbus.com/newsroom/press-releases/en/2017/10/cityairbus-demonstrator-passes-major-propulsion-testing-mileston.html, accessed on 9 February 2018.

G. Aldrich, 'Questions of International Law raised by the Seizure of the USS *Pueblo*' (1969) 63 *Proceedings of the American Society of International Law at Its Annual Meeting (1921–1969)* 2.

D. Alexander, 'Retired General cautions against overuse of "hated" drones', *Reuters*, 7 January 2013, available at www.reuters.com/article/us-usa-afghanistan-mcchrystal/retired-general-cautions-against-overuse-ofhated-drones-idUSBRE90608O20130107.

Allen Institute for Brain Science, available at www.alleninstitute.org/what-we-do/brain-science/.

B. R. Allenby, 'Are new Technologies Undermining the Laws of War?', *Bulletin of the Atomic Scientists*, 70(1) (2014), 21.

J. Altmann, *Military Nanotechnology: Potential Applications and Preventive Arms Control* (Routledge, 2006).

P. Alvelda, 'Neural Engineering System Design' (NESD), 2017, available at www.darpa.mil/program/neural-engineering-system-design.

W.-N. An, Z. Chen, et al., 'Remote Target Recognition and Tracking Based on Biometric Characteristics', 2017 Conference on Electrical Engineering and Automation Control, available at http://dpi-proceedings.com/index.php/dtetr/article/viewFile/10741/10291.

M. Andenas, 'Reassertion and Transformation: From Fragmentation to Convergence in International Law', 46 *Georgetown Journal of International Law* (2015) 685–734.

G. J. Annas, 'Protecting Soldiers from Friendly Fire: The Consent Requirement for Using Investigational Drugs and Vaccines in Combat', (1998) 24 *American Journal of Law & Medicine* 245, at 253.

C. Andrew and V. Mitrokhin, *The Mitrokhin Archive: The KGB in Europe and the West* (Alan Lane/The Penguin Press, 1999).

R. Armstrong, et al., *Bio-Inspired Innovation and National Security* (Washington: National Defence University Press, 2010),

I. Arreguin-Toft, 'How the Weak Win Wars. A Theory of Asymmetric Conflict', *International Security*, 8 December 2005, 26: 1, 93–128.

M. Astor, 'Microchip Implants for Employees? One Company Says Yes', *New York Times*, 25 July 2017.

M. Atwood, 'Margaret Atwood: The Road to Ustopia', *The Guardian*, 14 October 2011.

E. Auchard, 'Macron campaign was target of cyber attacks by spy-linked group', *Reuters*, 24 April 2017, available at www.reuters.com/article/us-france-election-macron-cyberidUSKBN17Q200.

Australia, 'Intelligence, Surveillance and Reconnaissance, Space, Electronic Warfare and Cyber Security', *Defence White Paper*, 2016, at www.defence.gov.au/White paper/docs/ISR-Cyber.pdf.

Australian Government, Australia's Cyber Security Strategy (2016), p. 28, available at https://cybersecuritystrategy.dpmc.gov.au/assets/img/PMC-Cyber-Strategy.pdf.

Australian Government Department of Defence, 'Department of Defence extends Optus Satellite contract', 1 June 2017, available at www.minister.defence.gov.au/minister/marise-payne/media-releases/department-defence-extnds-optus-satellite-contract.

A. Backstrom and I. Henderson, 'New Capabilities in Warfare: An Overview of Contemporary Technological Developments and the Associated Legal and Engineering Issues in Article 36 Weapons Reviews', *International Review of the Red Cross*, 94 (2012), 483.

D. Baird and A. Shew, 'Probing the History of Scanning Tunneling Microscopy', in D. Baird, A. Nordmann, and J. Schummer (eds.), *Discovering the Nanoscale* (IOS Press, 2004), 145–156.

B. Baker, 'Booming Biometrics: How Human Data Is Transforming the Battlefield', *Army Technology*, 18 March 2014.

J. M. Balbus, et al., 'Meeting Report: Hazard Assessment for Nanoparticles – Report from an Interdisciplinary Workshop', *Environmental Health Perspectives*, 115(11) (2007), 1654.

BBC News, 'Anthrax Jab Side-Effects Withheld', *BBC News* (21 February 2004), available at http://news.bbc.co.uk/2/hi/asia-pacific/3509037.stm.

BBC News, 'Music Torture: How Heavy Metal Broke Manuel Noriega', *BBC News*, 30 May 2017, available at www.bbc.co.uk/news/world-latin-america-40090809
J. Bean, 'Enhanced Blast Weapons and Forward Medical Treatment', *US Army Medical Department Journal*, (April–June 2004), 48–51.
M. Beard, J. Galliott and S. Lynch, 'Soldier Enhancement: Ethical Risks and Opportunities', *Australian Army Journal* XII:1 (Winter 2016), 13.
G. Bell, Talk entitled 'Managing the Machines: Building a New Applied Science for the 21st Century' (2018), available at https://cecs.anu.edu.au/3a-institute.
J. B. Bellinger III, legal adviser to the State Department and William J. Haynes II, Department of Defense General Counsel, 3 November 2006, available at www.defenselink.mil/news/newsarticle.aspx?id=3308.
S. M. Bellovin, M. Blaze, S. Clark and S. Landau, 'Lawful Hacking: Using Existing Vulnerabilities for Wiretapping on the Internet,' *Northwestern Journal of Technology and Intellectual Property* (2014); 12(1): 3–64.
B. Bender, 'Pentagon Considers Using Electricity to Stimulate Troops' Brains', *Boston Globe*, 18 February 2014.
L. Bennett Moses, 'Sui Generis Rules', in G. E. Marchant, B. R. Allenby and J. R. Heckert (eds.), *The Growing Gap between Emerging Technologies and Legal-Ethical Oversight: The Pacing Problem*, Vol. 7, International Library of Ethics, Law and Technology (New York: Springer, 2011).
L. Bennett Moses and J. Chan, 'Using Big Data for Legal and Law Enforcement Decisions: Testing the New Tools' (2014), 37(2), *University of New South Wales Law Journal* 643.
L. L. Bergeson and J. E. Plamondon, 'TSCA and Engineered Nanoscale Substances', *Nanotechnology Law & Business*, 4 (2007), 51.
D. M. Berube, *Nano-Hype: the Truth behind the Nanotechnology Buzz* (Amherst, NY: Prometheus Books, 2006).
C. Bing, 'NSA Technical Director: Sharing Hacker Information Isn't Enough, We Need a Shared Response', 30 March 2017, *Cyberscoop*, available at www.cyberscoop.com/nsa-technical-director-sharing-hacker-information-isnt-enough-need-shared-response/.
G. Binnig, et al., 'Surface Studies by Scanning Tunneling Microscopy', *Physical Review Letters*, 49(1) (1982), 57.
G. Binnig and H. Rohrer, 'In Touch with Atoms', *Reviews of Modern Physics*, 71(2) (1999), S324–330, S324.
'Scanning Tunneling Microscopy', *IBM Journal of Research and Development*, 44(1/2) (2000), 279.
J. Bird and G. Bird, 'Human Rights and the Military: The "Chemical Soldier" (2005) 30 *Alternative Law Journal* 81, 81–85.
M. Birnbaum, 'Russian Submarines are Prowling Around Vital Undersea Cables. It's Making NATO Nervous', *The Washington Post*, available at www.washingtonpost.com/world/europe/russian-submarines-are-prowling-around-vital-undersea-cables-its-making-nato-nervous/2017/12/22/d4c1f3da-e5d0-11e7-927a-e72eac1e73b6_story.html?utm_term=.c8392188f79e.
Blue Brain Project, *Digital Reconstruction*, available at https://bluebrain.epfl.ch/cms/lang/en/pid/59963.

R. G. Boire, 'Civil Liberties for the Mind: At the Crossroads of Civil Liberties and the Law', *Mind Matters (Centre for Cognitive Liberty & Ethics)* 1:2 (2004).

J. Boone, 'US Army Amasses Biometric Data in Afghanistan', *The Guardian*, 27 October 2010, at www.theguardian.com/world/2010/oct/27/us-army-biometric-data-afghanistan.

W. H. Boothby, *The Law of Targeting* (Oxford: Oxford University Press, 2012).

'How Will Weapons Reviews Address the Challenges Posed by New Technologies?', *Military Law and the Law of War Review*, 52(1) (2013), 58.

Weapons and the Law of Armed Conflict, 2nd edn (Oxford: Oxford University Press, 2016).

J. Borger, 'Chinese Warship Seizes US Underwater Drone in International Waters', *The Guardian*, 16 December 2016, available at www.theguardian.com/world/2016/dec/16/china-seizes-us-underwater-drone-south-china-sea.

P. J. A. Borm, et al., 'The Potential Risks of Nanomaterials: A Review Carried Out for ECETOC', *Particle and Fibre Toxicology*, 3(1) (2006).

N. Bostrom and A. Sandberg, 'Cognitive Enhancement: Methods, Ethics, Regulatory Challenges', *Science and Engineering Ethics* 15:3 (September 2009), 312.

N. E. Boudette and B. Vlasic, 'Tesla Self-Driving System Faulted by Safety Agency in Crash', *New York Times*, 12 September 2017, available at www.nytimes.com/2017/09/12/business/self-driving-cars.html.

V. Boulanin, *Implementing Article 36 Weapon Reviews in the Light of Increasing Autonomy in Weapon Systems* (November 2015) SIPRI Insights on Peace and Security, available at www.sipri.org/sites/default/files/files/insight/SIPRIInsight1501.pdf.

D. R. Boverhof, et al., 'Comparative Assessment of Nanomaterial Definitions and Safety Evaluation Considerations', *Regulatory Toxicology and Pharmacology*, 73(1) (2015), 137–150.

D. M. Bowman, 'More than a Decade On: Mapping today's Regulatory and Policy Landscape following the publication of Nanoscience and Nanotechnologies: Opportunities and Uncertainties', *NanoEthics*, 11(2) (2017), 169–186.

D. M. Bowman and J. Gatof, 'Reviewing the Regulatory Barriers for Nanomedicine: Global Questions and Challenges', *Nanomedicine*, 10(21) (2015), 3275.

D. M. Bowman and G. Hodge, 'A Small Matter of Regulation: An International Review of Nanotechnology Regulation', *Columbia Science and Technology Law Review*, 8 (2007), 5.

S. Bronitt and J. Stellios, 'Regulating Telecommunications Interception and Access in the Twenty-first Century: Technological Evolution or Legal Revolution?' (2006) 24 *Prometheus: Critical Studies in Innovation* 413.

S. Bronitt and B. McSherry, *Principles of Criminal Law*, 3rd edn (Sydney: Lawbook Co), 2010, 6.

R. Brownsword and M. Goodwin, *Law and the Technologies of the Twenty-First Century: Text and Materials* (Cambridge: Cambridge University Press, 2012).

H. Brueck, *This Tiny Robot Team Could Help Stop the No. 1 Killer in America* (12 January 2016) Fortune, available at http://fortune.com/2016/01/12/artery-clearing-nanobot/.

M. Brundage, et al., *The Malicious Use of Artificial Intelligence: Forecasting, Prevention, and Mitigation* (February 2018), available at https://img1.wsimg.com/blobby/go/3d82daa4–97fe-4096–9c6b-376b92c619de/downloads/1c6q2kc4v_50335.pdf.

W. Buhrow, *Biometrics in Support of Military Operations: Lessons from the Battlefield* (Boca Raton: Taylor and Francis, 2017), 64–65.

M. Burgess, 'Wikileaks Drops "Dark Matter": Part Two of Its Vault 7 CIA leaks', *Wired*, 24 March 2017.

O. Burkeman and R. Norton-Taylor, 'US Pilots Blame Drug for Friendly Fire Deaths', *The Guardian* 4 January 2003.

L. Bushak, 'Robot Surgeon Succeeds Without Help From Human Doctors', *Medical Daily*, 13 May 2016, available at www.medicaldaily.com/robot-surgeon-smart-tissue-autonomous-386079.

C. Buzea, Ivan P. and K. Robbie, 'Nanomaterials and Nanoparticles: Sources and Toxicity', *Biointerphases*, 2(4) (2007), 17–71.

CAA, 'How is UK Airspace Structured?', available at www.caa.co.uk/Consumers/Guide-to-aviation/Airspace/How-is-UK-airspace-structured-/.

J. Caldwell, et al., 'Modafinil's Effects on Simulator Performance and Mood in Pilots during 37 Hours without Sleep', *Aviation, Space and Environmental Medicine* 75:9 (September 2004), 777–784.

L. Caiyu, 'China's First Quantum Information Network Will Prevent Hacking: Expert', *Global Times*, 9 July 2017.

Canada Communications Security Establishment, Cyber Threats to Canada's Democratic Process, 7 June 2017, available at www.cse-cst.gc.ca/sites/default/files/cse-cyber-threat-assessment-e.pdf.

L. Carter, D. Burnett, S. Drew, G. Marle, L. Hagadorn, D. Bartlett-McNeill and N. Irvine, *Submarine Cables and the Oceans: Connecting the World*, at 8 (UNEP-WCMC Biodiversity Series No. 31, ICPC / UNEP-WCMC (2009), available at www.iscpc.org/documents/?id=132.

P. G. Carozza, 'Human Dignity', in D. Shelton (ed.) *The Oxford Handbook of International Human Rights Law* (Oxford: Oxford University Press, 2013) 345–359, 354.

A. B. Carter, M. Lettre and S. Smith, 'Keeping the Technological Edge', 129. See also H. Kleinberg, 'On War in Space', (2007) 5 *Astropolitics*, 19.

S. Casey-Maslen, N. Corney and A. Dymond-Bass, 'The Review of Weapons under International Humanitarian Law and Human Rights Law', in S. Casey-Maslen (ed.), *Weapons under International Human Rights Law* (Cambridge: Cambridge University Press, 2014).

S. Casey-Maslen and S. Connolly, *Police Use of Force under International Law* (New York: Cambridge University Press, 2017).

J.-L. Chameau, et al., *Emerging and Readily Available Technologies and National Security: A Framework for Addressing Ethical, Legal, and Societal Issues* (Washington: National Academies Press , 2014) 72.

C. Chambers, 'Neuro-Enhancement in the Military: Far-Fetched or an Inevitable Future', *The Guardian*, 7 October 2016.

E. Champion, 'Thirty-five Arguments against Google Glass', *Reluctant Habits* (14 March 2013), available at www.edrants.com/thirty-five-arguments-against-google-glass/ accessed 17 February 2018.

E. Chan, 'The Food and Drug Administration and the Future of the Brain-Computer Interface: Adapting FDA Device Law to the Challenges of Human-Machine Enhancement,' *John Marshall Journal of Computer and Information Law* (2007); 25: 117–164.

A. Chatterjee, 'The Ethics of Neuroenhancement', *Handbook of Clinical Neurology*, 118 (2013), 323.

Q. Chaudhry, et al., *Final Report: A Scoping Study to Identify Gaps in Environmental Regulation for the Products and Applications of Nanotechnologies* (2006) Safe Nano www.safenano.org/media/109893/DEFRA2005-Review_of_gaps_in_environmental_regulation_for_the%20products_and_applications_of_nanotechnologies.pdf.

L. Chengzhi, 'The Chinese GNSS System development and policy analysis', (2013) 29 *Space Policy*, 9.

Chemical Watch, 'EU Consults on Nanomaterial Definition: Revision of Recommendation Three Years Late', Chemical Watch, 19 September 2017, available at https://chemicalwatch.com/59016/eu-consults-onnanomaterial-definition.

International Committee of the Red Cross, 'Rule 154. Obedience to Superior Orders', Customary International Humanitarian Law Database, available at https://ihl-databases.icrc.org/customary-ihl/eng/docs/v1_rul_rule154.

E. S. Cho, et al., 'Graphene Oxide/Metal Nanocrystal Multilaminates as the Atomic Limit for Safe and Selective Hydrogen Storage', *Nature Communications*, (2016) www.nature.com/articles/ncomms10804.pdf.

Jens Clausen, et al., 'Help, Hope, and Hype: Ethical Dimensions of Neuroprosthetics' *Science*, 30 Jun 2017:. 356, 6345, 1338–1339, DOI: 10.1126/science.aam7731.

CNN, 'Bush presents Congressional Medal of Honor', 16 July 2001, available at http://edition.cnn.com/TRANSCRIPTS/0107/16/se.02.html.

D. R. Coats, 'Worldwide Threat Assessment of the US Intelligence Community', 11 May 2017, available at www.dni.gov/files/documents/Newsroom/Testimonies/SSCI%20Unclassified%20SFR%20-%20Final.pdf.

C. Coglianese and D. Lehr, 'Regulating by Robot: Administrative Decision Making in the Machine-Learning Age,' *Georgetown Law Journal* (2017); 105: 1147–1223.

R. Cohen Kadosh, 'Using Transcranial Electrical Stimulation to Enhance Cognitive Functions in the Typical and Atypical Brain", 4 (March 2013), *Translational Neuroscience*, 20–33.

R. H. Cole, *Operation Just Cause: The Planning and Execution of Joint Operations in Panama, February 1988-January 1990* (Joint History Office, Office of the Chairman of the Joint Chiefs of Staff, Washington 1995) 59–60.

P. Collins, 'The Value of Respect: What Does It Mean for an Army?', *Journal of Military Ethics* 16:1–2, 2.

Committee on Opportunities in Neuroscience for Future Army Applications, *Opportunities in Neuroscience for Future Army Applications* (2009), 117, available at www.ncbi.nlm.nih.gov/books/NBK207983/.

Committee on Toxicity of Chemicals in Food, Consumer Products and the Environment, *The Toxicology of Nanoparticles Used in Healthcare Products* (2006) HM Government, available at https://cot.food.gov.uk/sites/default/files/cot/cotstatementnanomats200701.pdf.

A. Conte, 'Drones with facial recognition technology will end anonymity, everywhere', *Business Insider*, 27 May 2013, available at www.businessinsider.com/facial-recognition-technology-and-drones-2013-5?IR=T.

J. M. Cosumano Jr, 'A Day without Space: Ensuring It Doesn't Happen', Summer 2002, *Army Space Journal*, 2–3.

Z. Corbyn 'Crispr: Is It A Good Idea to "Upgrade" Our DNA?' *The Observer (London)* 10 May 2015.

G. Corera, 'NHS Cyber attack was "launched by North Korea"', *BBC News*, 16 June 2017, available at www.bbc.co.uk/news/technology-40297493.

G. Corfield, 'BAe Systems' Autonomous Research Aircraft Flies Itself To Scotland', *The Register*, 5 December 2016, available at www.theregister.co.uk/2016/12/05/bae_systems_autonomous_flying/.

Lt Col R. Cornum, Dr J. Caldwell and Lt Col K. Cornum, 'Stimulant Use in Extended Flight Operations', Airpower Journal (Spring 1997), 57, available online: https://pdfs.semanticscholar.org/8128/64ff145b4a3916b824311ed1dbb554384178.pdf.

M. Crowley, 'The Use of Incapacitants in Law Enforcement', in Casey-Maslen (ed.), *Weapons under International Human Rights Law*, 357– 381, 358.

Daily Telegraph, 'Google Removes 640 Videos From Youtube Promoting Terrorism', 18 June 2012, available at www.telegraph.co.uk/technology/google/9337993/Google-removes-640-videos-from-YouTube-promoting-terrorism.html.

M. Dando, 'Find the Time to Discuss New Bioweapons', *Nature*, 535 (2016), 9.

I. Daoust, R. Coupland and R. Ishoey, 'New Wars, New Weapons? The Obligation of States to Assess the Legality of Means and Methods of Warfare' (2002) 846 *IRRC* 345, 348.

DARPA Launches Biological Technologies Office, 1 April 2014, available at www.darpa.mil/news-events/2014-04-01.

L. Dearden, 'ISIS Videos Targeted by Artificial Intelligence That Can Detect Propaganda Before It's Uploaded', The Independent, 13 February 2018, available at www.independent.co.uk/news/uk/home-news/isis-videos-artificial-intelligence-propaganda-ai-homeoffice-islamic-state-radicalisation-asi-data-a8207246.html.

A. Deeks, 'Confronting and Adapting: Intelligence Agencies and International Law,' *Virginia Law Review* (2016); 102(3): 599–685.

Defense Advanced Research Projects Agency (DARPA), Systems of Neuromorphic Adapted Plastic Scalable Electronics (SyNAPSE), available at www.darpa.mil/program/systems-of-neuromorphic-adaptive-plastic-scalable-electronics.

Defense Advanced Research Projects Agency (DARPA), 'Boosting Synaptic Plasticity to Accelerate Learning', 16 March 2016, available at www.darpa.mil/newsevents/2016-03-16.

Defense Advanced Research Projects Agency, 'Atoms to Product: Aiming to make Nanoscale Benefits Life-Sized' (Media Release, 22 August 2014), available at www.darpa.mil/news-events/2014-08/22.

Defence Science Institure, Symposium, 'Future Technologies for Personal Protection', 7–8 November 2013, available at www.defencescienceinstitute.com/2012/09/10/dsi-dstosymposium-future-technologies-personal-protection/.

N. Denette, 'Supreme Court Rejects Federal Government's Argument That Omar Khadr Is an Adult Offender', *National Post*, 14 May 2015, at http://nationalpost.com/news/canada/supreme-court-rejects-tory-argument-that-omar-khadr-was-adult-offender.

Department of Defense USA, 'Department of Defense Directive' (3000.03E), 25 April 2013, revised 27 September 2017, available at www.esd.whs.mil/Portals/54/Documents/DD/issuances/dodd/300003p.pdf?ver=2017-09-27-125836-647.

M. de Zwart, 'GOOGLE in Space? How Will Space Governance Accommodate Non-State Actors?', (May 27, 2016), RUMLAE Research Paper No. 16–21, available at SSRN: https://ssrn.com/abstract=2809409.

M. de Zwart and D. Stephens, 'Non-Military Space Testing in the Woomera Prohibited Area: Opportunities for the Australian Space Industry?', (2017) 45 *Federal Law Review*, 39.

J. J. DiCarlo, 'To Advance Artificial Intelligence, Reverse-Engineer the Brain,' *Wired*, 3 February 2018.

Diehl BGT Mutual Active Protection System, described at www.defense-update.com/20110112_maps.html.

C. Dillow, 'A Synthetic Brain Synapse Is Constructed from Carbon Nanotubes,' *Popular Science*, 22 April 2011.

H. Dinniss and J. K. Kleffner, 'Soldier 2.0: Military Human Enhancement and International Law', *International Law Studies* 92:432 (2016).

Y. Dinstein, *The Conduct of Hostilities under the Law of International Armed Conflict*, 3rd edn, (2016), 237–238.

District Court for the District of Columbia, 16 February 2018; Case 1:18-cr-00032-DLF, available at www.justice.gov/file/1035477/download.

A. Dolnik and R. Pilch, 'The Moscow Theater Hostage Crisis: The Perpetrators, their Tactics and the Russian Response' (2003) 8 *International Negotiation* 577, 600–601.

A. R. Dombe, 'Biometric Target Recognition', 19 March 2017, available at www.israeldefense.co.il/node/28881.

K. Donaldson, 'The UK Would Respond to a Russian Cyber Attack with the Weapon of Its Own Choice', *Bloomberg*, 18 December 2017.

A. W. Dorn, *Keeping Watch: Monitoring, Technology and Innovation in UN Peace Operations* (Tokyo: United Nations University Press, 2011), 2.

L. Doswald-Beck (ed.) *San Remo Manual on International Law Applicable to Armed Conflicts at Sea* (Cambridge/San Remo: Cambridge University Press/International Institute of Humanitarian Law, 1995).

'Humanitarian Law in Future Wars', in M. Schmitt and L. Green (eds.), *The Law of Armed Conflict: Into the Next Millennium* (Newport, RI: Naval War College, 2006) 139.

G. Dugué, W. Akemann and T. Knöpfel, 'A Comprehensive Concept of Optogenetics', in T. Knöpfel and E. Boyden (eds.), *Optogenetics: Tools for Controlling and Monitoring Neuronal Activity* (New York: Elsevier, 2012).

A. Dymond-Bass and N. Corney, 'The Use of "Less-Lethal" Weapons in Law Enforcement', in S. Casey-Maslen (ed.), *Weapons under International Human Rights Law* (New York: Cambridge University Press, 2014).

M. Eckstein, 'Navy Releases Details of New FFG(X) Guided Missile Frigate Program in Request to Industry', 10 July 2017, available at https://news.usni.org/2017/07/10/navy-releases-details-of-new-ffgx-guided-missile-frigate-program-in-request-to-industry.

D. Eigler, 'From the Bottom up: Building Things with Atoms', in G. Timp (ed.), *Nanotechnology* (Springer, 1999), 425–435.

A. Eisenberg, 'When Robotic Surgery Leaves Just a Scratch', *New York Times*, 17 November 2012.

A. C. Emery, 'Zero-Day Responsibility: The Benefits of a Safe Harbor for Cyber Security Research', 57 *Jurimetrics* 4, *Summer* 2017, 483, 488–489.

C. Enemark, '"Non-Lethal" Weapons and the Occupation of Iraq: Technology, Ethics and Law' (2008) 21(2) *Cambridge Review of International Affairs* 199, 201.

Environmental Protection Agency, *Nanotechnology white paper* (2007) US Government, available at www.epa.gov/sites/production/files/2015-01/documents/nanotechnology_whitepaper.pdf.

A. Estes, 'The Pentagon is Putting Its Big Brother Data Collection in the Cloud', *Gizmodo*, 01 June 2013, at www.gizmodo.com.au/2013/06/the-pentagon-is-putting-its-big-brother-data-collection-in-the-cloud/.

A. Estrada, 'Modafinil as a Replacement for Dextroamphetamine for Sustaining Alertness in Military Helicopter Pilots', *Aviation, Space and Environmental Medicine*, Vol. 83, No 6 (June 2012): 557.

European Commission, *Regulatory Aspects of Nanotechnologies: SEC(2008) 2036* (17 June 2008), available at https://ec.europa.eu/research/industrial_technologies/pdf/policy/comm_2008_0366_en.pdf.

Second Regulatory Review on Nanomaterials (COM(2012) 572 final) (2012) , available at http://eur-lex.europa.eu/LexUriServ/LexUriServ.do?uri%3DCOM:2012:0572:FIN:en:PDF.

Draft Recommendation for tackling illegal material online, dated February 2018, available at https://edri.org/files/illegal_content_ec_draft_recommendation_201802.pdf.

European Court on Human Rights, *Guide on Article 8 of The European Convention on Human Rights: Right to Respect for Private and Family Life* (30 April 2017), available at www.echr.coe.int/Documents/Guide_Art_8_ENG.pdf.

European Data Protection Supervisor, Opinion 9/2016, 'EDPS Opinion on Personal Information Management Systems: Towards more user empowerment in managing and processing personal data', 20 October 2016, available at https://secure.edps.europa.eu/EDPSWEB/webdav/site/mySite/shared/Documents/Consultation/Opinions/2016/16–10–20_PIMS_opinion_EN.pdf.

FAA Advisory Circular 21–2 dated 21 June 2016, available at www.faa.gov/uas/media/AC_107-2_AFS-1_Signed.pdf.

S. Fan, 'New Bionic Arm Blurs Line between Self and Machine for Wearers' *Singularity Hub*, 4 April 2018, available at https://singularityhub.com/2018/04/04/new-bionic-arm-blurs-line-between-self-and-machine-for-wearers/#sm.0000aeon81dp1ecg115s6ehz1nqub.

N. A. Farahany, 'A Neurological Foundation for Freedom,' *Stanford Technology Law Review* (7 December 2011): 11–53.

B. J. Feder and T. Zeller Jr, 'Identity Badge Worn under Skin Approved for Use in Health Care', 14 October 2014, www.nytimes.com/2004/10/14/technology/identity-badge-worn-under-skin-approved-for-use-in-health-care.html?_r=0.

G. Fein, 'Surface Navy 2018: Saab's NuMNS Extends MCM Operations', 11 January 2018, available at www.janes.com/article/77005/surface-navy-2018-saab-s-mumns-extends-mcm-operations.

W. J. Fenrick, 'The Conventional Weapons Convention: A Modest but Useful Treaty', 279 *IRRC* 498 (1990) 500.

W. Fenrick, 'The Law Applicable to Targeting and Proportionality after Operation Allied Force: A View from the Outside' (2000) *YIHL* 51.

D. P. Fidler, 'The International Legal Implications of "Non-Lethal" Weapons' (1999) 21 *Michigan Journal of International Law* 51, 63.

'The Meaning of Moscow: "Non-Lethal"'Weapons and International Law in the 21st Century', (2005) 859 *IRRC* 525 at pages 540–544.

'Requiem for the Internet Freedom Strategy,' *Net Politics*, 17 October 2017.

J. Fioretti, 'French Privacy Watchdog Raps WhatsApp over Facebook Data Sharing', *Reuters*, 18 December 2017, available at www.reuters.com/article/us-whatsapp-privacy-france/french-privacy-watchdog-raps-whatsapp-over-facebookdata-sharing-idUSKBN1EC285.

Food and Drug Administration, *Is the Product a Medical Device?*, available at www.fda.gov/MedicalDevices/DeviceRegulationandGuidance/Overview/ClassifyYourDevice/ucm051512.htm.

Nanotechnology: a Report of the US Food And Drug Administration Nanotechnology Task Force (2007), US Government, available at www.fda.gov/ScienceResearch/SpecialTopics/Nanotechnology/ucm2006659.htm.

Food Standards Agency, *Report of the FSA regulatory review* (2008) HM Government, available at www.food.gov.uk/sites/default/files/multimedia/pdfs/nanoregreviewreport.pdf.

S. Foss Hansen, et al., 'Late Lessons from Early Warnings for Nanotechnology', *Nature Nanotechnology*, 3(8) (2008), 444–447.

J. Foust, 'Debating a Code of Conduct for Space' *The Space Review* (online) 7 March 2011, www.thespacereview.com/article/1794/1.

'United Arab Emirates to Establish Human Spaceflight Program', *Space News* (online) http://spacenews.com/united-arab-emirates-to-establish-human-spaceflight-program, 2 October 2017.

R. A. Friedman, 'Why Are We Drugging Our Soldiers?', *New York Times*, 21 April 2012.

L. Fuller, *The Morality of Law* (New Haven, CT: Yale University Press, 1969).

G7 Group of States, Declaration on Responsible States [*sic*] Behaviour in Cyberspace, full text available at www.esteri.it/mae/resource/doc/2017/04/declaration_on_cyberspace.pdf.

J. Gallego, 'New Dust-Sized Implants May be the Key to Monitoring Your Internal Health', *Futurism*, 5 August 2016.

P. Ganapati, 'Surgical Robots Operate with Precision', *Wired*, 9 November 2009, available at www.wired.com/2009/09/surgical-robots/.

C. M. Garlic, 'Guidance by Clicks: The Unauthorized Practice of Law in a Technology Driven World,' *Trinity Law Review* (2014); 19: 267–307.

C. Gavaghan and J Moore, *A Review of the Adequacy of New Zealand's Regulatory Systems to Manage the Possible Impacts of Manufactured Nanomaterials* (2011), available at www.academyconferences.com/pdf/nznanoreview.pdf.

J. Giles, 'Beta-blockers Tackle Memories of Horror', *Nature* (28 July 2005), 436, 448–449.

Global Commission, meeting at Las Vegas on 27 July 2017 as described at https://cyberstability.org/news/645–2/.

J. Goldblatt, *Agreements for Arms Control: A Critical Survey* (Stockholm International Peace Research Institute, 1982), 89.

R. Gonzalez, 'Neuroscientists Just Launched an Atlas of the Developing Human Brian,' *Wired*, 7 December 2017, available at www.wired.com/story/neuroscientists-just-launched-an-atlas-of-the-developing-human-brain/?mbid=BottomRelatedStories.

H. T. Greely, 'Regulating Human Biological Enhancements: Questionable Justifications and International Complications', *Santa Clara Journal of International Law* 4:2 (2006), 88.

M. P. Green and D. R. Burnett, 'Security of International Submarine Cable Infrastructure – Time to Rethink?', in M. H. Nordquist, et al. eds., *Legal Challenges in Maritime Security*, (Dordrecht/Boston: Martinus Nijhoff, 2008).
A. Greenfield, *Radical Technologies* (New York: Verso, 2017).
C. Greenwood, 'The Advisory Opinion on Nuclear Weapons and the Contribution of the International Court to International Humanitarian Law' (1997), 316 *IRRC* 65, 73.
A. Grigsby, 'Year in Review: Militaries Got More Cyber in 2016,' *Net Politics*, 26 December 2016.
M. L. Gross, 'Bioethics and Armed Conflict: Mapping the Moral Dimensions of Medicine and War', *Hastings Centre Report* 34, no 6 (2004), 23.
L. Grush, 'SpaceX Makes Aerospace History with Successful Launch and Landing of a Used Rocket', 30 March 2017, *theverge.com* (online), www.theverge.com/2017/3/30/15117096/spacex-launch-reusable-rocket-success-falcon-9-landing.
The Guardian, 'Mystery of Sonic Weapon Attacks at US Embassy in Cuba Deepens', *The Guardian*, 15 September 2017, available at www.theguardian.com/world/2017/sep/14/mystery-of-sonic-weaponattacks-at-us-embassy-in-cuba-deepens.
The Guardian, 'Russia Names Gas Used to End Siege', *The Guardian*, 30 October 2002, available at www.theguardian.com/world/2002/oct/30/russia.
L. F. Haber, *The Poison Cloud: Chemical Warfare in the First World War* (Oxford: Oxford University Press, 1986).
D. Hambling, 'Ships Fooled in GPS Spoofing Attack Suggest Russian Cyberweapon', 10 August 2017, available at www.newscientist.com/article/2143499-ships-fooled-in-gps-spoofing-attack-suggest-russian-cyberweapon/.
The Hamlyn Symposium 2017, Surgical Robot Challenge 2017, available at http://hamlyn.doc.ic.ac.uk/hsmr/events/surgicalrobot-challenge-2017
J. Hansen, 'Female Sex Hormone May Save Injured Soldiers on the Battlefield', University of Alabama News Portal, 21 October 2015: www.uab.edu/medicine/news/latest/item/874-female-sex-hormone-may-save-injured-soldiers-on-the-battlefield.
S. Hansen, et al., 'A Critical Analysis of the Environmental Dossiers from the OECD Sponsorship Programme for the Testing of Manufactured Nanomaterials', *Environmental Science: Nano*, 4(2) (2017), 282–291.
H. Hanson, 'China's Censorship Fingers Reach into Our Cyberspace', *The Weekend Australian*, 24 June 2017, available at www.theaustralian.com.au/news/inquirer/chinas-censorship-fingers-reach-into-our-cyberspace/news-story/846944042e1e4227782611979a211cae.
J. Haque and O. Bashir, 'Banned Outfits in Pakistan Operate Openly on Facebook', 13 June 2017, *Dawn*.
L. Hardesty, 'Explained: Neural Networks,' *MIT News*, 14 April 2017, available at http://news.mit.edu/2017/explained-neural-networks-deep-learning-0414.
D. J. Harris, M. O'Boyle, E. P. Bates and C. M. Buckley, *Law of the European Convention on Human Rights* 3rd edn (Oxford: Oxford University Press, 2014).
G. Harris, J. Hirshfeld Davis and E. London, 'U.S. Expels 15 Cuban Diplomats, in Latest Sign Détente May Be Ending', *New York Times*, 3 October 2017.
R. G. Harrison 'Unpacking the Three C's: Congested, Competitive, and Contested Space' (2013) 11.3 *Astropolitics*. 123.
T. Harrison, 'Chaos and Uncertainty: The FYI 2014 Defense Budget and Beyond', *Centre for Strategic and Budgetary Assessments*, October 2013, 11–12.

H. A. Harrison Dinniss and J. K. Kleffner 'Soldier 2.0: Military Human Enhancement Technologies' (2016) 92 *International Law Studies* 432, available at http://digital-commons.usnwc.edu/ils/vol92/iss1/14/.

H. L. A. Hart, *The Concept of Law* (Oxford: Oxford University Press, 1961).

C. Haynes, 'Report to the UN General Assembly of the Special Rapporteur on Extrajudicial, Summary or Arbitrary Executions,' A/HRC/23/47 dated 9 April 2013, available at www.ohchr.org/Documents/HRBodies/HRCouncil/RegularSession/Session23/A-HRC-23-47_en.pdf.

J. Hawkins, 'What Intelligent Machines Need to Learn from the Neocortex', IEEE Spectrum, 2 June 2017, available at https://spectrum.ieee.org/computing/software/what-intelligentmachines-need-to-learn-from-the-neocortex.

W. Hays Parks, 'Air War and the Law of War' (1990) 32 (1) *The Air Force Law Review* 1, 113–116.

'Conventional Weapons and Weapons Reviews', 8 *Yearbook of International Humanitarian Law* (2005) 55.

'Means and Methods of Warfare, Symposium Issued in Honour of Edward R. Cummings', 38 *George Washington International Law Review* (2006) 511, 536 at note 25.

C. 'Sid' Heal, 'The Quest for the "Magic Bullet" 'in M. Dando (ed.) *Non-lethal Weapons: Technological and Operations Prospects* (London: Jane's Information Group, 2000) 32–42, 35–36.

Health and Safety Executive, *Review of the Adequacy of Current Regulatory Regimes to Secure Effective Regulation of Nanoparticles Created by Nanotechnology: the Regulations Covered by HSE* (2006) HM Government K. www.hse.gov.uk/nanotechnology/regulatoryreview.pdf.

K. Hearn, 'Future Soldiers Could Get Enhanced Minds' (19 March 2001) *United Press International,* available at www.geocities.ws/marksrealm/project457.html.

C. R. Hema, A. Elakkiya and M. P. Paulraj, 'Biometric Identification Using Electro-encephalography', 106 *International Journal of Computer Applications* 15 (November 2014), 17.

J.-M. Henckaerts and L. Doswald-Beck, 'Customary International Humanitarian Law, Vol 1: Rules' *ICRC* (2005) (Customary Law study), 237, Rule 70.

Hepburn, Col. M. 'Wound Stasis System', available at www.darpa.mil/program/wound-stasis-system.

G. Heppner and R. Gillmann, 'Autonomy of Mobile Robots'. In W. Heintschel von Heinegg, et al., eds., *Dehumanization of Warfare*, 77, 80–81.

S. Hobe, 'Legal Aspects of Space Tourism', (2007–2008) 86 *Nebraska Law Review*, 439.

P. H. Hoet, I. Brüske-Hohlfeld and O. V. Salata, 'Nanoparticles–Known and Unknown Health Risks', *Journal of Nanobiotechnology*, 2(1) (2004).

G. Hogben (ed.), 'ENISA Briefing: Behavioural Biometrics', 18 February 2010, available at www.enisa.europa.eu/publications/behavioural-biometrics.

C. Hope, 'Google: We'll make new tools to tackle terrorists online', *The Daily Telegraph*, 30 March 2017, available at www.telegraph.co.uk/news/2017/03/30/google-facebook-agree-setnew-industry-board-accelerate-removal/.

House of Commons Foreign Affairs Committee, 'Global Security: Non-Proliferation', Fourth Report of Session 2008–09 (3 June 2009), available at https://publications.parliament.uk/pa/cm200809/cmselect/cmfaff/222/222.pdf.

Z. Huang and K. Mačák, 'Towards the International Rule of Law in Cyberspace: Contrasting Chinese and Western Approaches', 16 *Chinese Journal of International Law* (2017) 271.

M. Hull, 'Nanotechnology Risk Management and Small Business: a Case Study of the Nanosafe Framework', in M. Hull and D. M. Bowman (eds.), *Nanotechnology Environmental Health and Safety: Risks, Regulation and Management* (New York: Elsevier, 2010), 247–295.

Human Rights Watch, 'Losing Humanity: The Case Against Killer Robots' (2012), available at www.hrw.org/sites/default/files/reports/arms1112ForUpload_0_0.pdf.

Hunton and Williams, 'New Data Processing Requirements Take Effect in Russia', *Hunton Privacy Blog*, 29 August 2017, available at www.huntonprivacyblog.com/2017/08/29/new-data-processing-notice-requirements-take-effect-russia/.

ICRC, 'Toxic Chemicals as Weapons for Law Enforcement: A Threat to Life and International Law?', September 2012, available at www.icrc.org/en/document/toxic-chemicals-weapons-lawenforcement-threat-life-and-international-law.

IEEE, 'Special Report: Can We Copy the Brain?', *IEEE Spectrum* (2017), available at https://spectrum.ieee.org/static/special-report-can-we-copy-the-brain.

M. Ienca and R. Andorno, 'Towards New Human Rights in the Age of Neuroscience and Neurotechnology' *Life Sciences, Society and Policy* (2017) 13:5.

India Today, 'SC asks govt, others to mull ways to check sex offence videos', *India Today*, 20 March 2017, available at http://indiatoday.intoday.in/story/sc-asks-govt-others-to-mullways-to-check-sex-offence-videos/1/908579.html.

Interfax-Ukraine, Ukraine with NATO's support prepares creation of cyber security centre, KyivPost, 5 April 2017, available at www.kyivpost.com/ukraine-politics/ukraine-natos-support-prepares-creation-cyber-security-centre.html.

International Committee of the Red Cross, 'Biological Weapons Review Conference: ICRC Statement', ICRC, 8 November 2016, available at www.icrc.org/en/document/biological-weapons-reviewconference-statement-icrc.

International Committee of the Red Cross, 'Rule 154. Obedience to Superior Orders', Customary International Humanitarian Law Database, available at https://ihl-databases.icrc.org/customary-ihl/eng/docs/v1_rul_rule154.

Investigatory Powers Act 2016, available at www.legislation.gov.uk/ukpga/2016/25/contents/enacted.

M. Jackson, 'UK Government Confirm Move to Force ISPs into Blocking "Adult" Sites', *ISP Review*, 21 November 2016, available at www.ispreview.co.uk/index.php/2016/11/uk-government-confirm-move-force-isps-blocking-adult-sites.html.

A. Jacobsen, 'Engineering Humans for War', *The Atlantic*, 23 September 2015.

A. K. Jain and A. Kumar, 'Biometric Recognition: An Overview', in E. Mordini and D. Tzovaras (eds.), *Second Generation Biometrics: The Ethical, Legal and Social Context*, The International Library of Ethics, Law and Technology, vol. 11. (Dordrecht: Springer, 2012), 49–78.

Janus Program, discussed at www.iarpa.gov/index.php/research-programs/janus.

J. Jie, 'Personal Information Leakage on the Rise in China', *People's Daily Online*, 31 March 2017, available at http://en.people.cn/n3/2017/0331/c90000–9197748.html.

J. Joannopoulos and F. Davis, *Institute for Soldier Nanotechnologies* (18 June 2015), Massachusetts Institute of Technology, available at http://web.mit.edu/annualreports/pres15/2015.18.06.pdf.

M. John-Hopkins , 'Regulating the Conduct of Urban Warfare: Lessons from Contemporary Asymmetric Armed Conflicts', *International Review of the Red Cross* 92:878 (June 2010), 471.

D. Johnson, *Nanoparticles Found in Brains Come from External Sources* (8 September 2016) IEEE Spectrum http://spectrum.ieee.org/nanoclast/semiconductors/materials/nanoparticles-found-in-brains-comes-from-external-sources.

H. Johnston, et al., 'Engineered Nanomaterial Risk. Lessons Learnt from Completed Nanotoxicology Studies: Potential Solutions to Current and Future Challenges', *Critical Reviews in Toxicology*, 43(1) (2013), 1–20.

I. Johnston, 'Tiny Implant Could Connect Humans and Machines Like Never Before', 3 August 2016, www.independent.co.uk/news/science/neural-dust-implant-sensor-brain-nerves-humans-machines-prosthetics-berkeley-a7170251.html.

E. Juengst, 'Can Enhancement Be Distinguished from Prevention in Genetic Medicine?', *The Journal of Medicine and Philosophy* 22 (1997), 125–142.

'The Meaning of Enhancement', in E. Parens (ed.), *Enhancing Human Traits: Ethical and Social Implications* (Washington, DC: Georgetown University Press, 1998), 29–47.

T. Junod, 'Elon Musk: Triumph of His Will' 15 November 2012, *Esquire* (online), available at www.esquire.com/news-politics/a16681/elon-musk-interview-1212/.

F. Kalshoven, 'Belligerent Reprisals Revisited' (1990) 21 *Netherlands Yearbook of International Law* 43.

Ł. Kamieński, *Shooting Up: A History of Drugs and War* (New York: Oxford University Press, 2016).

T. Kanunnikova, 'Russia Offers to Draft International Convention on Combating Cybercrime', 23 March 2017, available at http://russianconstruction.com/news-1/26888-russia-offers-to-draft-international-convention-on-combating-cybercrime.html.

A. O. Kaspersky 'Lab, Lazarus under the Hood', 3 April 2017, available at https://securelist.com/blog/sas/77908/lazarus-under-the-hood/.

R. D. Katz, 'Friendly Fire: The Mandatory Military Anthrax Vaccination Program, *Duke Law Journal* 50 (2001), 1835–1865.

J. Kelly, *Is Medical Ethics in Armed Conflict Identical to Medical Ethics in Times of Peace?* (Newcastle: Cambridge Scholars Publishing, 2013).

J. Kahn, 'Nano's Big Future' (2006) 209(6) *National Geographic* 98.

'Gas Warfare in International Law', *Military Law Review*, 1 (1960), 3–5.

J. Keuter 'Rules of the Road in Space: Does a Code of Conduct Improve US Security?' quoted in J. Beard 'Soft Law's Failure on the Horizon: The International Code of Conduct for Outer Space Activities' (2017) 38.2 *Pennsylvania Journal of International Law* 335 at 376.

U. Khandeparkar, 'Behind the Nano Hype', *Wall Street Journal*, (19 January 2009).

P. Khanna and A. Khanna, *The Pleasure and Danger of Augmented Reality*, BIG THINK, http://bigthink.com/hybrid-reality/the-pleasure-and-danger-of-augmented-reality (accessed 13 September 2018).

A. Kharpal, 'Apple vs FBI: All You Need to Know', 29 March 2016, available at www.cnbc.com/2016/03/29/apple-vs-fbi-all-you-need-to-know.html.

T. Kidd and S. Ward, 'A New Century for Unmanned Maritime Systems, October-December 2012, available at www.doncio.navy.mil/chips/ArticleDetails.aspx?ID=4195.

J. Kim, Georgetown University Otolaryngology Department, cited in Zachary Cohen, 'New Audio Adds to Mystery of Attacks on US Diplomats' (14 October 2017) *CNN*, available at http://edition.cnn.com/2017/10/13/politics/cuba-us-diplomats-acoustic-weapons/index.html.

K. Kjølberg and F. Wickson, 'Social and Ethical Interactions Wwth Nano: Mapping the Early Literature', *NanoEthics*, 1(2) (2007), 89–104.

N. Klein, *Maritime Security and the Law of the Sea*, . (Oxford: Oxford University Press, 2011).

A. J. Kobler, 'Will There Be a Neurolaw Revolution?,' *Indiana Law Journal* (2014); 89: 807–845.

C. Koch, 'To Keep Up with AI, We'll Need High-Tech Brains,' *Wall Street Journal*, 27 October 2017.

D. Koplow, 'Tangled up in Khaki and Blue: Lethal and Non-Lethal Weapons in Recent Confrontations' (2005) 36 *Georgetown Journal of International Law* 703–808, 708.

D. A. Koplow, *Non-Lethal Weapons: The Law and Policy of Revolutionary Technologies for the Military and Law Enforcement* (New York: Cambridge University Press, 2009).

B. Koranyi, 'G20 to Jointly Fight Bank Sector Hacking', Reuters, 17 March 2017, available at http://uk.reuters.com/article/us-g20-germany-cyber-idUKKBN16O2F0.

M. Kosal, *Nanotechnology for Chemical and Biological Defense* (New York: Springer, 2009).

J. Kosslyn and C. Yu, 'Fact Check Now Available in Google Search and News Round the World', Google Blog, 7 April 2017, available at https://blog.google/products/search/fact-check-now-available-google-search-and-news-around-world/.

L. Krabbenborg, 'DuPont and Environmental Defense Fund Co-Constructing a Risk Framework for Nanoscale Materials: An Occasion to Reflect on Interaction Processes in a Joint Inquiry', *NanoEthics*, 7(1) (2013), 45–54.

J. Kraska, *Maritime Power and the Law of the Sea* (Oxford: Oxford University Press, 2011).

V. Kumar Thakur and M. Kumari Thakur, *Handbook of Sustainable Polymers: Structure and Chemistry* (Abingdon: Taylor and Francis, 2016).

K. Kwang, Changes to Singapore's Cybercrime Law Passed, Channel News Asia, 3 April 2017, available at www.kyivpost.com/ukraine-politics/ukraine-natos-support-prepares-creation-cyber-security-centre.html.

A. Lalaj *Burning Secrets of the Corfu Channel Incident* (Woodrow Wilson International Center Cold War History Project Working paper #70, September 2014).

P. Larsen, 'Issues Relating to Civilian and Military Dual Uses of GNSS', (2001) 17 *Space Policy*, 111.

H. Lauterpacht, ed., *Oppenheim's International Law* Vol II (Reissue of 1952 edn, Longmans 1955), available at http://www.gutenburg.org/files/41047/41047-h/41047-h.htm).

K. Lawand, 'A Guide to the Legal Review of New Weapons, Means and Methods of Warfare: Measures to Implement Article 36 of Additional Protocol I of 1977', *ICRC* (2006).

R. Leheny, Statement submitted to the Subcommittee on Terrorism, Unconventional Threats and Capabilities, House Armed Services Committee, *United States House of Representatives*, 20 May 2009, 24, available at www.darpa.mil/attachments/TestimonyArchived%20(May%2020%202009).pdf.

J. Lehrer, 'The forgetting pill: How a new drug can erase your worst memories', *WIRED*, 17 February 2012, www.wired.com/2012/02/ff_forgettingpill/.

K. Leins, 'Shining a Regulatory Spotlight on New Lasers: Regulation of the Use of Nanolaser Technologies in Armed Conflict', *Jurimetrics*, 56 (2016), 261–278.

D. Letts, 'Naval Mines: Legal Considerations in Armed Conflict and Peacetime' (2016) 98/102 *International Review of the Red Cross* 543.

D. Letts and R. McLaughlin, 'Call-Out Powers for the Australian Defence Force in an Age of Terrorism: Some Legal Implications' (2017) No 85 *Australian Institute of Administrative Law Forum* 63, 75–79.

J. Levasseur-Moreau, et al., 'Non-Invasive Brain Stimulation Can Induce Paradoxical Facilitation. Are These Neuroenhancements Transferable and Meaningful to Security Services?', *Frontiers in Neuroscience* Vol. 7, 449 (August 2013), 2.

D. Levi-Faur and H. Comaneshter, 'The Risks of Regulation and the Regulation of Risks: The Governance of Nanotechnology', in G. Hodge, D. Bowman and K. Ludlow (eds.), *New Global Frontiers in Regulation: The Age of Nanotechnology* (Cheltenham: Edward Elgar, 2007), 149–165, 150.

S. Levy, 'Brain-Machine Interface Isn't Sci-Fi Anymore,' *Wired*, 13 September 2017, available at www.wired.com/story/brain-machine-interface-isnt-sci-fi-anymore/?mbid=BottomRelatedStories.

S. Z. Li, et al., 'Biometrics at a Distance', in M. Tistarelli, S. Z. Li and R. Chellappa (eds.), *Handbook of Remote Biometrics: for Surveillance and Security* (New York: Springer, 2009).

R. Liivoja, K. Leins and T. McCormack, 'Emerging Technologies of Warfare', in R. Liivoja and T. McCormack (eds.), *Routledge Handbook of the Law of Armed Conflict* (Routledge, 2016), 603, 620.

A. C. Lin, 'Size Matters: Regulating Nanotechnology', *Harvard Environmental Law Review*, 31 (2007), 349.

P. Lin, 'Ethical Blowback from Emerging Technologies', *Journal of Military Ethics* 9:4 (2010), 324–325.

'Could Human Enhancement Turn Soldiers into Weapons That Violate International Law? Yes', *The Atlantic*, 4 January 2013.

P. Lin and F. Allhoff, 'Against Unrestricted Human Enhancement', 18 *Journal of Evolution and Technology* 1, May 2008, 35, 4.

P. Lin, M. J. Mehlman and K. Abney, 'Enhanced Warfighters: Risk, Ethics and Policy', *Report Prepared for the Greenwall Foundation*, January 2013, 13.

B. K. Logan, 'Methamphetamine – Effect on Human Performance and Behaviour', *Forensic Science Review* 14:133 (2002), 136.

N. Lomas, 'Facebook Expands its Hate-fighting Counterspeech Initiative in Europe', 23 June 2017, available at https://techcrunch.com/2017/06/23/facebook-expands-its-hate-fighting-counterspeech-initiative-in-europe/.

'UK Kicks Off Driverless Car Law Review to Get Tech on the Road by 2021', *TechCrunch*, 7 March 2018, available at https://techcrunch.com/2018/03/07/uk-kicks-off-driverless-car-law-review-to-get-tech-on-the-road-by-2021/.

LRAD Corporation, 'Extended Range Communication & Scalable Escalation of Force', available at www.lradx.com/application/defense/.

B. Luber and S. H. Lisanby, 'Enhancement of Human Cognitive Performance Using Transcranial Magnetic Stimulation', *Neuroimage* 85: 03 (15 January 2014), 6–7.

K. Ludlow, D. M. Bowman and G. A. Hodge, 'A Review of Possible Impacts of Nanotechnology on Australia's Regulatory Framework: Final Report', Monash University Law, available at www.researchgate.net/publication/265997359_A_Review_of_Possible_Impacts_of_Nanotechnology_on_Australia's_Regulatory_Framework.

K. Ludlow, D. M. Bowman and D. D. Kirk, 'Hitting the Mark or Falling Short with Nanotechnology Regulation?', *Trends in Biotechnology*, 27(11) (2009), 615–620, 615.

K. McCann, 'Government "blocked" from Accessing Twitter Data to Help Spot Terrorist Plots', *The Telegraph*, 25 April 2017, available at www.telegraph.co.uk/news/2017/04/25/government-blocked-accessing-twitter-data-help-spot-terrorist/.

T. L. H. McCormack, 'A Non Liquet on Nuclear Weapons – The ICJ Avoids the Application of General Principles of International Humanitarian Law' (1997), 316 *IRRC* 76.

W. S. McCulloch and W. Pitts, 'A Logical Calculus of the Ideas Immanent in Nervous Activity,' *Bulletin of Mathematical Biophysics* (1943); 5: 115–133.

L. McGregor, 'Are Declaratory Orders Appropriate for Continuing Human Rights Violations? The Case of Khadr v Canada' (2010) 10:3 *Human Rights Law Review* 487.

R. McLaughlin, 'The Use of Lethal Force by Military Forces on Law Enforcement Operations – Is There a "Lawful Authority"?' (2009) 37:3 *Federal Law Review* 441.

'Unmanned Naval Vehicles at Sea: USVs, UUVs, and the Adequacy of the Law' (2011) 21:2 *Journal of Law Information and Science* 100.

'Dangerous Waters and International Law: The Corfu Channel Case, Warships, and Sovereignty Irritants', in K. Bannelier, T.e Christakis and S. Heathcote (eds.), *The ICJ and the Evolution of International Law: The Enduring Impact of the Corfu Channel Case* (Abingdon: Routledge, 2012).

'State Responsibility for Third Country Deployed/Embedded Military Personnel Engaged in Armed Conflict', *EJIL Talk*!, 10 September 2015, at www.ejiltalk.org/state-responsibility-for-third-country-deployed-embedded-military-personnel-engaged-in-armed-conflict/.

R. McLaughlin and D. Stephens, 'International Humanitarian Law in the Maritime Context: Conflict Characterization in Judicial and Quasi-Judicial Contexts', in D. Jinks, J. Maogoto and S. Solomon (eds.), *Applying International Humanitarian Law in Judicial and Quasi-Judicial Bodies: International and Domestic Aspects* (The Hague: TMC Asser Press, 2014).

A. Mack, 'Why Big Nations Lose Small Wars: The Politics of Asymmetric Conflict', *World Politics* 27:2 (January 1975), available online at: https://web.stanford.edu/class/polisci211z/2.2/Arreguin-Toft%20IS%202001.pdf.

B. Maher, et al., 'Magnetite Pollution Nanoparticles in the Human Brain', *Proceedings of the National Academy of Sciences of the United States of America*, 113(39) (2016), 10797–10801.

P. D. Marasco, et al., 'Illusory Movement Perception Improves Motor Control for Prosthetic Hands' (2018) 10(432) *Science Translational Medicine*, available at http://stm.sciencemag.org/content/10/432/eaao6990.

T. Marauhn, 'Meaningful Human Control – And the Politics of International Law' in W Heintschel von Heinegg, et al. (eds.) *Dehumanization of Warfare*, 2018, 207, 210–212.

G. E. Marchant, B. R. Allenby and J. R. Herkert (eds.), vol. 7, *The Growing Gap between Emerging Technologies and Legal-Ethical Oversight: The Pacing Problem* (New York: Springer, 2011).

D. Marcus, 'Famine Crimes in International Law', *American Journal of International Law*, 97(2) (2003), 245, 262–264.

G. Markoff, 'Explanation of Position at the Conclusion of the 2016–2017 Group of Governmental Experts on Developments in the Field of Information and Telecommunications in the Context of International Security', US Department of State, 23 June 2017, available at www.state.gov/s/cyberissues/releasesandremarks/272175.htm.

A. Marshall, 'War by Poison' (1934) *Nature* 952–953.

'Congress Unites (Gasp) to Spread Self-Driving Cars Across America', *Wired* 6 September 2017.

H. Maslen, T. Douglas, R. Cohen Kadosh, N. Levy and J. Savulescu, 'Mind Machines. The Regulation of Cognitive Enhancement Devices: Extending the Medical Model", *Oxford Martin Policy Paper*, 2014.

E. Massingham, 'Conflict Without Casualties ... A Note of Caution: Non-Lethal Weapons and International Humanitarian Law' (2012) 886 *International Review of the Red Cross* 673, 678.

M. J. Matheson, 'The United States Position on the Relation of Customary International Law to the 1977 Protocols Additional to the 1949 Geneva Conventions' (1987) 2 *AU JILP* 419.

R. Mathews, 'Chemical and Biological Weapons', in R. Liivoja and T. McCormack (eds.), *Routledge Handbook of the Law of Armed Conflict* (London: Routledge, 2016), 229–230.

J. Matsuura, *Nanotechnology Regulation and Policy Worldwide* (Boston and London: Artech House, 2006).

G. Matthieu and B. Williams-Jones, 'Examining the National Regulatory Environment of Medical Devices: Major Issues in the Risk-Benefit Assessment of High-Risk Devices,' *McGill Journal of Law and Health* (2015–2016); 9: 17–67.

A. D. Maynard, et al., 'Safe Handling of Nanotechnology', *Nature*, 444(7117) (2006), 267–269.

'Don't Define Nanomaterials' *Nature*, 475(7354) (2011), 31.

A. Maynard, 'Frankenstein Foods, Nanotech and the Trouble with Communicating Technology', World Economic Forum, 10 October 2016, available at www.weforum.org/agenda/2016/10/the-trouble-with-technology-innovation.

A. Mayor, *Greek Fire, Poison Arrows, and Scorpion Bombs: Biological and Chemical Weapons in the Ancient World* (New York: Overlook Press, 2009).

S. Medvedev, 'Data Protection in Russian Federation: Overview', Thomson Reuters, 2017, available at https://uk.practicallaw.thomsonreuters.com/2–502–2227?transitionType=Default&contextData=(sc.Default)&firstPage=true.

M. J. Mehlman, P. Lin and K. Abney, 'Enhanced Warfighters: A Policy Framework", in M. L. Gross and D. Carrick (eds.), *Military Medical Ethics for the 21st century*, 121–124.

M. W. Meier, 'Lethal Autonomous Weapon Systems (LAWS): Conducting a Comprehensive Weapons Review', 30 *Temple Int'l & Comp L J* 1 (2016) 119, 120.

C. Meili and M. Widmer, 'Voluntary Measures in Nanotechnology Risk Governance: The Difficulty of Holding the Wolf by the Ears', in G. Hodge, D. M. Bowman and A. Maynard (eds.), *International Handbook on Regulating Nanotechnologies* (Cheltenham: Edward Elgar, 2010), 446–461.

J. Menn, 'A Scramble at Cisco Exposes Uncomfortable Truths about US Cyber Defense', Reuters, 30 March 2017, available at http://uk.reuters.com/article/us-usa-cyber-defense-idUKKBN17013U.

L. D. Mermelstein, 'Ethics Update: Lawyers Must Keep Up with Technology,' American Bar Association (2013), available at www.americanbar.org/publications/blt/2013/03/keeping_current.html.

T. Meron, 'The Martens Clause, Principles of Humanity and Dictates of Public Conscience', *The American Journal of International Law*, 94(1) (2000), 78.

J. M. Meyer, 'Tearing Down the Façade: A Critical Look at the Current Law on Targeting the Will of the Enemy and Air Force Doctrine' (2001) 51 *Air Force Law Review* 143.

P. Meyer, 'Dark Forces Awaken: The Prospects for Cooperative Space Security', (2016) *The Nonproliferation Review*, 23 3–4, 495, 496.

MForce website, available at www.hrsid.com/product-mforce.

Microsoft Corporation, Policy Papers, available at https://mscorpmedia.azureedge.net/mscorpmedia/2017/04/Policy-Paper-Attribution-Organization.pdf.

J. W. Miller, 'US Steel Accuses China of Hacking', *Wall Street Journal*, 28 April 2016.

W. Miller, 'The British Experience in Northern Ireland: A Model for Modern Peacemaking Operations', US Army Command and General Staff College, Fort Leavenworth, 1993, 40, at www.dtic.mil/dtic/tr/fulltext/u2/a264419.pdf.

S. Mohanty and R. Bhatia, 'India Court's Privacy Ruling Is Blow to Government', Reuters, 24 August 2017, available at www.reuters.com/article/us-india-court-privacy/indian-courts-privacy-ruling-is-blow-to-government-idUSKCN1B40CE?utm_campaign=trueAnthem:+Trending+Content&utm_content=599e6b6204d3015a2054937c&utm_medium=trueAnthem&utm_source=twitter.

R. Mola 'How Things Work: Space Fence: The New Early-Warning System to Protect Spacecraft from Orbiting Junk', *Air & Space Magazine*, February 2016, www.airspacemag.com/space/how-things-work-space-fence-180957776/.

J. Monica, G. Van Calster and P. Agapi (2014), 'A Nanotechnology Legal Framework', in M. Hull and D. M. Bowman (eds.), *Nanotechnology Environmental Health and Safety: Risks, Regulation and Management*, 2nd edn, (New York: Elsevier), pp.265–312.

P. Moriuchi and W. Ladd, 'China Altered Public Vulnerability Data to Conceal MSS Influence, Recorded Future', 9 March 2018, available at www.recordedfuture.com/chinese-vulnerability-data-altered/.

E. Morozov 'The Perils of Perfection' *New York Times*, 3 March 2013.

L. B. Moses, 'Report of the Symposium on Ethical, Legal and Social Implications of Emerging Military Technologies, 29–30 May 2017', Melbourne Law School, August 2017.

NASA, 'Artificial Intelligence, NASA Data Used to Discover Eighth Planet Circling Distant Star', 14 December 2017, available at www.nasa.gov/press-release/artificial-intelligence-nasa-dataused-to-discover-eighth-planet-circling-distant-star.

H. Nasu, 'Nanotechnology and Challenges to International Humanitarian Law: A Preliminary Legal Assessment', *International Review of the Red Cross*, 94 (2012), 653–672.

H. Nasu and T. Faunce, 'Nanotechnology and the International Law of Weaponry: Towards International Regulation of Nano-Weapons' (2009) 20 *Journal of Law, Information and Science* 21, 34–43.

'Nanotechnology and the International Law of Weaponry: Towards International Regulation of Nano-Weapons', Journal of Law, Information and Science, 22 (2009), 21–54, available at https://law.anu.edu.au/sites/all/files/users/u9705219/nanoweapons_int_law_0.pdf.

H. Nasu and R. McLaughlin (eds.), *New Technologies and the Law of Armed Conflict* (The Hague: Asser Press, 2014) 145.

National Academies of Sciences, Engineering, and Medicine, *Thriving on Our Changing Planet: A Decadal Strategy for Earth Observation from Space* (2018), available at www.nap.edu/catalog/24938/thriving-on-our-changing-planet-a-decadal-strategy-for-earth.

National Research Council, *Opportunities in Neuroscience for Future Military Applications* (Washington: National Academies Press, 2009).

NATO, 'Policy on non-lethal weapons', *NATO*, (13 October 1999, updated 6 November 2008) para 3, available at www.nato.int/cps/ic/natohq/official_texts_27417.htm?selectedLocale=en.

S. Nelliss and C. Cadell, 'Apple Moves To Store Icloud Keys In China, Raising Human Rights Fears', Reuters 24 February 2018, available at www.reuters.com/article/us-china-apple-icloud-insight/apple-moves-to-store-icloud-keys-in-china-raising-human-rights-fears-idUSKCN1G8060.

New York Times, 'Prosthetic Limbs, Controlled by Thought', *New York Times*, 20 May 2015, available at www.nytimes.com/2015/05/21/technology/a-bionic-approach-to-prosthetics-controlled-by-thought.html (accessed 9 March 2018).

D. Ngo, 'Pinterest is Now Blocked by China's Great Firewall', C/Net, 16 March 2017, available at www.cnet.com/uk/news/pinterest-blocked-in-china-great-firewall/.

Z. Niemtus, 'The Disrupters – Humans 2.0: Meet the Entrepreneur Who Wants to Put a Chip in Your Brain,' *The Guardian*, 14 December 2017.

Nobel Peace Prize in Chemistry 2016, Nobelprize.org, available at www.nobelprize.org/nobel_prizes/chemistry/laureates/2016.

A. Norris, *Legal Issues Relating to Unmanned Maritime Systems*, available at www.iqpc.com/media/1002182/50661.pdf.

V. Novakovic, et al., 'Brain Stimulation in Posttraumatic Stress Disorder', *European Journal of Psychotraumatology* 2 (2011).

D. Oberhaus, 'The First Spy Satellites Had to Drop Gigantic Buckets of Film Back to Earth', 22 April 2017, *Motherboard* (online), https://motherboard.vice.com/en_us/article/jpzpkg/the-first-spy-satellites-had-to-drop-gigantic-buckets-of-film-back-to-earth.

R. L. O'Connell, *Of Arms and Men* (Oxford: Oxford University Press, 1989).

S. O'Neill, 'British Spy in IRA Arrested on Murder and Torture Charges', *The Times*, 31 January 2018, 5.

L. Oppenheim, *International Law*, Vol. II, 7th edn *et seq.* (7th ed. by Hersch Lauterpacht, London 1952).

R. O'Rourke, 'Navy Frigate (FFX[X]) Program: Background and Issue for Congress', 8 December 2017, available at https://fas.org/sgp/crs/weapons/R44972.pdf; www.naval-technology.com/projects/standard-missile-6/.

OECD, *OECD Chemical Studies Show Way Forward for Nanomaterial Safety* (9 June 2015), available at www.oecd.org/chemicalsafety/nanosafety/news-nanomaterial-safety.htm.

Office of Director of National Intelligence, 'IARPA launches "Odin" Program to harden biometric technology against attacks', News Release 22–17 dated 19 October 2017, available at www.dni.gov/index.php/newsroom/press-releases/item/1808-iarpa-launches-odin-program-to-harden-biometric-technology-against-attacks.

Office of the United Nations High Commissioner for Human Rights, 'Egypt extends its assault on freedom of expression by blocking dozens of websites UN experts warn', 30 August 2017, available at www.ohchr.org/EN/NewsEvents/Pages/DisplayNews.aspx?NewsID=22008&LangID=E.

Office of the Under Secretary of Defense for Acquisition, Technology, and Logistics, *Study on Next-Generation Unmanned Undersea Systems* (Washington, DC, October 2016), available at www.acq.osd.mil/dsb/reports/2010s/Next-Generation_Unmanned_Undersea_Systems.pdf.

H. Osborne and J. Elgot, 'MPs Summon Mark Zuckerberg and Accuse Facebook of Misleading Them', *The Guardian*, 20 March 2018.

J. Ostrower, 'Airbus "Flying Car" Takes to The Skies for the First Time', CNN Tech, 1 February 2018, available at http://money.cnn.com/2018/02/01/technology/airbus-vahana-firstflights/index.html.

B. Oswald, H. Durham and A. Bates (eds.), *Documents on the Law of UN Peace Operations*, (Oxford: Oxford University Press, 2010).

K. Owens, 'Wireless Underwater Charging Tech Will Bring a New Level of Autonomy for UUVs', *September* 1, 2017, available at https://defensesystems.com/articles/2017/09/01/navy-drone-uuv.aspx.

T. Paine 'Bombs in Orbit? Verification and Violation under the Outer Space Treaty', *The Space Review*, 19 March 2018.

R. Panditl, 'New Joint Doctrine Stresses Space, Cyber, Special Ops in War-Fighting', *Times of India*, 25 April 2017.

R. (Pete) Pedrozo, 'Military Activities in the Exclusive Economic Zone: East Asia Focus', 90 *International Law Studies* (2014), 514–543, 525 *et seq*.

J. Pelegrín-Borondo, E. Reinares-Lara and C. Olarte-Pascual, 'Assessing the Acceptance of Technological Implants (the Cyborg): Evidences and Challenges', (2017) 70 *Computers in Human Behaviour*, 104–112, 105.

N. Perlroth, and C. Krauss, 'A Cyber Attack in Saudi Arabia Had a Deadly Goal: Experts Fear Another Try', *New York Times*, 15 March 2018, available at www.nytimes.com/2018/03/15/technology/saudi-arabia-hacks-cyberattacks.html.

J. B. Perkins, 'Operation Praying Mantis: The Surface View' (1989) 115:5 *US Naval Institute Proceedings* 1035, at www.usni.org/magazines/proceedings/1989-05/operation-praying-mantis-surface-view.

The PEW Charitable Trusts, 'Project on Emerging Nanotechnologies', available at www.pewtrusts.org/en/archived-projects/project-on-emerging-nanotechnologies.

M. Pham, 'Vietnam Urges Firms to Stop Youtube and Facebook Ads in Protest over Fake Content', Reuters, 16 March 2017, available at www.reuters.com/article/us-vietnam-google-idUSKBN16N110.

E. Phoenix, 'NI State Papers: Concern over Use of Plastic Bullets', *BBC*, 28 December 2013, at www.bbc.com/news/uk-northern-ireland-25425779.

M. Piesing, 'Medical Robotics: Would you Trust a Robot with a Scalpel?', *The Guardian*, 10 October 2014.

E Pilkington, 'Judge Approves Use of "Truth Serum" on Accused Aurora Shooter James Holmes', *The Guardian* (12 March 2013).

W. Pincus, 'Senate Boosts Funding for Laser Weapons', *Washington Post*, 22 September 2008.

R. K. Pitman, K. M. Sanders, R. M. Zusman, et al., 'Pilot Study of Secondary Prevention of Posttraumatic Stress Disorder with Propranolol', *Biological Psychiatry* 51:2 (2002), 189–192.

C. A. Poland, et al., 'Carbon Nanotubes Introduced into the Abdominal Cavity of Mice Show Asbestos-Like Pathogenicity in a Pilot Study', *Nature Nanotechnology*, 3(7) (2008), 423–428.

S. Polavaram and G. Ascoli, 'Neuroinformatics,' *Scholarpedia*, 2015, doi: 10(11):1312, available at www.scholarpedia.org/article/Neuroinformatics.

A. Pop, 'Autonomous Weapon Systems: A Threat to Human Dignity', ICRC Humanitarian Law and Policy Blog, 10 April 2018.

B. Popken, 'Twitter Deleted 200,000 Russian Troll Tweets', NBC News, 14 February 2018, available at www.nbcnews.com/tech/social-media/now-available-more-200-000-deleted-russian-troll-tweets-n844731.

L. Potts, 'The Police "Super-Recognisers" Putting Names to Faces', *BBC News*, 19 October 2015, available at www.bbc.co.uk/news/uk-england-34544199.

J. M. Prescott, 'Building the Ethical Cyber Commander and the Law of Armed Conflict,' *Rutgers Computer and Technology Law Journal* (2014); 40: 42–77.

A. Priddy, 'The Use of Weapons in Counterpiracy Operations', in Casey-Maslen (ed.), *Weapons under International Human Rights Law*, 2014, 122 – 160, 154.

Privacy, Data Protection and Cyber Security Law Review, Edition 3, Russia dated November 2016, available at http://thelawreviews.co.uk/edition/the-privacy-data-protection-and-cybersecurity-law-review-edition-3/1140166/russia.

Program on Humanitarian Policy and Conflict Research at Harvard University, *HPCR Manual on International Law Applicable to Air and Missile Warfare* (Cambridge: Cambridge University Press, 2013).

A. E. Pusateri, 'US Department of Defence Haemorrhage and Resuscitation Research and Development Program', presented at the Remote Damage Control Resuscitation Conference, Bergen, Norway, 19 June 2002, available at: http://rdcr.org/wp-content/uploads/symp2012pdf/US%20Department%20of%20Defense%20Hemorrhage%20and%20Resuscitation%20Research%20and%20Development%20Program%20-%20Anthony%20Pusateri.pdf.

I. Puscas, 'La Quête du Soldat Augmenté', *Le Monde Diplomatique*, September 2017, 3.

A. G. Quinn 'The New Age of Space Law: The Outer Space Treaty and the Weaponization of Space', (2008) 17 *Minnesota Journal of International Law* 475, 479–480.

H. Quzhi, 'The Outer Space Treaty in Perspective' (1997) 40 *Proceedings on the Law of Outer Space* 51, 53.

S. Radchenko, *The Soviet Union and the North Korean Seizure of the USS Pueblo: Evidence from Russian Archives*, Woodrow Wilson International Center Cold War History Project Working paper #47, 7 July 2011.

Radio Free Europe/Radio Liberty, 'Telegram App Under the Gun in Russia After Court Ruling', 20 March 2018, available at www.rferl.org/a/russia-telegram-messaging-app-under-pressure-fsb-supreme-court/29111309.html.

A. Ram, 'DeepMind Develops AI to Diagnose Eye Diseases,' *Financial Times*, 4 February 2018.

R. Ramey, 'Armed Conflict on the Final Frontier: The Law of War in Space', (2000) 48 *Air Force Review*, 1, 13.

K. Rasmussen, et al., 'Review of Achievements of the OECD Working Party on Manufactured Nanomaterials' Testing and Assessment Programme: From Exploratory Testing to Test Guidelines', *Regulatory Toxicology and Pharmacology*, 74 (2016), 147–160.

K. Ravilious, 'The Secrets of Sonic Weapons' (8 November 2005) *The Guardian* www.theguardian.com/science/2005/nov/08/g2.weaponstechnology.

G. Rayner, 'WhatsApp Accused of Giving Terrorists a Secret Place to Hide as It Refuses to Hand Over Private Messages', *Daily Telegraph*, 27 March 2017.

Regulation (EU) 2016/679 of 27 April 2016, available at http://eur-lex.europa.eu/legal-content/EN/TXT/PDF/?uri=CELEX:32016R0679&from=EN, (2016 LAWS Report), para13.

Report of the 2016 Informal Meeting of Experts on Lethal Autonomous Weapon Systems, submitted by the Chairperson, available at www.unog.ch/80256EDD006B8954/(httpAssets)/DDC13B243BA863E6C1257FDB00380A88/$file/ReportLAWS_2016_AdvancedVersion.pdf.

S. Reschke, J. B. F. van Erp, A. M. Brouwer and M. Grootjen, 'Neural and Biological Soldier Enhancement: From SciFi to Deployment', Paper Presented at the NATO RTO Human Factors and Medicine Panel, Sofia, 5–7 October 2009, http://publica.fraunhofer.de/documents/N-194631.html.

Reuters, 'Germany Adopts Self-Driving Vehicles Law', *Reuters*, 12 May 2017, available at www.reuters.com/article/us-germany-autos-self-driving/germany-adopts-self-driving-vehicles-lawidUSKBN1881HY

G. Reynolds, 'Environmental Regulation of Nanotechnology: Some Preliminary Observations', *Environmental Law Reporter*, 31(6) (2001), 10681.

'Nanotechnology and Regulatory Policy: Three Phases', *Harvard Journal of Law and Technology*, 17(1) (2003), 179, 185 citing E. Drexler, *Engines of Creation* (Oxford: Oxford University Press, 1990), 56–63.

J. Rice, 'Seaweb Acoustic Communication and Navigation Network', available at https://calhoun.nps.edu/bitstream/handle/10945/40396/rice_seaweb.pdf?sequence=1&isAllowed=y.

J. Richardson, 'Inside the Race to Hack the Human Brain,' *Wired*, 16 November 2017, available at www.wired.com/story/inside-the-race-to-build-a-brain-machine-interface/?mbid=BottomRelatedStories.

J. R. Riches, et al., 'Analysis of Clothing and Urine from Moscow Theatre Siege Casualties Reveals Carfentanil and Remifentanil Use' (2012) 36(9) *Journal of Analytical Toxicology* 647–656.

B. Riley-Smith, et al., 'Airports and Nuclear Power Stations on Terror Alert As Government Officials Warn of Credible Cyber Threat', *Daily Telegraph*, 2 April

2017, available at www.telegraph.co.uk/news/2017/04/01/airports-nuclear-power-stations-terror-alert-government-officials/.

A. Ritchie and J. Gibbons, 'Letter to Editor: Plastic Bullets in Northern Ireland' (1990) 301(6764) *British Medical Journal* 1332.

K. Roach, 'The Supreme Court at the Bar of Politics: The Afghan Detainee and Omar Khadr Cases' (2010) 28 *National Journal of Constitutional Law* 115.

J. A. Roach and R. W. Smith, *Excessive Maritime Claims*, 3rd edn (Leiden/Boston: Martinus Nijhoff, 2012).

A. Roberts, *Is International Law International?* (Oxford: Oxford University Press, 2017).

M. C. Roco, 'International Perspective on Government Nanotechnology Funding in 2005', *Journal of Nanoparticle Research*, 7(6) (2005), 707–712.

A. P. V. Rogers, *Law on the Battlefield*, 3rd edn (Manchester: Manchester University Press, 2012).

C. A. Rogers, 'A Slow March towards Thought Crime: How the Department of Homeland Security's Fast Program Violates the Fourth Amendment', 64 *American University Law Review* 2 (2014), 337, 384.

L. Rohter, 'The Noriega Case: Panama City; Papal Envoy Asserts Psychology, Not Ultimatum, Swayed Noriega', *New York Times*, 6 January 1990.

E. Rosenfeld, 'US-China Agree Not to Conduct Cybertheft of Intellectual Property', CNBC, 25 September 2015, available at www.cnbc.com/2015/09/25/us-china-agree-to-not-conductcybertheft-of-intellectual-property-white-house.html.

S. Rossi, et al., 'Safety, Ethical Considerations, and Application Guidelines for the Use of Transcranial Magnetic Stimulation in Clinical Practice and Research', *Clinical Neurophysiology* 120 (2009), 2018–2024.

P. J. Rowe, *Defence, the Legal Implications* (London: Brassey's 1987).

W. Rowell, 'Haemorrhage Control: Lessons from the Military', *Emergency Nurse* Vol. 21, No 7 (November 2013): 22.

The Royal Society, 'Brain Wave Module 3: Neuroscience, Conflict and Security', February 2012, available at https://royalsociety.org/~/media/Royal_Society.../brain-waves/2012–02–06-BW3.pdf.

The Royal Society, 'Neuroscience, Conflict and Security', February 2012, https://royalsociety.org/~/media/Royal_Society_Content/policy/projects/brain-waves/2012–02–06-BW3.pdf.

Royal Society and Royal Academy of Engineering, *Nanoscience and Nanotechnologies: Opportunities and Uncertainties* (Royal Society, 2004).

L. Ruan, 'Internet Censorship: How China Does It', *The Strategist*, 9 October 2017, available at www.aspistrategist.org.au/internet-censorship-how-china-does-it/.

M. Ryan and D. Lamothe, 'Pentagon: Chinese Naval Ship Seized an Unmanned US Underwater Vehicle in South China Sea', *Washington Post*, 17 December 2016, available at www.washingtonpost.com/news/checkpoint/wp/2016/12/16/defense-official-chinese-naval-ship-seized-an-unmanned-u-s-ocean-glider/?utm_term=.2dfia5ad523d.

Saab Corporation, 'Naval Laser Warning Systems', available at https://saab.com/naval/electronic-warfare/naval-laser-warning-systems/nlws/.

S. Saeed, 'German Cyber Security Chief: Army attacked over 284,000 times this year', Politico, 3 April 2017, available at www.politico.eu/article/german-cybersecurity-chief-army-attacked-over-284000-times-this-year/.

I. Sample, 'Wired Awake', *The Guardian*, 29 July 2014.

I. Sample and R. Evans, 'MoD Bought Thousands of Stay Awake Pills in Advance of War in Iraq', *The Guardian*, 29 July 2004.

Y. Sandoz, et al. (eds.), *Commentary on the Additional Protocols of 8 June 1977 to the Geneva Conventions of 12 August 1949* (Geneva: ICRC, 1987).

D. E. Sanger and N. Perlroth, '6 Chinese Men Indicted in Theft of Code from US Tech Companies', *New York Times*, 19 May 2015.

M. Sassoli, 'Legitimate Targets of Attacks under International Humanitarian Law', *HCPR Policy Brief* (January 2003).

M. Schaefer 'The Contours of Permissionless Innovation in the Outer Space Domain' (March 28, 2017). *University of Pennsylvania Journal of International Law*, Vol. 39, No. 1, 2017, available at https://ssrn.com/abstract=2942526.

P. Scharre, 'Human Judgment and Lethal Decision-Making in War', ICRC Humanitarian Law and Policy Blog, 11 April 2018.

P. Schindler, 'Expanded safeguards for advertisers', 21 March 2017, available at https://blog.google/topics/ads/expanded-safeguards-for-advertisers/.

M. N. Schmitt, 'War, Technology and International Humanitarian Law', *Harvard University HPCR Occasional Paper Series*, (Summer 2005), 48.

'International Law and Military Operations in Space', in A. von Bogdandy and R. Wolfrum (eds.), *Max Planck Yearbook of United Nations Law*, 2006, volume 10, 200.

'Military Necessity and Humanity in International Humanitarian Law: Preserving the Delicate Balance', *Virginia Journal of International Law* 50:4 (2010), 795–839.

'Tallinn Manual 2.0 on the International Law of Cyber Operations: What It Is and Isn't', *Just Security*, 9 February 2017, available at www.justsecurity.org/37559/tallinn-manual-2-0-international-law-cyber-operations/.

M. N. Schmitt and J. Biller, 'The NotPetya Cyber Operation as a Case Study in International Law', *EjilTalk*, 11 July 2017.

M. N. Schmitt and David S. Goddard, 'International Law and the Military Use of Unmanned Maritime Systems', 98 *International Review of the Red Cross* 1–26 (August 2016).

M. Schmitt and J. Thurner, 'Out of the Loop – Autonomous Weapon Systems and the Law of Armed Conflict', *Harvard National Security Journal*, 4 (2013), 231, 275.

M. Schmitt and L. Vihul, 'International Cyber Law Politicized: The UN GGE's Failure to Advance Cyber Norms', *Just Security*, 30 June 2017.

M. Schmitt, et al. (eds.), *Tallinn Manual 2.0 on the International Law Applicable to Cyber Operations* (Cambridge: Cambridge University Press, 2017).

K.-U. Schrogl and J. Neumann, '[Outer Space Treaty] Article IV', in S. Hobe, B. Schmidt-Tedd and K.-U. Schrogl (eds.), *Cologne Commentary on Space Law Volume 1; Outer Space Treaty*, Cologne: Carl Heymanns, 2009.

K. Schwab, 'The Fourth Industrial Revolution: What It Means, How to Respond', World Economic Forum, Geneva, 14 January 2016, available at www.weforum.org/agenda/2016/01/the-fourth-industrial-revolution-what-it-means-and-how-to-respond/.

M. Sechrist, *Cyberspace in Deep Water: Protecting Undersea Communication Cables By Creating an International Public–Private Partnership* (Harvard Kennedy School, 23 March 2010), available at http://belfercenter.hks.harvard.edu/files/PAE_final_draft_-_043010.pdf.

K. Sennaar, 'Facial Recognition Applications: Security, Retail and Beyond', *Techemergence*, 16 November 2017.

N. Shachtman, 'Squirrel=Super Soldier', *WIRED* 3 December 2007, available at www.wired.com/2007/03/soldiers-if-you/.

A. Shahani, *Microsoft President Urges Nuclear-Like Limits on Cyberweapons* (16 May 2017) National Public Radio www.npr.org/sections/alltechconsidered/2017/05/16/528555400/microsofts-president-reflects-on-cyberattack-helping-pirates-and-the-nsa.

T. Shanker and M. Richtel, 'In New Military, Data Overload Can Be Deadly' *New York Times*, 16 January 2011.

T. Shaohui (ed.), 'International Strategy of Cooperation on Cyberspace', 1 March 2017, available at http://news.xinhuanet.com/english/china/2017–03/01/c_136094371.htm.

F. X. Shen, 'Law and Neuroscience 2.0', *Arizona State Law Journal* (2016).

'The Overlooked History of Neurolaw,' *Fordham Law Review* (2016); 85: 667–695.

'Neurolegislation: How U.S. Legislators Are Using Brain Science', *Harvard Journal of Law & Technology* (2016).

B. Shepard, 'UAB to Start First Human Trials of Drug to Prevent Death Due to Severe Blood Loss', 21 October 2015, available at https://uab.edu/news/innovation/item/6624-uab-to-start-first-human-trials-of-drug-to-prevent-death-due-to-severe-blood-loss.

R. Sherstobitoff, 'Hidden Cobra Targets Turkish Financial Sector with New Bankshot Implant', 8 March 2018, available at https://securingtomorrow.mcafee.com/mcafee-labs/hidden-cobra-targets-turkish-financial-sector-new-bankshot-implant/.

P. Sherwell, 'Blind Man Describes Joy at Seeing Wife for First Time in Decade Thanks to 'Bionic Eye', *The Telegraph*, 24 February 2015.

J. Z. Siegel and M. Crockett, 'How Serotonin Shapes Moral Judgment and Behavior', *Annals of the New York Academy of Sciences, Issue: Sociability, Responsibility and Criminality: From Lab to Law* (2013), 44.

M. Simon, 'A Robot Ear Surgeon Drills into the Future of Medicine', Wired, 15 March 2017 available at www.wired.com/2017/03/robotear-surgeon-drills-future-medicine/.

Sky, 'Websites we've blocked under order of the High Court', available at www.sky.com/help/articles/websites-blocked-under-order-of-the-high-court.

P. Smaglik, 'Building for the Future', *Nature*, 408(6812) (2000), 622; *IBM Research – Nanotechnology* (2002) IBM www.research.ibm.com/pics/nanotech/defined.shtml.

B. Smith, 'The Need for a Digital Geneva Convention', Microsoft Blog dated 14 February 2017, available at https://blogs.microsoft.com/on-the-issues/2017/02/14/need-digital-geneva-convention/#sm.00001e7re9mi1sd80wxcuwmn80906.

R. Smith, *The Utility of Force* (London: Allen Lane, 2005).

C. Som, et al., 'The Importance of Life Cycle Concepts for the Development of Safe Nanoproducts', *Toxicology*, 269 (2010), 160–169.

J. M. Spaight, *Aircraft in War* (London: Macmillan, 1914).

R. Sparrow, 'The Social Impacts of Nanotechnology: An Ethical and Policy Analysis', in A. Maynard and J. Stilgoe (eds.), *The Ethics of Nanotechnology, Geoengineering and Clean Energy* (Routledge, 2017), 179.

'Ethics as a Source of Law: The Martens Clause and Autonomous Weapons' on *International Committee of the Red Cross Blog* (14 November 2017).

S. Stadlmeier, 'CPU and Keyboard: Weapons of Mass Disruption?' in W. Heintschel von Heinegg, R. Frau and T. Singer (eds.), *Dehumanization of Warfare* (Springer, Cham, Switzerland, 2018).

E. Stanley, 'Neuroplasticity, Mind Fitness, and Military Effectiveness', in R. Armstrong, *Bio-Inspired Innovation*, 259–269.

H. Stark, 'A Bipartisan Bill to Strengthen Cyber Security for the Internet of Things', 20 August 2017, available at www.forbes.com/sites/haroldstark/2017/08/20/a-bipartisan-bill-to-strengthen-cybersecurity-for-the-internet-of-things/#3f51e1a95a5f.

State Coroner of NSW, *Inquest into the Deaths Arising from the Lindt Café Siege: Findings and Recommendations*, May 2017, ch14, especially paras 5–47, at www.lindtinquest.justice.nsw.gov.au/Documents/findings-and-recommendations.pdf.

C. Steer, *Translating Guilt: Identifying Leadership Liability for Mass Atrocity Crimes* (The Hague: Asser Press, 2017).

D. Stephens and M. Fitzpatrick, 'Legal Aspects of Contemporary Naval Mine Warfare' (1999) 21:4 *Loyola of Los Angeles International and Comparative Law Journal* 553.

P. N. Stockton and N. Golabek-Goldman, 'Curbing the Market for Cyber Weapons', 32 *Yale Law and Policy Review* 239, (2013) 248.

E. Stokes, 'Nanotechnology and the Products of Inherited Regulation', *Journal of Law and Society*, 39(1) (2012), 93–112.

E. Stokes and D. M. Bowman, 'Regulatory Inheritance and Emerging: Technologies: the Case of Nanotechnologies and Synthetic Biology', *European Journal of Risk Regulation*, 2 (2012), 235–241.

D. Stout, 'US Pilots Avoid Prosecution for Bombing Canadian Troops', *New York Times*, 19 June 2003.

C. Stupp, 'France Eyes EU Law to Crack Down on Terrorists' Use of Social Media,' *Euractiv*, 16 February 2018.

Subsea World News, 'Persistent Littoral Undersea Surveillance (PLUS) System Completes Initial in-Water Testing', *Subsea World News*, 28 October 2013, available at https://subseaworldnews.com/2013/10/28/persistent-littoral-undersea-surveillance-plus-system-completes-initial-in-water-testing/.

A. M. Sukumar, 'The UN GGE Failed: Is International Law in Cyberspace Doomed as Well?,' *Lawfare*, 4 July 2017, available at https://lawfareblog.com/un-gge-failed-international-law-cyberspace-doomed-well.

H. Surden, 'Machine Learning and the Law,' *Washington Law Review* (2014); 89: 87–115.

R. Susskind, *The Future of Law: Facing the Challenges of Information Technology* (Oxford: Clarendon Press 1996).

R. L. Swanson II, et al., 'Neurological Manifestations Among US Government Personnel Reporting Directional Audible and Sensory Phenomena in Havana, Cuba', *Journal of the American Medical Association Preliminary Communication*, 15 February 2018.

E. Sweeney, '"Troubles" Legacy: Derry Used as Testing Ground for Plastic Bullets', *Derry Journal*, 26 April 2016, at www.derryjournal.com/news/troubles-legacy-derry-used-as-testing-ground-for-plastic-bullets-1-7348893.

M. Tafirenyika, 'Drones are Effective in Protecting Civilians – Herve Ladsous', *Africa Renewal*, April 2016, at www.un.org/africarenewal/magazine/april-2016/drones-are-effective-protecting-civilians.

D. Talpalariu, 'Airport Speech and Brain Scans Identify the Terrorist in You', *Softpedia*, 3 December 2008.

T. Tether, Subcommittee on Terrorism, Unconventional Threats and Capabilities, House Armed Services Committee, United States House of Representatives,

27 March 2003, available at www.darpa.mil/attachments/TestimonyArchived (March%2027%202003).pdf.

The Economist, 'Know Your Enemy: How to Make Soldiers' Brains Better at Noticing Threats', 27 July 2017, available at www.economist.com/news/science-and-technology/21725543-target-recognition-warfare-how-make-soldiers-brains-better-noticing. (accessed 13 February 2018).

The Local, 'German government wants "backdoor" access to every digital device': Report, 1 December 2017, available at www.thelocal.de/20171201/german-government-wants-backdoor-access-to-every-digital-device-report.

The Scientific Committee on Emerging and Newly Identified Health Risks, *Nanomaterials* (2006), available at http://ec.europa.eu/health/scientific_committees/opinions_layman/en/nanotechnologies/l-3/5-nanoparticles-consumer-products.htm#opo.

S. Thielman, 'Yahoo Hack: 1bn Accounts Compromised by Biggest Data Breach in History'; the Guardian 15 December 2016, available at www.theguardian.com/technology/2016/dec/14/yahoo-hack-security-of-one-billion-accounts-breached.

D. Thomas, 'Facebook to Tackle Fake News with Education Campaign', *BBC News*, 6 April 2017, available at www.bbc.com/news/technology-39517033.

M. L. Thomas and M. B. Russo, 'Neurocognitive Monitors: Toward the Prevention of Cognitive Performance Decrements and Catastrophic Failures in the Operational Environment', *Aviation Space and Environmental Medicine* 78:5 (May 2007), 144–145.

J. Thurnher at J. S. Thurnher, 'Feasible Precautions in Attack and Autonomous Weapons', in W. Heintschel von Heinegg, et al., (eds.), *Dehumanization of Warfare* , 99, 108–114.

E. Tikk and M. Kerttunen, 'The Alleged Demise of the UN GGE: An Autopsy and Eulogy', Cyber Policy Institute, 2017, available at http://cpi.ee/wp-content/uploads/2017/12/2017-Tikk-Kerttunen-Demise-of-the-UN-GGE-2017–12–17-ET.pdf.

'Cyber Treaty Is Coming: Что делать?', Cyber Policy Institute (2018), 5, available at http://cpi.ee/wp-content/uploads/2018/02/Cyber-Treaty-is-Coming-Tikk-Kerttunen.pdf.

A. Tovey, 'Rise of The Machines: How Robot Brickies Could Be on Site Soon', *Daily Telegraph*, 27 June 2017.

M. Townsend and T. Helm, 'Threat of Russian Cyber Reprisal Puts UK Finance, Power and Water on High Alert', *The Guardian*, 18 March 2018, available at www.theguardian.com/world/2018/mar/17/uk-finance-power-water-on-high-alert-threat-russian-cyber-reprisal-grows.

A. Travis and C. Arthur, 'EU Court Backs Right To Be Forgotten: Google Must Amend Results on Request', the *Guardian*, 13 May 2014.

F. Tronchetti, 'Legal Aspects of the Military Uses of Outer Space', in F. von der Dunk and F. Tronchetti (eds.), *Handbook of Space Law*, (Cheltenham: Edward Elgar, 2015.

M. Troy, 'USS *John McCain* Destroyer: Who's At Fault When Ships Collide?', *ABC News*, 21 August 2017, available at www.abc.net.au/news/2017-08-21/uss-john-mccain-destroyer:-whos-at-fault-when-ships-collide/8827516.

J. B. Tucker (ed.), *Innovation, Dual Use, and Security: Managing the Risks of Emerging Biological and Chemical Technologies* (Cambridge, MA: Massachusetts Institute of Technology Press, 2012).

R. W. Tucker, *The Law of War and Neutrality at Sea*, p. (Washington, DC, 1955).

A. Turing, 'Intelligent Machinery: A Report,' National Physical Laboratory (1948), available at www.alanturing.net/intelligent_machinery/.

UK Department for Digital, Culture, Media and Sport, Secure by Design: Improving the cyber security of consumer Internet of Things Report, available at www.gov.uk/government/uploads/system/uploads/attachment_data/file/686089/Secure_by_Design_Report_.pdf.

UK Government Press Release 'Rocketing to success - Space Industry Act to unlock billions for the UK economy', 16 March 2018, available at www.gov.uk/government/news/rocketing-to-success-space-industry-act-to-unlock-billions-for-the-uk-economy.

UNIDIR, 'Framing Discussions on the Weaponisation of Increasingly Autonomous Technologies', (2014), available at www.unidir.org/files/publications/pdfs/framing-discussionson-the-weaponization-of-increasingly-autonomous-technologies-en-606.pdf.

United States Navy MK 15 - Phalanx Close-In Weapons System, described at www.navy.mil/navydata/fact_display.asp?cid=2100&tid=487&ct=2.

University of Canterbury, 'Simulating friendly fire to learn how to avoid it', 05 April 2016, available at www.canterbury.ac.nz/news/2016/simulating-friendly-fire-to-learn-how-to-avoid-it.html

University of Southern California Neuroimaging and Informatics Institute, 'The Human Connectome Project', available at www.humanconnectomeproject.org

K. R. Urban, 'Performance Enhancement at the Cost of Potential Brain Plasticity: Neural Ramifications of Nootropic Drugs in the Healthy Developing Brain', *Frontiers in Systems Neuroscience*, Vol. 8, No 38 (May 2014), 2.

USAF Judge Advocate General's Department, *Air Force Operations and the Law, A Guide for Air and Space Forces* (2002), 47.

US Defense Advanced Research Projects Agency (DARPA), Towards a High-Resolution, Implantable Neural Interface, available at www.darpa.mil/news-events/2017-07-10.

US Department of Defense, 'Description of Medals', http://valor.defense.gov/Description-of-Awards/.

US Department of Defense, Non-Lethal Weapons Program, 'Active Denial Technology', available at http://jnlwp.defense.gov/Future-Non-Lethal-Weapons/Active-Denial-Technology/.

'Distributed Sound and Light Array', available at http://jnlwp.defense.gov/Developing-Non-Lethal-Weapons/Distributed-Sound-and-Light-Array/.

US Department of Homeland Security, Future Attribute Screening Technology Fact-sheet, 2 March 2016, available at www.dhs.gov/publication/future-attribute-screening-technology.

US Department of Transportation, 23 February 2018, available at www.transportation.gov/AV.

US Director of National Intelligence website, available at www.iarpa.gov/index.php/research-programs/best.

US Food and Drug Administration, Computer-Assisted Surgical Systems, last updated 29 July 2015, available at www.fda.gov/MedicalDevices/ProductsandMedicalProcedures/SurgeryandLifeSupport/ComputerAssistedSurgicalSystems/default.htm.

USNI, 'Report to Congress on Navy Laser, Railgun and Hypervelocity Projectiles', 26 February, 2018, available at https://news.usni.org/2017/12/05/report-congress-navy-laser-railgun-hypervelocity-projectiles-2.

'Report to Congress on US Navy Destroyer Programs', 5 October 2017, available at https://news.usni.org/2017/10/05/report-congress-u-s-navy-destroyer-programs-2.
US National Research Council, *An Assessment of Non-Lethal Weapons Science and Technology* (Washington, DC: The National Academies Press, 2003).
H. Van de Meent, M. T. Hopman and J. P. Frölke, 'Walking Ability and Quality of Life in Subjects with Transfemoral Amputation: A Comparison of Osseointegration with Socket Prostheses', *Archives of Physical Medicine and Rehabilitation* (2013),94: 2174–2178.
A. Vance, *Elon Musk, Tesla, SpaceX and the Quest for a Fantastic Future*, (Harper Collins, 2015).
J. J. Vasterling, et al., 'Neuropsychological Outcomes of Army Personnel Following Deployment to the Iraq War', *Journal of the American Medical Association* 296, 5 (August 2006), 526–528.
B. Viswanath and S. Kim, 'Influence of Nanotoxicity on Human Health and Environment: The Alternative Strategies', *Review of Environmental Contamination and Toxicology*, 242 (2017), 61–104.
I. A. Vlasic, 'Space Law and the Military Applications of Space Technology', in N. Jasentuliyana (ed.) *Perspectives on International Law*, (Kluwer Law International, London, 1995).
G. Voelz, 'Catalysts of Military Innovation: A Case Study of Defense Biometrics', 23 *Defense ARJ* 2, *April* 2016, 178, 185.
W. Heintschel von Heinegg, 'The International Law of Mine Warfare at Sea', 23 *Israel Yearbook of Human Rights*, 53–76 (1993).
'Protecting Critical Submarine Cyber Infrastructure: Legal Status and Protection of Submarine Communications Cables under International Law', in K. Ziolkowski (ed.), *Peacetime Regime for State Activities in Cyberspace* (Tallinn: NATO, CCD COE, 2013).
'Unmanned Maritime Systems: Does the Increasing Use of Naval Weapon Systems Present a Challenge for IHL?', in W. Heintschel von Heinegg, R. Frau and T. Singer (eds.), *Dehumanization of Warfare: Legal Implications of New Weapon Technologies* (New York: Springer, 2018).
J. Wagner, 'China's Cybersecurity Law: What You Need to Know,' *The Diplomat*, 1 June 2017, available at http://thediplomat.com/2017/06/chinas-cybersecurity-law-what-you-need-to-know/.
E. Waldrop 'Integration of Military and Civilian Space Assets: Legal and National Security Implications' (2004) 55 *Air Force Review* 157.
G. Walker (ed.), *Definitions for the Law of the Sea*, p. 241 (Leiden/Boston: Martinus Nijhoff, 2012).
A. E. Wall, 'Legal and Ethical Lessons of NATO's Kosovo Campaign', *US Naval War College International Law Studies*, Vol. 78, Parts III and IV.
D. Wallace, 'Trends in Traumatic Limb Amputation in Allied Forces in Iraq and Afghanistan,' *Journal of Military and Veterans Health* (2012).
D. Wallace and S. Reeves, 'Modern Weapons and the Law of Armed Conflict', in G. S. Corn, R. E. VanLandingham, and S. R. Reeves (eds.), *U.S. Military Operations: Law, Policy, and Practice* (New York: Oxford University Press, 2016) 41, 58–59.

J. Wang and P. Dortmans, 'Nanotechnology: Selected Topics and Potential Military Application' (Paper presented at Land Warfare Conference, Melbourne, Australia, 27–30 September 2004) http://dspace.dsto.defence.gov.au/dspace/handle/1947/2559 69–81.

Washington Post Staff, 'Full Transcript: FBI Director James Comey Testifies on Russian Interference in 2016', *The Washington Post*, 20 March 2017, available at www.washingtonpost.com/news/post-politics/wp/2017/03/20/full-transcript-fbi-director-james-comey-testifies-on-russian-interference-in-2016-election/?utm_term=.6b6ef8d4c68d.

K. Watkin, 'Chemical Agents and "Expanding" Bullets: Limited Law Enforcement Exceptions and Unwarranted Handcuffs?', in A. M. Helm (ed.), *The Law of War in the 21st Century: Weaponry and the Use of Force* (US Naval War College International Law Studies, Vol. 82, 2006) 193, 196–197.

T. M. Wettach, 'Rebalancing Cyber Security in Asia', *The Diplomat*, 16 December 2016, available at http://thediplomat.com/2016/11/rebalancing-cybersecurity-in-asia/.

G. Webster, 'Observations on China's New International Cyberspace Cooperation Strategy', Lawfare blog, 7 March 2017, available at www.lawfareblog.com/observations-chinas-newinternational-cyberspace-cooperation-strategy.

L. Weymouth, 'Russia Is a Threat. . .to All of Europe', *Foreign Policy*, 24 March 2017, available at http://foreignpolicy.com/2017/03/24/russia-lithuania-nato-grybauskaite-putin-trump-interview/.

M. Wheelis, 'Will the "New Biology" Lead to New Weapons?', *Arms Control Today*, 34(6) (2004), 6.

M. White, 'New Tech Helps Facebook Remove 99% of IS Posts Before Users Flag It Up', Sky News, 29 November 2017, available at https://news.sky.com/story/new-tech-helpsfacebook-remove-99-of-is-posts-before-users-flag-it-up-11148386.

White House, *The National Security Strategy of the United States of America* (September 2002), available at www.state.gov/documents/organization/63562.pdf.

White House, www.whitehouse.gov/briefings-statements/statement-press-secretary-25/.

D. Whitehouse,. 'First dog in space died within hours', *BBC News*, 28 October 2002, available at http://news.bbc.co.uk/2/hi/science/nature/2367681.stm.

T. Widler, et al., 'Organisational Risk Management of Nanomaterials Using SUNDS: The Contribution of CENARIOS®' in F. Murphy, E. M. McAlea and M. Mullins (eds.), *Managing Risk in Nanotechnology* (New York: Springer International, 2016), 219–235.

P. Wiles, 'Metropolitan Police's Use of Facial Recognition Technology at the Notting Hill Carnival, 2017', Press Release, 23 August 2017.

G. Williams, *Textbook of Criminal Law*, 2nd edn (London: Stevens and Son, 1983).

I. Williams, *Rum: A Social and Sociable History* (New York: Nation Books, 2005).

A. Willoughby, 'Biometric Surveillance and the Right to Privacy', 36 *IEEE Technology and Society Magazine* 3, September 2017, available at http://ieeexplore.ieee.org/document/8038135/.

R. F. Wilson, 'Nanotechnology: The Challenge of Regulating Known Unknowns', *The Journal of Law, Medicine & Ethics*, 34(4) (2006), 704–713.

K. Wong, 'Australia Seeks Guidance on Quantum Technologies', 11 January 2018, available at www.janes.com/article/77006/australia-seeks-guidance-on-quantum-technologies.

'Sea Sentinels: Chinese Unmanned Maritime Systems Gain Traction', available at www.janes.com/images/assets/219/77219/Sea_sentinels_Chinese_unmanned_mari time_systems_gain_traction.pdf.

K. Wong, 'Australia Seeks Guidance on Quantum Technologies', 11 January 2018, available at www.janes.com/article/77006/australia-seeks-guidance-on-quantum-technologies.

J. Woodward, Jr, 'Using Biometrics to Achieve Identity Dominance in the Global War on Terrorism' (September–October 2005) LXXXV No.5 *Military Review* 30 at 31.

N. Woolf, 'DDoS Attack That Disrupted Internet Was Largest of Its Kind in History, Experts Say', *The Guardian*, 26 October 2016.

L. R. Wrathall, 'The Vulnerability of Subsea Infrastructure to Underwater Attack: Legal Shortcomings and the Way Forward', 12 *San Diego Int'l Law Journal* (2010–2011), 223–261, at 228.

D. Wroe, 'New Sabotage Laws for Cyber Attacks on Australia's Critical Infrastructure', *Sydney Morning Herald*, 8 December 2017, available at www.smh.com.au/federal-politics/political-news/new-sabotage-laws-for-cyber-attacks-on-australias-critical-infrastructure-20171207-h000c2.html.

S. S. Wu and M. Goodman, 'Neural Devices Will Change Humankind: What Legal Issues Will Follow?,' *American Bar Association SciTech Lawyer* (Winter 2012).

S. Xia, 'China Establishes Its First Cyber Court in Hangzhou: Thank You Alibaba', China Law Blog, 16 August 2017, available at www.chinalawblog.com/2017/08/china-establishes-its-firstcyber-court-in-hangzhou-thank-you-alibaba.html.

M. Xuequan, 'China asks online group, WeChat account admins to take responsibility', Xinhuanet, 19 September 1917, available at http://news.xinhuanet.com/english/2017–09/08/c_136592568.htm.

G. Z. Yang, et al., 'Medical robotics: Regulatory, Ethical And Legal Considerations for Increasing Levels of Autonomy', 2 *Science Robotics* 4, 15 March 2017, available at http://robotics.sciencemag.org/content/2/4/eaam8638.full.

H. Yin, et al., 'Structure-Guided Chemical Modification of Guide RNA Enables Potent Non-Viral in Vivo Genome Editing', *Nature Biotechnology* (2017), nature.com/articles/doi:10.1038/nbt.4005.

M. Young, 'Neural Network Finds 8-Planet System,' *Sky & Telescope*, 15 December 2017.

D. Zaidi, 'Inside China's Growing Use of Facial Recognition Technology' *Venture Beat*, 15 February 2018.

J. P. Zanders, 'The Chemical Weapons Convention and Universality: A Question of Quality over Quantity?', *Disarmament Forum*, 4 (2002), 23.

'International Norms against Chemical and Biological Warfare: An Ambiguous Legacy', (2003) 8(2) *JCSL* 391, 407–408.

G. Zorpette, 'The Benefits of Building an Artificial Brain,' *IEEE Spectrum*, 31 May 2017, available at https://spectrum.ieee.org/computing/hardware/can-we-copy-the-brain.

L. Zyga, *Microbots can clean up polluted water* (11 April 2016) Phys Org, available at https://phys.org/news/2016–04-microbots-polluted.html.

Index